145

OXFORD STUDIES IN NUCLEAR PHYSICS

GENERAL EDITOR
P.E. HODGSON

OXFORD STUDIES IN NUCLEAR PHYSICS
General editor: P E HODGSON

1. J.L. McL. Emmerson: *Symmetry principles in particle physics* (1972)
2. J.M. Irvine: *Heavy nuclei, superheavy nuclei, and neutron stars* (1975)
3. I.S. Towner: *A shell-model description of light nuclei* (1977)
4. P.E. Hodgson: *Nuclear heavy ion reactions* (1978)
5. R.D. Lawson: *Theory of nuclear shell model* (1980)
6. W.E. Frahn: *Diffractive purposes in nuclear physics* (1985)
7. S.S.M. Wong: *Nuclear statistical spectroscopy* (1986)
8. N. Ullah: *Matrix ensembles in the many-nucleon problem* (1987)
9. A.N. Antonov, P.E. Hodgson, and I.Zh. Petkov: *Nucleon nomentum and density distributions in nuclei* (1988)
10. D. Bonatsos: *Interacting boson models of nuclear structure* (1988)
11. H. Ejiri and M.J.A. deVoigt: *Gamma-ray and electron spectroscopy in nuclear physics* (1989)
12. B. Castel and I.S. Towner: *Modern theories of nuclear physics moments* (1990)
13. R.F. Casten: *Nuclear structure* (1990)
14. I.Zh. Petkov and M.V. Stoisov: *Nuclear density function theory* (1991)
15. I. Gadioli and P.E. Hodgson: *Pre-equilibrium nuclear reactions* (1992)
16. J.D. Walecka: *Theoretical nuclear and subnuclear physics* (1995)
17. D.N. Poenaru and W. Greiner (eds): *Handbook of nuclear properties* (1995)
18. P. Fröbrich and R. Lipperheide: *Theory of nuclear reactions* (1996)
19. Fl. Stancu: *Group theory in subnuclear physics* (1996)
20. S. Boffi, C. Giusti, F.D. Pacati, M. Radici: *Electromagnetic response of atomic nuclei* (1996)

Group Theory in Subnuclear Physics

Fl. STANCU

Institute of Physics, University of Liège

CLARENDON PRESS · OXFORD
1996

Oxford University Press, Great Clarendon Street, Oxford OX2 6DP
Oxford New York
Athens Auckland Bangkok Bogota Bombay Buenos Aires
Calcutta Cape Town Dar es Salaam Delhi Florence Hong Kong
Istanbul Karachi Kuala Lumpur Madras Madrid Melbourne
Mexico City Nairobi Paris Singapore Taipei Tokyo Toronto
and associated companies in
Berlin Ibadan

Oxford is a trade mark of Oxford University Press

Published in the United States
by Oxford University Press Inc., New York

© Fl. Stancu, 1996

All rights reserved. No part of this publication may be
reproduced, stored in a retrieval system, or transmitted, in any
form or by any means, without the prior permission in writing of Oxford
University Press. Within the UK, exceptions are allowed in respect of any
fair dealing for the purpose of research or private study, or criticism or
review, as permitted under the Copyright, Designs and Patents Act, 1988, or
in the case of reprographic reproduction in accordance with the terms of
licences issued by the Copyright Licensing Agency. Enquiries concerning
reproduction outside those terms and in other countries should be sent to
the Rights Department, Oxford University Press, at the address above.

This book is sold subject to the condition that it shall not,
by way of trade or otherwise, be lent, re-sold, hired out, or otherwise
circulated without the publisher's prior consent in any form of binding
or cover other than that in which it is published and without a similar
condition including this condition being imposed
on the subsequent purchaser.

A catalogue record for this book is available from the British Library

Library of Congress Cataloging in Publication Data
Stancu, Fl.
Group theory in subnuclear physics / Fl. Stancu.
(Oxford studies in nuclear physics; 19)
Includes bibliographical references and index.
1. Symmetry (Physics) 2. Nuclear physics. 3. Group theory.
4. Representations of groups. I.Title. II. Series.
QC793.3.S9S69 1996   539.7'25–dc20   96-14251
ISBN 0 19 851742 4

Typeset by EXPO Holdings, Malaysia

Printed in Great Britain by
Bookcraft (Bath) Ltd,
Midsomer Norton, Avon

To Alex
To my parents

Plus ça change, plus c'est la même chose

PREFACE

Group theory and representation theory are the mathematical tools needed to study the symmetry transformations of quantum systems. Recent progress, especially in theoretical high-energy physics, has brought group theory to occupy a crucial role in the present development of fundamental aspects of physics.

There already exist quite a large number of books on group theory, some of them famous. However, the subject is difficult and over the years I myself have encountered many frustrations in trying to learn group theory.

The aim of this book is to provide advanced undergraduates or graduate students, as well as professional physicists, with a practical introduction to concepts of group theory and some of their applications, especially to subnuclear physics. This is due to the growing interest in the quark structure of hadrons and in theories of particle interactions based on the principle of gauge symmetries.

The book has grown out of a series of lectures given over the past ten years at the University of Liège. The original notes have been reorganized and the sections on permutation groups considerably amplified with the purpose of discussing applications to multiquark systems as described in the last chapter. However, the book is far from being exhaustive because group theory is a very wide and extremely diverse subject. Intricate mathematical proofs have been avoided and references are given to books where such proofs can be found.

In exchange, we include intermediate steps which are usually omitted by experts, in order to make the book more accessible to non-experts. To help in acquiring a more solid foundation in group theory, many solved exercises are included and several others suggested for serious students. The reader should be familiar with quantum mechanics. Some exposure to elementary group theory is also desirable.

The book is organized as follows. Chapter 1 introduces the types of symmetries discussed. For self-consistency, in Chapters 2 and 3 the basics of group theory and representation theory, respectively, are reviewed. Chapter 4 is devoted to the symmetric groups and is one of the core chapters. It gives a detailed description of the Young diagrams associated to symmetric groups. Inner and outer products are defined and their use in many-particle physics is illustrated with examples, especially in relation to baryons, viewed as three-quark states. A simple introduction to the tensor method is presented to show the interrelation between symmetric and linear Lie groups. The last section of Chapter 4 gives an introduction to the theory of braid groups, a generalization of symmetric groups, which is now being developed.

In Chapter 5, Lie groups are described in a rather traditional manner. After introducing generators and associated Lie algebras, examples are given and the Cartan classification is reviewed. General properties of semi-simple groups are presented and it is shown how to label the basis vectors of an invariant subspace by using weights and invariant operators. The tensor method is revisited as a method of constructing irreducible representations of semi-simple Lie groups. The chapter finishes with a short introduction to quantum groups.

The subsequent chapters deal with a detailed description of some of the Lie groups used in physics. Chapter 6 is devoted to the orthogonal group and in particular to O(3) and O(4). Chapter 7 gives a short description of the Lorentz group because relativistic physics is becoming more and more important. Besides Chapter 4, the other core chapter is Chapter 8, devoted to unitary groups. Use is made again of Young diagrams to describe the irreducible representations of unitary groups. Applications to the classification of elementary particles are presented.

To keep up with the present theories of fundamental interactions, a short introduction to gauge groups is given in Chapter 9. The book contains a number of applications to multiquark systems in Chapter 10. The list of references is not exhaustive. I apologize to the large number of authors whose work is not quoted.

I have also included a few appendices: Appendix A illustrates Noether's theorem, Appendix B contains a proof of Schur's lemmas, Appendix C introduces invariant integration, and Appendix D derives the formula for the dimension of an SU(n) irreducible representation. In the table of contents, those sections which may safely be omitted at a first reading are marked with an asterisk.

An important goal of this book is to provide help in understanding and manipulating Young diagrams, in the context of both symmetric and unitary groups. These diagrams offer a powerful method for determining Clebsch–Gordan series. They were originally introduced by Young (1928) to describe the representations of the symmetric group, but later Weyl (1946) showed that they play a dual role due to the reciprocity between a linear group and a symmetric group in a given tensor space. This reciprocity is illustrated by a simple example related to S_3 and SU(3) in Section 4.5.

For further reading, I can recommend some of the books which helped me to learn group theory. The basic role of symmetries, together with applications to elementary particle physics, can be found in the book by T.D. Lee (1981). The classic textbook on group theory is Hamermesh's book (1962). It describes many aspects of representation theory and it contains chapters with applications to atomic, molecular, and nuclear physics. The book of Elliott and Dawber (1979) incorporates additional applications to elementary particle physics and solid state physics.

The book by Chen (1989) describes representation theory in an original and consistent way. Several examples considered in the present book and treated in a traditional manner are also discussed in the book by Chen. To complete the present reading with more mathematical proofs, a useful book is that of Cornwell (1984).

Of the people who have aided me at the beginning of my career, I wish to record my gratitude to my former colleagues at the Institute of Atomic Physics in Bucharest,

Mircea Iosifescu and Mircea Micu, who interested me in group theory and helped me to understand its basic concepts.

I am particularly grateful to David Brink who contributed friendly discussions, careful reading and criticism of much of the manuscript, and many valuable suggestions for improving it. Florin Constantinescu helped in bringing accuracy to some mathematical aspects and I am also much indebted to him for introducing me to the latest mathematical developments in braid groups and quantum groups. I am also grateful to Dennis Bonatsos for a critical reading of the section on quantum groups.

I thank Stéphane Pepin for reading parts of the final version of the manuscript and correcting a number of inadvertent mistakes and also for help in checking the proofs. I wish to express my gratitude to the students attending my lectures, whose active interest helped in clarifying some ideas and their presentation.

I am very grateful to Marie-Louise Joiris for transforming an untidy manuscript into an elegant and accurate typescript.

Liège Fl.S.
March 1995

CONTENTS

1 Symmetries in quantum mechanics   **1**
     1.1   The role of group theory   1
     1.2   Types of symmetries   2
     1.3   Invariance and conservation laws   6

2 Elements of group theory   **9**
     2.1   Definition of a group (of transformations)   9
     2.2   Subgroups   16
     2.3   Isomorphism and homomorphism   18
     2.4   Cayley's theorem   20
     2.5   Classes   22
     2.6   Simple and semi-simple groups   25
     2.7   Direct product groups and semi-direct product groups   27

3 Linear representations of a group   **29**
     3.1   Linear vector spaces   29
     3.2   Definition of a group representation   31
     3.3   Matrix representations   32
     3.4   Equivalent representations   32
     3.5   Unitary representations   33
     3.6   Characters   34
     3.7   Examples of representations   35
     3.8   Irreducible representations   39
     3.9   Direct products of representations   42
     3.10   Schur's lemmas and the orthogonality theorem   43
     3.11   The regular representation   48
     3.12   Construction of character tables   53
     3.13   Clebsch–Gordan series   57

4 Permutation group S_n   **59**
     4.1   General remarks   59
     4.2   Irreducible representations   60
     4.3   Basis functions of S_n 62
     4.4   Matrices of irreducible representations   89
     4.5   The tensor method   103
     4.6   Outer products   108
     4.7   Inner products. Clebsch–Gordan series and coefficients   112
     4.8   More about Clebsch–Gordan coefficients of S_n 119
     *4.9   The K-matrix (isoscalar factor)   128

	*4.10	The $\overline{K}$-matrix	131
	4.11	Baryons as three-quark states	140
	*4.12	Braid groups and new developments	144

5 Lie groups 148

5.1	Infinitesimal transformations	149
5.2	Structure constants	151
5.3	Generators	154
5.4	Simple and semi-simple groups	158
5.5	Simple and semi-simple Lie algebras	162
5.6	Examples of Lie groups	163
5.7	Compactness	169
*5.8	Direct and semi-direct sums of Lie algebras	170
5.9	Classification of semi-simple groups	171
5.10	Representations of semi-simple groups	192
5.11	The tensor method (revisited)	198
*5.12	Quantum groups	204

6 The orthogonal group 209

6.1	The rotation group R_3 or $SO(3)$	210
6.2	The group $O(4)$	230
6.3	The Euclidean groups	239

7 The Poincaré group and the Lorentz group 241

7.1	Notation	241
7.2	Lorentz transformations	242
7.3	Infinitesimal transformations	248
7.4	The spin of a Dirac particle	250
7.5	Irreducible representations	254
7.6	The Poincaré group	256

8 Unitary groups 262

8.1	General properties	263
8.2	The group $SU(2)$	265
8.3	The homomorphism of $SU(2)$ with R_3	266
8.4	Multiplets	269
8.5	G-parity	271
8.6	The group $SU(3)$	274
8.7	Beyond $SU(3)$	315
8.8	Heavy flavours	321
8.9	The adjoint representation	337
8.10	The tensor method	340
8.11	Clebsch–Gordan coefficients for $SU(3)$	344

*9	**Gauge groups**	**347**
	9.1 Abelian gauge symmetry	347
	9.2 Non-abelian gauge symmetry	349
*10	**Multiquark systems**	**355**
	10.1 The dynamics	356
	10.2 The baryons	360
	10.3 Diquonia	372
	10.4 Six-quark systems	383

Appendix A. Conservation laws — 398

Appendix B. The rearrangement theorem, Schur's lemmas, and the orthogonality theorem — 402

Appendix C. Invariant integration — 407

Appendix D. Dimension of an $SU(n)$ irrep — 410

References — 413

Index — 419

1
SYMMETRIES IN QUANTUM MECHANICS

1.1 THE ROLE OF GROUP THEORY

Group theory is a branch of mathematics appropriate for studying symmetries of physical systems. A symmetry is a natural attribute of the physical world and therefore is the starting point of any physical law. The foundation of dynamics is in Lagrangian theory or equivalently in Hamiltonian theory. A symmetry of a physical system is expressed by the invariance of its Lagrangian or Hamiltonian, or ultimately of its equations of motion, under certain transformations. When these transformations form groups it is convenient and useful to study symmetries through group theory.

Group theory can explain observed regularities and helps to simplify and unify the physics of apparently different systems. It is a valuable tool that helps in predicting the behaviour of systems for which a symmetry has been found.

Among all symmetries, those related to geometrical transformations are the most obvious. They correspond to groups of transformations such as, for example, the group of translations, the group of rotations, the group of reflections (inversions) in a point or in a plane, the Lorentz group, the Poincaré group, or the groups of symmetries in a crystal.

In quantum mechanics, there are additional symmetries. First of all, for identical particles, there is a permutation symmetry, associated with the symmetric group. The role of the symmetric group is to ensure that wave functions incorporate correctly the property of indistinguishability. Quantum systems also have symmetries related to *internal* degrees of freedom, such as spin, isospin, flavour, or colour. These symmetries involve unitary groups. In Section 1.2, we give a list of the main types of symmetries in quantum mechanics.

The first ideas of group theory as a branch of mathematics date from the early nineteenth century. At the end of the nineteenth century and the beginning of the twentieth, major developments were made by Frobenius, Schur, Lie, and Cartan. The essential role of group theory in physics was realized only after 1920 in relation to the development of the representation theory of groups, which is intimately connected with quantum mechanics (see Chapter 3). These achievements are contained in classical works, such as those of H. Weyl[†] (1946, 1950) and E. Wigner (1959).

[†] The bibliographies of Weyl's books contain important historical notes. The preface to the first German edition of *The theory of groups and quantum mechanics* dates from 1928. The application of the representation theory of the rotation group to atomic spectra was first considered by Wigner (1927).

At present, in the modern developments of physics, and in particular high-energy physics, symmetries play a crucial and indispensable role. Of great importance are the gauge symmetries, initiated by Weyl in 1929. Nowadays, it is believed that all fundamental interactions are described by some form of gauge theory, i.e. theories with gauge symmetries. Another aspect is the extension of the theory of Lie groups to supersymmetries. These concepts are currently applied to particle physics, quantum field theory, and gravity in the form of string and superstring theories.

1.2 TYPES OF SYMMETRIES

Here we enumerate the commonly encountered symmetries in quantum mechanics by dividing them into five main categories. Table 1.1 lists some of them, together with their consequences.

(1) *Discrete permutation symmetries*

In quantum mechanics, the expectation value of any physical observable remains unchanged on the permutation of identical particles. The permutations form a group called the symmetric group S_n (Chapter 4). A role of the symmetric group is to ensure that wave functions incorporate correctly the property of indistinguishability (Bose–Einstein and Fermi–Dirac statistics) (Schiff 1968, Chapter 10).

(2) *Continuous space–time symmetries*

These are related to:
  (a) Space translations

$$r' = r + \rho. \tag{1.1}$$

In this case the symmetry is based on the assumption of the homogeneity of space. This means that one can arbitrarily choose the origin of coordinates; in other words, there is no absolute position. The statement is valid for an isolated system (no external field) and it implies that the interaction potential acting between two particles does not depend on the origin of the coordinate system. The symmetry group is the translation group T_3 (Chapter 6) and a consequence of this symmetry is the conservation of linear momentum.
  (b) Time translation

$$t' = t + t_0. \tag{1.2}$$

This symmetry is based on the assumption of homogeneity of time, which means that the time origin is arbitrary or, equivalently, that the same physical phenomenon can be reproduced at any time. In other words, there is no absolute time. This symmetry is encountered in so-called conservative systems which are either isolated or located in a time-independent external field. The Lagrangian or Hamiltonian of such a system

TYPES OF SYMMETRIES

Table 1.1

Theoretical assumption	Symmetry transformation	Consequence
Indistinguishability of identical particles	permutation	Bose–Einstein or Fermi–Dirac statistics
Space homogeneity	Space translation $r' = r + \rho$.	Conservation of linear momentum
Time homogeneity	Time translation $t' = t + t_0$	Conservation of energy
Space isotropy	Rotation $x'_i = R_{ij} x_j \quad (i, j = 1, 2, 3)$	Conservation of angular momentum
No absolute uniform velocity	Lorentz transformation $x'^\mu = \Lambda^\mu_{\ \nu} x^\nu$ $(\mu, \nu = 0, 1, 2, 3)$	Conservation of relativistic angular momentum tensor

does not depend explicitly on time and an important consequence of this is energy conservation (Appendix A).

(c) Rotations in a three-dimensional space

$$x'_i = R_{ij} x_j \quad (i, j = 1, 2, 3) \tag{1.3}$$

where x_i are the components of a vector and R_{ij} is the rotation matrix. Rotational symmetry results from the assumption of space isotropy or lack of an absolute spatial direction for isolated systems. Rotational symmetry indicates the fact that the properties of a system are independent of its orientation in space. The matrix rotations R_{ij} form the rotation group R_3 (Chapter 6) and, from rotational invariance, one obtains the conservation law of angular momentum.

(d) Lorentz transformations

$$x'^\mu = \Lambda^\mu_{\ \nu} x^\nu \quad (\mu, \nu = 0, 1, 2, 3) \tag{1.4}$$

where x^ν is a point $x^\nu = (ct, x^i)$ in Minkowski space–time. The transformations (1.4) are either rotations as defined above, or Lorentz transformations between two systems moving uniformly with respect to each other, which preserve the velocity of light. In special relativity, the physical laws are formulated to remain identical for all inertial reference frames.[†] In the non-relativistic limit physical laws remain invariant under Galilean transformations. This is equivalent to the assumption that there is no absolute uniform velocity. The transformations (1.4) define the Lorentz group

---

[†] According to Einstein, an inertial system is defined as one in which an isolated particle, free from interactions, remains at rest or moves uniformly. The system of coordinates in which the local stars are fixed is a good approximation for the ideal inertial system (Goldstein 1980, Section 4–10).

(Chapter 7) and invariance under such transformations leads to conservation laws connected to the generators of the Lorentz group.

(3) *Discrete space–time symmetries*

(a) Space inversions (or reflections), P, where

$$Pr = r' = -r. \tag{1.5}$$

Applying an inversion twice, one returns to the original point, therefore $P^2 = I$. The unit operator I together with P form a very simple discrete group with two elements. However, in quantum mechanics, space inversion defined by a unitary operator introduces an important quantum number called parity, which is conserved by most interactions in nature. An exception is the weak interaction.

(b) Time reversal, T, where

$$t \to -t. \tag{1.6}$$

This is a change in the direction of the flow of time. It was introduced in quantum mechanics by Wigner in 1932 (see also Wigner 1959). This transformation is effected in quantum mechanics by an anti-unitary operator called T. The physical laws are generally symmetric with respect to T with some exceptions, for example, neutral kaon decay (Lee 1981, Chapter 9 or 15).

(c) Symmetry transformations of the point groups. These are transformations where at least one point of a finite body remains fixed and the body occupies the same space after the transformation as it did before. Each such transformation is one or a combination of two fundamental operations: rotation by a fixed angle and mirror reflection in a plane. For an infinite medium (crystal lattice) translations by a fixed segment should also be included in order to obtain the basic symmetries of solid state or molecular physics (Hamermesh 1962; Tinkham 1964; Landau and Lifshitz 1965, Chapter 12).

(4) *Internal continuous symmetries*

These are related to transformations acting in the space of internal degrees of freedom of particles, as, for example, spin, flavour, or colour. The flavour F is the collective name for isospin I, hypercharge Y, charm C, beauty B, and topness T degrees of freedom. The internal symmetries are more difficult to deal with because it is not obvious in advance which symmetry group is appropriate. Experience shows that the relevant transformations form unitary groups. In particular, the invariance under transformations described by the unitary group U(1) leads to charge or particle number conservation (leptons, baryons). The isospin symmetry of the strong interactions, described by $SU_I(2)$, is a manifestation of the proton and neutron mass (near) degeneracy. An alternative form is the $SU_F(2)$ symmetry expressing the (near) mass degeneracy of the up and down quarks. The flavour $SU_F(3)$ symmetry, of the strong interaction, is also a manifestation of the lightness of up, down, and strange quarks on the typical strong interaction energy scale. (See Chapter 8).

(5) *Internal discrete symmetries*

(a) Charge conjugation, C. By this transformation, the sign of an electric charge changes from positive to negative or vice versa. This is the particle–antiparticle symmetry. The operator is denoted by C. In the framework of the Dirac equation, it is an anti-unitary operator, but in field theory is unitary. The experiments confirming parity non-conservation in weak interactions also gave evidence of C-violation, or particle–antiparticle asymmetry (Lee 1981, Chapter 9).

(b) G-parity. The transformation related to this symmetry represents charge conjugation combined with a rotation of π in the isospin space of a particle. In strong interactions, G-parity is conserved (Chapter 8).

The validity of a symmetry rests on the theoretical assumption of 'non-observability' (Table 1.1, column 1). Violation of a symmetry appears when the non-observable turns out to be an observable. Some non-observables are fundamental and then the symmetry is called exact. Among these are all continuous space–time transformations, the U(1) symmetries leading to particle or charge conservation, and the internal continuous symmetries related to the colour degree of freedom and described by $SU_C(3)$. Some other symmetries are less fundamental, i.e. are violated to some extent. The amount of violation depends on the ability to find proper tests and the possibility of precise measurements. The present studies conclude that P, T, C, and also CP, PT, and TC, are violated or broken. However, the product CPT remains, so far, an exact symmetry.

The permutation symmetry associated with identical particles is also thought to be exact.

The $SU_I(2)$ symmetry is broken by the electromagnetic interaction. The $SU_F(n)$ symmetry, where n is the number of flavours, is broken especially by the heavy quarks (Chapter 8).

It is useful to mention that there exist so-called *spontaneously broken symmetries*, as for example *chiral* symmetry. This means that the physical vacuum is not invariant under the symmetry group of its Hamiltonian. In the case of chiral symmetry, the physical manifestation is the presence of a set of near-massless bosons (the pions), or more generally, of the pseudoscalar meson octet.

All continuous symmetries mentioned above are *global* symmetries. They are defined in terms of a finite number of parameters a^ρ which are independent of space–time (Chapter 5). There are also symmetry transformations where the parameters are space–time dependent. These are called *local* symmetries or gauge symmetries (Weyl 1929). They are used to generate the gauge interactions (Chapter 9). It is now believed that all fundamental interactions are described by gauge theories.

The gauge group of quantum electrodynamics (QED) is an abelian group. The extension of the gauge principle to non-abelian groups gave the Yang–Mills theories. They have been applied to the weak–electromagnetic interaction and to quantum chromodynamics (QCD). The electroweak theory is a gauge theory with spontaneous symmetry breaking. Recent proposals try to unify strong and electroweak interactions into a single theory. Theories attempting to do this unification are called 'grand unified theories' or GUTs for short.

1.3 INVARIANCE AND CONSERVATION LAWS

The concept of symmetry in quantum mechanics is analogous to that in classical mechanics. In particular, the geometrical transformations (translation, rotation, Lorentz transformation) are defined in the same way in both classical and quantum mechanics and the symmetries play the same role. As already mentioned, a symmetry is expressed by the invariance of a Lagrangian or of a Hamiltonian or, equivalently, of the equations of motion, with respect to some group of transformations. Irrespective of a classical or a quantal treatment, an invariant Lagrangian or Hamiltonian associated with a *continuous* symmetry implies a set of conservation laws. In Table 1.1, we list some examples.

Differences from classical mechanics arise from the fact that in quantum mechanics the Hamiltonian and the group of transformations are operators. Then a Hamiltonian is said to be invariant if it commutes with all operators defining a group of transformations.

In a non-relativistic or relativistic quantum mechanical Hamiltonian formalism, a conservation law appears most obviously in the Heisenberg picture, where the state vectors are time independent and the equation of motion is (Schiff 1968, p. 171)

$$\frac{d\Omega}{dt} = \frac{\partial \Omega}{\partial t} + \frac{1}{i\hbar}[\Omega, H], \qquad (1.7)$$

where Ω represents an observable operator, H is the Hamiltonian of the system, and $[\Omega, H] = \Omega H - H\Omega$. If $\frac{\partial \Omega}{\partial t} = 0$ and

$$[\Omega, H] = 0 \qquad (1.8)$$

it follows that $\frac{d\Omega}{dt} = 0$ or equivalently $\frac{d}{dt}\langle\Omega\rangle = 0$, since in the Heisenberg picture the state vectors are time independent. Note that $\langle\ \rangle$ designates an expectation value. Then, the eigenvalues of Ω are time independent, i.e. conserved (Table 1.1, column 3). Let us consider a few cases:

(1) $\Omega = H$. If the system remains invariant under a time translation one has $\frac{\partial H}{\partial t} = 0$. Hence, the eigenvalues of H are constant, i.e. the energy is conserved.

(2) $\Omega = P$, the parity operator, which transforms any vector $\mathbf{r}$ into $\mathbf{r}' = -\mathbf{r}$, or alternatively the coordinate system x, y, z into $-x, -y, -z$. Obviously $\frac{\partial P}{\partial t} = 0$.

Suppose that the system described by H contains no element which can distinguish between the two systems of coordinates, i.e.

$$[H, P] = 0 \tag{1.9}$$

thus $<P>$ is constant in time. Then if $\psi(r)$ is a non-degenerate eigenstate corresponding to an eigenvalue E the function $P\psi(r) = \psi(-r)$, which refers to the transformed coordinate system, must correspond to the same eigenvalue due to (1.9):

$$HP\psi(r) = PH\psi(r) = EP\psi(r). \tag{1.10}$$

Thus, $\psi(r)$ and $P\psi(r)$ cannot be linearly independent. In fact, they are proportional, which can be written

$$P\psi(r) = \psi(-r) = \pi\psi(r) \tag{1.11}$$

with π as the proportionality coefficient. But one can also transform $-r$ into r which gives

$$P\psi(-r) = \psi(r) = \pi\psi(-r). \tag{1.12}$$

From (1.11) or (1.12), it follows that the eigenvalues of P are $\pi = \pm 1$ and any non-degenerate eigenfunction has the property

$$P\psi(r) = \psi(r) \quad \text{or} \quad P\psi(r) = -\psi(r) \tag{1.13}$$

and the parity quantum number π is conserved.

(3) $\Omega = J_i (i = 1, 2, 3)$, the three components of the angular momentum. In terms of group theory, J_i are the generators of the rotation group R_3. If H is invariant under rotations it has to commute with all possible rotations in three dimensions, or equivalently with the generators of R_3. Hence, its invariance under rotations leads to the conservation of each of the components of the angular momentum.

In a Lagrangian formalism (classical mechanics, field theory) the connection between invariance and the corresponding conservation law is established by Noether's theorem (Goldstein 1980, p. 588).

Noether's theorem can be formulated either in terms of classical generalized coordinates q_i or in terms of field amplitudes ϕ_i, inasmuch as in classical field theory ϕ_i is the analogue of q_i for a continuous mechanical system. Both ϕ_i and q_i are independent functions of space r and time t, taken as parameters in the Lagrangian. In the discussion below, we shall formulate the theorem in terms of fields because in this form it is related to modern physics. For continuous systems the equations of motion are derived from a Lagrangian density $\mathscr{L}$. Introducing the covariant space–time coordinates x_μ (Chapter 7) the Lagrangian density appears as a function of the form

$$\mathscr{L} = \mathscr{L}(\phi_i, \partial_\mu \phi_i) \tag{1.14}$$

where ∂_μ are the covariant derivatives. One should also note that Noether's theorem refers to *continuous* transformations only, i.e. of types (2) and (4) of the previous section. This theorem says that for a classical system described by the Lagrangian

$$L = \int d^3x \, \mathscr{L}(\phi_i, \partial_\mu \phi_i) \tag{1.15}$$

and satisfying the equation of motion

$$\partial_\mu \frac{\delta \mathscr{L}}{\delta(\partial_\mu \phi_i)} - \frac{\delta \mathscr{L}}{\delta \phi_i} = 0, \tag{1.16}$$

any continuous transformation which leaves the action

$$S = \int L \, \mathrm{d}t \tag{1.17}$$

invariant implies the existence of a conserved current.[†] The volume integral of the time components of the current is called *charge* and as a consequence of the conservation law the charge is a constant of motion. Examples of charge are the electric charge itself, and the baryon or lepton number. In Appendix A, we illustrate Noether's theorem in the case of internal continuous symmetries of type (4) introduced in the previous section.

Apart from conservation laws, other important consequences of symmetries are: (a) degeneracies in energy, (b) selection rules, and (c) factorization of matrix elements related to observables, etc. These consequences are important both for understanding the physics of various processes and for simplifying calculations.

The degeneracy in energy is immediately apparent. Let U be a transformation belonging to a symmetry group which leaves the Hamiltonian invariant:

$$[U, H] = 0. \tag{1.18}$$

This transformation can connect two eigenstates $|i\rangle$ and $|j\rangle$ of the Hamiltonian

$$U|i\rangle = |j\rangle. \tag{1.19}$$

It follows that

$$H|j\rangle = HU|i\rangle = UH|i\rangle = E_i U|i\rangle = E_i |j\rangle, \tag{1.20}$$

i.e. $|i\rangle$ and $|j\rangle$ are degenerate.

The other consequences will be discussed later in the book. They are based on various elements of group theory, such as irreducible representations, irreducible tensors, the orthogonality theorem, invariant operators, and so on.

[†] It may happen that the conservation of a current at a classical level fails to hold at a quantum level. In quantum field theory these situations are called *anomalies*. They have been discovered in current algebra studies (see, for example, Adler (1969)).

2

ELEMENTS OF GROUP THEORY

The purpose of this chapter is to familiarize the reader with the principal concepts of group theory. In order to prevent the presentation from becoming too abstract, examples of very simple groups are given. In later chapters these concepts are used for a detailed study of a selection of groups currently used in physical applications. For important topics of group theory not covered in this book the reader is referred to other books or detailed articles (see references).

2.1 DEFINITION OF A GROUP (OF TRANSFORMATIONS)

A set G of transformations g forms a group if:
(i) the result of applying successively any two transformations
$$g_1 \in G, \quad g_2 \in G$$
is also a transformation of the set:
$$g_1 g_2 = g \in G; \tag{2.1}$$
the above relation is called a *product* or law of composition;
(ii) the product is *associative*: for all $g_1, g_2, g_3 \in G$,
$$(g_1 g_2) g_3 = g_1 (g_2 g_3); \tag{2.2}$$
(iii) one of the elements e is the *unity* or *identity* element, i.e. $e \in G$ such that
$$ge = eg = g; \tag{2.3}$$
(iv) when a transformation g belongs to the set, its *reciprocal* or *inverse* g^{-1} also does, i.e. $g^{-1} \in G$ is such that
$$gg^{-1} = g^{-1}g = e. \tag{2.4}$$

Although associativity is required by one of the axioms, commutativity is not a general property. There are, however, groups whose elements commute with each other. These are called *abelian* groups. Examples are translations $x \to x + \rho$ or rotations about a fixed axis.

A *finite* group is a group with a finite number of elements N. This number is also called the *order* of a finite group.

For example, the space reflections (or inversions) form a group of order $N = 2$. One element P is the transformation $x \to -x$. Two successive reflections P^2 give the identity element $x \to x$.

The group obtained from an element g and its higher powers g, g^2, ... is called a *cyclic* group. Its order $N = n$ is the least integer n such that $g^n = e$. Obviously the group of inversions is cyclic. Also the complex n-roots of unity

$$x_{k,n} = \cos\frac{2\pi}{n}k + i\sin\frac{2\pi}{n}k \quad (k = 1, 2, \ldots, n) \tag{2.5}$$

form a cyclic group. Here putting $k = n$ gives the identity element.

An *abstract* group is a set of elements satisfying the group axioms (i)–(iv). The concrete examples are called *realizations* of that abstract group. Note that, in general, the law of composition does not necessarily imply ordinary multiplication.

Exercise 2.1 Show that the abstract group of order 4 with the elements a, b, c, e is either cyclic, or abelian but not cyclic. In the latter case, find the multiplication table. Here e is the identity element.

Note that a cyclic group is abelian but the opposite is not necessarily true.

Solution One has two possibilities:

$$\text{(A)} \quad ab = e \quad \text{and} \quad \text{(B)} \quad ab = c. \tag{1}$$

Case (A)

One alternative is

$$ab = e \quad \text{and} \quad ac = e. \tag{2}$$

This is impossible because it leads to $b = c$ i.e a group of order 3 instead of 4.
Another alternative is

$$ab = e \quad \text{and} \quad ac = b. \tag{3}$$

Then we get $ab^2 = b = ac$ or $b^2 = c$, i.e. we have a cyclic group of elements

$$a = b^{-1}, \quad b, \quad c = b^2, \quad e. \tag{4}$$

Case (B)

We have the group of elements a, b, ab, e. The product $a(ab) = a^2b$ can be either (1) $a^2b = e$ or (2) $a^2b = b$.

Alternative (1) $a^2b = e \Rightarrow b = a^{-2}$ or $a^2 = b^{-1}$. This gives $ba = a^{-2}a = a^{-1}$ or equivalently $ba^2 = e$. Hence $b(ab) = bab = a^{-1}a^{-2} = a^{-3}$. The product bab must be equal to one of the remaining group elements e or a. If $bab = e$ it follows that $a^3 = e$ or $b^{-1}a = e$ or $a = b$ which is impossible. If $bab = a$ it means $a^{-3} = a$ or $a^4 = e$ or $b^{-1}a^2 = e$ or $b = a^2$. Hence this alternative is the cyclic group

DEFINITION OF A GROUP

$$a, \quad b = a^2, \quad c = a^3, \quad e = a^4 \qquad (5)$$

Alternative (2) $a^2b = b$ implies $a^2 = e$ or $a = a^{-1}$.

If $bab = e$ it means $ba = b^{-1}$ or $a = b^{-2}$ or $b^2 = a^{-1} = a$ and we find again a cyclic group in the sequence

$$b, \quad a = b^2, \quad c = b^3, \quad e = b^4. \qquad (6)$$

If $bab = a$ it means $ba = ab^{-1}$ or $aba = a^2b^{-1} = b^{-1}$ or $ab = (ab)^{-1}$. But $ab = c$ hence $c^2 = e$. We can now easily see that $ba = ab$. The other alternatives are excluded. Then we have $aba = a^2b$ and $abab = b^2$ or $c^2 = b^2 = e$. We therefore deal with an abelian group having the following multiplication table

	e	a	b	c
e	e	a	b	c
a	a	e	c	b
b	b	c	e	a
c	c	b	a	e

(7)

The abstract group of order 4 with this structure is called the four-group or *Vierergruppe* or simply the V group.

A realization of the cyclic structure of the abstract group of order 4 is the group C_4 of four rotations of $\frac{1}{2}\pi, \pi, \frac{3}{2}\pi$ and 2π around a fixed axis. A realization of the structure (7) of the four-group is the group D_2 formed of rotations of π around three perpendicular axes meeting at the same point.

Point groups

The groups C_4, D_2 introduced in Exercise 2.1 are examples of C_n and D_n ($n \geq 2$). These are finite groups called *point groups*. These groups are related to symmetries of bodies where at least one point remains fixed during the symmetry transformation. For D_2, for example, this is the point where the three perpendicular axes meet. The symmetry transformations of the point groups are defined such that they bring the body back into itself. They preserve distances and can be built up from three basic transformations:

(1) rotation of a given angle about a given axis; if the angle is $\frac{2\pi}{n}$ the transformation is denoted by C_n ($n \geq 2$);
(2) mirror reflection in a plane, for which the symbol σ is customarily used;
(3) translations, this symmetry occurs only for an infinite body which is an extrapolation of a real one.

The point groups are important both in physics and chemistry. For a detailed discussion of their properties see, for example, Hamermesh (1962), Chapter 2 or Tinkham (1964), Chapter 4.

The symmetric group S_n

An example of a finite group of particular importance in quantum mechanics is the *symmetric* group. Its role in quantum mechanics follows from the fact that particles are indistinguishable or identical, which means that the expectation value of any physical observable remains unaltered by the exchange of particles forming the system. The exchange is expressed as a permutation of the particle labels. All possible permutations of particle labels form a group called the symmetric group. A system of identical particles is described by a wave function which is either symmetric for bosons, or antisymmetric for fermions at the exchange of two particles. The role of the symmetric group is to provide wave functions which incorporate correctly the property of indistinguishability by taking into account all degrees of freedom of the system.

The symmetric group is also important for understanding properties of other finite groups through Cayley's theorem discussed in Section 2.4.

The symmetric group will be extensively discussed in Chapter 4. Here we give some definitions and notations only. The group itself is denoted by S_n where n is the *degree* of the group and represents the number of objects under permutation.

Let us consider the n objects in the order $1, 2, ..., n$. A permutation is a transition from this order to another, say $a_1, a_2, ..., a_n$. The notation is

$$P_a = \begin{pmatrix} 1 & 2 \ldots n \\ a_1 & a_2 \ldots a_n \end{pmatrix}. \tag{2.6}$$

The columns can in fact be interchanged in P_a, which means that we can start from any initial order but we always make the transformation $i \mapsto a_i$. To illustrate the action of a permutation operator on a wave function we take for example

$$P_a = \begin{pmatrix} 1 & 2 & 3 & 4 & 5 \\ 5 & 3 & 2 & 1 & 4 \end{pmatrix}. \tag{2.7}$$

Applying it to a five-particle wave function one gets

$$P_a \psi(1, 2, 3, 4, 5) = \psi(5, 3, 2, 1, 4). \tag{2.8}$$

The permutations of n objects form a group of order $N = n!$. Its identity element is

$$P_e = \begin{pmatrix} 1 & 2 \ldots n \\ 1 & 2 \ldots n \end{pmatrix}. \tag{2.9}$$

The inverse of any element P_a is

$$P_a^{-1} = \begin{pmatrix} a_1 & a_2 \ldots a_n \\ 1 & 2 \ldots n \end{pmatrix}. \tag{2.10}$$

Applying successively P_a and

$$P_b = \begin{pmatrix} a_1 & a_2 & \ldots & a_n \\ b_{a_1} & b_{a_2} & \ldots & b_{a_n} \end{pmatrix}, \tag{2.11}$$

DEFINITION OF A GROUP

one gets another permutation

$$P_c = P_b P_a = \begin{pmatrix} 1 & 2 & \cdots & n \\ b_{a_1} & b_{a_2} & \cdots & b_{a_n} \end{pmatrix}. \tag{2.12}$$

For example

$$\begin{aligned} P_c &= \begin{pmatrix} 1 & 2 & 3 & 4 & 5 \\ 4 & 3 & 5 & 1 & 2 \end{pmatrix} \begin{pmatrix} 1 & 2 & 3 & 4 & 5 \\ 5 & 3 & 2 & 1 & 4 \end{pmatrix} \\ &= \begin{pmatrix} 5 & 3 & 2 & 1 & 4 \\ 2 & 5 & 3 & 4 & 1 \end{pmatrix} \begin{pmatrix} 1 & 2 & 3 & 4 & 5 \\ 5 & 3 & 2 & 1 & 4 \end{pmatrix} \\ &= \begin{pmatrix} 1 & 2 & 3 & 4 & 5 \\ 2 & 5 & 3 & 4 & 1 \end{pmatrix} \end{aligned} \tag{2.13}$$

where, after the second equality, we wrote P_b with its columns interchanged so as to get its initial order identical to the final order of P_a.

It is easy to see by reordering the columns that any permutation can be written as a product of closed cycles without common elements, i.e. circular permutations. For example

$$P_c = \begin{pmatrix} 1 & 2 & 5 \\ 2 & 5 & 1 \end{pmatrix} \begin{pmatrix} 3 \\ 3 \end{pmatrix} \begin{pmatrix} 4 \\ 4 \end{pmatrix} = (125)(3)(4) \tag{2.14}$$

where the final form is a shorthand notation. In this notation, (125) stands for $\begin{pmatrix} 1 & 2 & 5 \\ 2 & 5 & 1 \end{pmatrix}$ which means that 1 is replaced by 2, 2 by 5, and 5 by 1 according to the sequence in the bracket. Note that the last number turns into the first again such as to close the cycle. The notation (3) or (4) is straightforward. The permutation (2.14) contains a 3-cycle and two 1-cycles. The number of elements ℓ in an ℓ-cycle gives the *length* of that cycle. Keeping in mind the degree of a permutation, the 1-cycles can be omitted. Also a circular permutation can be written in several identical forms. Hence equivalent notations for P_c are

$$P_c = (125) = (251) = (512). \tag{2.15}$$

In terms of cycles the 3! elements of S_3 are

$$\begin{aligned} e &= \begin{pmatrix} 1 & 2 & 3 \\ 1 & 2 & 3 \end{pmatrix}, \quad (12) = \begin{pmatrix} 1 & 2 & 3 \\ 2 & 1 & 3 \end{pmatrix}, \quad (13) = \begin{pmatrix} 1 & 2 & 3 \\ 3 & 2 & 1 \end{pmatrix}, \\ (23) &= \begin{pmatrix} 1 & 2 & 3 \\ 1 & 3 & 2 \end{pmatrix}, \quad (123) = \begin{pmatrix} 1 & 2 & 3 \\ 2 & 3 & 1 \end{pmatrix}, \quad (132) = \begin{pmatrix} 1 & 2 & 3 \\ 3 & 1 & 2 \end{pmatrix}. \end{aligned} \tag{2.16}$$

where, on the left hand side, the elements are identified by their shorthand notation, as explained above. One can easily verify that they give a multiplication table, shown in Table 2.1.

Table 2.1 Multiplication table for S_3.

	e	(12)	(13)	(23)	(123)	(132)
e	e	(12)	(13)	(23)	(123)	(132)
(12)	(12)	e	(132)	(123)	(23)	(13)
(13)	(13)	(123)	e	(132)	(12)	(23)
(23)	(23)	(132)	(123)	e	(13)	(12)
(123)	(123)	(13)	(23)	(12)	(132)	e
(132)	(132)	(23)	(12)	(13)	e	(123)

A 2-cycle is also called a *transposition*. Any cycle can be written as a product of transpositions having elements in common. For example, using the above table one finds that

$$(123) = (13)(12) = (12)(23) = (23)(13). \qquad (2.17)$$

Notice that the order of factors is significant in writing (2.17). In general the order of factors is important whenever the transpositions are *adjoint*, i.e. have elements in common.

A permutation can be written as a product of transpositions in many ways. Two possible rules are:

$$(abc\ldots ef) = (af)(ae)\ldots(ac)(ab) \qquad (2.18a)$$
$$= (ab)(bc)\ldots(ef). \qquad (2.18b)$$

However for a given permutation the number q of transpositions is always *even* or *odd*, independent of the way they are written, and one can define the parity of a permutation P as $\delta_P = (-1)^q$ because, by definition, a transposition has parity -1 and the total parity is the product of transposition parities.

A simple way to calculate the parity is to introduce the decrement $n - k$ of a permutation P when n is the number of objects and k the number of cycles. The parity is then

$$\delta_P = (-1)^{n-k}. \qquad (2.19)$$

As a transposition has parity -1, from (2.18) it follows that an n-cycle has a parity $(-1)^{n-1}$. To see that (2.19) is reasonable let us denote by n_i the number of objects in an n_i-cycle. A permutation of k cycles will have a parity

$$\delta_P = (-1)^{\sum_{i=1}^{k}(n_i-1)} = (-1)^{n-k}$$

which proves (2.19). The product of two even permutations is even, while the product of two odd permutations is also even. The even permutations of degree n form a group called the *alternating* group A_n. Its order is $\frac{1}{2}n!$.

DEFINITION OF A GROUP

Exercise 2.2 Write all the elements of S_5 in terms of cycles.

Solution The group has 120 elements which can be separated into seven different categories in terms of their cyclic structure.

1. e   1 element
2. (12), (13), (14), (15), (23), (24), (25), (34), (35), (45)   10 elements
3. (12) (34), (12) (35), (12) (45), (13) (24), (13) (25), (13) (45),(14) (23), (14) (25), (14) (35), (15) (23), (15) (24), (15) (34), (24) (35), (25) (34), (23) (45)   15 elements
4. (123), (132), (124), (142), (134), (143), (125), (152), (135), (153), (145), (154), (234), (243), (235), (253), (245), (254), (345), (354)   20 elements
5. (123) (45), (132) (45), (124) (35), (142) (35), (134) (25), (143) (25),(125) (34), (152) (34), (135) (24), (153) (24), (145) (23), (154) (23), (234) (15), (243) (15), (235) (14), (253) (14), (245) (13), (254) (13),(345) (12), (354) (12)   20 elements
6. (1234), (1243), (1324), (1342), (1423), (1432), (1235), (1245), (1325), (1345), (1425), (1435), (1254), (1253), (1354), (1352), (1453), (1452), (1534), (1543), (1524), (1542), (1523), (1532), (2345), (2354), (2435), (2453), (2534), (2543)   30 elements
7. (12345), (12435), (13245), (13425), (14235), (14325), (12354), (12453), (13254), (13452), (14253), (14352), (12543), (12534), (13542), (13524), (14532), (14523), (15342), (15432), (15243), (15423), (15234), (15324)   24 elements

Infinite groups

There are also groups with an infinite number of elements. These fall into two categories. In one category, the elements are discrete, like those of finite groups. These are called *discrete* groups of infinite order. An example is the infinite set of integers, where the law of composition is addition. Zero is the identity; the inverse of n is $-n$. There are many examples of infinite discrete groups and they never need more than one discrete parameter to label the elements. This parameter can always be chosen to take positive integer values (see Hamermesh 1962, Section 8.2).

There is another category of groups with an infinite number of elements but with a continuous change from one element to another. These are called *continuous* groups. Here we shall be concerned only with those groups where the elements can be labelled by a *finite* set $a_1, a_2, ..., a_r$ of continuously varying parameters. The continuous transformations (2) and (4) of Section 1.2, i.e. translations, rotations, Lorentz transformations, and internal continuous symmetries, are examples. If the group has its elements labelled by $r \geq 1$ continuously varying *real* independent parameters, the group is said to be an *r*-parameter continuous group or a finite continuous group. In physics one deals with Lie groups, named after the Norwegian mathematician Sophus Lie. These are finite continuous groups where the transformations are analytic functions of the real parameters; hence they can be reduced to the study of infinitesimal transformations. Through this simplification the study of a group with an infinite number of transformations can be reduced to the study of commutation relations between a finite number r of operators. These relations form a Lie algebra and r is called the *order* of the finite continuous group. Chapter 5 is devoted to the study of general properties of Lie groups.

Exercise 2.3 Find the order of the special unitary group SU(2) whose elements are 2×2 matrices

$$u = \begin{pmatrix} a & b \\ c & d \end{pmatrix}, \tag{1}$$

where $a, b, c, d, \in \mathbb{C}$, with the properties of being unitary

$$uu^+ = u^+u = 1 \tag{2}$$

and having

$$\det u = +1. \tag{3}$$

Solution a, b, c, d are complex parameters; therefore, in the absence of constraints, there are eight real parameters. The condition (3) gives one constraint

$$ad - bc = 1.$$

The unitarity property is explicitly

$$\begin{pmatrix} a & b \\ c & d \end{pmatrix} \begin{pmatrix} a^* & c^* \\ b^* & d^* \end{pmatrix} = \begin{pmatrix} 1 & 0 \\ 0 & 1 \end{pmatrix}$$

which gives the following four distinct constraints

$$|a|^2 + |b|^2 = 1$$
$$|c|^2 + |d|^2 = 1$$
$$\text{Re}(ac^* + bd^*) = 0$$
$$\text{Im}(ac^* + bd^*) = 0.$$

There are eight parameters and five constraints, i.e. three free parameters. Therefore the group order is $r = 3$.

2.2 SUBGROUPS

From the elements of a discrete or continuous group G one can select a subset H and write

$$H \subset G \quad \text{or} \quad G \supset H \tag{2.20}$$

to symbolize that the subset H is contained in G. If the subset H itself forms a group, i.e. the product of any two elements of H is an element of H, under the same law of composition as that of G, H is called a subgroup of G. It is essential to find the subgroups of a group both for mathematical studies and for physical applications. Starting from a given G one can find a chain of subgroups

$$G \supset H_1 \supset H_2 \supset \cdots$$

and it may happen that a given G has several different chains. An important task is to find out the appropriate chain for a specific physical problem. This will be discussed on several occasions through the book.

Every group has two subgroups called improper subgroups. One consists of the identity element alone, the other is the whole group itself. Any other subgroup is called a proper subgroup. In general, if a finite group G of order N has a subgroup H of order N_h one has

$$N = hN_h \tag{2.21}$$

where h is a positive integer called the *index* of the subgroup H. This result is known as Lagrange's theorem.

Cosets

If g is an element of G one can form the set of elements gH by multiplying g with every element of the subgroup H. There is a one-to-one correspondence between H and gH. If $g \in H$ then gH is the subgroup itself. But if $g \in G$ but is not contained in H, then gH is not a group because it does not contain the identity element. It is called the *left coset* of H. In an analogous way, one can define the *right coset* Hg. Any element of G belongs to either H or to one of its cosets.

Let us take again the example of Exercise 2.1, i.e. the abstract group of order 4 with elements e, a, b, c. The multiplication table (7) shows that this group has three subgroups each with two elements. These are: H_1: e, a, H_2: e, b, and H_3: e, c.

From Table 2.1 one can see that S_3 contains the following four subgroups

$$\begin{align} H_1 & \quad e, (12) \\ H_2 & \quad e, (13) \\ H_3 & \quad e, (23) \\ H_4 & \quad e, (123), (132) \end{align} \tag{2.22}$$

Notice that the H_4 is a cyclic group because $(123)^2 = (132)$.

The group S_4 contains permutations on four symbols and has therefore 24 elements. In the group S_5, there are five S_4 subgroups containing the symbols: 1234, 1235, 1245, 1345 and 2345. One can make a choice, say 1234, and use Exercise 2.2 to find out the cyclic structure of S_4 by selecting out those elements where the symbol 5 is absent in cycles with two or more symbols. This gives the following five cyclic structures for S_4.

1. e
2. $(12), (13), (14), (23), (24), (34)$
3. $(12)(34), (13)(24), (14)(23)$ (2.23)
4. $(123), (132), (124), (142), (134), (143), (234), (243)$
5. $(1234), (1243), (1324), (1342), (1423), (1432)$

S_4 itself has several subgroups. These are the four S_3 subgroups S_3 (1,2,3), S_3 (1,2,4), S_3(1,3,4), S_3 (2,3,4) and their own subgroups, the four-group

$$e, (12)(34), (13)(24), (23)(14),$$

the cyclic groups isomorphic with C_4 of the type

$$e, (1234), (1234)^2 = (13)(24), (1234)^3 = (1432),$$

and the alternating group A_4 formed of permutations of even parity

$$e, (12)(34), (13)(24), (14)(23),$$
$$(123), (132), (124), (142), (134), (143), (234), (243).$$

From the example of S_4, one can see that, besides the identity element, the subgroups of a group can have other elements in common.

In general, a group S_n has the following chain of subgroups

$$S_n \supset S_{n-1} \supset S_{n-2} \supset \cdots \supset S_2. \qquad (2.24)$$

In Chapter 4, we shall see that this chain is extremely important in the study of the symmetric group. Also, in general, the alternating group A_n, formed of permutations of even parity, is a subgroup of S_n:

$$S_n \supset A_n. \qquad (2.25)$$

Its index as defined by (2.21) is $h = 2$.

Let us illustrate now the notion of subgroup for continuous groups. The rotation group of three dimensions, R_3, has an infinity of subgroups R_2 (consisting of all rotations about a fixed axis) because an infinite number of axes or rotation can be chosen. In each R_2, two successive rotations of angles φ_1 and φ_2 about the same axis give a rotation of $\varphi_1 + \varphi_2$. The identity element is a rotation of $\varphi = 0$ or $\varphi = 2\pi$. The inverse of a rotation of an angle φ is a rotation of an angle $-\varphi$ about the same axis.

All translations along a fixed axis form a subgroup of the group of translations in three dimensions.

The group R_3 is a subgroup of the Lorentz group, as will appear in Chapter 7.

2.3 ISOMORPHISM AND HOMOMORPHISM

Two groups G and G' are called *isomorphic* if there exists a one-to-one correspondence between their elements, i.e. for each $g \in G$ there is one and only

ISOMORPHISM AND HOMOMORPHISM

one $g' \in G'$, the correspondence $g \leftrightarrow g'$ being preserved under the multiplication law.

Isomorphic groups have the same structure, i.e. the same abstract group. Hence two realizations of the same abstract group are isomorphic. For example, D_2 and a subgroup of S_4 formed of elements e, $(12)(34)$, $(13)(24)$, and $(14)(23)$ are isomorphic as follows:

$$\begin{aligned}
\text{rotation of } \pi \text{ about the } x\text{-axis} &\leftrightarrow (12)(34) \leftrightarrow a \\
\text{rotation of } \pi \text{ about the } y\text{-axis} &\leftrightarrow (13)(24) \leftrightarrow b \\
\text{rotation of } \pi \text{ about the } z\text{-axis} &\leftrightarrow (14)(23) \leftrightarrow c \\
\text{product of all three rotations} &\leftrightarrow (1)(2)(3)(4) \leftrightarrow e
\end{aligned} \quad (2.26)$$

where a, b, c, e are the elements of the abstract group of order 4.

Another example of isomorphism is the group S_3 and the group D_3 which is the point group (see Section 2.1) of the equilateral triangle. The three rotations in the plane of the triangle

of angle $\frac{2}{3}\pi n$ ($n = 1, 2, 3$) correspond to permutations e, (123), and (132), respectively, of the corners. Each rotation of π (out of the triangle plane) about the perpendicular from one corner to its opposite side corresponds to a transposition (12), (13), or (23) of the triangle corners.

A group G is *homomorphic* to G' if, to any $g \in G$, there corresponds a $g' \in G'$, and to each g' at least one g such that for $g_1 g_2 = g$ one has $g'_1 g'_2 = g'$. The notation is $G \to G'$.

As an example, we shall prove later (Section 8.3) that the group SU(2) of unitary 2×2 matrices u of det $u = +1$ is homomorphic to the rotation group in three dimensions, R_3. In this case the homomorphism SU(2) $\to R_3$ means that both u and $-u$ correspond to the same rotation. Another example is the homomorphism SL(2,$\mathbb{C}$) $\to$ proper orthochronous Lorentz group, where SL(2,$\mathbb{C}$) is the special (S) linear (L) group in two dimensions (see Chapter 5). This relation will be encountered in Chapter 7 and represents the relativistic generalization of the homomorphism SU(2) $\to R_3$.

Kernel of a homomorphic mapping

A *mapping* ϕ of G onto G' represents a rule by which each element g of G corresponds to some element $g' = \phi(g)$ of G'. By a mapping, each element of G' becomes the *image* of at least one element of G in the case of a homomorphic mapping.

The set of elements of G which are mapped onto the identity element of G' form the *kernel* H of the mapping. It follows that H is an invariant subgroup of G (Cornwell 1984, Chapter 2). The definition of an invariant subgroup is given in Section 2.6.

2.4 CAYLEY'S THEOREM

Aside from its importance in quantum mechanics the symmetric group plays a dominant role in the study of finite groups due to Cayley's theorem.

Theorem Every finite group of order n is isomorphic to a subgroup of the symmetric group S_n.

The proof is based on the so-called rearrangement theorem (see Appendix B) which states that multiplication of all the elements of a group G by a given element of G rearranges the group elements in a different order so that to each element $g \in G$ there corresponds a permutation P_g

$$g \leftrightarrow P_g = \begin{pmatrix} g_1 & \cdots & g_n \\ gg_1 & \cdots & gg_n \end{pmatrix}. \tag{2.27}$$

This correspondence is one-to-one, since a given permutation P_g cannot be produced by another element g' because this would imply $g\,g_i = g'\,g_i$, hence $g = g'$. Moreover, it is easy to see that the correspondence (2.27) is preserved under the multiplication law, that is, if $g \leftrightarrow P_g$ and $g' \leftrightarrow P_{g'}$ then $g'g \leftrightarrow P_{g'} P_g = P_{g'g}$. For this purpose we can write the permutation $P_{g'}$ starting from the final order of P_g as initial order of $P_{g'}$

$$P_{g'} = \begin{pmatrix} gg_1 & \cdots & gg_n \\ g'gg_1 & \cdots & g'gg_n \end{pmatrix}, \tag{2.28}$$

from which one gets

$$\begin{aligned} P_{g'}P_g &= \begin{pmatrix} gg_1 & \cdots & gg_n \\ g'gg_1 & \cdots & g'gg_n \end{pmatrix} \begin{pmatrix} g_1 & \cdots & g_n \\ gg_1 & \cdots & gg_n \end{pmatrix} \\ &= \begin{pmatrix} g_1 & \cdots & g_n \\ g'gg_1 & \cdots & g'gg_n \end{pmatrix} = P_{g'g}. \end{aligned} \tag{2.29}$$

It follows that the permutations P_g corresponding to elements g of a finite group G of order n form a subgroup of S_n isomorphic to G.

The subgroups of S_n isomorphic in this way to finite groups of order n have some special properties (see Hamermesh 1962, Section 1.3). Except for the identity element they contain only permutations which leave no symbol unchanged. If the permutations are decomposable into cycles with no common elements these cycles must be of equal length. Permutations with these properties are called *regular* and

CAYLEY'S THEOREM

form a *regular subgroup* of S_n. For example, using the list (2.23) one can find that the group S_4 has the following regular subgroups:

$$e, (12)(34), (13)(24), (14)(23)$$
$$e, (1234), (1234)^2 = (13)(24), (1234)^3 = (1432)$$
$$e, (1243), (1243)^2 = (14)(23), (1243)^3 = (1342) \quad (2.30)$$
$$e, (1324), (1324)^2 = (12)(34), (1324)^3 = (1423).$$

The group S_4 also contains permutations with cycles of different lengths. They are not regular because they leave symbols unchanged. One case is the permutation (123) $\equiv$ (123)(4) which leaves the symbol (4) unchanged. Its two cycles have different lengths, 3 and 1, respectively.

Cayley's theorem has two important consequences.

(a) From a practical point of view the theorem can be used to find subgroups of a group, as in the example of S_4 given above, or to understand the structure of some groups. An interesting case is the finite group of order n where n is a prime number. The subgroup of S_n isomorphic with it must contain regular permutations with cycles of equal length, i.e. n must be a multiple m of a cycle length ℓ, $n = m\ell$. Since n is prime one must have $\ell = 1$ or n. Therefore a regular subgroup of S_n must contain permutations having one n-cycle, i.e. a permutation and its powers up to n. One of the powers is the identity element e (for length $\ell = 1$). Therefore a finite group of order n is necessarily a cyclic group if n is prime.

(b) As a general fact, it follows that the number of non-isomorphic groups of order n is finite because S_n is finite and therefore it has a finite number of subgroups. In particular any point group is isomorphic to a subgroup of S_n.

Exercise 2.4 Find the regular subgroups of S_5.

Solution From the definition of regular permutations it follows that only cyclic structures (1) and (7) of Exercise 2.2 are acceptable. The cyclic structure (1) gives the identity element. The 24 elements of (7) can be distributed among six cyclic groups of order 5 obtained as follows. Let us take for starting the element (12345). Its powers give

$$(12345)^2 = \begin{pmatrix} 1 & 2 & 3 & 4 & 5 \\ 2 & 3 & 4 & 5 & 1 \end{pmatrix} \begin{pmatrix} 1 & 2 & 3 & 4 & 5 \\ 2 & 3 & 4 & 5 & 1 \end{pmatrix} = \begin{pmatrix} 1 & 2 & 3 & 4 & 5 \\ 3 & 4 & 5 & 1 & 2 \end{pmatrix}$$
$$= (13524)$$
$$(12345)^3 = \begin{pmatrix} 1 & 2 & 3 & 4 & 5 \\ 2 & 3 & 4 & 5 & 1 \end{pmatrix} \begin{pmatrix} 1 & 2 & 3 & 4 & 5 \\ 3 & 4 & 5 & 1 & 2 \end{pmatrix} = \begin{pmatrix} 1 & 2 & 3 & 4 & 5 \\ 4 & 5 & 1 & 2 & 3 \end{pmatrix}$$
$$= (14253)$$

$$(12345)^4 = \begin{pmatrix} 1 & 2 & 3 & 4 & 5 \\ 2 & 3 & 4 & 5 & 1 \end{pmatrix} \begin{pmatrix} 1 & 2 & 3 & 4 & 5 \\ 4 & 5 & 1 & 2 & 3 \end{pmatrix} = \begin{pmatrix} 1 & 2 & 3 & 4 & 5 \\ 5 & 1 & 2 & 3 & 4 \end{pmatrix}$$
$$= (15432)$$
$$(12345)^5 = \begin{pmatrix} 1 & 2 & 3 & 4 & 5 \\ 2 & 3 & 4 & 5 & 1 \end{pmatrix} \begin{pmatrix} 1 & 2 & 3 & 4 & 5 \\ 5 & 1 & 2 & 3 & 4 \end{pmatrix} = \begin{pmatrix} 1 & 2 & 3 & 4 & 5 \\ 1 & 2 & 3 & 4 & 5 \end{pmatrix}$$
$$= e$$

Proceeding in a similar way for (12435), (13245), (13425), (14235), and (14325) one gets another five cyclic groups. The six regular subgroups of S_5 obtained in this way are:

1. e, (12345), (13524), (14253), (15432)
2. e, (12435), (14523), (13254), (15342)
3. e, (13245), (12534), (14352), (15423)
4. e, (13425), (14532), (12354), (15243)
5. e, (14235), (12543), (13452), (15324)
6. e, (14325), (13542), (12453), (15234)

As a consequence of Cayley's theorem it follows that the finite group of order $n = 5$ is isomorphic to any of the above regular subgroups and is therefore cyclic, which happens whenever n is a prime number.

2.5 CLASSES

Two elements a and b of G are *conjugate* to each other if there is a third element $x_0 \in G$ such that

$$b = x_0 a x_0^{-1} \quad \text{or} \quad a = x_0^{-1} b x_0. \tag{2.31}$$

If two elements a and b are both conjugate to c they are all conjugate to each other. This can easily be seen because if a is conjugate to c there is an x_0 such that

$$a = x_0 c x_0^{-1} \quad \text{or} \quad c = x_0^{-1} a x_0$$

and if b is conjugate to c there is a y_0 such that

$$b = y_0 c y_0^{-1} = y_0 x_0^{-1} a x_0 y_0^{-1} = (y_0 x_0^{-1}) a (y_0 x_0^{-1})^{-1}$$

which shows that a and b are conjugate through $y_0 x_0^{-1} \in G$. It follows that any element is conjugate to itself.

A *conjugacy class* or simply a *class* is a set of elements conjugate to a given element by all elements of the group. From the above discussion one can see that all elements in a class are mutually conjugate. The elements of a group are divided into distinct classes. If we denote the classes of a group G by C_i ($i = 1, 2, ..., K$) the group can be written as the union of its classes

$$G = C_1 \cup C_2 \cup \cdots \cup C_K \tag{2.32}$$

where $K \leq N$ for a finite group of order N. For an abelian group each element is also a class, i.e. it has $K = N$.

For continuous groups related to space transformations, classes can be understood geometrically. For example, take a rotation of angle ω about an axis n_0. Any other rotation of ω but about an axis n different from n_0 belongs to the same class because one can always find a rotation R to bring the rotation axis n into n_0 and its inverse R^{-1} to take n_0 into n such that

$$R_{n_0}(\omega) = RR_n(\omega)R^{-1}, \tag{2.33}$$

i.e. first R^{-1} rotates n_0 into n, R_n produces the rotation of angle ω about n, and R rotates n back into n_0, which is equivalent to a rotation of ω about n_0.

Classes play an important role in the study of groups and their representations; this will be seen in the next chapter.

Let us consider the group S_3 and choose an element, say (123). Using Table 2.1 and the fact that a transposition is identical to its inverse, one obtains the class of (123) as follows

$$\begin{align}e(123)e &= (123)\\(12)(123)(12) &= (12)(13) = (132)\\(13)(123)(13) &= (13)(23) = (132)\\(23)(123)(23) &= (23)(12) = (132)\\(123)(123)(123)^{-1} &= (123)\\(132)(123)(132)^{-1} &= (132)(123)(123) = (132)(132) = (123)\end{align} \tag{2.34}$$

This shows that the class of (123) has two elements: (123) and (132). Using Table 2.1, one can get the other two classes of S_3. One is (e) itself and the other is formed of (12), (13), and (23). For any group the identity element is in a class by itself. By working out other examples one can find that in general a group S_n has as many classes as it has cyclic structures, and therefore all elements in a class have the same cyclic structure. From list (2.23) and Exercise 2.2 it follows that S_4 has five classes and S_5 has seven classes.

Partition

We have seen that each permutation can be written as a product of closed cycles, without common elements. Suppose that in a permutation of n objects an i-cycle appears k_i times. Then one must get

$$k_1 + 2k_2 + \cdots + nk_n = n \tag{2.35}$$

where $k_i \geq 0$. As each cyclic structure represents a class, it means that each set of integers k_i satisfying (2.35) corresponds to a class of S_n.

One can introduce the integers

$$\begin{aligned} \lambda_1 &= k_1 + k_2 + \cdots + k_n \\ \lambda_2 &= k_2 + \cdots + k_n \\ &\vdots \\ \lambda_n &= k_n. \end{aligned} \qquad (2.36)$$

Then equation (2.35) becomes

$$\lambda_1 + \lambda_2 + \cdots + \lambda_n = n \qquad (2.37)$$

where, by definition, $\lambda_1 \geq \lambda_2 \geq \ldots \geq \lambda_n \geq 0$.

The set $\lambda = [\lambda_1, \lambda_2, \ldots, \lambda_n]$ is called a *partition*. One can also express k_i in terms of λ_i

$$\begin{aligned} k_1 &= \lambda_1 - \lambda_2 \\ k_2 &= \lambda_2 - \lambda_3 \\ &\vdots \\ k_n &= \phantom{\lambda_2 - {}} \lambda_n. \end{aligned} \qquad (2.38)$$

In other words, there is a one-to-one correspondence between the sets $[k_1, k_2, \ldots, k_n]$ and $[\lambda_1, \lambda_2, \ldots, \lambda_n]$. This means that there is a one-to-one correspondence between a partition λ and a cyclic structure or a class, and that the number of partitions of n is equal to the number of classes of S_n. For example, for S_3, equation (2.37) takes one of the following three forms

$$\begin{aligned} 3 &= 3 + 0 + 0 \\ 3 &= 2 + 1 + 0 \\ 3 &= 1 + 1 + 1. \end{aligned}$$

Usually one omits the zeroes and the three partitions of 3 read [3], [21], and [111] $\equiv$ [1^3], respectively. As we have seen there are also three cyclic structures or classes in S_3: one is the identity element, another is made up of transpositions, and the third is formed of the permutations (123) and (132).

Inequalities

Let $\lambda = [\lambda_1, \lambda_2, \ldots, \lambda_n]$ and $\mu = [\mu_1, \mu_2, \ldots, \mu_n]$ be two partitions of n. By definition:

(1) $\lambda = \mu$ if $\lambda_i = \mu_i$ for $i = 1, 2, \ldots, n$;
(2) $\lambda > \mu$ if the first non-zero term in the sequence $\lambda_1 - \mu_1, \lambda_2 - \mu_2, \ldots$ is positive;
(3) $\lambda < \mu$ if the first non-zero term in the sequence $\lambda_1 - \mu_1, \lambda_2 - \mu_2, \ldots$ is negative.

Example For $n = 6$ the following inequalities hold

$$[6] > [51] > [42] > [33] > [321] > [311] > [222] > [21^4] > [1^6].$$

Exercise 2.5 Give the partitions of $n = 5, 6$.

Solution

$n = 5$ $[5], [41], [32], [31^2], [2^21], [21^3], [1^5]$

$n = 6$ $[6], [51], [42], [41^2], [3^2], [321], [31^3], [2^3], [2^21^2], [21^4], [1^6]$

i.e. S_5 (Exercise 2.2) and S_6 have seven and eleven cyclic structures or classes, respectively.

One can calculate the number of permutations contained in a class of S_n (Hamermesh 1962, Section 1.5). If the structure of a permutation is

$$k_1 \text{ 1-cycle,} \quad k_2 \text{ 2-cycle,} \quad \ldots, \quad k_n \text{ n-cycle}$$

the number of distinct permutations of such a structure is

$$n_{(k_1,\ldots,k_n)} = \frac{n!}{k_1! k_2! \cdots k_n! 2^{k_2} 3^{k_3} \cdots n^{k_n}}. \quad (2.39)$$

The explanation is as follows. The total number of permutations is $n!$ but not all are distinct. That is why one has to divide $n!$ by $k_1! k_2! \ldots k_n!$ where $k_i!$ is the number of times i-cycles can be permuted among themselves. The division by $2^{k_2} 3^{k_3} \ldots n^{k_n}$ is explained by the fact that in a cycle of n objects one can make n^{k_n} circular permutations which are not distinct, i.e. (123) is the same as (231) or (312).

We can check this formula with Exercise 2.2. For example, the category 7 of elements of S_5 contains 5-cycle objects only. This means

$$k_1 = k_2 = k_3 = k_4 = 0 \quad k_5 = 1.$$

Hence

$$n_{(0,0,0,0,1)} = \frac{5!}{5} = 24$$

as indicated in the exercise.

Remark

Any symmetric group has the property that the inverse of an element belongs to the same class as the element itself. This follows immediately from the definition (2.10) of an inverse: the initial and final orders of P_a^{-1} are interchanged with respect to those of P_a but the cyclic structure remains the same. Hence they belong to the same class.

2.6 SIMPLE AND SEMI-SIMPLE GROUPS

Let H be a subgroup of G, $H \subset G$. Then H is an invariant *subgroup* of G if it contains all its conjugate elements, i.e. if

$$ghg^{-1} \in H \quad \text{for all} \quad g \in G \text{ and } h \in H. \tag{2.40}$$

This is equivalent to writing

$$gH = Hg.$$

Alternative names for H are *self-conjugate subgroup, normal subgroup* or *normal divisor*. A characteristic of an invariant subgroup is that its elements belong to one or several complete classes. In other words, those classes of the union (2.32) which form a group give the invariant subgroup H of G.

If H is an invariant subgroup, its index h given by equation (2.21), is the order of the so-called *factor* or *quotient* group, denoted by G/H. This group has as elements the invariant subgroup and its cosets. For example, the subgroup A_n of S_n has only one coset, the set of odd permutations. Hence, the quotient group of S_n has two elements, one being A_n itself and the other the set of odd permutations.

Example $S_3/A_3 = \{A_3, B\}$ is a factor group of order 2 with

$$A_3 = ((e), (123), (132)), \quad B = ((12), (13), (23)).$$

This group is homomorphic to S_2:

$$A_3 \to e, B \to (12).$$

A group is called *simple* if it is not abelian and does not have any proper invariant subgroups. Since every subgroup of an abelian group is invariant, an abelian group is simple if and only if it has no proper subgroups (for abelian Lie groups see Section 5.5)

A group is called *semi-simple* if none of its invariant subgroups is abelian. Obviously a simple group is also semi-simple.

Using Table 2.1, one finds that the subgroup H_4 of S_3, defined by (2.22), is an Abelian invariant subgroup. Therefore, S_3 is not semi-simple. This subgroup of S_3, formed of elements e, (123), and (132) is precisely the alternating group A_3 of S_3. For $n = 4$, the reader is left to prove that besides A_4 the subgroup

$$e, (12)(34), (13)(24), (23)(14)$$

isomorphic to the four-group, is also an invariant subgroup. The four-group is an Abelian subgroup (see Exercise 2.1). Therefore, S_4 is not a semi-simple group either. For $n > 4$, the only invariant subgroup of S_n is the alternating group A_n. Therefore, S_n with $n > 4$ is not simple, but it is semi-simple.

As we shall see in Chapter 5, the property of being simple or semi-simple is essential in the classification of the Lie groups.

2.7 DIRECT PRODUCT GROUPS AND SEMI-DIRECT PRODUCT GROUPS

One can define a group G as a *direct product* of two other groups H_1 and H_2 if:

(a) all elements of H_1 commute with those of H_2;
(b) H_1 and H_2 have only the identity element in common;
(c) an element of G can be uniquely written as a product of $h_1 \in H_1$ and $h_2 \in H_2$:

$$g = h_1 h_2 = h_2 h_1. \tag{2.41}$$

The notation used here for the direct product group is[†]

$$G = H_1 \times H_2 = H_2 \times H_1. \tag{2.42}$$

In an enlarged definition, G is a direct product group if it is isomorphic with $H_1 \times H_2$. Naturally, the direct product can be generalized to more than two factors provided all H_i ($i = 1, 2, \ldots, n$) commute among themselves. All these groups have to be distinct and their only common element is the identity element e.

An important property is that each H_i is an invariant subgroup of G. Let us discuss this property in the case where G is a product of two factors. Let us choose an element $\bar{h}_1$ of H_1. Its conjugate

$$g\bar{h}_1 g^{-1} = h_1 h_2 \bar{h}_1 h_2^{-1} h_1^{-1} = h_1 h_2 h_2^{-1} \bar{h}_1 h_1^{-1} = h_1 \bar{h}_1 h_1^{-1}$$

obviously belongs to H_1, that is H_1 is an invariant subgroup of G. The same can be shown for H_2.

There are groups which occur in physical applications and have the structure of a direct product. One example is the group O(3) isomorphic to SO(3) $\times$ I where O(3) and SO(3) are orthogonal groups (Chapter 6) and I is the group of inversions consisting of two elements e ($x \to x$) and P ($x \to -x$) (see Chapter 3). Another example is the group SO(4) = SO(3) $\times$ SO(3) (Chapter 5).

A group G is a *semi-direct product* if it has two subgroups H_1 and H_2 such that

(a) H_1 is an invariant subgroup of G;
(b) H_1 and H_2 have only the identity element in common; and
(c) any element of G can be written uniquely as a product of $h_1 \in H_1$ and $h_2 \in H_2$.

The notation for the semi-direct product group used here is

$$G = H_1 \wedge H_2 \tag{2.43}$$

[†] Another notation used in the literature is $G = H_1 \otimes H_2$.

Examples of (2.43) are the Euclidean groups (Section 6.3) and the Poincaré group (Chapter 7).

SUPPLEMENTARY EXERCISES

2.6 Find the element (g) of S_6 satisfying the relation

$$(36) = (46)(34)(g).$$

2.7 The elements

$$e, (12)(34), (13)(24), (14)(23)$$

of the group S_4 form a subgroup. Indicate the classes of this subgroup.

2.8 A finite group G is generated by taking all possible products and powers of the matrices

$$h_1 = \begin{pmatrix} 0 & -1 \\ 1 & 0 \end{pmatrix}, \quad h_2 = \begin{pmatrix} -1 & 0 \\ 0 & 1 \end{pmatrix}.$$

(a) Find all matrices in G. Which is the identity element e?
(b) Find the order of each element in G. (The order of an element g is the smallest integer n for which $g^n = e$.)
(c) Partition the elements of G into classes.

3

LINEAR REPRESENTATIONS OF A GROUP

In quantum mechanics we are interested in the properties of eigenstates under various transformations (Chapter 1). Group theory offers a systematic way of finding these properties from the symmetries of the Hamiltonian. The eigenstates form linear vector spaces which in terms of group theory provide matrix representations of the group of transformations G under consideration. Let us denote by $D(g)$ such a representation where g is an element of G.

A trivially simple case of a representation is obtained for the group of inversions where the matrices are 1×1. The two elements of the group are the identity element $e(x \to x)$ and $P(x \to -x)$. For any even parity state, $\pi = +1$ (see Chapter 1), the representation is

$$D(e) = 1 \quad D(P) = 1. \tag{3.1}$$

The odd parity states, $\pi = -1$, give rise to a different representation

$$D(e) = 1 \quad D(P) = -1. \tag{3.2}$$

A representation is formed of as many matrices as there are elements in the group.

We give below a general definition of a linear representation. First we recall the definition of a linear vector space and a linear operator.

3.1 LINEAR VECTOR SPACES

A set of elements $x, y, z, \ldots$ form a *linear vector space* L if:

(1) to any pair x, y there corresponds a third element z obtained by *addition* of x and y:

$$z = x + y; \tag{3.3}$$

(2) to any x and any complex number α, there corresponds another vector $y = \alpha x \in L$ obtained by *multiplication* of x by α.

The operations of *addition* and *multiplication* have to satisfy the following axioms:

(a) the addition must be commutative

$$x + y = y + x \tag{3.4}$$

and associative

$$(x + y) + z = x + (y + z); \tag{3.5}$$

(b) the set L must contain a zero vector $\mathbf{0}$ such that

$$x + \mathbf{0} = \mathbf{0} + x = x \quad \text{for all } x; \tag{3.6}$$

(c) for any x there exists one and only one element $(-x)$ called the *inverse* of x such that

$$x + (-x) = \mathbf{0}; \tag{3.7}$$

(d)

$$1x = x; \tag{3.8}$$

(e) the multiplication must satisfy

$$(\alpha\beta)x = \alpha(\beta x) \tag{3.9}$$

$$(\alpha + \beta)x = \alpha x + \beta x \tag{3.10}$$

$$\alpha(x + y) = \alpha x + \alpha y \tag{3.11}$$

where the multiplication coefficients are real or complex numbers.

Realizations of the abstract definition of a linear vector space are the space of Euclidian geometry, the set of polynomials of degree $\leq n$, the set of $n \times n$ matrices, the set of all functions defined on a given set, etc.

In quantum mechanics the wave functions ψ form a linear vector space. They are assumed to have a finite norm which requires

$$\int \psi^*(r)\psi(r) \mathrm{d}^3 r < \infty. \tag{3.12}$$

If the Hamiltonian, which is a Hermitian operator H, is bounded from below, but not from above, which means that for any real constant c there exists a state vector ψ such that

$$\frac{<\psi|H|\psi>}{<\psi|\psi>} > c \tag{3.13}$$

then the set ψ_i of all eigenvectors of H is complete and infinite. The space spanned by such complete set is called Hilbert space (Courant and Hilbert 1953, Vol. 1, p. 55; Lee 1981, Chapter 1). The integral (3.12) is a special case of the scalar product of two state vectors

$$<\psi_i|\psi_j> = \int \psi_i^*(r)\psi_j(r)\mathrm{d}^3 r \qquad (3.14)$$

and when this product vanishes the states are called orthogonal.

An n-dimensional linear vector space L can be formed from a set of functions ψ_1, ψ_2, ..., ψ_n which correspond to the same eigenvalue of a Hamiltonian H.

Any system of linearly independent vectors forms a basis in the space L. Any vector can be expressed uniquely as a linear combination of the basis vectors.

In a linear vector space one can define linear operators acting in this space. Suppose that to any x there corresponds a vector x'. Then one can define an operator A which maps x into x'

$$x' = Ax. \qquad (3.15)$$

The operator A is *linear* if it satisfies

$$A(x+y) = Ax + Ay$$
$$A\alpha x = \alpha Ax \qquad (3.16)$$

for any vectors x, y and complex numbers α.

If A induces a one-to-one correspondence one can also write

$$x = A^{-1}x' \qquad (3.17)$$

where A^{-1} is the inverse of A, i.e.

$$AA^{-1} = A^{-1}A = 1. \qquad (3.18)$$

3.2 DEFINITION OF A GROUP REPRESENTATION

A group Γ of linear operators defined in a finite dimensional linear vector space L is called a linear representation of an arbitrary group G if G is homomorphic to Γ.

Let us call $S(g) \in \Gamma$ the operator corresponding to $g \in G$. Then one has

$$S(g_1)S(g_2) = S(g_1 g_2) \qquad (3.19)$$

$$S(e) = e \qquad (3.20)$$

$$S(g^{-1}) = S^{-1}(g). \qquad (3.21)$$

The third relation follows from the fact that the operator $S(g)$ has to be non-singular, i.e. it has an inverse $S^{-1}(g)$ in order to fulfil the group axioms. Then one can write

$$S(g)S^{-1}(g) = e. \qquad (3.22)$$

On the other hand, $g\, g^{-1} = e$ implies

$$S(g)S(g^{-1}) = S(e) = e. \qquad (3.23)$$

The comparison of (3.22) and (3.23) leads to (3.21).

If G and Γ are isomorphic the representation is called *faithful*.

3.3 MATRIX REPRESENTATIONS

If the dimension of L is n the representation has *degree n* or is an *n*-dimensional representation. In fact the operators $S(g)$ generate $n \times n$ matrices acting on the basis vectors $|1>, |2>, ..., |n>$ of L:

$$S(g)|k> = \sum D^{\mu}_{ik}(g)|i> \qquad (3.24)$$

The matrices $D^{\mu}(g)$ form a *matrix representation* of the group G. They usually carry a superscript index μ related to the dimension of the representation. The common notations for a matrix representation are Γ (see e.g. Tinkham 1964, Cornwell 1984) or D (see e.g. Wigner 1959, Hamermesh 1962). We shall use the letter D.

Let us take a set of functions $\psi_1, ..., \psi_n$ which correspond to the same eigenvalue of a Hamiltonian H. Suppose H is invariant to the group of transformations Γ, i.e.

$$[H, S(g)] = 0. \qquad (3.25)$$

Then any new function $S(g)\psi_i$ corresponds to the same eigenvalue. The matrix transformation from ψ_i to $S(g)\psi_i$ is a linear representation of G. This is a particular case of what is called in group theory an invariant subspace. In a linear space L it may be possible to find a subspace L' of basis vectors ψ_i having the property that a transformed vector $S(g)\psi_i$ also belongs to L'. Then L' is called an *invariant subspace* if this property is maintained for all transformations $S(g)$ of the representation Γ.

3.4 EQUIVALENT REPRESENTATIONS

Let us consider two representations $S(g)$ and $S'(g)$ in L and L', respectively. If L and L' have the same dimension and one can find a non-singular linear operator M transforming L and L' into each other such that

$$MS(g) = S'(g)M \qquad (3.26)$$

for any g then the two representations are called *equivalent*. In other words, if we change the basis in the space L by a matrix M the representation will become

$$S'(g) = MS(g)M^{-1}. \qquad (3.27)$$

Any transformation of a matrix having the form (3.27) is called a *similarity transformation*.

3.5 UNITARY REPRESENTATIONS

In quantum mechanics a wave function is a state vector or a linear combination of state vectors, i.e. of eigenstates ψ_i of the Hamiltonian. As mentioned above, the ψ_i form a Hilbert space, in which a scalar product is defined by

$$<\psi_i|\psi_j> = <\psi_j|\psi_i>^*. \qquad (3.28)$$

The basis vectors ψ_i can be chosen to be orthonormal:

$$<\psi_i|\psi_j> = \delta_{ij}. \qquad (3.29)$$

An operator U is called unitary if

$$<U\phi, U\psi> = <\phi, \psi>.$$

For an orthonormal basis this implies

$$<U\psi_i|U\psi_j> = <\psi_i|\psi_j> = \delta_{ij} \quad \text{for all } i,j. \qquad (3.30)$$

The matrix associated to U in an orthonormal basis is a unitary matrix

$$UU^+ = U^+U = 1. \qquad (3.31)$$

If the operators of a representation $S(g)$ of a group G are unitary, or equivalently the matrices $D(g)$ are unitary, the representation is called *unitary*. The unitary matrices have useful properties, which simplify calculations, as we shall see in the following chapters. But there are representations which are not unitary or are not equivalent to a unitary representation. This happens only for some infinite groups. Most groups of interest in physics possess only unitary representations or representations that can be transformed into unitary ones.

Theorem Every representation is equivalent to a unitary representation for finite or compact Lie groups.

Proof For finite groups there is a standard proof, followed below.

One can introduce an invariant sum operator as

$$H^2 = \frac{1}{N} \sum_{h \in G} S^+(h)S(h) \qquad (3.32)$$

where N is the group order and the sum runs through all elements h of G. This is a Hermitian operator and one can show that its eigenvalues are real and positive (see Cornwell 1984, Appendix C). It is then legitimate to define its square root

$$H = (H^2)^{\frac{1}{2}}. \qquad (3.33)$$

The operator H gives the equivalent representation

$$S'(g) = HS(g)H^{-1} \tag{3.34}$$

which we want to prove is unitary. Let us first show that

$$\begin{aligned} S^+(g)H^2S(g) &= \frac{1}{N}\sum_{h\in G} S^+(g)S^+(h)S(h)S(g) \\ &= \frac{1}{N}\sum_{h\in G} S^+(hg)S(hg) \\ &= \frac{1}{N}\sum_{h'\in G} S^+(h')S(h') \\ &= H^2 \end{aligned} \tag{3.35}$$

where $h' = hg$ runs also through all elements of G, rearranged by multiplication on the right, with the fixed element g. It is in the sense of the equality (3.35) that H^2 is an invariant operator. Multiplying (3.35) by H^{-1} on the left and $S^{-1}H^{-1}$ on the right one gets

$$H^{-1}S^+(g)H = HS^{-1}H^{-1} \tag{3.36}$$

or alternatively, by using the fact that H is Hermitian,

$$\left(HS(g)H^{-1}\right)^+ = \left(HS(g)H^{-1}\right)^{-1} \tag{3.37}$$

or using the definition (3.34)

$$S'^+(g) = S'^{-1}(g) \tag{3.38}$$

which proves that the equivalent representation S' is unitary.

The proof can be generalized to compact Lie groups (Chapter 5) through the use of an invariant integration over the group elements (see Appendix C) instead of the invariant sum (3.32). The proof is then analogous to that for finite groups.

3.6 CHARACTERS

The character of an element g in a matrix representation $D(g)$ is defined as

$$\chi^\mu(g) = \sum_i D^\mu_{ii}(g). \tag{3.39}$$

The character of the group G in the representation $D(g)$ is the set $\chi^\mu(g)$ where g runs through all elements of G. This is also called the character of the representation.

The following two results follow from the definition of the character.

(A) Two equivalent representations D' and D have the same character because,

EXAMPLES OF REPRESENTATIONS

omitting μ, one has

$$\chi'(g) = \sum_i (MD(g)M^{-1})_{ii} = \sum_i (M^{-1}MD(g))_{ii} = \sum D_{ii}(g) = \chi(g) \qquad (3.40)$$

for any g.

(B) Two elements g and g' belonging to the same class of a group G have the same character. If g and g', of the same class, are conjugate to each other by h, i.e. $g' = hgh^{-1}$, using the properties of a representation one has

$$D(g') = D(h)D(g)D^{-1}(h) \qquad (3.41)$$

or, by circular permutations of matrices in a trace, as above, one gets

$$\chi(g') = \chi(g). \qquad (3.42)$$

If the classes of G are labelled by $C_1, C_2, \ldots, C_K$, a representation D^μ can be described by the characters $\chi_1^\mu, \chi_2^\mu, \ldots, \chi_K^\mu$ which is a set of numbers. Tables of characters are available for all finite groups of practical use. The entries are $\chi_1^\mu, \chi_2^\mu, \ldots, \chi_K^\mu$ for each μ. A discussion of how to find character tables is given in Section 3.12.

3.7 EXAMPLES OF REPRESENTATIONS

Examples of representations will appear in connection with the study of various groups starting with Chapter 4. Here we consider two simple examples which will serve to illustrate the meaning of matrix representations.

Example 3.1 The group R_z of rotations in two dimensions is a continuous group of one parameter θ, the angle of rotation about the z-axis. Let us call its elements $R_z(\theta)$ and consider a rotation (Fig. 3.1) of a vector $\mathbf{r}$ of components

$$x = r\cos\alpha, \qquad y = r\sin\alpha. \qquad (3.43)$$

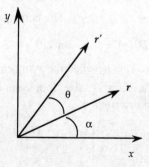

Figure 3.1 Rotation of angle θ of the vector $\mathbf{r} = (r\cos\alpha, r\sin\alpha)$ about the z-axis.

Under the transformation $R_z(\theta)$ the vector $\boldsymbol{r}$ becomes

$$\boldsymbol{r}' = R_z(\theta)\boldsymbol{r} \tag{3.44}$$

with $|\boldsymbol{r}'| = |\boldsymbol{r}|$ and components

$$x' = r\cos(\alpha + \theta), \qquad y' = r\sin(\alpha + \theta). \tag{3.45}$$

This transformation can be written in matrix form as

$$\begin{pmatrix} x' \\ y' \end{pmatrix} = \begin{pmatrix} \cos\theta & -\sin\theta \\ \sin\theta & \cos\theta \end{pmatrix} \begin{pmatrix} x \\ y \end{pmatrix} \tag{3.46}$$

which gives a 2×2 matrix representation to R_z:

$$D_z(\theta) = \begin{pmatrix} \cos\theta & -\sin\theta \\ \sin\theta & \cos\theta \end{pmatrix}. \tag{3.47}$$

An equivalent representation can be found by using the matrix

$$M = \begin{pmatrix} 1 & -i \\ 1 & i \end{pmatrix}, \qquad M^{-1} = \frac{1}{2}\begin{pmatrix} 1 & 1 \\ i & -i \end{pmatrix} \tag{3.48}$$

which brings $D_z(\theta)$ to diagonal form

$$MD_z(\theta)M^{-1} = \begin{pmatrix} e^{-i\theta} & 0 \\ 0 & e^{i\theta} \end{pmatrix} \tag{3.49}$$

for any θ.

Example 3.2 The group S_3 of permutations of three objects has been defined by Table 2.1. Its representations can be introduced as follows.

1. Any symmetric function f_S with respect to permutations of indices 1, 2, 3 form an invariant subspace (see Section 3.3) of S_3 because all permutations leave it unchanged :

$$g f_S = f_S \quad \text{for any } g \in S_3.$$

Examples of f_S are $x_1 + x_2 + x_3$ or $x_1 x_2 x_3$. The corresponding representation is

$$D^S(g) = +1 \quad \text{for all } g \in S_3 \tag{3.50}$$

i.e. a 1×1 matrix for all g. This is called the symmetric representation of S_3.

2. Any antisymmetric function f_A also forms an invariant subspace. Such a function can be written as a determinant

$$f_A = \begin{vmatrix} x_1 & x_2 & x_3 \\ y_1 & y_2 & y_3 \\ z_1 & z_2 & z_3 \end{vmatrix}. \tag{3.51}$$

EXAMPLES OF REPRESENTATIONS

Application of elements of S_3 on f_A gives $\pm f_A$, depending on the parity of the permutation. Then the representation is

$$D^A(g) = \begin{cases} +1 & \text{for } e, (123), \text{ or } (132) \\ -1 & \text{for } (12), (23), \text{ or } (13). \end{cases} \quad (3.52)$$

This is the antisymmetric representation of S_3.

3. A convenient simple basis for finding a third representation is given by the internal coordinates of a system of three particles of equal mass (Fig. 3.2).

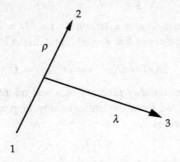

Figure 3.2 Internal Jacobi coordinates for three identical particles.

In three-dimensional space these are

$$\boldsymbol{\rho} = \frac{1}{\sqrt{2}}(\boldsymbol{r}_1 - \boldsymbol{r}_2)$$

$$\boldsymbol{\lambda} = \frac{1}{\sqrt{6}}(\boldsymbol{r}_1 + \boldsymbol{r}_2 - 2\boldsymbol{r}_3). \quad (3.53)$$

A third vector giving the centre of mass of the system is necessary for completeness:

$$\boldsymbol{R} = \frac{1}{\sqrt{3}}(\boldsymbol{r}_1 + \boldsymbol{r}_2 + \boldsymbol{r}_3). \quad (3.54)$$

Note that the transformation matrix of coordinates (3.53) and (3.54) gives a Jacobian

$$\frac{\partial(\boldsymbol{r}_1, \boldsymbol{r}_2, \boldsymbol{r}_3)}{\partial(\boldsymbol{\rho}, \boldsymbol{\lambda}, \boldsymbol{R})} = 1 \quad (3.55)$$

and $\boldsymbol{\rho}, \boldsymbol{\lambda}$, and $\boldsymbol{R}$ are called Jacobi coordinates (for n particles, see for example Moshinski 1969). The internal coordinates $\boldsymbol{\rho}$ and $\boldsymbol{\lambda}$ span an invariant subspace with respect to all six elements of S_3. In particular, any transposition (ij) acting on $\begin{pmatrix} \boldsymbol{\lambda} \\ \boldsymbol{\rho} \end{pmatrix}$ yields, as a result, a combination of $\boldsymbol{\lambda}$ and $\boldsymbol{\rho}$

$$(ij)\begin{pmatrix}\lambda\\\rho\end{pmatrix} = D(ij)\begin{pmatrix}\lambda\\\rho\end{pmatrix} \quad i,j = 1,2,3. \tag{3.56}$$

A straightforward calculation gives the following matrices

$$D(12) = \begin{pmatrix} 1 & 0 \\ 0 & -1 \end{pmatrix} \quad D(13) = \begin{pmatrix} -\frac{1}{2} & -\frac{\sqrt{3}}{2} \\ -\frac{\sqrt{3}}{2} & \frac{1}{2} \end{pmatrix}$$
$$D(23) = \begin{pmatrix} -\frac{1}{2} & \frac{\sqrt{3}}{2} \\ \frac{\sqrt{3}}{2} & \frac{1}{2} \end{pmatrix}. \tag{3.57}$$

Note that these matrices are symmetric, i.e. $D = D^T$ which is natural for transpositions. In that case, one has $g = g^{-1}$, and, using Exercise 3.1 below, one finds

$$D(g) = D(g^{-1}) = D^{-1}(g) = D^T(g).$$

The last equality is the unitarity property of a real matrix for which $D^+ = D^T$. The unitarity property follows from the orthogonality of ρ and λ in a space where the scalar product is defined as

$$<r_i|r_j> = \delta_{ij}. \tag{3.58}$$

Its meaning will be better understood in the next chapter, Section 4.3. The property $D = D^T$ is valid for any transposition of any S_n. It therefore simplifies the construction of representations of large dimensions (see Exercise 4.4).

The matrix associated with e is

$$D_e = \begin{pmatrix} 1 & 0 \\ 0 & 1 \end{pmatrix} \tag{3.59}$$

and those associated with 3-cycles are obtained through multiplications of matrices in one-to-one correspondence with elements of S_3 as shown in Table 2.1:

$$D(123) = D(13)D(12) = \begin{pmatrix} -\frac{1}{2} & \frac{\sqrt{3}}{2} \\ -\frac{\sqrt{3}}{2} & -\frac{1}{2} \end{pmatrix} \tag{3.60}$$

$$D(132) = D(12)D(13) = \begin{pmatrix} -\frac{1}{2} & -\frac{\sqrt{3}}{2} \\ \frac{\sqrt{3}}{2} & -\frac{1}{2} \end{pmatrix}.$$

Note that if we wish to obtain the matrices $D(123)$ and $D(132)$ directly by acting on $\begin{pmatrix}\lambda\\\rho\end{pmatrix}$ we have to use the permutations (132) and (123), respectively. This inversion is necessary due to the summation convention in (3.24). For a real matrix one has

$$S(g)|k> = \Sigma D_{ik}(g)|i> = \Sigma D^T_{ki}(g)|i> = \Sigma D^{-1}_{ki}(g)|i>$$
$$= \Sigma D_{ki}(g^{-1})|i>$$

IRREDUCIBLE REPRESENTATIONS

where use has been made of Exercise 3.1 below in the last equality. The last summation is over the second index, as in the multiplication of an $n \times n$ matrix with a column matrix of n rows. As the permutations (123) and (132) are inverse of each other the inversion mentioned above is now understood.

The set of six 2×2 matrices (3.57), (3.59), and (3.60) gives the mixed representation D^M of S_3. The character of the group in this representation is

$$\chi^M(g) = \begin{cases} 2 & \text{for } (e) \\ 0 & \text{for } (12), (13), \text{ and } (23) \\ -1 & \text{for } (123), \text{ and } (132) \end{cases} \quad (3.61)$$

which is consistent with statement (B) of Section 3.6, i.e. elements of the same class have the same character.

In Chapter 4, where we shall study the representations of the permutation group in a more systematic way, it will be shown that these are the only three non-equivalent representations of S_3.

Exercise 3.1 If $D(g)$ is a matrix representation of a group prove the equality

$$D(g^{-1}) = D^{-1}(g).$$

Solution Using the definition of a representation we get

$$e = gg^{-1} \to D(e) = D(g)D(g^{-1}) = e. \quad (1)$$

On the other hand, one has

$$D(g)D^{-1}(g) = e \quad (2)$$

from which the desired equality follows immediately.

3.8 IRREDUCIBLE REPRESENTATIONS

Irreducibility is a very important property of the representations of a group. In physics we use groups of symmetries especially through their irreducible representations. Degenerate states of a Hamiltonian can provide basis functions for irreducible representations. The role of irreducibility will become clearer in the following section.

If, in a linear vector space L, one can find a basis in which the matrices $D(g)$ of an n-dimensional representation can simultaneously be put in the form

$$D(g) = \begin{vmatrix} D^1(g) & C(g) \\ 0 & D^2(g) \end{vmatrix} \quad (3.62)$$

for all elements g of the group G, the representation $D(g)$ is called *reducible*. The block matrices are here two square matrices D^1 and D^2 of dimensions n_1 and n_2, respectively and two rectangular matrices, the one in the lower left corner having all elements equal to zero. Such a form implies the existence of an invariant subspace L^1 of dimension n_1 (see Section 3.3). Let us call $\begin{pmatrix} X^1 \\ 0 \end{pmatrix}$ the vectors belonging to L^1 only. Then one has

$$\begin{pmatrix} D^1 & C \\ 0 & D^2 \end{pmatrix} \begin{pmatrix} X^1 \\ 0 \end{pmatrix} = \begin{pmatrix} D^1 X^1 \\ 0 \end{pmatrix} \tag{3.63}$$

i.e. the transformed vectors also belong to L^1. Let us take now a vector $\begin{pmatrix} 0 \\ X^2 \end{pmatrix}$ belonging to L^2. Then the result is

$$\begin{pmatrix} D^1 & C \\ 0 & D^2 \end{pmatrix} \begin{pmatrix} 0 \\ X^2 \end{pmatrix} = \begin{pmatrix} C X^2 \\ D^2 X^2 \end{pmatrix} \tag{3.64}$$

i.e. a vector belonging to the whole space. Hence, in order to have L^2 as an invariant subspace too, one has to take $C = 0$. If the matrix C is zero the representation D is called *fully reducible*. Then there is a second invariant subspace L^2 of dimension n_2, and the whole space L can be written as the direct sum

$$L = L^1 + L^2 \tag{3.65}$$

and the representation D is the sum

$$D = D^1 + D^2. \tag{3.66}$$

We have encountered such a situation in Example 3.1 of Section 3.7. There, by using the similarity transformation M, the matrix D_z has been brought to a diagonal form (Equation (3.49)), i.e. it has been *reduced* to two one-dimensional representations which obviously cannot be reduced further.

If, for a given representation D, there is no similarity transformation which brings the matrices $D(g)$ into block diagonal form simultaneously for all $g \in G$ the representation is called *irreducible*. An irreducible representation is sometimes called an *irrep* for brevity.

In the case of fully reducible representations the matrices D^1 and D^2 may be reduced further to sums of matrices of smaller dimensions which are irreducible. Some of the irreducible representations may be equivalent. They do not count as distinct and we can introduce a multiplicity $m_r \geq 1$ to write

$$D = \sum_r m_r D^r. \tag{3.67}$$

The matrices D^r can have different dimensions from each other and the summation means that D is composed of square matrices strung along its diagonal

IRREDUCIBLE REPRESENTATIONS

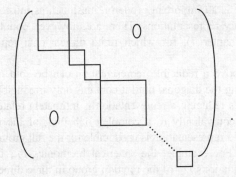

Actually, to make a distinction between the sum (3.67) and ordinary matrix addition, the symbol $\oplus$ should be used in (3.67). For example, if one takes $r \leq K$ and $m_r = 1$ for any r, (3.67) reads

$$D = D^1 \oplus D^2 \oplus \cdots \oplus D^K. \qquad (3.67')$$

From the definition of the characters (Section 3.6) it follows immediately that the character of D is given by

$$\chi(g) = \Sigma m_r \chi^r(g). \qquad (3.68)$$

For a unitary representation (Section 3.5) reducibility implies full reducibility. Let us take an orthonormal basis $e_i^1 \in L^1, e_i^2 \in L^2$:

$$(e_i^1, e_j^1) = (e_i^2, e_j^2) = \delta_{ij}; \qquad (e_i^1, e_j^2) = 0. \qquad (3.69)$$

The invariance of the subspace L^1 means

$$S(g)e_i^1 = \sum_{j=1}^{n_1} D_{ji}^1 e_j^1, \qquad (3.70)$$

while, for a vector in L^2, we have

$$S(g)e_j^2 = \sum_{\ell=1}^{n_1} C_{\ell j}^1 e_\ell^1 + \sum_{k=1}^{n_2} D_{kj}^2 e_k^2. \qquad (3.71)$$

The orthogonality relations (3.69) give

$$C_{ij} = (e_i^1, S(g)e_j^2). \qquad (3.72)$$

But by definition S is a unitary operator which leads to

$$C_{ij} = <S^{-1}(g)e_i^1, e_j^2> = <S(g^{-1})e_i^1, e_j^2> = 0 \qquad (3.73)$$

i.e. the matrix C in (3.64) has all its elements equal to zero. Hence, L^2 is also an invariant subspace.

Full reducibility is an important property, inasmuch as most groups of interest in physics have unitary representations. There are, however, counterexamples, like the Lorentz group (Chapter 7), for which finite dimensional representations are not unitary.

As discussed above a reducible representation can be split again and again into block matrices along the diagonal until it contains only irreducible representations. In quantum mechanics reducing a representation is intimately related to the elimination of degeneracy or equivalently to a complete labelling of the wave function. This happens because if a representation is irreducible for the full group it may be reducible for its subgroups. For example, the spherical harmonics $Y_{\ell m}$ form a basis for the irreducible representations D^{ℓ} of the rotation group in three dimensions, R_3, and they are degenerate eigenstates of a spherically symmetric system. The representations D^{ℓ} of R_3 can be reduced to diagonal forms for the subgroup R_2 of R_3 containing rotations about the z-axis, like the matrix (3.49). Each irreducible representation of R_2 is labelled by a given value of m. Also the eigenvalues of a system invariant under R_2 depend on m, i.e. the degeneracy is lifted and the index m gives a complete labelling of the space of the irreducible representations of R_3. The group R_3 will be discussed in Chapter 6. Other examples of complete labelling will be encountered throughout the book, as for instance the symmetric or the unitary groups.

3.9 DIRECT PRODUCTS OF REPRESENTATIONS

From two matrix representations, $D^{\mu_1}(g)$ and $D^{\mu_2}(g)$, of dimensions n_1 and n_2, respectively, one can construct a representation $D^{\mu}(g)$ as a *direct* or Kronecker product of the two matrices. This is an $n_1 \times n_2$ matrix with elements labelled by a double suffix

$$D^{\mu}_{ik,j\ell}(g) = D^{\mu_1}_{ij}(g) D^{\mu_2}_{k\ell}(g). \quad (3.74)$$

Symbolically one can write

$$D^{\mu} = D^{\mu_1} \times D^{\mu_2}. \quad (3.75)$$

The matrix D^{μ} describes the transformation properties of the product functions $\psi^1_j \psi^2_\ell$ if one performs simultaneously the same transformation g on the coordinates of ψ^1 and ψ^2. These functions can, for example, describe two different particles or generally independent parts of the same system. Separately we have

$$S(g)\psi^1_j = \Sigma D^{\mu_1}_{ij} \psi^1_i; \qquad S(g)\psi^2_\ell = \Sigma D^{\mu_2}_{k\ell} \psi^2_k$$

and for the composite system we get

$$S(g)\psi^1_j \psi^2_\ell = \Sigma D^{\mu_1}_{ij} D^{\mu_2}_{k\ell} \psi^1_i \psi^2_k$$
$$= \Sigma D^{\mu}_{ik,j\ell} \psi^1_i \psi^2_k.$$

The direct product offers a way to generate new representations from given ones. If D^{μ_1} and D^{μ_2} are irreducible the product matrix D^μ is generally reducible. One is interested in finding which are the irreducible representations which occur inside D^μ. This mathematical problem has important implications in physics. For example, in the coupling of two angular momenta, j_1 and j_2 the resulting j values, $|j_1 - j_2| \le j \le j_1 + j_2$, are related to the irreducible representations of the rotation group obtained from the direct product of two irreducible representations corresponding to $\mu_1 = j_1$ and $\mu_2 = j_2$.

The character of D^μ defined by Equation (3.74) can easily be calculated:

$$\chi(g) = \sum_{i,j} D_{ij,ij}(g) = \sum_i D_{ii}^{\mu_1}(g) \sum_j D_{jj}^{\mu_2}(g) = \chi^{\mu_1}(g)\chi^{\mu_2}(g) \tag{3.76}$$

i.e. it is equal to the product of the characters of the factor representations.

3.10 SCHUR'S LEMMAS AND THE ORTHOGONALITY THEOREM

For any group it is essential to discover which irreducible representation the group can have and to analyse a reducible representation in terms of irreducible ones. An important role in this goal is played by the orthogonality theorem for representations and its implication for the properties of characters. The orthogonality theorem is a corollary of two theorems named Schur's lemmas. Below we give the content of these lemmas and a statement of the orthogonality relations. Their proof is given in Appendix B.

Lemma 1 Let S and S' be two irreducible representations of a group G defined in the linear vector spaces L and L' of dimensions n and n', respectively. If a linear operator A transforming vectors from L to L' satisfies

$$S'(g)A = A S(g) \tag{3.77}$$

for any $g \in G$, then either A gives a one-to-one correspondence between L and L' or $A = 0$.

Remark If A gives a one-to-one correspondence between L and L' it means it has an inverse and S and S' are equivalent:

$$S'(g) = A S(g) A^{-1}.$$

Lemma 2 If $S(g)$ is a finite dimensional irreducible representation of a group G and A is an arbitrary linear operator such that

$$S(g)A = A S(g) \tag{3.78}$$

for all $g \in G$, then A is a multiple of the unit operator.

This lemma can be proved either independently or as a consequence of Lemma 1. In Appendix B we use the latter approach. Both lemmas can also be formulated in terms of matrices instead of operators (see, for example, Cornwell 1984, p. 80). In matrix form the relation (3.77) reads

$$D^\mu(g)A = AD^\nu(g) \qquad (3.79a)$$

where A is now an $n' \times n$ rectangular matrix and the relation (3.78) becomes

$$D^\mu(g)A = AD^\mu(g) \qquad (3.79b)$$

with A being an $n \times n$ square matrix.

Lemma 2 is a useful test of irreducibility. If one finds that A satisfying (3.79b) is not a multiple of the unit operator then $D^\mu(g)$ is reducible.

Exercise 3.2 Use Schur's Lemma 2 to prove that any irreducible representation of an Abelian group is one-dimensional.

Solution For an Abelian group, any two elements g_1 and g_2 commute giving

$$g = g_1 g_2 = g_2 g_1 \qquad (1)$$

which, for any representation $D(g)$, implies

$$D(g) = D(g_1)D(g_2) = D(g_2)D(g_1). \qquad (2)$$

According to Lemma 2, if a matrix A satisfies

$$AD(g) = D(g)A \qquad (3)$$

for all $D(g)$ of an irrep, the matrix A is a multiple λ of the unit matrix, i.e. $A = \lambda\,I$. From (2) and (3) one gets

$$AD(g_1)D(g_2) = D(g_1)D(g_2)A = D(g_2)D(g_1)A = D(g_2)AD(g_1)$$

i.e. $A\,D(g_1)$ commutes with $D(g_2)$. If g_2 runs through all elements of the group, Lemma 2 requires

$$AD(g_1) = \lambda_1 I$$

hence

$$D(g_1) = \frac{\lambda_1}{\lambda} I.$$

Since D is an irreducible representation it follows that it must be one-dimensional.

The orthogonality theorem for representations of a finite or a compact Lie group is an orthogonality relation for the matrix elements of the representations of that group. It involves a summation over the group elements of a finite group which becomes an *invariant integration* in the case of compact Lie groups, where although the number of elements is infinite they are defined in terms of a finite number of parameters taking values in a closed and bounded domain (see Chapter 5).

For such groups every finite dimensional representation is equivalent to a unitary representation (Section 3.5) and therefore the theorem refers to unitary representations directly.

Orthogonality theorem for finite groups

Theorem Let D^μ and D^ν be two unitary irreducible representations of a finite group G of order N. If D^μ and D^ν are not equivalent one has

$$\frac{1}{N} \sum_{g \in G} D^\mu_{ij}(g)[D^\nu_{k\ell}(g)]^* = 0 \tag{3.80}$$

and if they are equivalent one has

$$\frac{1}{N} \sum_{g \in G} D^\mu_{ij}(g)[D^\nu_{k\ell}(g)]^* = \frac{\delta_{\mu\nu}}{n_\mu} M_{ik}(M^{-1})_{\ell j} \tag{3.81}$$

where n_μ is the dimension of D^μ and M is a similarity transformation (Section 3.4), which transforms the matrices of the representation μ into those of ν or vice versa.

Note that if the two representations are identical one has $M = I$ i.e. $M_{ik} = \delta_{ik}$ and $(M^{-1})_{\ell j} = \delta_{\ell j}$. The first and second parts of the theorem follow from Schur's Lemmas 1 and 2, respectively.
Proof See Appendix B.

A geometrical interpretation of the theorem is that the matrix elements D^ν_{ij} for fixed ν, i, j form a vector in an N-dimensional space, where N is the group order. Each irreducible representation provides n^2_ν such vectors and the total number of vectors must satisfy

$$\sum_\nu n^2_\nu \leq N \tag{3.82}$$

since there are no more than N orthogonal vectors available. One can show that the equality sign always holds. To prove this one needs the concept of *regular representation* (see the next section).

Let us consider D^μ and D^ν to be two identical representations and take $i = j$ and $k = \ell$ in (3.81). This gives

$$\sum_g D^\mu_{ii}(g)[D^\nu_{jj}(g)]^* = \frac{N}{n_\mu} \delta_{\mu\nu} \delta_{ij}.$$

Summing over i and j one gets

$$\sum_g \chi^\mu(g)[\chi^\nu(g)]^* = N\delta_{\mu\nu}. \tag{3.83}$$

Moreover, since all group elements in the same class have the same character we can rewrite (3.83) as

$$\sum_C N_C \chi^\mu(C)[\chi^\nu(C)]^* = N\delta_{\mu\nu} \qquad (3.84)$$

where C represents a class and N_C the number of its elements, hence $\sum_C N_C = N$. For given μ, $(N_C)^{1/2} \chi^\mu(C)$ can be interpreted as the components of a vector in a K-dimensional space where K is the number of classes. The vectors obtained from two non-equivalent representations are orthogonal. One can have at most K such vectors. In fact for a finite group one can prove that the number of non-equivalent irreducible representations is equal to the number of classes (see the next section).

Such a result is very important since the initial step in representation theory is the construction of a complete list of non-equivalent irreducible representations of a group.

Example 3.3 We illustrate the above discussion with the permutation group S_3 of order $N = 6$. Its representations have been introduced in Example 3.2, Section 3.7. This group has three different representations: symmetric (S), mixed (M) and antisymmetric (A). It also has three different classes:

$$\begin{aligned} C_1 &: e & N_1 &= 1 \\ C_2 &: (12), (13), (23) & N_2 &= 3 \\ C_3 &: (123), (132) & N_3 &= 2. \end{aligned}$$

Hence the number of irreducible representations is equal to the number of classes. One can also prove the validity of Equation (3.84) for any μ. The characters $\chi(C_i)$ are given by the matrices (3.50) and (3.52) for $\mu = S$ and A respectively, and are explicitly exhibited in equation (3.61) for $\mu = M$.

$$\mu = S \qquad \chi(C_i) = 1; \qquad n_S = 1$$
$$1 \times 1 + 3 \times 1 + 2 \times 1 = 6$$

$$\mu = M \quad \chi(C_1) = 2, \quad \chi(C_2) = 0, \quad \chi(C_3) = -1; \quad n_M = 2$$
$$1 \times 2^2 + 3 \times 0 + 2 \times (-1)^2 = 6$$

$$\mu = A \quad \chi(C_1) = 1, \quad \chi(C_2) = -1, \quad \chi(C_3) = 1; \quad n_A = 1$$
$$1 \times 1^2 + 3 \times (-1)^2 + 2 \times 1^2 = 6$$

Note also that the orthogonality between different χ^μ holds and that

$$n_S^2 + n_M^2 + n_A^2 = 6$$

i.e. the equality sign in (3.82) holds for the representations of S_3.

Orthogonality theorem for compact Lie groups

The compact Lie groups have many of the properties of finite groups. An example is the theorem of Section 3.5. The extension to compact Lie groups is based on the fact that a sum over the group elements of a finite group can be replaced by an integral called an invariant integral

$$\frac{1}{N}\sum_{g\in G}f(g) \to \int_G f(g)\,\mathrm{d}g.$$

This integral is defined in terms of a measure $\mathrm{d}g$ on the group.

For a finite group, the sum $\sum_{g\in G} f(g)$ is an invariant

$$\sum_{g\in G}f(g) = \sum_{g\in G}f(g'g) = \sum_{g\in G}f(gg')$$

i.e. it does not change its value if g is replaced by gg' or $g'g$, where g' is fixed but arbitrary, because gg' or $g'g$ runs also through all elements of G (the rearrangement theorem, Appendix B). In the same way, for compact Lie groups the measure $\mathrm{d}g$ is unique up to a multiplicative constant and must be such that

$$\int_G f(g'g)\,\mathrm{d}g = \int_G f(gg')\,\mathrm{d}g = \int_G f(g)\,\mathrm{d}g.$$

One can write

$$\int_G f(g)\,\mathrm{d}g = \int_{a_1}^{b_1}\mathrm{d}\epsilon_1 \int_{a_2}^{b_2}\mathrm{d}\epsilon_2 \cdots \int_{a_r}^{b_r}\mathrm{d}\epsilon_r\, f(g(\epsilon_1,\epsilon_2,\cdots,\epsilon_r))\rho(\epsilon_1,\epsilon_2,\cdots,\epsilon_r) \quad (3.85)$$

where $f(g)$ is a continuous function of $\epsilon_1,\epsilon_2,\ldots,\epsilon_r$, and ρ is a weight function of the parameters ϵ_i, $a_i \leq \epsilon_i \leq b_i$ ($i=1,\ldots,r$) of the group. This integral exists and is finite for any continuous f if G is a compact Lie group (Peter and Weyl 1927). The function ρ can be chosen such that

$$\int_G \mathrm{d}g = \int_{a_1}^{b_1}\mathrm{d}\epsilon_1 \int_{a_2}^{b_2}\mathrm{d}\epsilon_2 \cdots \int_{a_r}^{b_r}\mathrm{d}\epsilon_r\, \rho(\epsilon_1,\epsilon_2,\cdots,\epsilon_r) = 1. \quad (3.86)$$

Examples of compact Lie groups will be considered in Chapter 5 and in a few cases the weight ρ can be found as in Appendix C.

The transcription of the relations (3.80) and (3.81) for compact Lie groups is

$$\int_G \mathrm{d}g\, D^{\mu}_{ij}(g) D^{\nu}_{k\ell}(g) = 0 \quad (3.87)$$

for non-equivalent representations and

$$\int_G \mathrm{d}g\, D^{\mu}_{ij}(g)[D^{\nu}_{k\ell}(g)]^* = \frac{\delta_{\mu\nu}}{n_\mu} M_{ik}(M^{-1})_{\ell j} \quad (3.88)$$

for equivalent ones. For each compact Lie group one should use definition (3.85) to express the left-hand side of (3.87) or (3.88) in terms of the group parameters. An example is provided in Section 6.1 for the rotation group R_3.

In a similar way one can write the equivalent of (3.83) for a compact Lie group

$$\int_G dg \chi^\mu(g)[\chi^\nu(g)]^* = \delta_{\mu\nu}. \tag{3.89}$$

Orthogonality relations for characters are useful in finding some fundamental properties of irreducible representations. For applications to unitary groups (Chapter 8) see, for example, Itzykson and Nauenberg (1966).

3.11 THE REGULAR REPRESENTATION

The regular representation of a finite group, together with Schur's lemmas, play a crucial role in the representation theory of groups. The principal properties of the representations follow from them.

The regular representation itself can be easily understood through the rearrangement theorem (Appendix B) which states that multiplying the group $g_1, g_2, \ldots, g_N$, say, on the left, by any element g_s, results in the same group but with the elements permuted among themselves. One can therefore represent an element g_s in the following way

$$D^{\text{reg}}(g_s)_{k\ell} = \begin{cases} 1 & \text{if } g_s g_\ell = g_k \\ 0 & \text{if } g_s g_\ell \neq g_k \end{cases} \tag{3.90}$$

with $D^{\text{reg}}(g_s)$ an $N \times N$ matrix where in any row only one matrix element is different from zero (equal to 1). The diagonal matrix elements of all matrices are zero except for the matrix representing the identity element. The relation (3.90) implies

$$g_s g_\ell = \sum_{k=1}^{N} D^{\text{reg}}(g_s)_{k\ell} g_k. \tag{3.91}$$

It is easy to prove that the matrices (3.90) form an $N \times N$ representation. Take $g_s = g_r g_t$ in (3.91). Then

$$(g_r g_t) g_\ell = \sum_{k=1}^{N} D^{\text{reg}}(g_r g_t)_{k\ell} g_k$$

$$= g_r (g_t g_\ell)$$

$$= \sum_{m=1}^{N} g_r D^{\text{reg}}(g_t)_{m\ell} g_m$$

$$= \sum_{k,m=1}^{N} D^{\text{reg}}(g_r)_{km} D^{\text{reg}}(g_t)_{m\ell} g_k$$

Hence

$$D^{\text{reg}}_{k\ell}(g_r g_t) = \sum_{m=1}^{N} D^{\text{reg}}(g_r)_{km} D^{\text{reg}}(g_t)_{m\ell}. \tag{3.92}$$

Example 3.4 The group S_3. Use the multiplication Table 2.1 and label the matrix of the regular representation in the same order. Take, for example, $g_s = (12)$ and search for the row $k = 3$ of D^{reg}. This corresponds to the element $g_k = (13)$. By using the second row of Table 2.1 one gets

$$D^{\text{reg}}_{3\ell}(g_s) = \begin{cases} 1 & \text{for } \ell = 6 \\ 0 & \text{for } \ell \neq 6 \end{cases}$$

because $(12)(132) = (13)$. The whole matrix is

$$D^{\text{reg}}(12) = \begin{pmatrix} 0 & 1 & 0 & 0 & 0 & 0 \\ 1 & 0 & 0 & 0 & 0 & 0 \\ 0 & 0 & 0 & 0 & 0 & 1 \\ 0 & 0 & 0 & 0 & 1 & 0 \\ 0 & 0 & 0 & 1 & 0 & 0 \\ 0 & 0 & 1 & 0 & 0 & 0 \end{pmatrix}. \tag{3.93}$$

From the definition (3.90) it follows immediately that

$$\chi^{\text{reg}}(g) = \begin{cases} 0 & \text{for } g \neq e \\ N & \text{for } g = e. \end{cases} \tag{3.94}$$

One can express the regular representation in terms of the non-equivalent irreducible representations D^ν of the group. Then for each class C one can write

$$\chi^{\text{reg}}(C) = \sum_\nu a_\nu \chi^\nu(C) \tag{3.95}$$

where a_ν are coefficients to be found. In particular for the class C_1 of the identity element one has

$$\chi^{\text{reg}}(e) = \chi^{\text{reg}}(C_1) = N \tag{3.96}$$
$$\chi^\nu(e) = \chi^\nu(C_1) = n_\nu \tag{3.97}$$

where n_ν is the dimension of D^ν. With this notation (3.95) becomes

$$N = \sum a_\nu n_\nu. \tag{3.98}$$

Using the orthogonality relation (3.84) one gets from (3.95)

$$a_\nu = \frac{1}{N} \sum_C N_C [\chi^\nu(C)]^* \chi^{\text{reg}}(C).$$

According to (3.94) in this sum only the class of the identity element contributes. For this class $N_C = 1$ and hence

$$a_\nu = n_\nu = \chi^\nu(C_1) \tag{3.99}$$

which shows that the regular representation contains each irreducible representation a number of times equal to the degree of that representation. Using (3.99), the relation (3.98) becomes

$$N = \sum_\nu n_\nu^2 \tag{3.100}$$

i.e. the equality sign holds in Equation (3.82). Substituting (3.99) in (3.95) one gets

$$\chi^{\text{reg}}(C) = \sum_\nu \chi^\nu(C_1)\chi^\nu(C)$$

which, together with the definition (3.94), leads to the orthogonality relation

$$\sum_\nu \chi^\nu(C_1)\chi^\nu(C) = N\delta_{C_1 C}. \tag{3.101}$$

For continuous groups one can also define a regular representation. In group theory this is called the *adjoint* representation and denoted by Ad (Cornwell 1984, Chapter 11, Section 5). The adjoint representation of SU(n) is presented in Section 8.9.

We return now to the discussion following Equation (3.84) where, for fixed μ, the character $\chi^\mu(C)$ of a class C has been interpreted as the Cth component of a vector in a K-dimensional space, where K is the number of classes of a finite group. Each non-equivalent irreducible representation μ gives rise to such a vector and their total number N_{irr} is then at most equal to K, i.e.

$$N_{\text{irr}} \leq K. \tag{3.102}$$

Below we shall prove that the equality sign always holds. This is the content of the following theorem.

Theorem The number N_{irr} of non-equivalent irreducible representations of a group is equal to the number K of its classes.

Proof One needs to introduce the product $C_i C_j$ of two classes C_i and C_j of a group G. This is a set of elements of G obtained by multiplying each element of C_i by each element of C_j. This gives $N_i N_j$ elements of G but some of them may be identical. From the definition of a class (Section 2.5) it follows that

$$hC_i C_j h^{-1} = hC_i h^{-1} hC_j h^{-1} = C_i C_j \tag{3.103}$$

for any $h \in G$, which means that $C_i C_j$ consists of complete classes. Thus one can write

$$C_i C_j = \sum c_{ij}^k C_k \tag{3.104}$$

where the coefficients c_{ij}^k give the multiplicity of the class C_k in the product $C_i C_j$. The relation (3.104) implies that there is the same number of elements on both sides of (3.104),

$$N_i N_j = \sum c_{ij}^k N_k. \qquad (3.105)$$

In group representation theory the coefficients c_{ij}^k determine the simple characters of the group G and are called the *structure constants* of a finite group. Moreover one can write

$$C_i C_j = C_j C_i \qquad (3.106a)$$

which can be easily understood. Take $h \in C_i$ and $g \in C_j$. Then $hg = hgh^{-1}h$, where $hgh^{-1} \in C_j$, hence $hg \in C_j C_i$ too. This relation leads to the symmetry property

$$c_{ij}^k = c_{ji}^k. \qquad (3.106b)$$

Let us consider an irreducible representation D^ν of G and introduce the matrix

$$D_i^\nu = \sum_{g \in C_i} D^\nu(g) \qquad (3.107)$$

where the sum runs over all elements of a class C_i. From the class property $h C_i h^{-1} = C_i$ and Exercise 3.1 it follows that

$$D^\nu(h) D_i^\nu [D^\nu(h)]^{-1} = D_i^\nu \qquad (3.108)$$

for any $h \in G$, because the matrix product on the left-hand side of equation (3.108) has only rearranged the terms in the sum (3.107) defining D_i^ν. Relation (3.108) means that D_i^ν commutes with all matrices $D^\nu(h)$ of the irreducible representation D^ν chosen for discussion. It follows from Schur's Lemma 2 that D_i^ν must be a multiple of the unit matrix

$$D_i^\nu = \lambda_i^\nu I. \qquad (3.109)$$

The matrices $D^\nu(g)$ in the sum (3.107) have the same character $\chi^\nu(C_i)$ and if there are N_i elements in the class C_i the relation (3.109) implies

$$N_i \chi^\nu(C_i) = \lambda_i^\nu n_\nu = \lambda_i^\nu \chi^\nu(C_1) \qquad (3.110)$$

where equation (3.97) has been used in the second equality. Due to (3.109) the relation (3.104) taken in the representation D^ν leads to

$$\lambda_i^\nu \lambda_j^\nu = \sum c_{ij}^k \lambda_k^\nu.$$

Replacing the value of λ_i^ν resulting from (3.110) one gets

$$N_i N_j \chi^\nu(C_i) \chi^\nu(C_j) = \sum c_{ij}^k N_k \chi^\nu(C_1) \chi^\nu(C_k). \qquad (3.111)$$

The above relation is valid for any irreducible representation ν of the group.

Now one can make a particular choice of classes. Let us take C_j to be the class $C_{i^{-1}}$ of the inverses g^{-1} of the elements of C_i. The product C_iC_j becomes

$$C_iC_{i^{-1}} = N_ie + \sum_{k \neq 1} c^k_{ii^{-1}} C_k \qquad (3.112)$$

where we write separately the identity element e which is the class C_1 of its own. Its corresponding coefficient is (see Example 3.5 at the end of this section)

$$c^1_{ij} = N_i\delta_{j,i^{-1}}. \qquad (3.113)$$

Summing (3.111) over all possible representations ν and using (3.101) one obtains

$$N_iN_j \sum_\nu^{N_{\text{irr}}} \chi^\nu(C_i)\chi^\nu(C_j) = \sum_k c^k_{ij} N_k N \delta_{1k}$$

$$= c^1_{ij} N$$

because $N_1 = 1$. Hence using (3.113) one obtains

$$\sum_\nu^{N_{\text{irr}}} \chi^\nu(C_i)\chi^\nu(C_j) = \frac{N}{N_i}\delta_{j,i^{-1}} \qquad (3.114)$$

because $N_{i^{-1}} = N_i$, i.e. a class C_i and the class of its inverses have the same number of elements. For a unitary representation one has

$$D(g^{-1}) = D^{-1}(g) = D^+(g)$$

from which it follows that the class $C_{i^{-1}}$ has the property

$$\chi(C_{i^{-1}}) = \chi^*(C_i)$$

which, substituted into (3.114), leads to

$$\sum_\nu^{N_{\text{irr}}} \chi^\nu(C_i)\chi^{\nu*}(C_j) = \frac{N}{N_i}\delta_{ij}. \qquad (3.115)$$

Note that this is the generalization of (3.101) to any class and represents another orthogonality relation for the characters besides (3.84). The characters can again be interpreted as mutually orthogonal vectors but now in an N_{irr}-dimensional space. The number of these vectors is K and there are at most N_{irr} such vectors

$$K \leq N_{\text{irr}}.$$

This relation, together with (3.102), gives

$$N_{\text{irr}} = K \qquad (3.116)$$

which proves the theorem.

An immediate consequence of this theorem is that the irreducible representations of an abelian group are one-dimensional. For an abelian group one has

$$N = K = N_{\text{irr}}.$$

Hence the sum in (3.100) runs from 1 to N for which the only possible solution is

$$n_v = 1$$

consistent with Exercise 3.2.

Example 3.5 Denote the classes of S_3 by

$$C_1 : e$$
$$C_2 : (12), (23), (13)$$
$$C_3 : (123), (132).$$

Each transposition of C_2 is identical to its inverse, so $C_{2^{-1}}$ is identical to C_2. In C_3 the permutations (123) and (132) are inverses of each other so $C_{3^{-1}}$ and C_3 have the same elements but ordered differently. From the multiplication Table 2.1 of S_3 one easily gets

$$C_2 C_{2^{-1}} = C_{2^{-1}} C_2 = 3C_1 + 3C_3$$

i.e.

$$c^1_{22^{-1}} = c^3_{22^{-1}} = 3; \qquad c^2_{22^{-1}} = 0$$

and

$$C_3 C_{3^{-1}} = C_{3^{-1}} C_3 = 2C_1 + C_3$$

i.e

$$c^1_{33^{-1}} = 2; \qquad c^2_{33^{-1}} = 0; \qquad c^3_{33^{-1}} = 1.$$

Remark The equality (3.106b) is satisfied.

3.12 CONSTRUCTION OF CHARACTER TABLES

It is customary to construct character tables, indicating the character $\chi^v(C_i)$ for each class C_i and irreducible representation v. From the equality (3.116) it follows that the character table of a group is a square of size N_{irr}. Usually the columns label the classes and the rows the non-equivalent irreducible representations. Example 3.3 leads to the character table for S_3 shown in Table 3.1.

Table 3.1 Character table of S_3.

v \ C_i	C_1	C_2	C_3
S	1	1	1
M	2	0	−1
A	1	−1	1

54   LINEAR REPRESENTATIONS OF A GROUP

The relation (3.84) represents the orthogonality of rows with the weighting factor N_C. The relation (3.115) represents the orthogonality of columns. Alternatively the orthogonality relations (3.84) and (3.115) allow us to associate to each character table a unitary matrix X of dimension $N_{irr} \times N_{irr}$ having as matrix elements

$$X_{vi} = \left(\frac{N_i}{N}\right)^{1/2} \chi^v(C_i). \tag{3.117}$$

For a group of low order N one can determine the character table directly by using the relations (3.100), (3.116), and the orthogonality relations (3.84) and (3.115). From these relations one can see that the only information required is the group order N, the number K of its classes, and the size N_C of each class. See, for example, Exercise 3.1.

For larger groups one has to make use of the multiplication table of the group. Tables of characters are available for most groups of practical applications in Physics. For point groups, see for example Hamermesh 1962, Chapter 4. For S_n there is a general formula due to Frobenius (Hamermesh 1962, Chapter 7). Tables of characters can be found in Hamermesh (1962) for groups up to S_7 and in Littlewood (1958) up to S_{10}.

Exercise 3.3   Find the character table of S_4

Solution   The result is given in Table 3.2 and is obtained as follows. The group S_4 is of order $N = 24$ and the elements of its five classes are given by (2.23):

$C_1 : e$
$C_2 : (12), (13), (14), (23), (24), (34)$
$C_3 : (12)(34), (13)(24), (14)(23)$
$C_4 : (123), (132), (124), (142), (134), (143), (234), (243)$
$C_5 : (1234), (1243), (1324), (1342), (1423), (1432).$

Therefore

$$N_1 = 1, \quad N_2 = 6, \quad N_3 = 3, \quad N_4 = 8, \quad N_5 = 6. \tag{1}$$

From the equality (3.116) it follows that the number of irreducible representations is

$$N_{irr} = 5 \tag{2}$$

and the relation (3.100) now reads

$$\sum_{v=1}^{5} n_v^2 = 24. \tag{3}$$

Let us denote the five irreducible representations by

$$v = S, M_1, M_2, M_3, A \tag{4}$$

CONSTRUCTION OF CHARACTER TABLES

where S stands for symmetric, A for antisymmetric and M_i ($i = 1, 2, 3$) for mixed. As for S_3 (see Example 3.2) one can write a symmetric function of four variables which remains unchanged under all permutations of S_4. This gives rise to the symmetric irreducible representation D^S of dimension $n_S = 1$. Hence

$$\chi^S(p) = D^S(p) = 1 \quad \text{for any } p \in S_4. \tag{5}$$

A determinant like (3.51) but of order 4 forms an invariant subspace which generates the antisymmetric irreducible representation D^A of dimension $n_A = 1$. It follows that

$$\chi^A(p) = D^A(p) = \delta_p$$

where δ_p is the parity of the permutation p as defined by equation (2.19). In detail this gives

$$\chi^A(C_1) = \chi^A(C_3) = \chi^A(C_4) = +1$$
$$\chi^A(C_2) = \chi^A(C_5) = -1. \tag{6}$$

These characters represent the elements of the last row of Table 3.2.

For reasons which will become obvious in the next chapter the other three representations of S_4, M_1, M_2, and M_3 are called mixed. With $n_S = 1$ and $n_A = 1$ the relation (3) gives

$$n_{M_1}^2 + n_{M_2}^2 + n_{M_3}^2 = 22. \tag{7}$$

The set of integers compatible with Equation (7) is $\{3, 3, 2\}$. As the M_i are not specified, and one doesn't need their definition at this level, we can take for example

$$n_{M_1} = 3, \quad n_{M_2} = 2, \quad n_{M_3} = 3. \tag{8}$$

Now for the class C_1 of the identity element one has

$$\chi^\nu(C_1) = n_\nu \tag{9}$$

which means that we have also found the first column of Table 3.2 because we know all the values of n_ν at this stage.

Any symmetric group S_n has the property that the inverse g^{-1} of an element g occurs in the same class as the element itself (cf. Section 2.5). Hence they have the same character. For any unitary representation ν of S_n it follows that

$$\chi^\nu(g) = \chi^\nu(g^{-1}) = \chi^{\nu^*}(g) \tag{10}$$

which means that the characters of S_n are real numbers inasmuch as we deal with unitary representations. Then for $i = 1$ and $j \neq 1$ the relation (3.115) becomes

$$\chi^S(C_j) + 3\chi^{M_1}(C_j) + 2\chi^{M_2}(C_j) + 3\chi^{M_3}(C_j) + \chi^A(C_j) = 0. \tag{11}$$

Use of (5) and (6) in (11) leads to

$$3\chi^{M_1}(C_2) + 2\chi^{M_2}(C_2) + 3\chi^{M_3}(C_2) = 0 \tag{12}$$

$$3\chi^{M_1}(C_3) + 2\chi^{M_2}(C_3) + 3\chi^{M_3}(C_3) = -2 \tag{13}$$

$$3\chi^{M_1}(C_4) + 2\chi^{M_2}(C_4) + 3\chi^{M_3}(C_4) = -2 \tag{14}$$

$$3\chi^{M_1}(C_5) + 2\chi^{M_2}(C_5) + 3\chi^{M_3}(C_5) = 0. \tag{15}$$

The relation (3.115) for $i = j = 2$ leads to

$$[\chi^{M_1}(C_2)]^2 + [\chi^{M_2}(C_2)]^2 + [\chi^{M_3}(C_2)]^2 = 2. \tag{16}$$

The only values of χ^{M_i} compatible with Equation (12) are

$$\chi^{M_1}(C_2) = 1, \qquad \chi^{M_2}(C_2) = 0, \qquad \chi^{M_3}(C_2) = -1 \tag{17}$$

or

$$\chi^{M_1}(C_2) = -1, \qquad \chi^{M_2}(C_2) = 0, \qquad \chi^{M_3}(C_2) = 1. \tag{18}$$

As the M_i are not specified we can make either choice. In the following, we shall use the solution (17) which, together with (5) and (6), completes the second column of Table 3.2.

Taking into account (5) and (6) the relation (3.115) for $i = j = 3$ leads to

$$[\chi^{M_1}(C_3)]^2 + [\chi^{M_2}(C_3)]^2 + [\chi^{M_3}(C_3)]^2 = 6. \tag{19}$$

This leads unambiguously to

$$\chi^{M_1}(C_3) = -1, \qquad \chi^{M_2}(C_3) = 2, \qquad \chi^{M_3}(C_3) = -1 \tag{20}$$

compatible with (13). The resulting column 3 of Table 3.2 is orthogonal to column 2.

The relation (3.115) for $i = j = 4$ gives

$$[\chi^{M_1}(C_4)]^2 + [\chi^{M_2}(C_4)]^2 + [\chi^{M_3}(C_4)]^2 = 1. \tag{21}$$

The only solution compatible with equation (14) is

$$\chi^{M_1}(C_4) = 0, \qquad \chi^{M_2}(C_4) = -1, \qquad \chi^{M_3}(C_4) = 0 \tag{22}$$

which leads to column 4.

Finally the relation (3.115) for $i=j=5$ gives the equation

$$[\chi^{M_1}(C_5)]^2 + [\chi^{M_2}(C_5)]^2 + [\chi^{M_3}(C_5)]^2 = 2. \tag{23}$$

Its solutions, compatible with equation (15) and the choice (17), are

$$\chi^{M_1}(C_5) = +1, \qquad \chi^{M_2}(C_5) = 0, \qquad \chi^{M_3}(C_5) = -1 \tag{24a}$$

$$\chi^{M_1}(C_5) = -1, \qquad \chi^{M_2}(C_5) = 0, \qquad \chi^{M_3}(C_5) = 1. \tag{24b}$$

Solution (24a) is excluded by (3.115) with $i = 2$ and $j = 5$.
This completes column 5.

Table 3.2 Character table of S_4.

 C_i ν	C_1	C_2	C_3	C_4	C_5
S	1	1	1	1	1
M_1	3	1	−1	0	−1
M_2	2	0	2	−1	0
M_3	3	−1	−1	0	1
A	1	−1	1	1	−1

We anticipate that by determining explicitly the matrix elements of the mixed representations of S_4 in the next chapter one can make the identification

$$M_1 \longleftrightarrow \quad \square\!\square\!\square / \square$$

$$M_2 \longleftrightarrow \quad \square\!\square / \square\!\square$$

$$M_3 \longleftrightarrow \quad \square\!\square / \square / \square$$

Note that the above table is also the character table of the point group T_d. For the latter see, for example, Table 4-7 of Hamermesh (1962) where some rows and columns have to be interchanged in order to recover Table 3.2 above. The identity of the table characters follows from the fact that T_d is isomorphic with S_4.

3.13 CLEBSCH–GORDAN SERIES

In Section 3.9, we defined direct products of representations D^{μ_1} and D^{μ_2}. The resulting matrix is generally reducible so, using the form (3.67), one can write

$$D^{\mu_1} \times D^{\mu_2} = \Sigma\, m_\nu\, D^\nu \qquad (3.118)$$

where m_ν is the multiplicity of the irrep D^ν appearing in the direct product. The expansion (3.118) is called a Clebsch–Gordan series.

From (3.76), it follows that

$$\chi^{\mu_1}(g)\chi^{\mu_2}(g) = \Sigma\, m_\nu \chi^\nu(g). \tag{3.119}$$

If the representations are unitary, one can use the orthogonality relation (3.83) to obtain m_ν as

$$m_\nu = \frac{1}{N}\sum_g \chi^{\mu_1}(g)\chi^{\mu_2}(g)[\chi^\nu(g)]^* \tag{3.120}$$

for a finite group of order N. In a similar way, the orthogonality relation (3.89) gives

$$m_\nu = \int_G dg\, \chi^{\mu_1}(g)\chi^{\mu_2}(g)[\chi^\nu(g)]^* \tag{3.121}$$

for a compact Lie group.

If the dimensions of D^{μ_1}, D^{μ_2}, and D^ν are d_{μ_1}, d_{μ_2} and d_ν, respectively, the relation (3.118) implies

$$d_{\mu_1} d_{\mu_2} = \Sigma\, m_\nu\, d_\nu. \tag{3.122}$$

Clebsch–Gordan series will be presented for the symmetric group (Section 4.7), the rotation group R_3 (Section 6.1), and the unitary group SU(3) (Section 8.11).

SUPPLEMENTARY EXERCISE

3.4 Use the basis (x^2, xy, y^2) to construct a three-dimensional representation of R_2. Reduce it to a sum of irreps. Check the result by using property (A), Section 3.6, of characters.

4

PERMUTATION GROUP S_n

4.1 GENERAL REMARKS

In Chapter 2 the symmetric group S_n was introduced together with its basic properties. Its role in quantum mechanics was also specified. In this chapter we shall analyse the irreducible representations of S_n by using Young diagrams (Young 1928). These diagrams play a very important role in group theory because they relate the symmetric group to some continuous groups which are also important in physics. The relation is such that one can use Young diagrams to describe representations of the symmetric group as well as of most Lie groups. The use of Young diagrams for Lie groups was initiated by Herman Weyl in 1931 (see Weyl 1946). The method works particularly well for $SU(\ell + 1)$ groups. For further discussion see Chapter 5.

Other basic properties of the symmetric group are due to Frobenius (1900) who gave the theory of characters and to Yamanouchi (1937) who derived the standard orthogonal representation. Further contributions to the mathematical properties of S_n will be quoted throughout the book.

The symmetric group has been applied extensively to problems of atomic and nuclear spectroscopy. There is a line developed by Racah (1942, 1943, 1949) (see also Racah 1965) and other later developments due to Jahn (1950, 1951), Flowers (1952) (see also Edmonds and Flowers 1952) and Elliott, Hope, and Jahn (1953). Racah based his studies on the theory of Lie groups, while in the second line of approach use has been made of Young's theory of substitutional analysis (see Rutherford 1948) and the theory of Schur's functions, named S-functions (see Littlewood 1958). The basic concepts in these developments are the outer and inner products and fractional parentage coefficients, which are introduced in this chapter.

The applications considered in this book refer especially to multiquark systems (Chapter 10). From the permutation group point of view the problem is analogous to atomic or nuclear physics inasmuch as one deals with a system of fermions. One aims at constructing a totally antisymmetric wave function in all cases, but for multiquark systems there is an extra degree of freedom, the colour. Therefore one can borrow and adapt the techniques developed in atomic and nuclear physics. The technique initiated by Jahn is particularly convenient because the fractional parentage coefficients can be derived separately for each degree of freedom (orbital, spin, isospin, colour) of the total wave function. This separation simplifies the calculations of the two-body matrix elements.

4.2 IRREDUCIBLE REPRESENTATIONS

General properties of the symmetric group S_n were introduced in Chapter 2 as follows. In Section 2.1, permutations of n objects, as elements of S_n were defined together with their cyclic structure and parity property. The group S_3 and its multiplication table were given as an example of the simplest non-trivial case. In Section 2.2 it was shown how to look for subgroups of S_n. A particular case, the regular subgroup, was defined and its relation to Cayley's theorem specified in Section 2.4. Below we recall the notion of partition and show its connection with Young diagrams.

Partition

A partition of an integer n is a set of positive numbers $(\lambda_1, \lambda_2, \cdots, \lambda_n)$ where

$$\lambda_1 \geq \lambda_2 \geq \ldots \geq \lambda_n \geq 0$$

$$\lambda_1 + \lambda_2 + \ldots + \lambda_n = n.$$

In Section 2.5 we showed that there is a one-to-one correspondence between a partition and the cyclic or class structure. Hence the number of partitions is equal to the number of classes for any permutation group. On the other hand, in Section 3.11 we proved that the number of classes of any finite group is equal to the number of its irreducible representations. It follows that the number of partitions is equal to the number of irreducible representations. This relation is reinforced by the introduction of Young diagrams which illustrate the one-to-one correspondence between a partition and an irreducible representation.

Young diagrams

A Young diagram is a collection of n boxes arranged in n rows with λ_1 boxes in the first row, λ_2 in the second, $\cdots \lambda_n$ in the nth row, such that

$$\lambda_1 \geq \lambda_2 \geq \ldots \lambda_n \geq 0$$
$$\lambda_1 + \lambda_2 + \ldots \lambda_n = n \tag{4.1}$$

An example is shown in Fig. 4.1

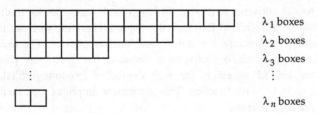

Figure 4.1 A Young diagram.

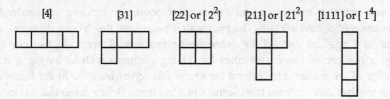

Figure 4.2 The partitions of 4 and their Young diagrams.

A shorthand notation for a diagram is

$$[\lambda] = [\lambda_1, \lambda_2, \ldots, \lambda_n]. \quad (4.2)$$

For the permutation group the irreducible representation which is in one-to-one correspondence with a Young diagram is usually designated by the letter f instead of λ. One writes

$$[f] = [f_1, f_2, \ldots, f_n]. \quad (4.2a)$$

Moreover, if starting from a given i the partition has

$$f_i = f_{i+1} = \ldots = f_n = 0$$

these zeroes are omitted in the shorthand notation (4.2a) which is then written as

$$[f] = [f_1, f_2, \ldots, f_k] \quad (4.2b)$$

where $k = i - 1$. For an illustration of this notation see Fig. 4.2.

In general, whenever k successive rows have the same length f_i a notation of the form $[\cdots f_i^k \cdots]$ can also be used, as illustrated with the above example.

Two partitions or Young diagrams are called conjugate if one is obtained from the other by the interchange of rows and columns, i.e. by a reflection with respect to a dashed line as drawn in the figure below. One says that the diagrams

are *conjugate* diagrams, while the diagram

is called *self-conjugate* because it transforms into itself.

It turns out that there is a one-to-one correspondence between an irreducible representation of S_n and a Young diagram with n boxes, and that Young diagrams can be used as a practical method for decomposing products of irreps. In fact one can develop a number of elementary rules for Young diagrams without having a deep knowledge of the mathematics behind the irreducible representations. In the following chapters we shall indicate these rules without proving them. Before doing that it is useful to have a better understanding of the relation between an irreducible representation of S_n and a Young diagram. This can be done by introducing the Young tableaux and writing the basis functions of some irreducible representations explicitly.

4.3 BASIS FUNCTIONS OF S_n

We first introduce the notion of a standard Young tableau. In the following subsections we shall build invariant subspaces for all the irreducible representations of S_2, S_3, and S_4, and for some representations of the groups S_5 and S_6 by using a simple technique. The main aim is to show explicitly in these examples that there is a one-to-one correspondence between a basis vector of an irreducible representation defined by a Young diagram and a standard Young tableau (see below) associated with that diagram. General formulae for the dimension of an irrep of S_n are given at the end of this section.

Young tableau

A Young *tableau* is a Young diagram of n boxes filled with the figures *1*, *2*, ..., *n* in any order, each figure being used only once. There are $n!$ ways of filling the boxes. A *standard* Young tableau is a tableau where the figures are arranged in increasing order when read from left to right in a row and from top to bottom in a column. Some examples are given in Fig. 4.3. They all correspond to the partition [211] of S_4 and are the only three possible standard Young tableaux for this partition. For each diagram or partition one has to find all possible standard Young tableaux because there is a one-to-one correspondence (cf. the end of this section) between a standard Young tableau and a basis vector of the irreducible representation associated to the given Young diagram. This is so because each Young tableau represents a particular process of symmetrization and antisymmetrization of n indices. Explicit examples will be given in the following subsections.

$$Y_1 = \begin{array}{|c|c|}\hline 1 & 2 \\\hline 3 \\\cline{1-1} 4 \\\cline{1-1}\end{array} \qquad Y_2 = \begin{array}{|c|c|}\hline 1 & 3 \\\hline 2 \\\cline{1-1} 4 \\\cline{1-1}\end{array} \qquad Y_3 = \begin{array}{|c|c|}\hline 1 & 4 \\\hline 2 \\\cline{1-1} 3 \\\cline{1-1}\end{array}$$

(a) (b) (c)

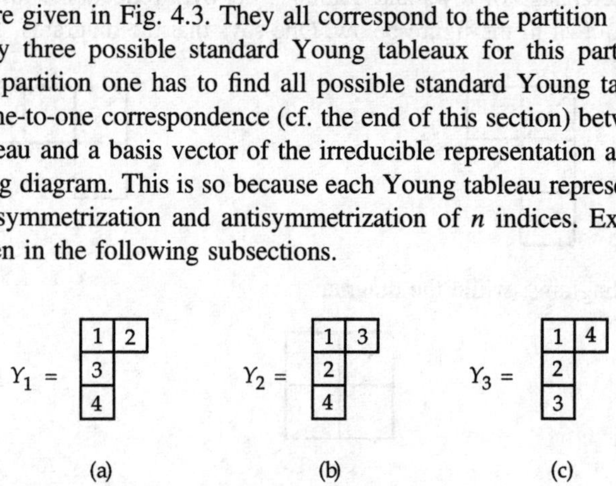

Figure 4.3 Young tableaux.

There are standard Young tableaux where the figures appear in order when one reads from left to right successively all the rows one after the other from top to bottom. Such particular tableaux are called *normal*. In Fig. 4.3 the case (a) is a normal Young tableau. From the normal Young tableau of a given partition one can obtain all its other standard tableaux by appropriate permutations. For example the tableau (b) can be obtained from (a) by using the transposition (23), and (c) from (a) through the permutation

$$(34)\,(23) = (243).$$

This reads:

$$(23)Y_1 = Y_2 \tag{4.3a}$$

$$(243)Y_1 = Y_3. \tag{4.3b}$$

A more compact way of specifying a Young tableau as in Fig. 4.3 is to use the so-called *Yamanouchi symbol* or *label* $Y = (r_n, r_{n-1}, \cdots, r_1)$ for each tableau. Here r_i represents the row of the figure i. The Yamanouchi symbols of the tableaux (a), (b), (c) of Fig. 4.3 are therefore

$$Y_1 = (3211) \qquad Y_2 = (3121) \qquad Y_3 = (1321)$$

In the case of Y_1, (3211) means the object 1 is in the first row, 2 is also in the first row, 3 is in the second row and 4 in the third row.

The Yamanouchi symbols are used to label wave functions of n particles from the symmetric group point of view. For example, suppose that ψ is a state of four particles each having angular momentum ℓ. If ψ has S_4 symmetry described by the tableau (a) of Figure 4.3 it can be labelled as follows

$$\psi(\ell^4[211]\xi\,|\,3211)$$

where ξ stands for other quantum numbers, used to specify the state completely. In general, for n particles one writes

$$\psi(\ell^n[f]\xi\,|\,r_n, r_{n-1}, \ldots, r_1)$$

where $[f]$ is the irreducible representation of S_n to which this state belongs. Here $r_n, r_{n-1}, \cdots, r_1$ is the Yamanouchi symbol, not to be confused with the space coordinates of the particles.

Symmetrizer and antisymmetrizer

For a system of identical particles the Hamiltonian is invariant under the permutation of particles. In a permutation one interchanges all coordinates (position, spin, etc.) of a particle with those of another particle. Let us call 1, 2, ..., n the coordinates of the particles 1, 2, ..., n, respectively. Then if $\psi(1, 2, ..., n)$ is an eigenfunction belonging to a given eigenvalue E, any permutation of particle coordinates

$$P_a = \begin{pmatrix} 1 & 2 & \cdots & n \\ a_1 & a_2 & \cdots & a_n \end{pmatrix}$$

will give

$$P_a \psi = \psi(a_1, a_2, \ldots, a_n) \tag{4.4}$$

where $P_a \psi$ is another eigenfunction belonging to the same eigenvalue if H is invariant under the permutation of particles. Quantum mechanics requires symmetric functions for bosons and antisymmetric functions for fermions. They can be obtained by using the symmetrizer operator $\mathscr{S}$ and the antisymmetrizer operator $\mathscr{A}$ which are defined as

$$\mathscr{S} = \frac{1}{n!} \sum_P P, \qquad \mathscr{A} = \frac{1}{n!} \sum_P \delta_P P. \tag{4.5}$$

In both operators the sum extends over all $n!$ elements P of the group S_n. In $\mathscr{A}$ each permutation is multiplied by its parity as defined in Equation (2.19). So if ψ is a product of single-particle states, $\mathscr{S}$ and $\mathscr{A}$ produce a *permanent* and a *determinant* respectively. These operators obviously have the property

$$P\mathscr{S} = \mathscr{S}P = \mathscr{S}, \qquad P\mathscr{A} = \mathscr{A}P = \delta_P \mathscr{A} \qquad \text{for any } P \in S_n, \tag{4.6}$$

that is, they commute with all elements of S_n; this follows immediately from the rearrangement theorem (Appendix B). They are particular examples of class sums. They also act as projection operators, and hence satisfy

$$\mathscr{S}^2 = \mathscr{S}, \qquad \mathscr{A}^2 = \mathscr{A}. \tag{4.7}$$

We shall sometimes use the notation $\mathscr{S}_{1,2,\cdots,n}$ and $\mathscr{A}_{1,2,\cdots,n}$ to specify the particles 1, 2, ..., n on which these operators act (see below).

Consider first the case $n = 2$ and introduce the two-particle states

$$\psi_N = u(1)\,d(2) \tag{4.8}$$

where u and d are orthonormal single-particle states. The symmetrizer

$$\mathscr{S}_{12} = \frac{1}{2}(e + P_{12}) \tag{4.9}$$

projects out the symmetric part ψ^S of ψ_N:

$$\psi^S = \mathscr{S}_{12}\,\psi_N = \frac{1}{2}[u(1)\,d(2) + d(1)\,u(2)] \tag{4.10}$$

and the antisymmetrizer

$$\mathscr{A}_{12} = \frac{1}{2}(e - P_{12}) \tag{4.11}$$

projects out the antisymmetric part ψ^A of ψ_N:

$$\psi^A = \mathscr{A}_{12}\psi_N = \frac{1}{2}[u(1)d(2) - d(1)u(2)]. \tag{4.12}$$

Starting from the single-particle states u and d one can also put two particles in the same state to get $u(1)u(2)$ or $d(1)d(2)$, which are obviously symmetric. In all, there are four linearly independent states among which three are symmetric and one antisymmetric. The representation of S_2 in the four-dimensional space defined above is reducible to four one-dimensional representations, three symmetric and one antisymmetric. All three symmetric states are characterized by the same Young tableau

$$\boxed{\begin{array}{|c|c|} \hline 1 & 2 \\ \hline \end{array}} \;\rightarrow\; \begin{cases} u(1)u(2) \\ \frac{1}{2}[u(1)d(2) + d(1)u(2)] \\ d(1)d(2) \end{cases} \tag{4.13}$$

and for the antisymmetric state one has the correspondence

$$\begin{array}{|c|} \hline 1 \\ \hline 2 \\ \hline \end{array} \;\rightarrow\; \frac{1}{2}[u(1)d(2) - d(1)u(2)]. \tag{4.14}$$

In other words, two boxes in a row correspond to a symmetric state and two boxes in a column correspond to an antisymmetric state.

Starting with $n = 3$ the situation gets more complicated because mixed symmetries can appear.

Before discussing this case we introduce another way of specifying permutations which involve permutations of states instead of particles.

State permutations and Weyl tableaux

The generalization of (4.8) to $n > 2$ particles occupying n distinct single-particle states $\alpha_1, \alpha_2, \cdots, \alpha_n$ is

$$\psi_N = \alpha_1(1)\alpha_2(2)\ldots\alpha_n(n) \tag{4.15}$$

where an initial order has been established both in the particles and in the particle states, such that the particle i occupies the state α_i. This is referred to as a *normal order* state. In the previous subsection an element P of S_n was defined as a permutation of the particles, as in the relation (4.4). But one can alternatively permute the states $\alpha_1, \alpha_2, \cdots, \alpha_n$ and keep the particle order fixed as 1, 2, ..., n, as in (4.10) and (4.12). All the permutations of the n states also form a group called the state permutation group. Let us denote it by $\mathscr{S}_n$ and its elements by $\mathscr{P}$. The two groups are isomorphic. For the normal order state (4.15) one has

$$P\psi_N = \mathscr{P}^{-1}\psi_N. \tag{4.16}$$

The meaning of this relation is that P acts in the space of numbers $1, 2, ..., n$ and $\mathscr{P}$ acts in the space of single-particle states $\alpha_1, \alpha_2, \cdots, \alpha_n$. In the following sections we shall use $\mathscr{S}_n$ instead of S_n to construct basis functions for the irreducible representations of the permutation groups S_n for n up to $n = 6$.

As the particle label and the state label appear on an equal footing this suggests that the boxes of a Young diagram can be filled with state labels instead of particle labels. This operation produces the so-called *Weyl tableaux*. It turns out that there is a relation between the structure of a Weyl tableau and the symmetries of the corresponding state. Two simple rules for constructing Weyl tableaux are the following.

(1) The same value of α_i does not appear twice in a column. A column represents an antisymmetric state or a Slater determinant and two identical rows or columns in a Slater determinant are forbidden by the Pauli principle.

(2) In a particular tableau, the values of α_i must be read in increasing order, first all α_1, then all α_2, and so on, as we read a row from left to right or a column from top to bottom. The initial order is specified in Equation (4.15). An example of a correct tableau is:

$$\begin{array}{|c|c|c|c|c|} \hline \alpha_1 & \alpha_1 & \alpha_1 & \alpha_2 & \alpha_3 \\ \hline \alpha_2 & \alpha_3 & \alpha_4 \\ \cline{1-3} \alpha_3 & \alpha_4 \\ \cline{1-2} \alpha_5 \\ \cline{1-1} \end{array}$$

In the case $n = 2$, discussed in the previous subsection, for each Young diagram there is only one Weyl tableau, where $\alpha_1 = u, \alpha_2 = d$:

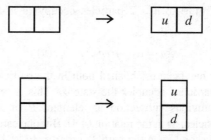

This is because a two-particle state can be either symmetric or antisymmetric. For $n > 2$ particles, one can construct states of mixed symmetries, and more than one Weyl tableau can be attached to each such state. Examples will appear in the following subsections.

Basis functions of S_3

We shall leave the Weyl tableaux for the moment. We first base our discussion of the $n = 3$ case on Young tableaux and return to Weyl tableaux at the end of this subsection. We start from the normal state with $\alpha_1 = u, \alpha_2 = d, \alpha_3 = s$:

$$\psi_N = u(1)\,d(2)\,s(3) \tag{4.17}$$

and assume that u, d and s are orthonormal functions. This notation is chosen because in applications to the quark model u, d and s stand for up, down, and strange quark states. By permuting the particle labels 1, 2, 3 or alternatively the state labels u, d, s by keeping the particle order fixed, one obtains 3! linearly independent functions. Following the arguments in the previous subsection we choose to permute the state labels and for simplicity we shall omit the particle labels which remain everywhere in the order 1, 2, 3. The six functions obtained by permutations can be recombined into another six linearly independent functions. We first write the symmetric ψ^S and the antisymmetric ψ^A functions obtained with the help of the operators $\mathscr{S}_{123}$ and $\mathscr{A}_{123}$ respectively:

$$\psi^S = \mathscr{S}_{123}\,u\,d\,s = \frac{1}{3!}(e + P_{12} + P_{13} + P_{23} + P_{123} + P_{132})\,u\,d\,s$$

$$= \frac{1}{3!}(uds + dus + sdu + usd + dsu + sud) \tag{4.18}$$

$$\psi^A = \mathscr{A}_{123}\,u\,d\,s = \frac{1}{3!}(e - P_{12} - P_{13} - P_{23} + P_{123} + P_{132})\,u\,d\,s$$

$$= \frac{1}{3!}(uds - dus - sdu - usd + dsu + sud). \tag{4.19}$$

The Young tableaux associated with these functions are

$$\psi^S \rightarrow \boxed{\begin{array}{|c|c|c|} \hline 1 & 2 & 3 \\ \hline \end{array}} \qquad \psi^A \rightarrow \boxed{\begin{array}{|c|} \hline 1 \\ \hline 2 \\ \hline 3 \\ \hline \end{array}} \tag{4.20}$$

We are left with building another four independent functions and these have to have a mixed symmetry. The procedure is as follows. We start again from (4.17) and symmetrize (or antisymmetrize) it with respect to two particles only and then antisymmetrize (or symmetrize) with respect to two other particles, one particle being common to the previous operation. This means we use the successive action of operators of type (4.9) and (4.11), i.e. $\mathscr{S}_{ij}$ and $\mathscr{A}_{ij}$. There is some arbitrariness in the order of $\mathscr{S}_{ij}$ and $\mathscr{A}_{ij}$ and in the choice of i and j. Of course i and j cannot be the same in $\mathscr{S}$ and $\mathscr{A}$ because

$$\mathscr{S}_{ij}\,\mathscr{A}_{ij} = 0. \tag{4.21}$$

We found it convenient to start with the functions (for other choices, see for example Lichtenberg 1970)

$$\psi_1 = \mathscr{A}_{13} \mathscr{S}_{12} \psi_N \qquad (4.22)$$
$$\psi_2 = \mathscr{A}_{23} \mathscr{S}_{12} \psi_N \qquad (4.23)$$
$$\psi_3 = \mathscr{S}_{13} \mathscr{A}_{12} \psi_N \qquad (4.24)$$
$$\psi_4 = \mathscr{S}_{23} \mathscr{A}_{12} \psi_N. \qquad (4.25)$$

The first two functions have $\mathscr{S}_{12}$ and the other two $\mathscr{A}_{12}$ as common operators. It is easy to see that ψ_1 and ψ_2 form an invariant subspace. Direct calculations give

$$(12)\psi_1 = \psi_2; \qquad (13)\psi_1 = -\psi_1$$
$$(12)\psi_2 = \psi_1; \qquad (13)\psi_2 = \psi_2 - \psi_1.$$

All the other group elements in S_3 can be obtained by taking products of (12) and (13) (cf. Table 2.1). In a similar way one can show that ψ_3 and ψ_4 also form an invariant subspace. One can see that by construction

$$\langle\psi_1,\psi_3\rangle = \langle\psi_1,\psi_4\rangle = \langle\psi_2,\psi_3\rangle = \langle\psi_2,\psi_4\rangle = 0 \qquad (4.26)$$

but

$$\langle\psi_1,\psi_2\rangle \neq 0, \qquad \langle\psi_3,\psi_4\rangle \neq 0 \qquad (4.27)$$

so that neither basis is orthogonal. However the sum and the difference

$$\psi'_1 \sim \psi_1 + \psi_2 \qquad (4.28)$$
$$\psi'_2 \sim \psi_1 - \psi_2 \qquad (4.29)$$

produce two orthogonal functions ψ'_1 and ψ'_2. Also

$$\psi'_3 \sim \psi_3 + \psi_4 \qquad (4.30)$$
$$\psi'_4 \sim \psi_3 - \psi_4 \qquad (4.31)$$

are orthogonal to each other.

In normalized form these four functions are

$$\psi'_1 = -\frac{1}{\sqrt{12}}(2uds + 2dus - sdu - sud - usd - dsu) \qquad (4.32)$$

$$\psi'_2 = -\frac{1}{2}(usd + dsu - sdu - sud) \qquad (4.33)$$

$$\psi'_3 = \frac{1}{\sqrt{12}}(2uds - 2dus + sdu - sud + usd - dsu) \qquad (4.34)$$

$$\psi'_4 = \frac{1}{2}(-sdu + sud + usd - dsu). \qquad (4.35)$$

The phase convention will be explained after Equation (4.40). The $\psi'_i (i = 1, 2, 3,$ and 4) are also orthogonal on ψ^S of (4.18) and ψ^A of (4.19). Hence we have a set of six linearly independent functions, as we needed. The ψ'_i have mixed symmetries as inferred by construction. The argument given after Equation (4.25) and the definitions of ψ'_i show that ψ'_1 and ψ'_4 are both symmetric with respect to the transposition (12), and ψ'_2 and ψ'_3 are both antisymmetric with respect to the same transposition. This can be immediately checked, together with the remark that they do not transform into themselves upon application of the transpositions (13) and (23). These properties give the Young tableaux

$$\begin{array}{|c|c|}\hline 1 & 2 \\\hline 3 \\\hline\end{array} \quad \text{for } \psi'_1 \text{ or } \psi'_4 \qquad \begin{array}{|c|c|}\hline 1 & 3 \\\hline 2 \\\hline\end{array} \quad \text{for } \psi'_2 \text{ or } \psi'_3 \qquad (4.36)$$

where 1, 2 in a row (column) indicates symmetry (antisymmetry). ψ'_1 and ψ'_2 form an invariant subspace (Section 3.3), i.e. any element of S_3 applied to ψ'_1 or ψ'_2 gives a linear combination of them. In the same way, ψ'_3 and ψ'_4 form another invariant subspace. Both subspaces give rise to the same irreducible representation of S_3, the mixed representation, the Young diagram of which is

$$\begin{array}{|c|c|}\hline & \\\hline \\\hline\end{array} \qquad (4.37)$$

In terms of matrices this diagram corresponds to the set of six matrices (3.57), (3.59), and (3.60), one matrix for each group element.

The action of any element of S_3 on ψ'_1 or ψ'_4 is the same as the action of S_3 on the internal coordinate λ of (3.53), and any element of S_3 acts on ψ'_2 or ψ'_3 in the same way as it acts on the internal coordinate ρ of (3.53). In fact, in ψ'_i taking for example

$$u(1) = r_1, \qquad u(2) = r_2, \qquad u(3) = r_3$$

and $d(n) = \text{const.}$, $s(n) = \text{const.}$ for any $n = 1, 2,$ and 3 one obtains

$$\psi'_1 \sim \psi'_4 \sim r_1 + r_2 - 2r_3$$
$$\psi'_2 \sim \psi'_3 \sim r_1 - r_2 \qquad (4.38)$$

which shows that λ is a particular case of ψ'_1 or ψ'_4 and ρ is a particular case of ψ'_2 or ψ'_3. For this reason one labels ψ'_1 or ψ'_4 with a superscript index λ, and ψ'_2 or ψ'_3 with a superscript index ρ. We shall adopt this notation in the rest of the book, i.e. the 2×2 matrices of the mixed representation (4.37) can be obtained in one of the two-dimensional spaces

$$\begin{pmatrix} \psi^\lambda \\ \psi^\rho \end{pmatrix} = \begin{pmatrix} \psi'_1 \\ \psi'_2 \end{pmatrix} \qquad (4.39)$$

or

$$\begin{pmatrix} \psi^\lambda \\ \psi^\rho \end{pmatrix} = \begin{pmatrix} \psi'_4 \\ \psi'_3 \end{pmatrix}. \qquad (4.40)$$

The phases in front of (4.32)–(4.35) are a matter of convention. They are consistent with the Yamanouchi phase convention (Section 4.4) and with the phase convention for the unitary group SU(3) (Section 8.6). We recall that Young diagrams are also used in unitary groups. Here we anticipate that the doublets (4.39) and (4.40) can describe the SU(3) flavour content of the baryons Σ^0 and Λ^0 respectively, each

Table 4.1 Mixed symmetry states of three quarks for the baryon octet. (For the phase convention between isomultiplets, see Section 8.6).

Baryon	$\phi^\lambda = \begin{array}{\|c\|c\|} \hline 1 & 2 \\ \hline 3 \\ \cline{1-1} \end{array}$	$\phi^\rho = \begin{array}{\|c\|c\|} \hline 1 & 3 \\ \hline 2 \\ \cline{1-1} \end{array}$	Weyl tableau
p	$-\frac{1}{\sqrt{6}}(udu + duu - 2uud)$	$\frac{1}{\sqrt{2}}(udu - duu)$	$\begin{array}{\|c\|c\|} \hline u & u \\ \hline d \\ \cline{1-1} \end{array}$
n	$\frac{1}{\sqrt{6}}(udd + dud - 2ddu)$	$\frac{1}{\sqrt{2}}(udd - dud)$	$\begin{array}{\|c\|c\|} \hline u & d \\ \hline d \\ \cline{1-1} \end{array}$
Σ^+	$\frac{1}{\sqrt{6}}(usu + suu - 2uus)$	$-\frac{1}{\sqrt{2}}(usu - suu)$	$\begin{array}{\|c\|c\|} \hline u & u \\ \hline s \\ \cline{1-1} \end{array}$
Σ^0	$-\frac{1}{\sqrt{12}}(2uds + 2dus$ $-sdu - sud - usd - dsu)$	$-\frac{1}{2}(usd + dsu - sdu - sud)$	$\begin{array}{\|c\|c\|} \hline u & d \\ \hline s \\ \cline{1-1} \end{array}$
Σ^-	$\frac{1}{\sqrt{6}}(dsd + sdd - 2dds)$	$-\frac{1}{\sqrt{2}}(dsd - sdd)$	$\begin{array}{\|c\|c\|} \hline d & d \\ \hline s \\ \cline{1-1} \end{array}$
Λ^0	$\frac{1}{2}(sud - sdu + usd - dsu)$	$\frac{1}{\sqrt{12}}(2uds - 2dus$ $+sdu - sud + usd - dsu)$	$\begin{array}{\|c\|c\|} \hline u & s \\ \hline d \\ \cline{1-1} \end{array}$
Ξ^0	$-\frac{1}{\sqrt{6}}(uss + sus - 2ssu)$	$-\frac{1}{\sqrt{2}}(uss - sus)$	$\begin{array}{\|c\|c\|} \hline u & s \\ \hline s \\ \cline{1-1} \end{array}$
Ξ^-	$-\frac{1}{\sqrt{6}}(dss + sds - 2ssd)$	$-\frac{1}{\sqrt{2}}(dss - sds)$	$\begin{array}{\|c\|c\|} \hline d & s \\ \hline s \\ \cline{1-1} \end{array}$

defined as a system of three quarks u, d, and s. So the flavour part of the total wave function of each of the Σ^0 and Λ^0 is a set ϕ^λ, ϕ^ρ. The identification is

$$\phi^\lambda_{\Sigma^0} = \psi'_1 \tag{4.41}$$
$$\phi^\rho_{\Sigma^0} = \psi'_2 \tag{4.42}$$
$$\phi^\lambda_{\Lambda^0} = \psi'_4 \tag{4.43}$$
$$\phi^\rho_{\Lambda^0} = \psi'_3. \tag{4.44}$$

These flavour components have to be combined with mixed symmetry spin functions to form a totally symmetric flavour–spin state of spin 1/2. In the ground state, where the orbital part is symmetric, the antisymmetry of the total wave function comes from the colour part. Mixed symmetric states of three particles with two identical state labels also exist. They can be derived starting with a ψ_N containing only two distinct state labels or obtained by taking $s = u, s = d$ or $u = d$ in (4.32)–(4.35) and normalizing again. Then ψ'_1 and ψ'_4 reduce to the same function. The same happens with ψ'_2 and ψ'_3. The results of various possibilities are displayed in Table 4.1. The corresponding baryon name is given in column 1 in each case. This correspondence will be explained in Chapter 8 where the meaning of the flavour degree of freedom in the context of unitary groups will be given. Each pair ϕ^λ, ϕ^ρ, with its Weyl tableau shown in the last column (see Chen 1989, Table 7.5-2), describes the flavour content of a baryon of spin 1/2. One can see that the Weyl tableaux of Σ^0 and Λ^0 are different. Hence they can be used as labels to distinguish between two equivalent irreps of S_3.

Symmetric states of three particles can be made of three different, two identical, or three identical, state labels. The last two cases can be derived independently or from the expression (4.18) by analogy with the case of mixed symmetry states, as explained above. The possible normalized functions are given in Table 4.2. Each function from this table can be associated to the flavour part of a baryon of spin 3/2 made out of three quarks. The detailed explanation will be given in Chapter 8.

Basis functions of S_4

The irreducible representations of S_4 are given by the following Young diagrams

[4]   [31]   [22]   [211]   [1⁴]

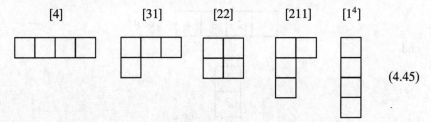

(4.45)

The most general normal state of four particles is

$$\psi_N = \alpha_1(1)\alpha_2(2)\alpha_3(3)\alpha_4(4) \equiv \alpha_1 \alpha_2 \alpha_3 \alpha_4 \tag{4.46}$$

Table 4.2 Symmetric states of three quarks for the baryon decuplet.

Baryon	ϕ^S	Weyl tableau
Δ^{++}	uuu	$u\,u\,u$
Δ^+	$\frac{1}{\sqrt{3}}(uud + udu + duu)$	$u\,u\,d$
Δ^0	$\frac{1}{\sqrt{3}}(udd + dud + ddu)$	$u\,d\,d$
Δ^-	ddd	$d\,d\,d$
Σ^{*+}	$\frac{1}{\sqrt{3}}(uus + usu + suu)$	$u\,u\,s$
Σ^{*0}	$\frac{1}{\sqrt{6}}(uds + dus + usd + sud + sdu + dsu)$	$u\,d\,s$
Σ^{*-}	$\frac{1}{\sqrt{3}}(sdd + dsd + dds)$	$d\,d\,s$
Ξ^{*0}	$\frac{1}{\sqrt{3}}(uss + sus + ssu)$	$u\,s\,s$
Ξ^{*-}	$\frac{1}{\sqrt{3}}(dss + sds + ssd)$	$d\,s\,s$
Ω^-	sss	$s\,s\,s$

with $\alpha_1 \neq \alpha_2 \neq \alpha_3 \neq \alpha_4$. The invariant subspaces of the representations [4] and [1^4] are the symmetric ψ^S and antisymmetric ψ^A functions, respectively defined as

$$\psi^S = \mathcal{S}_{1234}\,\psi_N \tag{4.47}$$
$$\psi^A = \mathcal{A}_{1234}\,\psi_N. \tag{4.48}$$

These are both one-dimensional representations and there is one standard Young tableau for each:

$$\boxed{1\,|\,2\,|\,3\,|\,4} \quad \text{for } \psi^S$$

and

$$\begin{array}{|c|}\hline 1 \\ \hline 2 \\ \hline 3 \\ \hline 4 \\ \hline \end{array} \quad \text{for } \psi^A$$

BASIS FUNCTIONS S_n

In order to facilitate the algebraic calculations involved in the construction of mixed symmetry states corresponding to [31], [22], and [211] we shall use simple forms of (4.46) in each case. There are states with two or more identical labels. However the procedure applied remains valid for the most general form of (4.46). In the following we shall consider the mixed representations one by one.

The standard Young tableaux associated with the representation [31] are

$$Y_1 = \begin{array}{|c|c|c|} \hline 1 & 2 & 3 \\ \hline 4 \\ \cline{1-1} \end{array} , \quad Y_2 = \begin{array}{|c|c|c|} \hline 1 & 2 & 4 \\ \hline 3 \\ \cline{1-1} \end{array} , \quad Y_3 = \begin{array}{|c|c|c|} \hline 1 & 3 & 4 \\ \hline 2 \\ \cline{1-1} \end{array}$$

(4.49)

i.e. this representation is three-dimensional. To build its invariant subspace we start with the functions

$$\psi_1 = \mathscr{A}_{14} \mathscr{S}_{123} \psi_N \qquad (4.50)$$
$$\psi_2 = \mathscr{A}_{24} \mathscr{S}_{123} \psi_N \qquad (4.51)$$
$$\psi_3 = \mathscr{A}_{34} \mathscr{S}_{123} \psi_N \qquad (4.52)$$

and for simplicity take $\alpha_1 = \alpha_2 = \alpha_3 = \alpha$, $\alpha_4 = \beta$ that is

$$\psi_N = \alpha \alpha \alpha \beta = \alpha^3 \beta \qquad (4.53)$$

with $<\alpha|\beta> = 0$. The ψ_i become

$$\psi_1 = \alpha \alpha \alpha \beta - \beta \alpha \alpha \alpha \qquad (4.54)$$
$$\psi_2 = \alpha \alpha \alpha \beta - \alpha \beta \alpha \alpha \qquad (4.55)$$
$$\psi_3 = \alpha \alpha \alpha \beta - \alpha \alpha \beta \alpha. \qquad (4.56)$$

These ψ_i are not orthogonal and we shall orthogonalize them by the Gram–Schmidt procedure. As in the case of S_3 we start with the sum of the available ψ_i

$$\psi'_1 = \psi_1 + \psi_2 + \psi_3 = 3\alpha\alpha\alpha\beta - \alpha\alpha\beta\alpha - \alpha\beta\alpha\alpha - \beta\alpha\alpha\alpha. \qquad (4.57)$$

This function is symmetric with respect to any permutation of 1, 2, and 3 and it corresponds to the tableau Y_1 of (4.49). Next we search for a function orthogonal to ψ'_1. There are two choices

$$\psi'_2 = \psi'_1 + a\,\psi_2 \qquad (4.58)$$

or

$$\psi'_2 = \psi'_1 + b\,\psi_3 \qquad (4.59)$$

with coefficients such that

$$<\psi'_2|\psi'_1> = 0. \qquad (4.60)$$

Performing the algebra it turns out that the interesting choice is (4.59) because it leads to a function with the property

$$P_{12}\,\psi'_2 = \psi'_2 \qquad (4.61)$$

required by the tableau Y_2 of (4.49). This function is

$$\psi'_2 = 2\alpha\alpha\beta\alpha - \alpha\beta\alpha\alpha - \beta\alpha\alpha\alpha \tag{4.62}$$

obtained from (4.59) with $b = -3$ required by (4.60).

Finally the third linear combination must be

$$\psi'_3 = \psi'_1 + c\psi'_2 + d\psi_2. \tag{4.63}$$

The orthogonality relations

$$<\psi'_1|\psi'_3> = <\psi'_2|\psi'_3> = 0 \tag{4.64}$$

lead to

$$\psi'_3 = \psi'_1 + \frac{1}{2}\psi'_2 - 3\psi_2 = \frac{3}{2}(\alpha\beta\alpha\alpha - \beta\alpha\alpha\alpha) \tag{4.65}$$

which has the property

$$P_{12}\psi'_3 = -\psi'_3 \tag{4.66}$$

consistent with the tableau Y_3 of (4.49) where the column indicates antisymmetry between 1 and 2.

The result is summarized in Table 4.3 where the functions are normalized. Their Yamanouchi symbol is also indicated. The relative phases between the normalized ψ'_1, ψ'_2, and ψ'_3 are consistent with the Yamanouchi phase convention. For completeness we also give the basis functions of the same representation in the configurations $\alpha^2\beta^2$ and $\alpha\beta^3$. Any of the configurations $\alpha^3\beta$, $\alpha^2\beta^2$ or $\alpha\beta^3$ leads to the same set of 4! matrices defining the irreps associated with the Young diagram [31]. The matrices corresponding to the adjacent transpositions (12), (23), and (34) are given in Table 4.5.

The next step is to find the invariant subspace of [22] which is two-dimensional because there are only two possible standard Young tableaux:

1	2
3	4

1	3
2	4

(4.67)

For simplicity we start with

$$\psi_N = \alpha\alpha\beta\beta = \alpha^2\beta^2 \tag{4.68}$$

and build the functions

$$\psi_1 = \mathcal{A}_{13}\mathcal{A}_{24}\mathcal{S}_{12}\mathcal{S}_{34}\psi_N \tag{4.69}$$

$$\psi_2 = \mathcal{A}_{14}\mathcal{A}_{23}\mathcal{S}_{12}\mathcal{S}_{34}\psi_N. \tag{4.70}$$

Their sum and difference give the following orthonormal basis states

$$\psi'_1 = \frac{1}{\sqrt{12}}(2\alpha\alpha\beta\beta + 2\beta\beta\alpha\alpha - \alpha\beta\alpha\beta - \beta\alpha\alpha\beta - \alpha\beta\beta\alpha - \beta\alpha\beta\alpha) \tag{4.71}$$

$$\psi'_2 = \frac{1}{2}(\alpha\beta\alpha\beta - \beta\alpha\alpha\beta + \beta\alpha\beta\alpha - \alpha\beta\beta\alpha) \tag{4.72}$$

Table 4.3 Basis functions of the irreducible representation [31] of S_4.

Young tableau + (Yamanouchi symbol)	Configuration		
	$\alpha^3\beta$	$\alpha^2\beta^2$	$\alpha\beta^3$
$\begin{array}{\|c\|c\|c\|}\hline 1 & 2 & 3 \\\hline 4 \\\cline{1-1}\end{array}$ (2111)	$\frac{1}{\sqrt{12}}(3\alpha\alpha\alpha\beta - \alpha\beta\alpha\alpha - \alpha\alpha\beta\alpha - \beta\alpha\alpha\alpha)$	$-\frac{1}{\sqrt{6}}(\alpha\alpha\beta\beta + \alpha\beta\alpha\beta + \beta\alpha\alpha\beta - \beta\beta\alpha\alpha - \beta\alpha\beta\alpha - \alpha\beta\beta\alpha)$	$\frac{1}{\sqrt{12}}(3\beta\beta\beta\alpha - \alpha\beta\beta\beta - \beta\alpha\beta\beta - \beta\beta\alpha\beta)$
$\begin{array}{\|c\|c\|}\hline 1 & 2 & 4 \\\hline 3 \\\cline{1-1}\end{array}$ (1211)	$\frac{1}{\sqrt{6}}(2\alpha\alpha\beta\alpha - \alpha\beta\alpha\alpha - \beta\alpha\alpha\alpha)$	$-\frac{1}{\sqrt{6}}(2\alpha\alpha\beta\beta - 2\beta\beta\alpha\alpha - \alpha\beta\alpha\beta - \beta\alpha\alpha\beta + \alpha\beta\beta\alpha + \beta\alpha\beta\alpha)$	$\frac{1}{\sqrt{6}}(2\beta\beta\alpha\beta - \alpha\beta\beta\beta - \beta\alpha\beta\beta)$
$\begin{array}{\|c\|c\|}\hline 1 & 3 & 4 \\\hline 2 \\\cline{1-1}\end{array}$ (1121)	$\frac{1}{\sqrt{2}}(\alpha\beta\alpha\alpha - \beta\alpha\alpha\alpha)$	$-\frac{1}{2}(\alpha\beta\alpha\beta - \beta\alpha\alpha\beta + \alpha\beta\beta\alpha - \beta\alpha\beta\alpha)$	$-\frac{1}{\sqrt{2}}(\alpha\beta\beta\beta - \beta\alpha\beta\beta)$

with the properties

$$P_{12}\psi'_1 = \psi'_1 \qquad P_{34}\psi'_1 = \psi'_1$$
$$P_{12}\psi'_2 = -\psi'_2 \qquad P_{34}\psi'_2 = -\psi'_2;$$
(4.73)

hence

$$\psi'_1 \rightarrow \begin{array}{|c|c|}\hline 1 & 2 \\\hline 3 & 4 \\\hline\end{array} \qquad \psi'_2 \rightarrow \begin{array}{|c|c|}\hline 1 & 3 \\\hline 2 & 4 \\\hline\end{array}$$

The last irreducible representation to consider for S_4 is [211] which is three-dimensional, as [31]. Its Young tableaux are

$$Y_1 = \begin{array}{|c|c|}\hline 1 & 4 \\\hline 2 \\\cline{1-1} 3 \\\cline{1-1}\end{array}, \quad Y_2 = \begin{array}{|c|c|}\hline 1 & 3 \\\hline 2 \\\cline{1-1} 4 \\\cline{1-1}\end{array}, \quad Y_3 = \begin{array}{|c|c|}\hline 1 & 2 \\\hline 3 \\\cline{1-1} 4 \\\cline{1-1}\end{array} \qquad (4.74)$$

To have an antisymmetric state of three particles as in Y_1 we need at least three distinct states. So we make the choice $\alpha_1 = \alpha$, $\alpha_2 = \beta$, $\alpha_3 = \alpha_4 = \gamma$, hence

$$\psi_N = \alpha(1)\beta(2)\gamma(3)\gamma(4) = \alpha\beta\gamma\gamma. \qquad (4.75)$$

The starting basis functions are built as follows

$$\begin{aligned}\psi_1 &= \mathscr{S}_{14}\mathscr{A}_{123}\psi_N \\ \psi_2 &= \mathscr{S}_{24}\mathscr{A}_{123}\psi_N \\ \psi_3 &= \mathscr{S}_{34}\mathscr{A}_{123}\psi_N.\end{aligned} \qquad (4.76)$$

There is a similarity between these functions and (4.50)–(4.52) for the [31] representation. This is because [31] and [211] are conjugate representations. The same technique as above gives the orthogonal functions

$$\begin{aligned}\psi'_1 &= \psi_1 + \psi_2 + \psi_3 \\ \psi'_2 &= \psi'_1 - 3\psi_3 \\ \psi'_3 &= \psi'_1 + \frac{1}{2}\psi'_2 - 3\psi_2\end{aligned} \qquad (4.77)$$

which, if normalized, become

$$\psi'_1 = \frac{1}{\sqrt{6}}(\alpha\beta\gamma\gamma - \beta\alpha\gamma\gamma - \gamma\beta\alpha\gamma - \alpha\gamma\beta\gamma + \beta\gamma\alpha\gamma + \gamma\alpha\beta\gamma) \qquad (4.78)$$

$$\psi'_2 = \frac{1}{\sqrt{48}}(-2\alpha\beta\gamma\gamma + 2\beta\alpha\gamma\gamma - \gamma\beta\alpha\gamma - \alpha\gamma\beta\gamma + \beta\gamma\alpha\gamma + \gamma\alpha\beta\gamma$$
$$+ 3\gamma\beta\gamma\alpha + 3\alpha\gamma\gamma\beta - 3\beta\gamma\gamma\alpha - 3\gamma\alpha\gamma\beta) \tag{4.79}$$

$$\psi'_3 = \frac{1}{4}(2\gamma\gamma\alpha\beta - 2\gamma\gamma\beta\alpha + \alpha\gamma\beta\gamma + \gamma\alpha\beta\gamma - \gamma\beta\alpha\gamma - \beta\gamma\alpha\gamma$$
$$+ \gamma\beta\gamma\alpha + \beta\gamma\gamma\alpha - \alpha\gamma\gamma\beta - \gamma\alpha\gamma\beta). \tag{4.80}$$

They have the properties

$$\begin{aligned} P\psi'_1 &= \delta_P \psi'_1 \quad \text{for any } P \in S_3(1,2,3) \\ P_{12}\psi'_2 &= -\psi'_2 \\ P_{12}\psi'_3 &= \psi'_3, \quad P_{34}\psi'_3 = -\psi'_3 \end{aligned} \tag{4.81}$$

so the correspondence with (4.74) is

$$\psi'_1 \rightarrow Y_1, \qquad \psi'_2 \rightarrow Y_2, \qquad \psi'_3 \rightarrow Y_3.$$

Properties like (4.73) or (4.81) stem from the requirement that a vector belonging to an invariant subspace of S_n must also belong to a definite representation of $S_{n-1}, S_{n-2}, \ldots, S_2$ (see Section 4.4). Such properties mean that consecutive numbers $i, i+1, \ldots, i+v$ in a row (column) represent symmetry (antisymmetry). If the numbers are not consecutive there is no definite permutation property. Based on these considerations one can abstract the pure permutation properties of the invariant subspaces constructed above by leaving out the single-particle structure. The simplified functions are obtained by taking all single-particle state functions as constants except for a few. The remaining ones are denoted as follows:

(a) $\quad \alpha_1(1) = x_1, \quad \alpha_1(2) = x_2, \quad \alpha_1(3) = x_3, \quad \alpha_1(4) = x_4 \tag{4.82}$
 in ψ^S of (4.47);

(b) $\quad \beta(1) = x_1, \quad \beta(2) = x_2, \quad \beta(3) = x_3, \quad \beta(4) = x_4 \tag{4.83}$
 in the expression (4.57), (4.62), and (4.65);

(c) $\quad \alpha(1) = x_1, \quad \alpha(2) = x_2, \quad \alpha(3) = x_3 \quad \alpha(4) = x_4 \tag{4.84}$
 in (4.71) and (4.72);

(d) $\quad \alpha(1) = x_1, \quad \alpha(2) = x_2, \quad \alpha(3) = x_3, \quad \alpha(4) = x_4$
 $\quad \beta(1) = y_1, \quad \beta(2) = y_2, \quad \beta(3) = y_3, \quad \beta(4) = y_4 \tag{4.85}$
 in (4.78)–(4.80);

(e) $\quad \alpha_1(1) = x_1, \quad \alpha_1(2) = x_2, \quad \alpha_1(3) = x_3, \quad \alpha_1(4) = x_4$
 $\quad \alpha_2(1) = y_1, \quad \alpha_2(2) = y_2, \quad \alpha_2(3) = y_3, \quad \alpha_2(4) = y_4 \tag{4.86}$
 $\quad \alpha_3(1) = z_1, \quad \alpha_3(2) = z_2, \quad \alpha_3(3) = z_3, \quad \alpha_3(4) = z_4.$
 in ψ^A of (4.48).

The resulting (normalized) simplified expressions for the basis states are indicated next to their Young tableaux in Table 4.4. Note that the functions associated to the

Table 4.4 Simplest basis functions for the irreducible representations $[f]$ of S_4.

$[f]$	Young tableau	Basis status
$[4]$	`1 2 3 4`	$\frac{1}{\sqrt{4}}(x_1 + x_2 + x_3 + x_4)$
$[31]$	`1 2 3` / `4`	$\frac{1}{\sqrt{12}}(x_1 + x_2 + x_3 - 3x_4)$
	`1 2 4` / `3`	$\frac{1}{\sqrt{6}}(x_1 + x_2 - 2x_3)$
	`1 3 4` / `2`	$\frac{1}{\sqrt{2}}(x_1 - x_2)$
$[22]$	`1 2` / `3 4`	$\frac{1}{\sqrt{12}}[2x_1x_2 + 2x_3x_4 - (x_1+x_2)(x_3+x_4)]$
	`1 3` / `2 4`	$\frac{1}{2}(x_1 - x_2)(x_3 - x_4)$
$[211]$	`1 4` / `2` / `3`	$\frac{1}{\sqrt{6}}(x_1y_2 - y_1x_2 + x_2y_3 - y_2x_3 + x_3y_1 - y_3x_1)$
	`1 3` / `2` / `4`	$\frac{1}{\sqrt{48}}[(x_1-x_2)(y_3-3y_4) + 2x_1y_2 - (y_1-y_2)(x_3-3x_4) - 2y_1x_2]$
	`1 2` / `3` / `4`	$\frac{1}{4}[2x_3y_4 - 2y_3x_4 + x_1y_3 - y_1x_3 + x_2y_3 - y_2x_3$ $-(x_2y_4 - y_2x_4) - (x_1y_4 - y_1x_4)]$
$[1^4]$	`1` / `2` / `3` / `4`	$\frac{1}{\sqrt{24}}\begin{vmatrix} x_1 & y_1 & z_1 & 1 \\ x_2 & y_2 & z_2 & 1 \\ x_3 & y_3 & z_3 & 1 \\ x_4 & y_4 & z_4 & 1 \end{vmatrix}$

In Table 1C-1 of Bohr and Mottelson (1969), some terms are missing in the last two functions of the symmetry $[211]$.

representations [4] and [31] are proportional to the centre of mass and internal Jacobi coordinates, respectively, of a system of four particles with x_i considered as position vectors of the particles. This is the extension to four particles of the basis states of the representation [21] of S_3 defined in Section 3.7 as internal Jacobi coordinates.

For n particles a general formula of the centre of mass and internal Jacobi coordinates is (see Moshinski 1969)

$$\dot{x}^n = \frac{1}{\sqrt{n}} \sum_{t=1}^{n} x^t \qquad (4.87a)$$

$$\dot{x}^s = [s(s+1)]^{-1/2} \left[\sum_{t=1}^{s} x^t - sx^{s+1} \right] \qquad 1 \le s \le n-1. \qquad (4.87b)$$

In general one expects that (4.87a) and (4.87b) always form an invariant subspace for the representations [n] and [n − 1, 1] of S_n, respectively.

Exercise 4.1 Find the basis vectors of the representation [41] of S_5 in the configuration $\psi_N = \alpha^3 \beta^2$.

Solution There are four standard Young tableaux associated with the partition [41]. Their corresponding Yamanouchi symbols are

$$Y_1 = (21111), \qquad Y_2 = (12111), \qquad Y_3 = (11211), \qquad Y_4 = (11121).$$

Hence the invariant subspace of the irrep [41] of S_5 is four-dimensional. Its basis vectors can be worked out as linear combinations of the following functions:

$$\psi_1 = \mathscr{A}_{15} \mathscr{S}_{1234} \psi_N = \alpha\alpha\alpha\beta\beta + \alpha\alpha\beta\alpha\beta + \alpha\beta\alpha\alpha\beta \\ - \beta\alpha\alpha\beta\alpha - \beta\alpha\beta\alpha\alpha - \beta\beta\alpha\alpha\alpha \qquad (1)$$

$$\psi_2 = \mathscr{A}_{25} \mathscr{S}_{1234} \psi_N = \alpha\alpha\alpha\beta\beta + \alpha\alpha\beta\alpha\beta + \beta\alpha\alpha\alpha\beta \\ - \alpha\beta\alpha\beta\alpha - \alpha\beta\beta\alpha\alpha - \beta\beta\alpha\alpha\alpha \qquad (2)$$

$$\psi_3 = \mathscr{A}_{35} \mathscr{S}_{1234} \psi_N = \alpha\alpha\alpha\beta\beta + \alpha\beta\alpha\alpha\beta + \beta\alpha\alpha\alpha\beta \\ - \alpha\alpha\beta\beta\alpha - \alpha\beta\beta\alpha\alpha - \beta\alpha\beta\alpha\alpha \qquad (3)$$

$$\psi_4 = \mathscr{A}_{45} \mathscr{S}_{1234} \psi_N = \alpha\alpha\beta\alpha\beta + \alpha\beta\alpha\alpha\beta + \beta\alpha\alpha\alpha\beta \\ - \alpha\alpha\beta\beta\alpha - \alpha\beta\alpha\beta\alpha - \beta\alpha\alpha\beta\alpha. \qquad (4)$$

These functions are not orthogonal. The orthogonalization procedure is analogous to that of the representation [31] of S_4. The first step is to build

$$\psi'_1 = \psi_1 + \psi_2 + \psi_3 + \psi_4 \qquad (5)$$

which turns out to be symmetric under all adjacent transpositions of S_4. Hence ψ'_1 corresponds to Y_1. The second step is to search for

$$\psi'_2 = \psi'_1 + a_i \psi_i \qquad (6)$$

where $i = 2, 3$, or 4, such that

$$\langle \psi'_1 | \psi'_2 \rangle = 0. \tag{7}$$

In a similar way one finds ψ'_3 and ψ'_4. The resulting orthogonal basis is shown in the table below. The relative phases are consistent with the Yamanouchi phase convention (Section 4.4).

Young tableau	Basis functions of [41]
$\begin{array}{\|c\|c\|c\|c\|}\hline 1 & 2 & 3 & 4 \\ \hline 5 \\ \cline{1-1}\end{array}$	$\frac{1}{\sqrt{60}}(3\alpha\alpha\alpha\beta\beta + 3\alpha\alpha\beta\alpha\beta + 3\alpha\beta\alpha\alpha\beta + 3\beta\alpha\alpha\alpha\beta$ $- 2\beta\alpha\alpha\beta\alpha - 2\beta\alpha\beta\alpha\alpha - 2\beta\beta\alpha\alpha\alpha$ $- 2\alpha\beta\alpha\beta\alpha - 2\alpha\beta\beta\alpha\alpha - 2\alpha\alpha\beta\beta\alpha)$
$\begin{array}{\|c\|c\|c\|c\|}\hline 1 & 2 & 3 & 5 \\ \hline 4 \\ \cline{1-1}\end{array}$	$\frac{1}{6}(3\alpha\alpha\alpha\beta\beta - \alpha\alpha\beta\alpha\beta - \alpha\beta\alpha\alpha\beta - \beta\alpha\alpha\alpha\beta$ $+ 2\beta\alpha\alpha\beta\alpha - 2\beta\alpha\beta\alpha\alpha - 2\beta\beta\alpha\alpha\alpha$ $+ 2\alpha\beta\alpha\beta\alpha - 2\alpha\beta\beta\alpha\alpha + 2\alpha\alpha\beta\beta\alpha)$
$\begin{array}{\|c\|c\|c\|}\hline 1 & 2 & 4 & 5 \\ \hline 3 \\ \cline{1-1}\end{array}$	$\frac{1}{\sqrt{18}}(2\alpha\alpha\beta\alpha\beta + 2\alpha\alpha\beta\beta\alpha - 2\beta\beta\alpha\alpha\alpha$ $- \alpha\beta\alpha\alpha\beta - \beta\alpha\alpha\alpha\beta - \beta\alpha\alpha\beta\alpha - \alpha\beta\alpha\beta\alpha$ $+ \beta\alpha\beta\alpha\alpha + \alpha\beta\beta\alpha\alpha)$
$\begin{array}{\|c\|c\|c\|}\hline 1 & 3 & 4 & 5 \\ \hline 2 \\ \cline{1-1}\end{array}$	$\frac{1}{\sqrt{6}}(\alpha\beta\alpha\alpha\beta - \beta\alpha\alpha\alpha\beta + \alpha\beta\alpha\beta\alpha - \beta\alpha\alpha\beta\alpha$ $+ \alpha\beta\beta\alpha\alpha - \beta\alpha\beta\alpha\alpha)$

Exercise 4.2 Find the basis states of the representation [51] of S_6 in the configuration $\psi_N = \alpha^3 \beta^3$.

Solution Here the starting functions are

$$\psi_1 = \mathscr{A}_{16} \mathscr{S}_{12345} \psi_N \tag{1}$$
$$\psi_2 = \mathscr{A}_{26} \mathscr{S}_{12345} \psi_N \tag{2}$$
$$\psi_3 = \mathscr{A}_{36} \mathscr{S}_{12345} \psi_N \tag{3}$$
$$\psi_4 = \mathscr{A}_{46} \mathscr{S}_{12345} \psi_N \tag{4}$$
$$\psi_5 = \mathscr{A}_{56} \mathscr{S}_{12345} \psi_N. \tag{5}$$

The procedure of finding the orthogonal basis is entirely analogous to that used in Exercise 4.1 but a bit more tedious. As before, the first linear combination of interest is the sum of the linearly independent functions (1)–(5)

$$\psi'_1 = \psi_1 + \psi_2 + \psi_3 + \psi_4 + \psi_5 \quad \longrightarrow \quad \begin{array}{|c|c|c|c|c|}\hline 1 & 2 & 3 & 4 & 5 \\\hline 6 \\\cline{1-1}\end{array} \qquad (6)$$

The second basis vector is a function of type (6) of Exercise 4.1 but with $i = 2, 3, 4$, or 5. Using previous experience we can already make a straight choice. We take a linear combination of ψ'_1 and ψ_5 because this is the only one which preserves the permutation symmetry under S_4 and leads to a Young tableau containing $\boxed{1\,2\,3\,4}$ to which $\boxed{5}$ and $\boxed{6}$ are attached unambiguously such as to differ from ψ'_1. The orthogonality on ψ'_1 leads to

$$\psi'_2 = \psi'_1 - 5\psi_5 \quad \longrightarrow \quad \begin{array}{|c|c|c|c|c|}\hline 1 & 2 & 3 & 4 & 6 \\\hline 5 \\\cline{1-1}\end{array} \qquad (7)$$

The third basis vector can be obtained as a linear combination of ψ'_1, ψ'_2, and ψ_4. Such a combination is invariant to any permutation of S_3, which implies that its Young tableau should be built starting from $\boxed{1\,2\,3}$. The addition of $\boxed{4}$ must give $\begin{array}{|c|c|c|}\hline 1 & 2 & 3 \\\hline 4 \\\cline{1-1}\end{array}$ which leaves, for $\boxed{5}$ and $\boxed{6}$, no choice but the symmetric construction $\boxed{5\,6}$ to be added to the first row. By orthogonality one gets

$$\psi'_3 = \psi'_1 + \frac{1}{4}\psi'_2 - 5\psi_4 \quad \longrightarrow \quad \begin{array}{|c|c|c|c|c|}\hline 1 & 2 & 3 & 5 & 6 \\\hline 4 \\\cline{1-1}\end{array} \qquad (8)$$

which indeed satisfies

$$(56)\,\psi'_3 = \psi'_3. \qquad (9)$$

The linear combination of $\psi'_1, \psi'_2, \psi'_3$ and ψ_3 must give a function with the property

$$(12)\,\psi'_4 = \psi'_4. \qquad (10)$$

The orthogonality on ψ'_1, ψ'_2, and ψ'_3 leads to

$$\psi'_4 = \psi'_1 + \frac{1}{4}\psi'_2 + \frac{1}{3}\psi'_3 - 5\psi_3 \quad \longrightarrow \quad \begin{array}{|c|c|c|c|c|}\hline 1 & 2 & 4 & 5 & 6 \\\hline 3 \\\cline{1-1}\end{array} \qquad (11)$$

for which one can check that

$$(45)\psi'_4 = \psi'_4, \qquad (56)\psi'_4 = \psi'_4. \tag{12}$$

Finally, the linear combination of ψ'_1, ψ'_2, ψ'_3, ψ'_4, and ψ_2 orthogonal to ψ'_i ($i = 1, 2, 3$, and 4) should lead to a function ψ'_5 with the property

$$(12)\psi'_5 = -\psi'_5. \tag{13}$$

Such a function can also be directly built by observing that it must be symmetric for any permutation of particles 3, 4, 5, and 6. Actually it is just the product

$$\psi'_5 = \mathscr{A}_{12}(\alpha\beta)\mathscr{S}_{3456}(\alpha^2\beta^2). \tag{14}$$

The basis vectors of [51] obtained in this way are given in the table on p. 83 where again the Yamanouchi phase convention is used. Hence, from the above basis one can derive the matrices of the representation [51] shown in Table 4.5.

Dimension of an irreducible representation

At the beginning of this section we introduced the notion of standard Young tableau. We then gave examples of the explicit construction of some invariant subspaces of irreducible representations of S_n with values up to $n = 6$. These examples may serve as illustrations of some important general results relating standard Young tableaux to properties of the irreducible representations of S_n. These results are:

1. The dimension $d_{[f]}$ of the irreducible representation associated with a Young diagram of partition

   $$[f] = [f_1, f_2, \ldots, f_k]$$

   written in the notation (4.2b) is equal to the number of distinct standard Young tableaux obtained for that diagram.

2. A general formula for the dimension of an irreducible representation $[f]$ is

   $$d_{[f]} = \frac{n!}{\prod_{i=1}^{k}(f_i + k - i)!} \prod_{i<j\leq k}(f_i - f_j + j - i). \tag{4.88}$$

Frobenius derived this formula from the theory of characters. Actually Frobenius showed how to determine all the simple characters of the symmetric group (Frobenius 1900). Equation (4.88) then immediately follows because the dimension of an irreducible representation is equal to the character of the identity element in that representation, as shown in Equation (3.97). For a more detailed discussion of the results 1 and 2, see Hamermesh (1962), Chapter 7, or Boerner (1963), Chapter 4.

The Frobenius formula can be checked with the examples given earlier in this section. For a better understanding here is another example.

Young tableau	Basis functions of [51]
[[1,2,3,4,5],[6]]	$\frac{1}{\sqrt{20}}(\alpha\alpha\alpha\beta\beta\beta + \alpha\alpha\beta\alpha\beta\beta + \alpha\beta\alpha\alpha\beta\beta + \beta\alpha\alpha\alpha\beta\beta + \alpha\alpha\beta\beta\alpha\beta$ $+ \alpha\beta\beta\alpha\alpha\beta + \beta\beta\alpha\alpha\alpha\beta + \alpha\beta\alpha\beta\alpha\beta + \beta\alpha\beta\alpha\alpha\beta + \beta\alpha\alpha\beta\alpha\beta$ $- \beta\alpha\alpha\beta\beta\alpha - \beta\alpha\beta\alpha\beta\alpha - \beta\beta\alpha\alpha\beta\alpha - \beta\alpha\beta\beta\alpha\alpha - \beta\beta\beta\alpha\alpha\alpha$ $- \beta\beta\alpha\beta\alpha\alpha - \alpha\beta\alpha\beta\beta\alpha - \alpha\beta\beta\alpha\beta\alpha - \alpha\beta\beta\beta\alpha\alpha - \alpha\alpha\beta\beta\beta\alpha)$
[[1,2,3,4,6],[5]]	$\frac{1}{\sqrt{120}}(3\alpha\alpha\alpha\beta\beta\beta + 3\alpha\alpha\beta\alpha\beta\beta + 3\alpha\beta\alpha\alpha\beta\beta + 3\beta\alpha\alpha\alpha\beta\beta$ $- 3\beta\alpha\beta\beta\alpha\alpha - 3\beta\beta\beta\alpha\alpha\alpha - 3\beta\beta\alpha\beta\alpha\alpha - 3\alpha\beta\beta\beta\alpha\alpha$ $- 2\alpha\alpha\beta\beta\alpha\beta - 2\alpha\beta\beta\alpha\alpha\beta - 2\beta\beta\alpha\alpha\alpha\beta - 2\alpha\beta\alpha\beta\alpha\beta$ $- 2\beta\alpha\beta\alpha\alpha\beta - 2\beta\alpha\alpha\beta\alpha\beta$ $+ 2\beta\alpha\alpha\beta\beta\alpha + 2\beta\alpha\beta\alpha\beta\alpha + 2\beta\beta\alpha\alpha\beta\alpha + 2\alpha\beta\alpha\beta\beta\alpha$ $+ 2\alpha\beta\beta\alpha\beta\alpha + 2\alpha\alpha\beta\beta\beta\alpha)$
[[1,2,3,5,6],[4]]	$\frac{1}{\sqrt{72}}(3\alpha\alpha\alpha\beta\beta\beta - 3\beta\beta\beta\alpha\alpha\alpha + 2\alpha\alpha\beta\beta\alpha\beta - 2\alpha\beta\beta\alpha\alpha\beta - 2\beta\beta\alpha\alpha\alpha\beta$ $+ 2\alpha\beta\alpha\beta\alpha\beta - 2\beta\alpha\beta\alpha\alpha\beta + 2\beta\alpha\alpha\beta\alpha\beta + 2\beta\alpha\alpha\beta\beta\alpha - 2\beta\alpha\beta\alpha\beta\alpha$ $- 2\beta\beta\alpha\alpha\beta\alpha + 2\alpha\beta\alpha\beta\beta\alpha - 2\alpha\beta\beta\alpha\beta\alpha + 2\alpha\alpha\beta\beta\beta\alpha - \alpha\alpha\beta\alpha\beta\beta$ $- \alpha\beta\alpha\alpha\beta\beta - \beta\alpha\alpha\alpha\beta\beta + \beta\alpha\beta\beta\alpha\alpha + \beta\beta\alpha\beta\alpha\alpha + \alpha\beta\beta\beta\alpha\alpha)$
[[1,2,4,5,6],[3]]	$\frac{1}{6}(2\alpha\alpha\beta\alpha\beta\beta + 2\alpha\alpha\beta\beta\alpha\beta + 2\alpha\alpha\beta\beta\beta\alpha - 2\beta\beta\alpha\alpha\alpha\beta - 2\beta\beta\alpha\alpha\beta\alpha$ $- 2\beta\beta\alpha\beta\alpha\alpha + \alpha\beta\beta\alpha\alpha\beta + \alpha\beta\beta\alpha\beta\alpha + \alpha\beta\beta\beta\alpha\alpha + \beta\alpha\beta\alpha\alpha\beta$ $+ \beta\alpha\beta\alpha\beta\alpha + \beta\alpha\beta\beta\alpha\alpha - \beta\alpha\alpha\alpha\beta\beta - \beta\alpha\alpha\beta\alpha\beta - \beta\alpha\alpha\beta\beta\alpha$ $- \alpha\beta\alpha\alpha\beta\beta - \alpha\beta\alpha\beta\alpha\beta - \alpha\beta\alpha\beta\beta\alpha)$
[[1,3,4,5,6],[2]]	$\frac{1}{\sqrt{12}}(\alpha\beta\alpha\alpha\beta\beta + \alpha\beta\alpha\beta\alpha\beta + \alpha\beta\alpha\beta\beta\alpha$ $+ \alpha\beta\beta\alpha\alpha\beta + \alpha\beta\beta\alpha\beta\alpha + \alpha\beta\beta\beta\alpha\alpha$ $- \beta\alpha\alpha\alpha\beta\beta - \beta\alpha\alpha\beta\alpha\beta - \beta\alpha\alpha\beta\beta\alpha$ $- \beta\alpha\beta\alpha\alpha\beta - \beta\alpha\beta\alpha\beta\alpha - \beta\alpha\beta\beta\alpha\alpha)$

Table 4.5 Matrices of adjacent transpositions for the irreducible representations of S_2–S_5.

S_3:

$$
\begin{array}{c} \\ 211 \\ 121 \end{array}
\quad
\begin{array}{c}(12)\\ \begin{pmatrix} 1 & 0 \\ 0 & -1 \end{pmatrix} \end{array}
\quad
\begin{array}{c}(23)\\ \begin{pmatrix} -\frac{1}{2} & \frac{\sqrt{3}}{2} \\ \frac{\sqrt{3}}{2} & \frac{1}{2} \end{pmatrix}\end{array}
$$

S_4:

$$
\begin{array}{c} 2111 \\ 1211 \\ 1121 \end{array}
\quad
\begin{array}{c}(12)\\ \begin{pmatrix} 1 & 0 & 0 \\ 0 & 1 & 0 \\ 0 & 0 & -1 \end{pmatrix} \end{array}
\quad
\begin{array}{c}(23)\\ \begin{pmatrix} 1 & 0 & 0 \\ 0 & -\frac{1}{2} & \frac{\sqrt{3}}{2} \\ 0 & \frac{\sqrt{3}}{2} & \frac{1}{2} \end{pmatrix}\end{array}
\quad
\begin{array}{c}(34)\\ \begin{pmatrix} -\frac{1}{3} & \frac{\sqrt{8}}{3} & 0 \\ \frac{\sqrt{8}}{3} & \frac{1}{3} & 0 \\ 0 & 0 & 1 \end{pmatrix}\end{array}
$$

$$
\begin{array}{c} 3211 \\ 3121 \\ 1321 \end{array}
\quad
\begin{array}{c}(12)\\ \begin{pmatrix} 1 & 0 & 0 \\ 0 & -1 & 0 \\ 0 & 0 & -1 \end{pmatrix} \end{array}
\quad
\begin{array}{c}(23)\\ \begin{pmatrix} -\frac{1}{2} & \frac{\sqrt{3}}{2} & 0 \\ \frac{\sqrt{3}}{2} & \frac{1}{2} & 0 \\ 0 & 0 & -1 \end{pmatrix}\end{array}
\quad
\begin{array}{c}(34)\\ \begin{pmatrix} -1 & 0 & 0 \\ 0 & -\frac{1}{3} & \frac{\sqrt{8}}{3} \\ 0 & \frac{\sqrt{8}}{3} & \frac{1}{3} \end{pmatrix}\end{array}
$$

$$
\begin{array}{c} 2211 \\ 2121 \end{array}
\quad
\begin{array}{c}(12)\\ \begin{pmatrix} 1 & 0 \\ 0 & -1 \end{pmatrix} \end{array}
\quad
\begin{array}{c}(23)\\ \begin{pmatrix} -\frac{1}{2} & \frac{\sqrt{3}}{2} \\ \frac{\sqrt{3}}{2} & \frac{1}{2} \end{pmatrix}\end{array}
\quad
\begin{array}{c}(34)\\ \begin{pmatrix} 1 & 0 \\ 0 & -1 \end{pmatrix}\end{array}
$$

S_5:

$$
\begin{array}{c} 21111 \\ 12111 \\ 11211 \\ 11121 \end{array}
\quad
\begin{array}{c}(12)\\ \begin{pmatrix} 1 & 0 & 0 & 0 \\ 0 & 1 & 0 & 0 \\ 0 & 0 & 1 & 0 \\ 0 & 0 & 0 & -1 \end{pmatrix} \end{array}
\quad
\begin{array}{c}(23)\\ \begin{pmatrix} 1 & 0 & 0 & 0 \\ 0 & 1 & 0 & 0 \\ 0 & 0 & -\frac{1}{3} & \frac{\sqrt{3}}{2} \\ 0 & 0 & \frac{1}{2} & \frac{1}{2} \end{pmatrix}\end{array}
\quad
\begin{array}{c}(34)\\ \begin{pmatrix} 1 & 0 & 0 & 0 \\ 0 & -\frac{1}{3} & \frac{\sqrt{8}}{3} & 0 \\ 0 & \frac{\sqrt{8}}{3} & \frac{1}{3} & 0 \\ 0 & 0 & 0 & 1 \end{pmatrix}\end{array}
\quad
\begin{array}{c}(45)\\ \begin{pmatrix} -\frac{1}{4} & \frac{\sqrt{15}}{4} & 0 & 0 \\ \frac{1}{4} & \frac{1}{4} & 0 & 0 \\ 0 & 0 & 1 & 0 \\ 0 & 0 & 0 & 1 \end{pmatrix}\end{array}
$$

Table 4.5 *cont.*

$$
\begin{array}{c}
\begin{array}{cc}
& (12) \\
\end{array} \\
\begin{array}{c}
22111 \\ 21211 \\ 21121 \\ 12211 \\ 12121
\end{array}
\left(\begin{array}{ccccc}
1 & 0 & 0 & 0 & 0 \\
 & 1 & 0 & 0 & 0 \\
 & & -1 & 0 & 0 \\
 & & & 1 & 0 \\
 & & & & -1
\end{array}\right)
\end{array}
\qquad
\begin{array}{c}
(23) \\
\left(\begin{array}{ccccc}
1 & 0 & 0 & 0 & 0 \\
-\tfrac{1}{2} & \tfrac{\sqrt{3}}{2} & 0 & 0 & 0 \\
 & \tfrac{1}{2} & 0 & 0 & 0 \\
 & & & -\tfrac{1}{2} & \tfrac{\sqrt{3}}{2} \\
 & & & & \tfrac{1}{2}
\end{array}\right)
\end{array}
$$

$$
\begin{array}{c}
(34) \\
\left(\begin{array}{ccccc}
-\tfrac{1}{3} & \tfrac{\sqrt{8}}{3} & 0 & 0 & 0 \\
 & \tfrac{1}{3} & 0 & 0 & 0 \\
 & & 1 & 0 & 0 \\
 & & & 1 & 0 \\
 & & & & -1
\end{array}\right)
\end{array}
\qquad
\begin{array}{c}
(45) \\
\left(\begin{array}{ccccc}
1 & 0 & 0 & 0 & 0 \\
-\tfrac{1}{2} & 0 & \tfrac{\sqrt{3}}{2} & 0 \\
 & -\tfrac{1}{2} & 0 & \tfrac{\sqrt{3}}{2} \\
 & & \tfrac{1}{2} & 0 \\
 & & & & \tfrac{1}{2}
\end{array}\right)
\end{array}
$$

$$
\begin{array}{c}
\begin{array}{c}
(12) \\
\end{array} \\
\begin{array}{c}
32111 \\ 31211 \\ 31121 \\ 13211 \\ 13121 \\ 11321
\end{array}
\left(\begin{array}{cccccc}
1 & 0 & 0 & 0 & 0 & 0 \\
 & 1 & 0 & 0 & 0 & 0 \\
 & & -1 & 0 & 0 & 0 \\
 & & & 1 & 0 & 0 \\
 & & & & -1 & 0 \\
 & & & & & -1
\end{array}\right)
\end{array}
\qquad
\begin{array}{c}
(23) \\
\left(\begin{array}{cccccc}
1 & 0 & 0 & 0 & 0 & 0 \\
-\tfrac{1}{2} & \tfrac{\sqrt{3}}{2} & 0 & 0 & 0 & 0 \\
 & \tfrac{1}{2} & 0 & 0 & 0 & 0 \\
 & & & -\tfrac{1}{2} & \tfrac{\sqrt{3}}{2} & 0 \\
 & & & & \tfrac{1}{2} & 0 \\
 & & & & & -1
\end{array}\right)
\end{array}
$$

Table 4.5 cont.

$$
(34) \qquad\qquad (45)
$$

$$
\begin{pmatrix}
-\frac{1}{3} & \frac{\sqrt{8}}{3} & 0 & 0 & 0 & 0 \\
 & \frac{1}{3} & 0 & 0 & 0 & 0 \\
 & & 1 & 0 & 0 & 0 \\
 & & & -1 & 0 & 0 \\
 & & & & -\frac{1}{3} & \frac{\sqrt{8}}{3} \\
 & & & & & \frac{1}{3}
\end{pmatrix}
\qquad
\begin{pmatrix}
-1 & 0 & 0 & 0 & 0 & 0 \\
 & -\frac{1}{4} & 0 & \frac{\sqrt{15}}{4} & 0 & 0 \\
 & & -\frac{1}{4} & 0 & \frac{\sqrt{15}}{4} & 0 \\
 & & & \frac{1}{4} & 0 & 0 \\
 & & & & \frac{1}{4} & 0 \\
 & & & & & 1
\end{pmatrix}
$$

$$
(12) \qquad\qquad (23)
$$

$$
\begin{array}{r}
32211 \\
32121 \\
23211 \\
23121 \\
21321
\end{array}
\begin{pmatrix}
1 & 0 & 0 & 0 & 0 \\
 & -1 & 0 & 0 & 0 \\
 & & 1 & 0 & 0 \\
 & & & -1 & 0 \\
 & & & & -1
\end{pmatrix}
\qquad
\begin{pmatrix}
-\frac{1}{2} & \frac{\sqrt{3}}{2} & 0 & 0 & 0 \\
 & \frac{1}{2} & 0 & 0 & 0 \\
 & & -\frac{1}{2} & \frac{\sqrt{3}}{2} & 0 \\
 & & & \frac{1}{2} & 0 \\
 & & & & -1
\end{pmatrix}
$$

$$
(34) \qquad\qquad (45)
$$

$$
\begin{pmatrix}
1 & 0 & 0 & 0 & 0 \\
 & -1 & 0 & 0 & 0 \\
 & & -1 & 0 & 0 \\
 & & & -\frac{1}{3} & \frac{\sqrt{8}}{3} \\
 & & & & \frac{1}{3}
\end{pmatrix}
\qquad
\begin{pmatrix}
-\frac{1}{2} & 0 & \frac{\sqrt{3}}{2} & 0 & 0 \\
 & -\frac{1}{2} & 0 & \frac{\sqrt{3}}{2} & 0 \\
 & & \frac{1}{2} & 0 & 0 \\
 & & & \frac{1}{2} & 0 \\
 & & & & -1
\end{pmatrix}
$$

Table 4.5 *cont.*

	(12)	(23)	(34)	(45)
43211	$\begin{pmatrix} 1 & 0 & 0 & 0 \\ & -1 & 0 & 0 \\ & & -1 & 0 \\ & & & -1 \end{pmatrix}$	$\begin{pmatrix} -\frac{1}{2} & \frac{\sqrt{3}}{2} & 0 & 0 \\ \frac{1}{2} & & 0 & 0 \\ & & -1 & 0 \\ & & & -1 \end{pmatrix}$	$\begin{pmatrix} -1 & 0 & 0 & 0 \\ -\frac{1}{3} & \frac{\sqrt{8}}{3} & 0 \\ & \frac{1}{3} & 0 \\ & & & -1 \end{pmatrix}$	$\begin{pmatrix} -1 & 0 & 0 & 0 \\ & -1 & 0 & 0 \\ & & -\frac{1}{4} & \frac{\sqrt{15}}{4} \\ & & & \frac{1}{4} \end{pmatrix}$
43121				
41321				
14321				

The matrices of the elements (45), (15), (35), and (25) of the representation [32] of S_5 are incorrect in Table 7.3 of Hamermesh (1962). The correct values are given in Exercise 4.5.

Example 4.1 Calculate the dimension of the irreducible representation [321] of S_6.

$$f_1 = 3, \quad f_2 = 2, \quad f_3 = 1$$

For this representation $k = 3$, i.e. the last row is row 3. Then

$$d_{[321]} = \frac{6!(f_1 - f_2 + 1)(f_1 - f_3 + 2)(f_2 - f_3 + 1)}{(f_1 + 2)!(f_2 + 1)!f_3!} = 16. \tag{4.89}$$

There also exists a simpler formula popular among particle physicists, due to Robinson (1961), the formula is related to the concept of *hook graphs* (Frame, Robinson, and Thrall 1954). The hook length of a given box in a Young diagram is equal to $(p + q + 1)$ where p is the number of boxes to the right of that box in the same row and q is the number of boxes below the given box in the same column. For the representation [321] of S_6 considered in the above example the hook graph is

5	3	1
3	1	
1		

Let ℓ_i be the hook lengths. Here $\ell_1 = 5$, $\ell_2 = 3$, $\ell_3 = 1$, $\ell_4 = 3$, $\ell_5 = 1$, and $\ell_6 = 1$. Then the dimension of $[f]$ of S_n is

$$d_{[f]} = \frac{n!}{\prod_i^n \ell_i}. \tag{4.90}$$

Using this formula we again get $d_{[321]} = 16$. Tables of dimensions of irreducible representations of S_n can be found in Wybourne (1970) for $n \leq 14$.

Exercise 4.3 Using formula (4.89) calculate the dimension $d_{[f]}$ of all irreducible representations $[f]$ of S_5.

Solution the answer is given in the table below.

$[f]$	Hook graph	$d_{[f]}$
$[5]$	5 4 3 2 1	1
$[41]$	5 3 2 1 / 1	4
$[32]$	4 3 1 / 2 1	5
$[31^2]$	5 2 1 / 2 / 1	6
$[2^2 1]$	4 2 / 3 1 / 1	5
$[21^3]$	5 1 / 3 / 2 / 1	4
$[1^5]$	5 / 4 / 3 / 2 / 1	1

We may check our result with the formula (3.100)

$$\sum_{[f]} d^2_{[f]} = 5!$$

4.4 MATRICES OF IRREDUCIBLE REPRESENTATIONS

The practical method used in Section 4.3 to construct invariant subspaces of S_n is convenient for irreducible representations $[f]$ of rather small dimensions $d_{[f]}$ and for some simple configurations of single-particle states such as those chosen there. The results obtained serve two purposes: (a) they can be used in practical applications (see Chapter 10) and (b) they fit into a more general scheme for choosing basis vectors for irreducible representations of the symmetric group developed by Yamanouchi, and called the Young–Yamanouchi basis.

The construction of the irreducible representations of S_n can be approached in several ways. One standard method is based on the complete decomposition of the regular representation (Section 3.11) of S_n. This makes use of the notion of *group algebra* and *irreducible symmetrizers* or *Young operators* (Hamermesh 1962, Sections 7-9 and 7-10; Tung 1985, Chapter 5). The practical method used in the previous sections deals with operators which are inspired by the construction of Young operators. The method developed by Yamanouchi (1937) makes use of the Schur or S-functions. For a description, see Hamermesh 1962, Section 7.7. The S-functions are defined, for example, in Wybourne 1970, Chapter 3. Yamanouchi's method is rather lengthy but the result is simply summarized in what are called Yamanouchi matrix elements which are given below. Both methods can also be applied to continuous groups, as for example the general linear group $GL(m)$ and its subgroups, through tensor analysis. Such groups are called classical Lie groups (see Chapter 5). The argument is that S_n and a classical Lie group play dual roles on the space of n-rank tensors defined in an m-dimensional space. That is the basic reason why a Young diagram with n boxes can be associated both to S_n and a classical Lie group. For a very simple illustration, see Section 4.5.

Other less standard methods are also available (see Chen 1989, Section 4.20).

Young–Yamanouchi basis

In this section we describe properties of the Young–Yamanouchi basis. These properties are essential for understanding the standard form of irreducible representations of S_n introduced in the next subsection. The starting point is that an irreducible representation $[f]$ of S_n is generally reducible with respect to its subgroup S_{n-1}. The permutations of S_n contained in S_{n-1} leave the symbol n unchanged. So, for an element $g \in S_{n-1}$ the matrix $D^{[f]}(g)$ is reducible and can be written as

$$D^{[f]}(g) = \sum_{[f']} D^{[f']}(g). \tag{4.91}$$

This is a particular case of the general formula (3.67) where $r \equiv [f']$ and $m_{[f']} = 1$, which means that in the reduction each $[f']$ appears only once. This is illustrated by the example (4.92) shown below. The Young diagrams $[f']$ in the sum (4.91) are obtained by removing a box in the Young diagram $[f]$. For example the irreducible representation $[3^2 1]$ of S_7 contains the irreducible representations $[321]$ and $[3^2]$ of S_6 only once

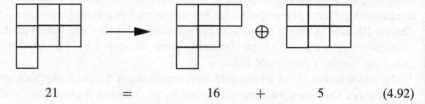

$$21 \quad = \quad 16 \quad + \quad 5 \tag{4.92}$$

This decomposition can be checked by dimensional arguments

$$d_{[f]} = \sum_{[f']} d_{[f']} \tag{4.93}$$

where the dimension d of each matrix is given by (4.88) or (4.90). Here it gives $21 = 16 + 5$.

The relation (4.91) can be written more explicitly in terms of partitions (4.2a) as

$$[f_1, f_2, \ldots, f_n] = \sum [f_1, f_2, \ldots, f_i - 1, \ldots, f_n] \tag{4.94}$$

when the partitions on the right-hand side must have

$$f_i - 1 \geq 0 \tag{4.95a}$$

and

$$f_i - 1 > f_{i+1} \tag{4.95b}$$

in order to obtain correct Young diagrams as defined by (4.1). Equation (4.94) is called the *branching law* of S_n.

The process of reduction can be continued further to S_{n-2} and then down to S_1. At this stage it is useful to discuss the basis vectors in a particular reduction of a representation of S_n. It is convenient to use Young tableaux because there is a one-to-one correspondence between a tableau and a basis vector, as explained in Section 4.3. This procedure is to remove first the box containing n, then the box containing $n - 1$, and so on. For example

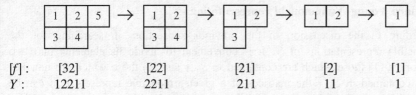

[f]:	[32]	[22]	[21]	[2]	[1]
Y :	12211	2211	211	11	1

This shows that the basis vector ([32] 12211) of S_5 is also a basis vector of the irreducible representations [22], [21], [2], and [1] of S_4, S_3, S_2, and S_1, respectively. At each stage of the reduction the first digit of the Yamanouchi symbol is removed as indicated above. In each case the process of reduction is unique. This is an indication that the labelling of basis vectors by Young tableaux or Yamanouchi symbols is complete. In general a basis vector of an irreducible representation is completely defined by [f] and Y, where Y is its Yamanouchi symbol. In other words a basis vector of S_n is *completely* defined by the chain

$$S_n \supset S_{n-1} \supset S_{n-2} \supset \ldots \supset S_2 \supset S_1 \quad (4.96)$$

where the last group S_1 can be omitted because of redundancy. Thus there is a one-to-one correspondence between a standard Young tableau labelled by [f] Y and a basis vector denoted by $|[f]Y>$. The distinction between them will be made clear below. The vectors $|[f]Y>$ form the *Young–Yamanouchi basis*. This is the basis in which the representations of S_3 and S_4 were derived in Section 4.3. We give below a general formula for matrix elements of any irreducible representation of S_n.

For clarity in what follows it is useful to recall or add the following notation:

[f]	a Young diagram of S_n
Y	a Yamanouchi label for a tableau associated with a given [f]
[f]Y	a standard Young tableau
|[f]Y >	a standard Young–Yamanouchi basis vector
$Y = (p_i q_{i-1})$	a Yamanouchi label in which the particle i is in row p and particle $i-1$ is in row q, the distribution of the other particles being irrelevant
$Y = (q_i p_{i-1})$	a Yamanouchi label as above but where the particle i is in row q and the particle $i-1$ in row p
$\mu_{i-1,i}$	*axial distance* from $i-1$ to i in the Young tableau Y. This is defined as follows: starting from $i-1$ we try to reach i by any rectangular route. Any time we cross a line going upwards (downwards) or to the right (left) we count $+1$ (-1). The number of steps is $\mu_{i-1,i}$ which is given by

$$\mu_{i-1,i} = c_i - c_{i-1} - (r_i - r_{i-1}) \quad (4.97)$$

where r_k and c_k are the row and the column of particle k. Examples are given in Exercise 4.4.

Standard Young–Yamanouchi representation

According to the discussion in the previous subsection, in searching for the irreducible representations of S_n it is convenient to divide its elements into two categories: (1) those which are contained in S_{n-1} and (2) those which are not. For those contained in S_{n-1} the matrices of a given irreducible representation can be obtained from the branching law (4.94) in terms of irreducible representations of S_{n-1} (see Exercise 4.4). Then if the representations of S_{n-1} are known the problem of finding all irreducible representations of S_n reduces to finding the matrix elements of the transposition $(n-1, n)$ because any permutation from the category (2) can be expressed as a product of transpositions, as in (2.18), one of them being $(n-1, n)$.

Also, any transposition can be expressed as a product of adjacent transpositions by the following formula (Chen 1989, Chapter 1)

$$(i, i+v) = (i+1, i+v)(i, i+1)(i+1, i+v). \tag{4.98}$$

Let us take, for example, the group S_4. With $i = 1$, $v = 2$ we get

$$(13) = (23)(12)(23)$$

and with $i = 2$, $v = 2$ the result is

$$(24) = (34)(23)(34)$$

i.e. both (13) and (24) are expressed in terms of adjacent transpositions. Furthermore, the set $i = 1$, $v = 3$ gives

$$(14) = (24)(12)(24) = (34)(23)(34)(12)(34)(23)(34)$$

where (14) is again expressed in terms of adjacent transpositions. For other examples see Exercise 4.5. So, in the end, the problem of finding the irreducible representations of S_n reduces to finding the matrices of the adjacent transpositions. These were obtained by Yamanouchi (1937) and the resulting irreducible representations form the *standard Young–Yamanouchi representation*. Yamanouchi's result is very simple:

1. If $i-1$ and i are in the same row or the same column of $|[f]Y>$ one has

$$(i-1, i)|[f]p_i q_{i-1}> = \mu |[f]p_i q_{i-1}>. \tag{4.99a}$$

2. If $i-1$ and i are neither in the same row nor in the same column

$$(i-1, i)|[f]p_i q_{i-1}> = \frac{1}{\mu}|[f]p_i q_{i-1}> + \frac{\sqrt{\mu^2-1}}{|\mu|}|[f]q_i p_{i-1}> \tag{4.99b}$$

where $\mu = \mu_{i-1,i}$ is the axial distance defined by (4.97). One can easily see that in the formula (4.99a) μ is either +1 or −1. The value +1 corresponds to particles in the same row and −1 corresponds to particles in the same column. The formula (4.99a) expresses the fact that a Young–Yamanouchi basis vector is symmetric (antisymmetric) with respect to any adjacent particles in the same row (column). For non-adjacent particles this property is no longer valid. For a better understanding

of formula (4.99b) let us give an example. Take $[f] = [31]$ of S_4 and $i = 4$, $c_4 = 1$, $r_4 = 2$, $c_3 = 3$, $r_3 = 1$. This gives
$$\mu = -3.$$

If the basis vectors are represented by their Young tableaux, the relation (4.99b) reads

$$(34)\left|\begin{array}{|c|c|c|}\hline 1 & 2 & 3 \\\hline 4 \\\cline{1-1}\end{array}\right\rangle = -\frac{1}{3}\left|\begin{array}{|c|c|c|}\hline 1 & 2 & 3 \\\hline 4 \\\cline{1-1}\end{array}\right\rangle + \frac{2\sqrt{2}}{3}\left|\begin{array}{|c|c|c|}\hline 1 & 2 & 4 \\\hline 3 \\\cline{1-1}\end{array}\right\rangle \quad (4.100)$$

Note that the left-hand side is the action of a transposition on a basis vector and not on a Young tableau. The action on the Young tableau gives (see Equations (4.3))

$$(34)\begin{array}{|c|c|c|}\hline 1 & 2 & 3 \\\hline 4 \\\cline{1-1}\end{array} = \begin{array}{|c|c|c|}\hline 1 & 2 & 4 \\\hline 3 \\\cline{1-1}\end{array} \quad (4.101)$$

It is important to note that a Young tableau and its associated basis vector are distinct. The distinction is made by writing

$$[f]Y \quad \text{for a Young tableau}$$
$$|[f]Y> \quad \text{for the basis vector associated to } [f]Y,$$

so in general

$$(i, i-1)|[f]Y> \neq (i, i-1)[f]Y. \quad (4.102)$$

The result (4.100) can be illustrated by using the basis functions of Table 4.3. For example in the configuration $\alpha^3\beta$ one has

$$\left|\begin{array}{|c|c|c|}\hline 1 & 2 & 3 \\\hline 4 \\\cline{1-1}\end{array}\right\rangle = \frac{1}{\sqrt{12}}(3\alpha\alpha\alpha\beta - \alpha\beta\alpha\alpha - \alpha\alpha\beta\alpha - \beta\alpha\alpha\alpha) \quad (4.103)$$

$$\left|\begin{array}{|c|c|c|}\hline 1 & 2 & 4 \\\hline 3 \\\cline{1-1}\end{array}\right\rangle = \frac{1}{\sqrt{6}}(2\alpha\alpha\beta\alpha - \alpha\beta\alpha\alpha - \beta\alpha\alpha\alpha) \quad (4.104)$$

Thus, the action of (34) is

$$(34)\left|\begin{array}{|c|c|c|}\hline 1 & 2 & 3 \\\hline 4 \\\cline{1-1}\end{array}\right\rangle = \frac{1}{\sqrt{12}}(3\alpha\alpha\beta\alpha - \alpha\beta\alpha\alpha - \alpha\alpha\alpha\beta - \beta\alpha\alpha\alpha) \quad (4.105)$$

where particles 3 and 4 have been interchanged, or alternatively their states have been interchanged as explained in Section 4.3. One can see that (4.103)–(4.105) satisfy (4.100).

The matrix elements of all irreducible representations of S_2 up to S_5 for the adjacent transpositions of these groups have been calculated from formulae (4.99) and are exhibited in Table 4.5. The matrices of the other elements of each group are obtained by multiplication of the matrices given in that table, by using either (2.18) or (4.98), and in this way one can obtain all matrices of a given irreducible representation. The relations (4.99) indicate that the matrices representing transpositions are real. As shown in Example 3.2, they are also symmetric.

The property that irreps of S_n are real is related to the property of characters being real. The latter has been proved in Exercise 3.1. If the character of a representation D is real, this representation is equivalent to its complex conjugate D^*, but cannot always be brought to a real form (see Section 5.11).

We should say a word about the phase convention. The off-diagonal matrix elements in (4.99b) are always chosen to be *positive* (see, for example, Hamermesh 1962, p. 219). This is called the *Yamanouchi phase convention*.

Exercise 4.4 Find the matrices of irreducible representations (a) [42] and (b) [33] for the adjacent transpositions of S_6.

Solution The adjacent transpositions are (12), (23), (34), (45), and (56). The formula (4.90) gives

$$d_{[42]} = 9, \qquad d_{[33]} = 5$$

which means we have to find five 9×9 matrices for [42] and five 5×5 matrices for [33]. Note that they are all symmetric matrices (see Example 3.2).

1. The nine basis vectors of [42] can be ordered following the conventional order of Table 4.8.

Row or column	Young tableau	Yamanouchi symbol	μ_{56}
1	1 2 3 4 / 5 6	221111	1
2	1 2 3 5 / 4 6	212111	−3
3	1 2 4 5 / 3 6	211211	−3

Row or column	Young tableau	Yamanouchi symbol	μ_{56}
4	$\begin{array}{\|c\|c\|c\|c\|}\hline 1&3&4&5\\\hline 2&6\\\cline{1-2}\end{array}$	211121	−3
5	$\begin{array}{\|c\|c\|c\|c\|}\hline 1&2&3&6\\\hline 4&5\\\cline{1-2}\end{array}$	122111	+3
6	$\begin{array}{\|c\|c\|c\|c\|}\hline 1&2&4&6\\\hline 3&5\\\cline{1-2}\end{array}$	121211	+3
7	$\begin{array}{\|c\|c\|c\|c\|}\hline 1&3&4&6\\\hline 2&5\\\cline{1-2}\end{array}$	121121	+3
8	$\begin{array}{\|c\|c\|c\|c\|}\hline 1&2&5&6\\\hline 3&4\\\cline{1-2}\end{array}$	112211	1
9	$\begin{array}{\|c\|c\|c\|c\|}\hline 1&3&5&6\\\hline 2&4\\\cline{1-2}\end{array}$	112121	1

The last column gives $\mu_{i-1,i}$ of Equation (4.99) for $i-1=5$, $i=6$.

The formula (4.99a) gives the matrix of the element (12). This matrix is therefore diagonal:

$$D^{[42]}(12) = \begin{pmatrix} 1 & & & & & & & \\ & 1 & & & & & & \\ & & 1 & & & & 0 & \\ & & & -1 & & & & \\ & & & & 1 & & & \\ & & & & & 1 & & \\ & & 0 & & & & -1 & \\ & & & & & & & 1 \\ & & & & & & & -1 \end{pmatrix}.$$

To construct the matrices of the transpositions (23), (34), and (45) one can use matrices from Table 4.5 because these transpositions belong also to the subgroups S_3,

S_4, and S_5 respectively. The matrices of Table 4.5 become submatrices of the [42] representation. For example, the symmetric [3] and the [21] representations of the element (23) of S_3 lead to

$$D^{[42]}(23) = \begin{pmatrix} 1 & 0 & 0 & 0 & 0 & 0 & 0 & 0 & 0 \\ & 1 & 0 & 0 & 0 & 0 & 0 & 0 & 0 \\ & & -\frac{1}{2} & \frac{\sqrt{3}}{2} & 0 & 0 & 0 & 0 & 0 \\ & & \frac{1}{2} & & 0 & 0 & 0 & 0 & 0 \\ & & & & 1 & 0 & 0 & 0 & 0 \\ & & & & & -\frac{1}{2} & \frac{\sqrt{3}}{2} & 0 & 0 \\ & & & & & \frac{1}{2} & & 0 & 0 \\ & & & & & & & -\frac{1}{2} & \frac{\sqrt{3}}{2} \\ & & & & & & & & \frac{1}{2} \end{pmatrix}.$$

One can see that both [3] and [21] appear with multiplicity 3.

In a similar way the representations [4], [31], and [22] of the element (34) of S_4 give rise to

$$D^{[42]}(34) = \begin{pmatrix} 1 & 0 & 0 & 0 & 0 & 0 & 0 & 0 & 0 \\ & -\frac{1}{3} & \frac{\sqrt{8}}{3} & 0 & 0 & 0 & 0 & 0 & 0 \\ & \frac{1}{3} & & 0 & 0 & 0 & 0 & 0 & 0 \\ & & & 1 & 0 & 0 & 0 & 0 & 0 \\ & & & & -\frac{1}{3} & \frac{\sqrt{8}}{3} & 0 & 0 & 0 \\ & & & & \frac{1}{3} & & 0 & 0 & 0 \\ & & & & & & 1 & 0 & 0 \\ & & & & & & & 1 & 0 \\ & & & & & & & & -1 \end{pmatrix}$$

and the representations [41] and [32] of the element (45) of S_5 produce

$$D^{[42]}(45) = \begin{pmatrix} -\frac{1}{4} & \frac{\sqrt{15}}{4} & 0 & 0 & 0 & 0 & 0 & 0 & 0 \\ & \frac{1}{4} & 0 & 0 & 0 & 0 & 0 & 0 & 0 \\ & & 1 & 0 & 0 & 0 & 0 & 0 & 0 \\ & & & 1 & 0 & 0 & 0 & 0 & 0 \\ & & & & 1 & 0 & 0 & 0 & 0 \\ & & & & & -\frac{1}{2} & 0 & \frac{\sqrt{3}}{2} & 0 \\ & & & & & & -\frac{1}{2} & 0 & \frac{\sqrt{3}}{2} \\ & & & & & & & \frac{1}{2} & 0 \\ & & & & & & & & \frac{1}{2} \end{pmatrix}.$$

This is a particular example of the branching law (4.94) which reads in this case

$$D^{[42]}(45) = D^{[41]}(45) + D^{[32]}(45).$$

With the help of (4.99a,b) and μ_{56} indicated in the above table, we get

$$D^{[42]}(56) = \begin{pmatrix} 1 & 0 & 0 & 0 & 0 & 0 & 0 & 0 & 0 \\ & -\frac{1}{3} & 0 & 0 & \frac{\sqrt{8}}{3} & 0 & 0 & 0 & 0 \\ & & -\frac{1}{3} & 0 & 0 & \frac{\sqrt{8}}{3} & 0 & 0 & 0 \\ & & & -\frac{1}{3} & 0 & 0 & \frac{\sqrt{8}}{3} & 0 & 0 \\ & & & & \frac{1}{3} & 0 & 0 & 0 & 0 \\ & & & & & \frac{1}{3} & 0 & 0 & 0 \\ & & & & & & \frac{1}{3} & 0 & 0 \\ & & & & & & & 1 & 0 \\ & & & & & & & & 1 \end{pmatrix}.$$

2. The five basis vectors of [33] are ordered as follows:

Row or column	Young tableau	Yamanouchi symbol	μ_{12}	μ_{56}
1	$\begin{array}{\|c\|c\|c\|}\hline 1 & 2 & 3 \\ \hline 4 & 5 & 6 \\ \hline\end{array}$	222111	1	1
2	$\begin{array}{\|c\|c\|c\|}\hline 1 & 2 & 4 \\ \hline 3 & 5 & 6 \\ \hline\end{array}$	221211	1	1
3	$\begin{array}{\|c\|c\|c\|}\hline 1 & 3 & 4 \\ \hline 2 & 5 & 6 \\ \hline\end{array}$	221121	−1	1
4	$\begin{array}{\|c\|c\|c\|}\hline 1 & 2 & 5 \\ \hline 3 & 4 & 6 \\ \hline\end{array}$	212211	1	−1
5	$\begin{array}{\|c\|c\|c\|}\hline 1 & 3 & 5 \\ \hline 2 & 4 & 6 \\ \hline\end{array}$	212121	−1	−1

The matrices of the elements (12) and (56) result from the formula (4.99a). They are both diagonal and their diagonal elements are given by columns 4 and 5 from the table above.

$$D^{[33]}(12) = \begin{pmatrix} 1 & & & & \\ & 1 & & & \\ & & -1 & & \\ & & & 1 & \\ & & & & -1 \end{pmatrix} \quad D^{[33]}(56) = \begin{pmatrix} 1 & & & & \\ & 1 & & & \\ & & 1 & & \\ & & & -1 & \\ & & & & -1 \end{pmatrix}.$$

The matrices of (23) and (34) are obtained in a way analogous to 1. They are

$$D^{[33]}(23) = \begin{pmatrix} 1 & 0 & 0 & 0 & 0 \\ & -\tfrac{1}{2} & \tfrac{\sqrt{3}}{2} & 0 & 0 \\ & \tfrac{1}{2} & & 0 & 0 \\ & & & -\tfrac{1}{2} & \tfrac{\sqrt{3}}{2} \\ & & & & \tfrac{1}{2} \end{pmatrix}$$

$$D^{[33]}(34) = \begin{pmatrix} -\frac{1}{3} & \frac{\sqrt{8}}{3} & 0 & 0 & 0 \\ & \frac{1}{3} & 0 & 0 & 0 \\ & & 1 & 0 & 0 \\ & & & 1 & 0 \\ & & & & -1 \end{pmatrix}.$$

The transposition (45) has the same matrix either in the representation [33] of S_6 or the representation [32] of S_5 because the additional particle 6 acts as an inert object. According to Table 4.5, one gets

$$D^{[33]}(45) = \begin{pmatrix} 1 & 0 & 0 & 0 & 0 \\ & -\frac{1}{2} & 0 & \frac{\sqrt{3}}{2} & 0 \\ & & -\frac{1}{2} & 0 & \frac{\sqrt{3}}{2} \\ & & & \frac{1}{2} & 0 \\ & & & & \frac{1}{2} \end{pmatrix}.$$

The matrix of any other transposition can be obtained from the matrix product of adjacent transpositions. The required product results from the formula (4.98). Taking for example $i = 4$, $v = 2$ one gets

$$(46) = (56)(45)(56)$$

or $i = 3$, $v = 3$ gives

$$(36) = (46)(34)(46) = (56)(45)(56)(34)(56)(45)(56).$$

By performing the multiplications one gets

$$D^{[33]}(46) = \begin{pmatrix} 1 & 0 & 0 & 0 & 0 \\ & -\frac{1}{2} & 0 & -\frac{\sqrt{3}}{2} & 0 \\ & & -\frac{1}{2} & 0 & -\frac{\sqrt{3}}{2} \\ & & & \frac{1}{2} & 0 \\ & & & & \frac{1}{2} \end{pmatrix}$$

and

$$D^{[33]}(36) = \begin{pmatrix} -\frac{1}{3} & -\frac{\sqrt{2}}{3} & 0 & -\sqrt{\frac{2}{3}} & 0 \\ & \frac{5}{6} & 0 & -\frac{\sqrt{3}}{6} & 0 \\ & & -\frac{1}{2} & 0 & \frac{\sqrt{3}}{2} \\ & & & \frac{1}{2} & 0 \\ & & & & \frac{1}{2} \end{pmatrix}.$$

Exercise 4.5 Find the matrices of the elements (15), (25), and (35) of the representation [32] of S_5.

Solution Use is made of the relation (4.98):

$$(i, i+v) = (i+1, i+v)(i, i+1)(i+1, i+v). \tag{1}$$

1. With $i = 3$, $v = 2$ we get

$$(35) = (45)(34)(45)$$

which implies that for any irreducible representation v one has

$$D^v(35) = D^v(45) D^v(34) D^v(45). \tag{2}$$

Using Table 4.5 we get, for $v = [32]$,

$$D^{[32]}(35) = \begin{pmatrix} -\frac{1}{3} & -\frac{\sqrt{2}}{3} & 0 & \sqrt{\frac{2}{3}} & 0 \\ & \frac{5}{6} & 0 & \frac{\sqrt{3}}{6} & 0 \\ & & -\frac{1}{2} & 0 & -\frac{\sqrt{3}}{2} \\ & & & \frac{1}{2} & 0 \\ & & & & \frac{1}{2} \end{pmatrix}. \tag{3}$$

2. The next step is to take $i = 2$, $v = 3$ in (1) which gives

$$(25) = (35)(23)(35).$$

Hence

$$D^v(25) = D^v(35) D^v(23) D^v(35). \tag{4}$$

Using the matrix (3) and Table 4.5 for the matrix of the transposition (23) we get

$$D^{[32]}(25) = \begin{pmatrix} -\frac{1}{3} & \sqrt{\frac{1}{18}} & -\sqrt{\frac{1}{6}} & -\sqrt{\frac{1}{6}} & \sqrt{\frac{1}{2}} \\ & -\frac{1}{6} & -\sqrt{\frac{1}{3}} & -\sqrt{\frac{1}{3}} & -\frac{1}{2} \\ & & \frac{1}{2} & -\frac{1}{2} & 0 \\ & & & \frac{1}{2} & 0 \\ & & & & \frac{1}{2} \end{pmatrix}. \quad (5)$$

3. The last step is to take $i = 1$, $v = 4$ in (1) which gives

$$(15) = (25)(12)(25).$$

In an analogous manner, we get

$$D^{[32]}(15) = \begin{pmatrix} -\frac{1}{3} & \sqrt{\frac{1}{18}} & \sqrt{\frac{1}{6}} & -\sqrt{\frac{1}{6}} & -\sqrt{\frac{1}{2}} \\ & -\frac{1}{6} & \sqrt{\frac{1}{3}} & -\sqrt{\frac{1}{3}} & \frac{1}{2} \\ & & \frac{1}{2} & \frac{1}{2} & 0 \\ & & & \frac{1}{2} & 0 \\ & & & & \frac{1}{2} \end{pmatrix}. \quad (5)$$

Diagonalized Young–Yamanouchi–Rutherford representation

From the standard Young–Yamanouchi representation one can obtain the so-called *diagonalized Young–Yamanouchi–Rutherford* representation (see Elliott, Hope, and Jahn 1953 or Harvey 1981a,b) where all basis functions of a representation have a property in common: they are either symmetrical or antisymmetrical in the last two particles $n - 1$ and n, even if n and $n - 1$ are neither in the same row nor in the same column. This property is suitable for evaluating matrix elements of two-body interaction operators because one can reduce the calculation of n-body matrix elements to matrix elements of the last two particles (Sections 4.10 and 10.4).

To understand how to obtain this property let us consider as an example the basis functions (4.103) and (4.104) of the representation [31] of S_4. By construction these are neither symmetrical nor antisymmetrical in particles 3 and 4. According to Table 4.5 one has

$$(34)|[31]1211> = \frac{1}{3}|[31]1211> + \frac{\sqrt{8}}{3}|[31]2111>$$
$$(34)|[31]2111> = -\frac{1}{3}|[31]2111> + \frac{\sqrt{8}}{3}|[31]1211>. \quad (4.106)$$

In this discussion the quantity of interest is the axial distance μ of the last two particles, as defined in Equation (4.97):

$$\mu = c_n - c_{n-1} + r_{n-1} - r_n.$$

This quantity can be either positive or negative. Here

$$\begin{aligned} \mu &= +3 \quad \text{for } |[31]1211> \\ \mu &= -3 \quad \text{for } |[31]2111>. \end{aligned} \quad (4.107)$$

One can see that the linear combinations

$$|[31]\overline{12}11> = \sqrt{\frac{\mu+1}{2\mu}}|[31]1211> + \sqrt{\frac{\mu-1}{2\mu}}|[31]2111> \quad (4.108a)$$

$$|[31]\widetilde{12}11> = \sqrt{\frac{\mu-1}{2\mu}}|[31]1211> - \sqrt{\frac{\mu+1}{2\mu}}|[31]2111> \quad (4.108b)$$

with $\mu = 3$ satisfy

$$(34)|[31]\overline{12}11> = |[31]\overline{12}11>$$
$$(34)|[31]\widetilde{12}11> = -|[31]\widetilde{12}11>$$

i.e. are symmetrical and antisymmetrical respectively in particles 3 and 4, which is the desired property. One can generalize (4.108) in the following way (see Rutherford 1948, Chapter 4, Section 27):

$$|[f]\overline{pq}y> = \sqrt{\frac{\mu+1}{2\mu}}|[f]pqy> + \sqrt{\frac{\mu-1}{2\mu}}|[f]qpy> \quad (4.109a)$$

$$|[f]\widetilde{pq}y> = \sqrt{\frac{\mu-1}{2\mu}}|[f]pqy> - \sqrt{\frac{\mu+1}{2\mu}}|[f]qpy> \quad (4.109b)$$

provided

$$p < q, \quad (4.110)$$

which ensures that μ is positive. Here, particle n is in row p and particle $n-1$ is in row q, or vice versa, and y is the Yamanouchi symbol of the remaining $n-2$ particles. In the left-hand side, $\overline{pq}$ represents a symmetric and $\widetilde{pq}$ an antisymmetric pair of particles, which means that

$$(n-1,n)|[f]\overline{pq}y> = |[f]\overline{pq}y> \quad (4.111a)$$
$$(n-1,n)|[f]\widetilde{pq}y> = -|[f]\widetilde{pq}y>. \quad (4.111b)$$

These functions are referred to as belonging to the diagonalized Young–Yamanouchi–Rutherford representation. Applications will be considered in Section 10.4.

4.5 THE TENSOR METHOD

This section gives a preliminary introduction to the tensor method used in representation theory and prepares the ground for the following sections. In a more elaborate form, the method is reconsidered in Section 5.11 in connection with Lie groups. The discussion given below can be understood without a detailed knowledge of the mathematics of Lie groups.

The tensor method is related to the notion of direct product of representations, introduced in Section 3.9. The direct product of two irreducible matrix representations of a group is in general reducible. The reduction of the direct product generates new irreducible representations. This procedure can be applied either to a finite group or to a Lie group.

Suppose we are given a continuous group G of linear transformations in an N-dimensional space. A vector x of this space

$$x = \begin{pmatrix} x_1 \\ x_2 \\ \vdots \\ x_N \end{pmatrix}$$

transforms into x' by

$$x' = ax \tag{4.112}$$

where $a \in G$ is an $N \times N$ matrix, the elements of which depend on the parameters of the group G under discussion.

The product of n vectors $x(1) \times x(2) \times \ldots \times x(n)$ forms a tensor space or a tensor of rank n in N-dimensional space. This tensor has N^n components defined by

$$T_{i_1 i_2 \ldots i_n} = x_{i_1}(1) x_{i_2}(2) \ldots x_{i_n}(n) \tag{4.113}$$

where the indices $i_1, i_2, \ldots, i_n$ run from 1 to N.

Under the action of any $a \in G$ the tensor components transform as

$$T'_{i_1 i_2 \ldots i_n} = a_{i_1 i'_1} a_{i_2 i'_2} \ldots a_{i_n i'_n} T_{i'_1 i'_2 \ldots i'_n}, \tag{4.114}$$

where a sum over repeated indices is implied. The $N^n \times N^n$ matrix

$$D_{i_1 i_2 \ldots i_n, i'_1 i'_2 \ldots i'_n} = a_{i_1 i'_1} a_{i_2 i'_2} \ldots a_{i_n i'_n}, \tag{4.115a}$$

which is a generalization of (3.74) to more than two factors, can be denoted in a more compact form by

$$D(a) = a \times a \times \ldots \times a. \tag{4.115b}$$

$D(a)$ is generally reducible with respect to both the group G and the group S_n. The involvement of S_n is related to the fact that the matrix (4.115) can satisfy certain symmetry properties with respect to permutation of its indices.

The arguments below are based on the fact that the components of a vector x which form a tensor of rank $n = 1$ correspond to the Young diagram $\square$. Let us first consider the second-rank tensor $T_{i_1 i_2}$. By permuting its indices, we obtain the component $T_{i_2 i_1}$, independent of $T_{i_1 i_2}$. From these, one can define the components $T_{i_1 i_2} \pm T_{i_2 i_1}$ which form the bases of the symmetric and the antisymmetric product representations described by the Young diagrams $\square\square$ and $\genfrac{}{}{0pt}{}{\square}{\square}$, respectively. Recall that $i_1, i_2, \ldots, i_n$ run from 1 to N. Then, the matrix $D(a) = a \times a$ reduces to an antisymmetric representation of dimension $C_N^2 = \dfrac{N(N-1)}{2}$ and a symmetric representation of dimension

$$N^2 - \frac{N(N-1)}{2} = \frac{N(N+1)}{2}.$$

The interchange of i_1 with i_2 is equivalent to applying the transposition $p = (12)$ to (4.113) where we take $n = 2$. According to (4.114), this gives

$$pT'_{i_1 i_2} = T'_{i_2 i_1} = a_{i_2 i'_2} a_{i_1 i'_1} T_{i'_2 i'_1}$$
$$= a_{i_1 i'_1} a_{i_2 i'_2} T_{i'_2 i'_1}$$
$$= a_{i_1 i'_1} a_{i_2 i'_2} p T_{i'_1 i'_2}$$

which shows that any transformation belonging to G commutes with p. This is due to the property of $a_{i_1 i'_1} a_{i_2 i'_2}$ being *bisymmetric*, which means that $a \times a$ remains invariant when the same permutation is applied to both $i_1 i_2$ and $i'_1 i'_2$.

The property remains valid for any n-rank tensor. Instead of a transposition of two indices, one has to consider any permutation (2.6) of S_n applied on (4.113). This gives

$$pT'_{i_1 i_2 \ldots i_n} = x_{i_1}(a_1) x_{i_2}(a_2) \ldots x_{i_n}(a_n)$$

or, knowing that the permutation of objects 1, 2, ..., n, is equivalent to the permutation of indices $i_1, i_2, \ldots, i_n$, one can write more compactly

$$pT_{(i)} = T_{p(i)}$$
$$pT'_{(i)} = T'_{p(i)} = D_{p(i)p(j)} T_{p(j)} = D_{p(i)p(j)}(pT)_{(j)} = D_{(i)(j)}(pT)_{(j)}. \qquad (4.115c)$$

The last equality results from the fact that $D(a)$ is bisymmetric.

The meaning of (4.115c) is that any permutation of S_n commutes with any bisymmetric transformation produced by an element of G in the tensor space. Then, those linear combinations which have a particular permutation symmetry of their indices transform among themselves. Moreover, they can be described by Young tableaux associated to the same Young diagram and generate an invariant subspace of S_n. The whole space of an n-rank tensor is therefore reducible into subspaces of tensors of different permutation symmetries. In conclusion, a tensor space can be reduced with respect to both G and S_n. According to Weyl, there is a duality

THE TENSOR METHOD

(reciprocity) between a linear group and a symmetric group in a tensor space (see Weyl 1946, Chapter 4).

To better understand the spirit of the tensor method, let us consider the case of a third-rank tensor in a physical context.

Let us call $x_1 = u$, $x_2 = d$, $x_3 = s$ the components of x so we have

$$x = \begin{pmatrix} u \\ d \\ s \end{pmatrix} \quad (4.116)$$

where u, d, and s are the amplitudes of three physical states corresponding to the up, down, and strange quarks, which transform among themselves under SU(3) transformations.

Suppose we are interested in producing states of n = 3 quarks starting from (4.116). These form a tensor space as defined by (4.113) of $3^n = 27$ components.

In terms of Young diagrams the 27 states can be grouped into three categories:

Category		T_{ijk}	Multiplicity
(1)	▭▭▭	u u u d d d s s s	3
		u u d u u s d d u d d s s s u s s d	6
		u d s	1
(2)	▭▭/▭	u u d u u s d d u d d s s s u s s d	6 × 2
		u d s	1 × 4
(3)	▯/▯/▯	u d s	1

In each category the multiplicity is defined by states of similar content (three identical quarks, two identical and one different, three non-identical quarks). The multiplicity of each of the states in category (2) can be understood from Section 4.3. Summarizing, the 3^3 matrix representation of S_3 is reduced to

$$\begin{array}{ll} \text{ten} & 1-\text{dimensional irreps of type } [3] \\ \text{eight} & 2-\text{dimensional irreps of type } [21] \\ \text{one} & 1-\text{dimensional irrep of type } [1^3]. \end{array} \quad (4.117)$$

The action of the group SU(3) in the same space also creates a reducible matrix. It turns out that the ten 1-dimensional representations of S_3 of type [3] form together a 10-dimensional irrep of SU(3), the eight 2-dimensional representations of type [21] form two 8-dimensional irreps of SU(3), and the antisymmetric representation $[1^3]$ of S_3 is the identity representation of SU(3). Hence the Young diagrams serve to label also the irreps of SU(3).

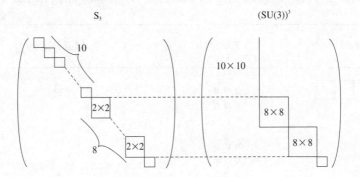

The dimensions can be checked if we anticipate that the dimension of an irreducible representation of SU(N) associated with the partition $[f_1, f_2, ..., f_N]$ is given by

$$d_{[f]}^{SU(N)} = \prod_{i<j}^{N} \frac{f_i - f_j + j - i}{j - i}. \quad (4.118)$$

From this formula we find

$[f]$	$d_{[f]}^{SU(3)}$
[3]	10
[21]	8
$[1^3]$	1

The numbers in the second column are identical to the multiplicities of (4.117). They satisfy the following general rule:

THE TENSOR METHOD   107

The multiplicity of the irreducible representation of [f] of S_n denoted by $m_{[f]}^{S_n}$ is equal to the dimension $d_{[f]}^{SU(N)}$ of the irreducible representation [f] of SU(N)

$$m_{[f]}^{S_n} = d_{[f]}^{SU(N)} \tag{4.119}$$

and, vice versa, the multiplicity $m_{[f]}^{SU(N)}$ is equal to the dimension $d_{[f]}^{S_n}$

$$m_{[f]}^{SU(N)} = d_{[f]}^{S_n} \tag{4.120}$$

in the same tensor space of dimension N^n.

This rule shows the dual role of $SU(N)$ and S_n in the same tensor space.

The rule given above can be checked by other examples (see Exercise 4.6). The simultaneous reduction of an N^n-dimensional tensor space by S_n and SU(N) helps give a better understanding of the inner products of S_n (Section 4.7). We shall return to the tensor method in Chapter 5.

Exercise 4.6 In the tensor space of 3^4 components defined by (4.113) with x_{i_k} taking the values u, d, or s, the representations of $SU(3)$ and S_4 are reducible. Find the multiplicities of the irreducible representations of both $SU(3)$ and S_4.

Solution The irreducible representations accepted in this space by both groups have the following Young diagrams

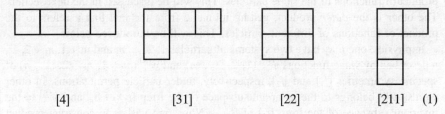

[4]   [31]   [22]   [211]   (1)

Note that the totally antisymmetric representation

(2)

does not appear because this is provided by a fourth-order Slater determinant, which cannot be constructed in this case because only three distinct components u, d, and s are available.

Using formulae (4.90) and (4.118) one can determine the dimensions $d_{[f]}^{S_4}$ and $d_{[f]}^{SU(3)}$ of the irreducible representations (1) of S_4 and SU(3), respectively. By using (4.119) and (4.120) we obtain the following multiplicities:

$[f]$	$m_{[f]}^{S_4}$	$m_{[f]}^{SU(3)}$
[4]	15	1
[31]	15	3
[22]	6	2
[211]	3	3

The result can be checked:

$$15 \times 1 + 15 \times 3 + 6 \times 2 + 3 \times 3 = 81, \tag{3}$$

which is precisely the dimension of the space to be reduced.

4.6 OUTER PRODUCTS

For the symmetric group one can introduce two types of product representations. One is the direct or Kronecker product known as the *inner* product, since it is related to the product of functions of the same particles. This will be discussed in the next section. The other is the *outer* product, getting its name from the fact that it refers to the product of functions of different particles. This will be introduced below.

In physics, one may have two systems of particles 1, 2, ..., m and $m+1, m+2, ...,$ n described by wave functions $\psi^{[f_1]}(1, 2, \ldots, m)$ and $\psi^{[f_2]}(m+1, m+2, \ldots, n)$ of specific symmetries $[f_1]$ and $[f_2]$, respectively, under particle permutations. In other words, $\psi^{[f_1]}$ belongs to the invariant subspace of the irrep $[f_1]$ of S_m and $\psi^{[f_2]}$ to the invariant subspace of the irrep $[f_2]$ of S_{n-m}. Now, one wishes to construct product states of the combined system $\psi^{[f_1]}(1, 2, \ldots, m)\, \psi^{[f_2]}(m+1, m+2, \ldots, n)$. The problem is to find the possible permutation symmetries $[f]$ of the total function $\psi^{[f]}(1, 2, \ldots, m, m+1, \ldots, n)$. The values of $[f]$ are given by the outer product. So, to perform an outer product of two irreps, $[f_1]$ of S_{n_1} and $[f_2]$ of S_{n_2}, means to decompose the resulting matrix, which is a reducible representation of $S_{n_1+n_2}$, into irreps $[f]$ of $S_{n_1+n_2}$ and to find their multiplicities $m_{[f]}$. Symbolically, one writes

$$[f_1] \otimes [f_2] = [f_2] \otimes [f_1] = \sum_{[f]} m_{[f]}[f] \tag{4.121}$$

where the symbol $\otimes$ has been used to distinguish the outer from the inner product (see next Section). However, in situations where there is no ambiguity, the sign $\times$ will be used. The first equality (4.121) shows that the outer product is commutative.

The rules which govern the outer product will not be proved mathematically but justified in terms of our knowledge of Young diagrams. Consider first the simplest case where one of the systems is formed of one particle only, i.e. take $[f_2] = [1]$. For the other system take as an example the irrep $[f_1] = [321]$ of S_6. The result must be a sum of irreducible representations of S_7. This is

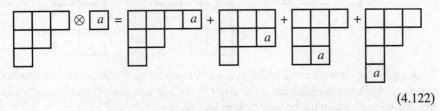

(4.122)

The diagrams on the right-hand side have been obtained from [321] by attaching to it an extrabox a in all possible ways so as to obtain correct Young diagrams (Section 4.2). The operation is the inverse of stripping off a box, as was done in (4.92), for example, and it is legitimate inasmuch as it generates standard Young tableaux (Section 4.3). For example, if we choose the Young tableau described by the Yamanouchi symbol $Y_1 = (312211)$ and label now the extra particle by 7 we can write (4.122) in an explicit form

$$\begin{array}{|c|c|c|}\hline 1&2&5\\\hline 3&4\\\cline{1-2} 6\\\cline{1-1}\end{array} \otimes \boxed{7} = \begin{array}{|c|c|c|c|}\hline 1&2&5&7\\\hline 3&4\\\cline{1-2} 6\\\cline{1-1}\end{array} + \begin{array}{|c|c|c|}\hline 1&2&5\\\hline 3&4&7\\\hline 6\\\cline{1-1}\end{array} + \begin{array}{|c|c|c|}\hline 1&2&5\\\hline 3&4\\\hline 6&7\\\hline\end{array} + \begin{array}{|c|c|c|}\hline 1&2&5\\\hline 3&4\\\cline{1-2} 6\\\cline{1-1} 7\\\cline{1-1}\end{array}$$

(4.123)

The game is somehow more complicated if both $[f_1]$ and $[f_2]$ contain more than one box, i.e. $n_1 \neq 1$ and $n_2 \neq 1$. We give here, without proof, the practical rule of reducing the product matrix into irreps $[f]$ of $S_{n_1+n_2}$ compatible with $[f_1]$ and $[f_2]$. The rule, sometimes named Littlewood's rule, is the following:

(A). Label the boxes in one of the diagrams $[f_1]$ or $[f_2]$ with a in the first row, with b in the second row, with c in the third row. Due to the property (4.121) one usually labels the boxes of the diagrams with the smaller number of particles. The remaining one is the *trunk* on which the representations contained in the product will be built. Let us choose the following example to illustrate the procedure

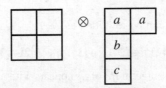

(B). First apply the boxes labelled with a to the trunk in all possible ways so as to obtain a standard Young tableau subject to the restriction that no two symbols a

appear in the same column. This gives

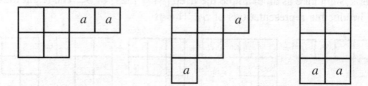

Then repeat the same procedure with b, c, etc. subject to the further restriction that the added symbols, when read from right to left in the first row, then from right to left in the second row, etc. must be such that at any stage

number of symbols $a \geq$ number of symbols $b \geq$ number of symbols $c \geq$ ..., etc.

By 'any stage' one means that the first b must be preceded by at least one a, the second b by at least two symbols a, ..., the first c by at least one a and one b, the second c by at least two symbols a and two symbols b, etc. For the example given above, the result is the following sum of irreps $[f]$ of S_8

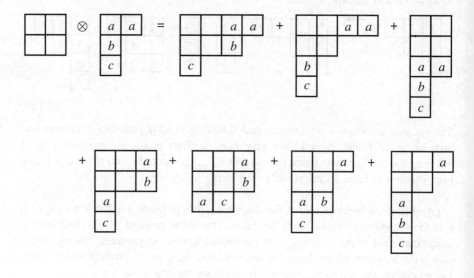

(4.124a)

In terms of partitions this can be rewritten as

$$[2^2] \otimes [21^2] = [431] + [421^2] + [2^3 1^2] + [3^2 1^2] + [3^2 2] + [3\,2^2 1] + [321^3]. \quad (4.124b)$$

In this example each representation $[f]$ of S_8 appears with multiplicity one. This is not always the case (see Exercise 4.7). It is possible to test the validity of the result of the reduction by using dimensionality arguments. If the dimension $d_{[f]}$ of an irrep $[f]$ of S_n is calculated according to (4.88) or (4.90) the following equality must be satisfied by (4.121):

$$d_{[f_1]} \times d_{[f_2]} \times \frac{(n_1+n_2)!}{n_1! n_2!} = \sum_{[f]} m_{[f]} d_{[f]}. \quad (4.125)$$

Using arguments based on dimensions of irreps, an alternative test can be established by considering that all Young diagrams, appearing on both left- and right-hand sides of (4.121), are associated to irreps of the same SU(N). With $d_{[f]}^{SU(N)}$ defined by (4.118) the following equality must be satisfied by (4.121)

$$d_{[f_1]}^{SU(N)} \times d_{[f_2]}^{SU(N)} = \sum_{[f]} m_{[f]} d_{[f]}^{SU(N)} \quad (4.126)$$

For the example (4.124) we need at least SU(5) in order to take into account all diagrams, because five is the largest number of rows appearing in some diagrams in the right-hand side of (4.124). Then, either by using the formula (4.118) or Table 8.7 (Chapter 8), we have for SU(5)

$$d_{[22]} = 50, \quad d_{[21^2]} = 45, \quad d_{[431]} = 1050, \quad d_{[332]} = 315$$
$$d_{[32^21]} = 175, \quad d_{[2^31^2]} = 10, \quad d_{[421^2]} = 450, \quad d_{[3^21^2]} = 210$$
$$d_{[321^2]} = 40$$

which fulfil (4.126):

$$50 \times 45 = 1050 + 315 + 175 + 10 + 450 + 210 + 40.$$

Tables of the outer product decomposition have been calculated by Itzykson and Nauenberg (1966) for S_n up to $n = 8$.

Exercise 4.7 Applying Littlewood's rule find the decomposition of the outer product $[21] \otimes [21]$ into irreducible representations of S_6

Solution

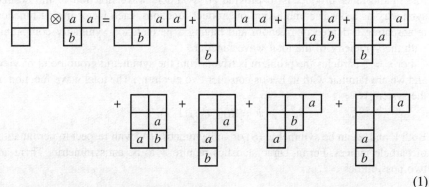

(1)

In terms of partitions the relation (1) reads

$$[21] \otimes [21] = [42] + [411] + [33] + 2[321] + [222] + [2211] + [3111]. \quad (2)$$

The position of the labels a and b is irrelevant in the reduced matrix representation. All irreducible representations appear with multiplicity one, with the exception of [321] which appears twice.

One can test the result with the formula (4.125) for example. According to (4.90) the dimensions required are

$$d_{[21]} = 2$$

on the left-hand side and

$$d_{[42]} = d_{[2211]} = 9, \quad d_{[33]} = d_{[222]} = 5, \quad d_{[321]} = 16$$
$$d_{[411]} = d_{[3111]} = 10$$

on the right-hand side. These values indeed satisfy the identity (4.125):

$$2 \times 2 \times \frac{6!}{3!\,3!} = 2 \times (9 + 16 + 10 + 5).$$

4.7 INNER PRODUCTS. CLEBSCH–GORDAN SERIES AND COEFFICIENTS

We have already mentioned that for the symmetric group the direct or Kronecker product of representations (Section 3.9) is called an *inner* product since it refers to the product of irreps of the S_n, as we shall see in the following.

The physical applications of the inner product can be easily understood. A microscopic particle is described by a wave function ψ which is usually expressed as a product of several functions, each representing a degree of freedom. For example, for a quark these are the space coordinates R, the spin χ, the flavour ϕ, and the colour C functions. We shall call R the *orbital* part of the wave function. For a system of n particles one can treat the permutation symmetry individually in each degree of freedom and construct afterwards a total wave function of a given S_n symmetry. The role of the inner product is to provide an n-particle wave function of the desired symmetry as a linear combination of product functions, each factor in this function representing a degree of freedom and having a permutation symmetry compatible with the symmetry of the total wave function.

For $n = 2$ particles the problem is trivial from the symmetric group point of view and we are familiar with it. Let us consider two electrons. The total wave function of the system is

$$\psi = R(r_1, r_2)\chi(1, 2).$$

Both R and χ can be symmetric (S) or antisymmetric (A) with respect to permutation of particle indices. Fermi–Dirac statistics require ψ to be antisymmetric. There are two possibilities

$$\psi^A = \begin{cases} R^S \chi^A \\ R^A \chi^S. \end{cases} \quad (4.127)$$

Let us now consider two nucleons. The total wave function is

$$\psi = R(r_1, r_2) \chi(1,2) \phi(1,2) \tag{4.128}$$

where the flavour function ϕ reduces to the isospin space function. We shall ignore R for the moment and study the product $\chi \phi$ only.

The spin states of a nucleon are

$$\chi_{1/2} = \begin{pmatrix} 1 \\ 0 \end{pmatrix} = \uparrow; \quad \chi_{-1/2} = \begin{pmatrix} 0 \\ 1 \end{pmatrix} = \downarrow. \tag{4.129}$$

They form a basis in the spin space which, from a group theory point of view, can be treated as an invariant subspace for SU(2) transformations: application of an arbitrary SU(2) transformation either on $\chi_{1/2}$ or $\chi_{-1/2}$ gives a linear combination of them. The same considerations are valid for the isospin and the notation to be used is

$$\phi_{1/2} = \begin{pmatrix} 1 \\ 0 \end{pmatrix} = u; \quad \phi_{-1/2} = \begin{pmatrix} 0 \\ 1 \end{pmatrix} = d. \tag{4.130}$$

To distinguish between spin and isospin, the groups will be denoted by $SU_S(2)$ and $SU_I(2)$, respectively, where S stands for spin and I for isospin.

Now, coupling the spin or isospin of two nucleons, one has

$$S = s_1 + s_2 \quad \text{and} \quad I = i_1 + i_2,$$

obtaining $S = 0$ or 1 and $I = 0$ or 1. The $S = 1$ or $I = 1$ states

$$\chi^S_{11} = \uparrow\uparrow, \quad \chi^S_{10} = \frac{1}{\sqrt{2}}(\uparrow\downarrow + \downarrow\uparrow), \quad \chi^S_{1-1} = \downarrow\downarrow$$
$$\phi^S_{11} = uu, \quad \phi^S_{10} = \frac{1}{\sqrt{2}}(ud + du), \quad \phi^S_{1-1} = dd \tag{4.131}$$

are symmetric with respect to particle permutations. They can both be described by the Young tableau (cf. (4.13))

1	2

The $S = 0$ or $I = 0$ states

$$\chi^A_{00} = \frac{1}{\sqrt{2}}(\uparrow\downarrow - \downarrow\uparrow)$$
$$\phi^A_{00} = \frac{1}{\sqrt{2}}(ud - du) \tag{4.132}$$

are antisymmetric under permutations, each having a Young tableau of the form (cf. (4.14))

1
2

In terms of Young diagrams the four possible states of a two-nucleon system read

$$S=1 \quad I=1 \quad (SI)=(11)$$

$$S=0 \quad I=0 \quad (SI)=(00)$$

$$S=1 \quad I=0 \quad (SI)=(10)$$

$$S=0 \quad I=1 \quad (SI)=(01) \tag{4.133}$$

Each of these states can be combined with R to give a totally antisymmetric state for (4.128). The states (4.133) are the simplest examples of inner products of representations. In general the inner product of two irreducible representations $[f']$ and $[f'']$ of S_n gives rise to a sum of irreducible representations of S_n

$$[f'] \times [f''] = \Sigma\, m_{[f]}[f]. \tag{4.134}$$

This is called the *Clebsch–Gordan series* of S_n. The relations (4.133) can be regarded as particular cases of (4.134) with only one term on the right-hand side The carrier (invariant) subspace of $[f]$ is spanned by vectors $|[f]Y>$ defined as the sum of products of $|[f']Y'>$ and $|[f'']Y''>$ by

$$|[f]Y> = \sum_{Y',Y''} S([f']Y'[f'']Y''\,|[f]Y')\,|[f']Y'>\,|[f'']Y''>. \tag{4.135}$$

The coefficients $S([f']Y'[f'']Y''\,|[f]Y)$ are the Clebsch–Gordan coefficients of S_n. They form an orthogonal matrix which gives the transformation between the bases $|[f]Y>$ and $|[f']Y'>|[f'']Y''>$. Using the orthogonality properties of this matrix, one can invert the relation (4.135) to give

$$|[f']Y'>\,|[f'']Y''> = \sum_{[f]Y} S([f']Y'[f'']Y''\,|[f]Y)\,|[f]Y>. \tag{4.136}$$

The first task is to find the Clebsch–Gordan series of an inner product. In the following we shall study products of S_3. For this purpose, it is convenient to use the

INNER PRODUCTS

notion of outer product defined in Section 4.6 and the dual role of SU(N) and S_n in the same tensor space, as explained in Section 4.5.

Let us first make some comments related to dimensions of irreps associated to the two particle cases discussed above, because these will be the kind of basic arguments we shall use in deriving Clebsch–Gordan series in practice.

Note that the two-dimensional spin and isospin spaces of the nucleon can produce a four-dimensional space by direct product. Any vector x in this space can be written as

$$x = \begin{pmatrix} u \uparrow \\ u \downarrow \\ d \uparrow \\ d \downarrow \end{pmatrix} \qquad (4.137)$$

and one can introduce the action of the group

$$SU(4) \supset SU_S(2) \times SU_I(2)$$

where the right-hand side is the direct product (Section 2.7) of $SU_S(2)$ acting in the spin space and $SU_I(2)$ acting in the isospin space. Any state in the right-hand side of (4.133) can be treated as a component of a tensor space of rank two as defined by Equation (4.113) with $x_1 = u \uparrow$, $x_2 = u \downarrow$, $x_3 = d \uparrow$, $x_4 = d \downarrow$. On the other hand the dimension of [2] and [11] as irreducible representations of SU(4) can be calculated by using the formula (4.118). This gives

$$d_{[2]}^{SU(4)} = 10, \qquad d_{[11]}^{SU(4)} = 6. \qquad (4.138)$$

The representations [2] and [11] of SU(4) decompose into irreps of SU(2) × SU(2) in the following way:

(4.139)

(4.140)

The numbers in brackets above each Young diagram gives the dimension of the SU(N) irrep indicated below the diagram. One can see that the dimensionality test is satisfied: $10 = 3 \times 3 + 1 \times 1$ and $6 = 3 \times 1 + 1 \times 3$.

For S_3, the inner products can be obtained as an extension of the technique displayed above. Let us assume that each of the three objects is a particle, the state of which is an SU(4) vector as above.

First we perform the outer product of [2] of S_2 and [1] of S_1:

$$[\square\square]_{SU(4)}^{(10)} \otimes [\square]_{SU(4)}^{(4)} = [\square\square\square]_{SU(4)}^{(20)} + \left[\begin{array}{c}\square\square\\ \square\end{array}\right]_{SU(4)}^{(20)} \quad (4.141)$$

This is also a direct product of two irreps of SU(4) which are indicated by their Young diagrams and their dimensions. They correspond to tensors of rank one ($\square$), two ($\square\square$) and, three ($\square\square\square$ or $\begin{array}{c}\square\square\\\square\end{array}$).

On the other hand, on the left-hand side one can use the relation (4.139) and

$$\square_{SU(4)} = \square_{SU(2)} \times \square_{SU(2)} \quad (4.142)$$

to get

$$\square\square_{SU(4)} \otimes \square_{SU(4)} = \left(\square\square_{SU(2)} \times \square\square_{SU(2)} + \begin{array}{c}\square\\\square\end{array}_{SU(2)} \times \begin{array}{c}\square\\\square\end{array}_{SU(2)}\right) \otimes \left(\square_{SU(2)} \times \square_{SU(2)}\right)$$

$$= \left(\square\square_{SU(2)} \otimes \square_{SU(2)}\right) \times \left(\square\square_{SU(2)} \otimes \square_{SU(2)}\right) + \left(\begin{array}{c}\square\\\square\end{array}_{SU(2)} \otimes \square_{SU(2)}\right) \times \left(\begin{array}{c}\square\\\square\end{array}_{SU(2)} \otimes \square_{SU(2)}\right)$$

$$= \left(\square\square\square_{SU(2)} + \begin{array}{c}\square\square\\\square\end{array}_{SU(2)}\right) \times \left(\square\square\square_{SU(2)} + \begin{array}{c}\square\square\\\square\end{array}_{SU(2)}\right) + \begin{array}{c}\square\square\\\square\end{array}_{SU(2)} \times \begin{array}{c}\square\square\\\square\end{array}_{SU(2)}$$

$$= \square\square\square_{SU(2)} \times \square\square\square_{SU(2)} + \begin{array}{c}\square\square\\\square\end{array}_{SU(2)} \times \square\square\square_{SU(2)} + \square\square\square_{SU(2)} \times \begin{array}{c}\square\square\\\square\end{array}_{SU(2)} +$$

$$+ 2 \begin{array}{c}\square\square\\\square\end{array}_{SU(2)} \times \begin{array}{c}\square\square\\\square\end{array}_{SU(2)} \quad (4.143)$$

INNER PRODUCTS

Note that only diagrams with at most two rows are allowed for SU(2). Now one can identify the two terms on the right-hand side of (4.141) as

$$\underset{SU(4)}{(20)\;\square\square\square\square} = \underset{SU(2)}{(4)\;\square\square\square} \times \underset{SU(2)}{(4)\;\square\square\square} + \underset{SU(2)}{(2)\;\begin{array}{c}\square\square\\\square\end{array}} \times \underset{SU(2)}{(2)\;\begin{array}{c}\square\square\\\square\end{array}}$$

(4.144)

and

$$\underset{SU(4)}{(20)\;\begin{array}{c}\square\square\square\\\square\end{array}} = \underset{SU(2)}{(4)\;\square\square\square} \times \underset{SU(2)}{(2)\;\begin{array}{c}\square\square\\\square\end{array}} + \underset{SU(2)}{(2)\;\begin{array}{c}\square\square\\\square\end{array}} \times \underset{SU(2)}{(4)\;\square\square\square}$$

$$+ \underset{SU(2)}{(2)\;\begin{array}{c}\square\square\\\square\end{array}} \times \underset{SU(2)}{(2)\;\begin{array}{c}\square\square\\\square\end{array}}$$

(4.145)

These are decompositions of irreps of SU(4) into irreps of SU(2) × SU(2). The first term on the right-hand side of (4.144) is naturally expected because a product of two symmetric states leads to a symmetric state. The addition of the second is the only alternative satisfying

$$20 = 4 \times 4 + 2 \times 2.$$

After this identification the remaining terms in (4.143) lead to (4.145).

Next we consider the outer product

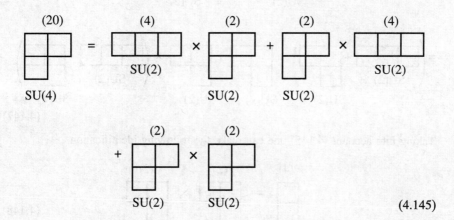

(4.146)

and use relations (4.140) and (4.142) on the left-hand side to get

$$\begin{aligned}&\underset{SU(4)}{\tiny\yng(1,1,1)}\otimes\underset{SU(4)}{\tiny\yng(1)}=\left(\underset{SU(2)}{\tiny\yng(2)}\times\underset{SU(2)}{\tiny\yng(1,1)}+\underset{SU(2)}{\tiny\yng(1,1)}\times\underset{SU(2)}{\tiny\yng(2)}\right)\otimes\left(\underset{SU(2)}{\tiny\yng(1)}\times\underset{SU(2)}{\tiny\yng(1)}\right)\\ &=\left(\underset{SU(2)}{\tiny\yng(2)}\otimes\underset{SU(2)}{\tiny\yng(1)}\right)\times\left(\underset{SU(2)}{\tiny\yng(1,1)}\otimes\underset{SU(2)}{\tiny\yng(1)}\right)+\left(\underset{SU(2)}{\tiny\yng(1,1)}\otimes\underset{SU(2)}{\tiny\yng(1)}\right)\times\left(\underset{SU(2)}{\tiny\yng(2)}\otimes\underset{SU(2)}{\tiny\yng(1)}\right)\\ &=\left(\underset{SU(2)}{\tiny\yng(3)}+\tiny\yng(2,1)\right)\times\left(\tiny\yng(2,1)+\tiny\yng(1,1,1)\right)\times\left(\underset{SU(2)}{\tiny\yng(3)}+\tiny\yng(2,1)\right)\end{aligned}$$

(4.147)

Taking into account (4.145) one can make the following identification

$$\underset{SU(4)}{\overset{(4)}{\tiny\yng(1,1,1,1)}}=\underset{SU(2)}{\overset{(2)}{\tiny\yng(2,1)}}\times\underset{SU(2)}{\overset{(2)}{\tiny\yng(2,1)}} \qquad (4.148)$$

From the S_3 point of view, one can use relations (4.144), (4.145), and (4.148) to get Clebsch–Gordan series. For example, if we search for the inner product $[21] \times [21]$, we find it appears once in each of the relations (4.144), (4.145), and (4.148). So, one can write

$$[21] \times [21] = [3] + [21] + [1^3]. \qquad (4.149)$$

By analogy, one obtains

$$[3] \times [3] = [3] \qquad (4.150)$$
$$[3] \times [21] = [21] \times [3] = [21]. \qquad (4.151)$$

For completeness one can add the following obvious Clebsch–Gordan series

$$[3] \times [1^3] = [1^3] \qquad (4.152)$$
$$[1^3] \times [1^3] = [3]. \qquad (4.153)$$

The first expresses the fact that the product of a symmetric and an antisymmetric function is an antisymmetric function and the latter states that the product of two antisymmetric functions is a symmetric function. We missed these trivial relations because SU(2) excludes the representation $[1^3]$. They can be recovered if we start with a group SU(N) larger than SU(2). Another, less trivial case is the Clebsch–Gordan series

$$[21] \times [1^3] = [21], \tag{4.154}$$

which is discussed below.

By using the same technique, one can go further and consider S_4 and so on. To obtain all Clebsch–Gordan series of S_4, one has to use at least SU(4). The Clebsch–Gordan series of S_4, S_5, and S_6 are given in Table 4.6.

For products at the left of Table 4.6 the multiplicities refer to the partitions at the top of the columns; for products on the right the multiplicities refer to partitions at the bottom.

Actually a relation like (4.152) is a particular case of a more general Clebsch–Gordan series which for any $[f]$ of a given S_n takes the form (Hamermesh 1962, Section 7-13)

$$[f] \times [1^n] = [1^n] \times [f] = [\tilde{f}] \tag{4.155}$$

where $[\tilde{f}]$ is the partition conjugate to $[f]$ (Section 4.2). The series (4.154) is a particular case of (4.155) because the partition [21] is self-conjugate.

Other useful general properties are

$$[f] \times [\tilde{g}] = [\tilde{f}] \times [g] \tag{4.156}$$
$$[f] \times [g] = [\tilde{f}] \times [\tilde{g}]. \tag{4.157}$$

They follow from the distributivity and commutativity properties of inner products (Hamermesh 1962, Section 7-13):

$$([f] \times [g]) \times [h] = [f] \times ([g] \times [h]) = ([f] \times [h]) \times [g]. \tag{4.158}$$

Taking $[h] = [1^n]$ and using (4.155) one obtains (4.156), and (4.157) results from (4.156) if one replaces $[f]$ by $[\tilde{f}]$.

Clebsch – Gordan series for S_n with $n = 2, 3, ..., 8$ can be found in Itzykson and Nauenberg (1966) and in Wybourne (1970) (Table B2) for S_n with $3 \leq n \leq 9$.

4.8 MORE ABOUT CLEBSCH–GORDAN COEFFICIENTS OF S_n

The Clebsch–Gordan (CG) coefficients of S_n have been defined by the equation (4.135). Here we discuss some of their properties in the Young–Yamanouchi basis. First we note that they are real quantities which is a characteristic of the symmetric group.

Table 4.6 Clebsch–Gordan series for S_n with $n = 3, 4, 5, 6$ (from Itzykson and Nauenberg (1966)).

$n = 3$

	[3]	[21]	[1³]	[3]×[1³]
[3]×[3]	1			
[3]×[21]		1		
[21]×[21]	1	1		

	[1³]	[21]	[3]
[1³]×[1³]			1
[1³]×[21]		1	

$n = 4$

	[4]	[31]	[2²]	[21²]	[1⁴]
[4]×[4]	1				
[4]×[31]		1			
[4]×[2²]			1		
[31]×[31]	1	1	1	1	
[31]×[2²]		1		1	
[2²]×[2²]	1		1		1

	[1⁴]	[21²]	[2²]	[31]	[4]
[4]×[1⁴]	1				
[4]×[21²]		1			
[31]×[21²]		1	1	1	
[4]×[1⁴]					
[1⁴]×[31]				1	
[31]×[21²]					

Table 4.6 cont.

$n = 5$

	[5]	[41]	[32]	[31²]	[2²1]	[21³]	[1⁵]	
[5]×[5]	1							[1⁵]×[1⁵]
[5]×[41]		1						[1⁵]×[21³]
[5]×[32]			1					[1⁵]×[2²1]
[5]×[31²]				1				[1⁵]×[31²]
[41]×[41]	1	1	1	1				[21³]×[21³]
[41]×[32]		1	1	1	1			[21³]×[2²1]
[41]×[31²]		1	1	1	1	1		[21³]×[31²]
[32]×[32]	1	1	1	1	1			[2²1]×[2²1]
[32]×[31²]		1	1	2	1	1		[2²1]×[31²]
[31²]×[31²]	1	1	2	1	2	1	1	[31²]×[31²]
	[1⁵]	[21³]	[2²1]	[31²]	[32]	[41]	[5]	

Table 4.6 *cont.*

$n = 6$

	[6]	[51]	[42]	[41²]	[3²]	[321]	[2³]	[31³]	[2²1²]	[21⁴]	[1⁶]		
[6]×[6]	1											[1⁶]×[1⁶]	[6]×[1⁶]
[6]×[51]		1										[1⁶]×[21⁴]	[6]×[21⁴]
[6]×[42]			1									[1⁶]×[2²1²]	[6]×[2²1²]
[6]×[41²]				1								[1⁶]×[31³]	[6]×[31³]
[6]×[3²]					1							[1⁶]×[3²]	[6]×[2³]
[6]×[321]						1						[1⁶]×[321]	
[51]×[51]	1	1	1	1								[21⁴]×[21⁴]	[51]×[1⁶]
[51]×[42]		1	1	1	1	1						[21⁴]×[2²1²]	[51]×[2²1²]
[51]×[41²]		1	1	1	1	1						[21⁴]×[31³]	[51]×[31³]
[51]×[3²]			1	1		1						[21⁴]×[3²]	[51]×[2³]
[51]×[321]		1	1	2	1	1	1	1	1			[21⁴]×[321]	
[42]×[42]	1	1	2	1	1	1	1					[2²1²]×[2²1²]	[42]
[42]×[41²]		1	2	1	1	2	1	1				[2²1²]×[31³]	[42]×[41²]
[42]×[3²]			1	1	1	1						[2²1²]×[3²]	[42]×[2³]
[42]×[321]		1	2	2	1	3	1	2	1			[2²1²]×[321]	
[41²]×[41²]	1	1	1	2	1	2	1	1	1			[31³]×[31³]	[41²]
[41²]×[3²]			1	1		1		1				[31³]×[3²]	[41²]×[2³]
[41²]×[321]		1	2	2	1	3	1	2	1			[31³]×[321]	
[3²]×[3²]	1		1		1	1						[2³]×[3²]	
[3²]×[321]		1	1	1	1	2	1	1				[2³]×[321]	
[321]×[321]	1	2	3	4	2	5	2	4	3	2	1		

	[1⁶]	[21⁴]	[2²1²]	[31³]	[2³]	[321]	[3²]	[41²]	[42]	[51]	[6]

For purposes of illustration of some of their properties we shall first derive the CG coefficients of the inner product of S_3

$$[21] \times [21] = [3] + [21] + [1^3].$$

Let us choose, to begin with, the one-dimensional representation [3]. Its invariant subspace is a totally symmetric function, which, normalized, must be

$$|[3]111\rangle = \frac{1}{\sqrt{2}}\left(\begin{array}{c}\boxed{\begin{array}{cc}1&2\\3\end{array}}\;\boxed{\begin{array}{cc}1&2\\3\end{array}} + \boxed{\begin{array}{cc}1&3\\2\end{array}}\;\boxed{\begin{array}{cc}1&3\\2\end{array}}\end{array}\right) \quad (4.159)$$

from which we can extract

$$S([21]211[21]211\,|\,[3]111) = S([21]121[21]121\,|\,[3]111) = \frac{1}{\sqrt{2}}. \quad (4.160)$$

The totally antisymmetric state, forming an invariant subspace of $[1^3]$, can be written as

$$|[1^3]321\rangle = \frac{1}{\sqrt{2}}\left(\boxed{\begin{array}{cc}1&2\\3\end{array}}\;\boxed{\begin{array}{cc}1&3\\2\end{array}} - \boxed{\begin{array}{cc}1&3\\2\end{array}}\;\boxed{\begin{array}{cc}1&2\\3\end{array}}\right) \quad (4.161)$$

The antisymmetry with respect to (12) is obvious and by using Table 4.5 one can see that it is also antisymmetric with respect to (23). From (4.161) one identifies

$$S([21]211[21]121\,|\,[1^3]321) = \frac{1}{\sqrt{2}}; \quad S([21]121[21]211\,|\,[1^3]321) = -\frac{1}{\sqrt{2}}. \quad (4.162)$$

The two basis vectors of the representation [21] can be obtained by orthogonality to (4.159) and (4.161). These are

$$|[21]211\rangle = \frac{1}{\sqrt{2}}\left(\boxed{\begin{array}{cc}1&2\\3\end{array}}\;\boxed{\begin{array}{cc}1&2\\3\end{array}} - \boxed{\begin{array}{cc}1&3\\2\end{array}}\;\boxed{\begin{array}{cc}1&3\\2\end{array}}\right) \quad (4.163)$$

$$|[21]121\rangle = -\frac{1}{\sqrt{2}}\left(\boxed{\begin{array}{cc}1&2\\3\end{array}}\;\boxed{\begin{array}{cc}1&3\\2\end{array}} + \boxed{\begin{array}{cc}1&3\\2\end{array}}\;\boxed{\begin{array}{cc}1&2\\3\end{array}}\right) \quad (4.164)$$

Hence

$$S([21]211[21]211 \mid [21]211) = -S([21]121[21]121 \mid [21]211) = \frac{1}{\sqrt{2}} \qquad (4.165)$$

$$S([21]211[21]121 \mid [21]121) = S([21]121[21]211 \mid [21]121) = -\frac{1}{\sqrt{2}}. \qquad (4.166)$$

We have adopted the phase convention of Hamermesh (1962), Chapter 7, which is consistent with (4.197) and with the discussion of Section 4.9.

The other non-zero CG coefficients of S_3 are equal to ± 1. They come out of the relations (4.150)–(4.154). For S_n with $n > 3$, but not large, the CG coefficients can be calculated directly as in the above example. The symmetry properties and orthogonality relations presented below simplify the task, reducing the number of coefficients to be found.

The relations (4.135) and (4.136) represent transformations between two orthonormal bases. Hence the CG coefficients satisfy the orthogonality relations

$$\sum_{Y'Y''} S([f']Y'[f'']Y'' \mid [f]Y) S([f']Y'[f'']Y'' \mid [f_1]Y_1) = \delta_{ff_1} \delta_{YY_1} \qquad (4.167)$$

$$\sum_{fY} S([f']Y'[f'']Y'' \mid [f]Y) S([f']Y_1'[f'']Y_1'' \mid [f]Y) = \delta_{Y'Y_1'} \delta_{Y''Y_1''}. \qquad (4.168)$$

Symmetry properties of CG coefficients with respect to permutations of their arguments can be found in Hamermesh (1962), Section 7-14. For $[f] \neq [f'] \neq [f'']$ one has

$$S([f']Y'[f'']Y'' \mid [f]Y) = S([f'']Y''[f']Y' \mid [f]Y) \qquad (4.169)$$

$$\frac{S([f']Y'[f'']Y'' \mid [f]Y)}{(d_{[f]})^{1/2}} = \frac{S([f]Y[f'']Y'' \mid [f']Y')}{(d_{[f']})^{1/2}} \qquad (4.170)$$

$$\frac{S([f']Y'[f'']Y'' \mid [f]Y)}{(d_{[f]})^{1/2}} = \frac{S([f']Y'[f]Y \mid [f'']Y'')}{(d_{[f'']})^{1/2}} \qquad (4.171)$$

where $d_{[f]}$, $d_{[f']}$, and $d_{[f'']}$ are given by (4.90).

In the case $[f'] = [f'']$ there is a subtlety related to the fact that the product of a representation with itself obtained in the space product $\phi_a \chi_b$ (where ϕ and χ are independent bases) is reducible into a sum of a symmetric product representation $[D^\mu \times D^\mu]$ and an antisymmetric product representation $\{D^\mu \times D^\mu\}$ (see Hamermesh 1962, Section 5-2). The subspace of the former is spanned by symmetric functions $\phi_a \chi_a$, $\phi_a \chi_b + \phi_b \chi_a$, $\phi_b \chi_b$ and of the latter by antisymmetric functions $\phi_a \chi_b - \phi_b \chi_a$. For this reason the property (4.169) must be modified by including a phase $\delta_{[f]}$ whenever $[f'] = [f'']$. The interchange of $[f']Y'$ with $[f']Y''$ gives in this case

$$S([f']Y'[f']Y'' \mid [f]Y) = \delta_{[f]} S([f']Y''[f']Y' \mid [f]Y) \qquad (4.172)$$

where $\delta_{[f]} = +1$ if $\mid [f]Y >$ is contained in the symmetrized product and $\delta_{[f]} = -1$ if it is contained in the antisymmetrized product.

The property (4.172) with $\delta_{[1^3]} = -1$ is illustrated by the coefficients of (4.162).

For $[f] \neq [f'] = [f'']$ the properties (4.170) and (4.171) remain valid because only the order of $[f]$ and $[f']$ or of $[f]$ and $[f'']$ is interchanged. In the right-hand side of (4.170) the interchange of $[f]Y$ and $[f'']Y''$ gives

$$\frac{S([f']Y'[f'']Y'' \mid [f]Y)}{(d_{[f]})^{1/2}} = \frac{S([f'']Y''[f]Y \mid [f']Y')}{(d_{[f']})^{1/2}}. \tag{4.173}$$

For example, using the first coefficient of (4.162) in (4.173) and taking $d_{[1^3]} = 1$ and $d_{[21]} = 2$ one gets

$$S([21]121[1^3]321 \mid [21]211) = +1. \tag{4.174}$$

Using (4.172) in the left-hand side of (4.173) leads to

$$\delta[f] \frac{S([f'']Y''[f']Y' \mid [f]Y)}{(d_{[f]})^{1/2}} = \frac{S([f'']Y''[f]Y \mid [f']Y')}{(d_{[f']})^{1/2}}. \tag{4.175}$$

Substituting now the first coefficient of (4.162) into (4.175), i.e. taking $Y'' = 211$, $Y' = 121$ and recalling that $\delta_{[1^3]} = -1$ one gets

$$S([21]211[1^3]321 \mid [21]121) = -1. \tag{4.176}$$

For the symmetry properties of CG coefficients in the case where $[f] = [f'] = [f'']$ we refer to Hamermesh (1962). The CG coefficients of S_3 are collected together in Table 4.7.

Construction of symmetric and antisymmetric functions as in (4.159) and (4.161), respectively, can be easily generalized to any S_n. The argument is that $[f] \times [f']$ contains the symmetric representation $[n]$ once if and only if $[f] = [f']$ and the antisymmetric representation $[1^n]$ once if and only if $[f'] = [\tilde{f}]$ (see Hamermesh 1962, Section 7.13).

The generalization of (4.159) reads

$$|[n]1> = \left(\frac{1}{d_{[f]}}\right)^{1/2} \sum_Y |[f]Y> |[f]Y> \tag{4.177}$$

where 1 is the abbreviation of the Yamanouchi symbol $Y = 11\ldots 1$. From the orthogonality property of the irreps $D^{[f]}_{YY'}$ of S_n it follows immediately that (4.177) is symmetric with respect to any permutation P of S_n

$$P|[n]1> = \left(\frac{1}{d_{[f]}}\right)^{1/2} \sum_{Y'Y''} \left(\sum_Y D^{[f]}_{Y'Y}(P) D^{[f]}_{Y''Y}(P)\right) |[f]Y'> [f]Y''>$$

$$= \left(\frac{1}{d_{[f]}}\right)^{1/2} \sum_{Y'} |[f]Y'> |[f]Y'>.$$

Hence

$$S([f]Y[f]Y \mid [n]1) = \left(\frac{1}{d_{[f]}}\right)^{1/2}. \tag{4.178}$$

For the generalization of (4.161) one has to introduce the basis functions $|[\tilde{f}]\tilde{Y}>$ of the conjugate partition $[\tilde{f}]$ of $[f]$. A Young tableau associated to $|[\tilde{f}]\tilde{Y}>$ is obtained

Table 4.7 Clebsch–Gordan coefficients for S_3.

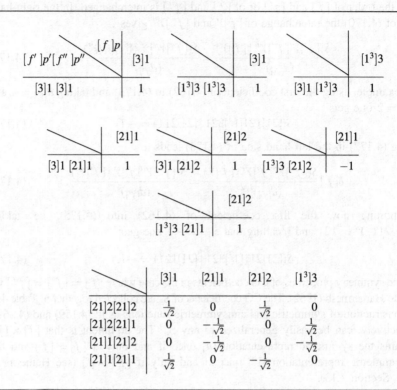

from $|[f]Y>$ by interchanging rows with columns as in the definition of $[\tilde{f}]$ (cf. Section 4.2). For example

$$|[f]Y> = \begin{array}{|c|c|c|} \hline 1 & 2 & 3 \\ \hline 4 & 5 \\ \cline{1-2} 6 \\ \cline{1-1} \end{array} \qquad |[\tilde{f}]\tilde{Y}> = \begin{array}{|c|c|c|} \hline 1 & 4 & 6 \\ \hline 2 & 5 \\ \cline{1-2} 3 \\ \cline{1-1} \end{array} \qquad (4.179)$$

We also need to recall that any permutation can be written as a product of adjacent transpositions (Section 4.4) therefore it is enough to discuss the action of a single adjacent transposition. From formula (4.99b) and the examples given in Table 4.5 one can see that

$$D^{[\tilde{f}]}_{YY} = -D^{[f]}_{YY} \qquad (4.180)$$

and

$$D^{[\tilde{f}]}_{Y'Y} = D^{[f]}_{Y'Y} \qquad \text{if } Y' \neq Y. \qquad (4.181)$$

For a given Y there is only one Y' for which the matrix element (4.181) is non-zero and this is just the Yamanouchi symbol obtained from Y by a transposition of adjacent numbers. The action of a transposition $(i-1, i)$ on the product $|[f]Y> |[\tilde{f}]\tilde{Y}>$ can then be written as

$$(i-1,i)|[f]Y>|[\tilde{f}]\tilde{Y}> = -\left(D_{YY}^{[f]}\right)^2 |[f]Y>|[\tilde{f}]\tilde{Y}> + \left(D_{Y'Y}^{[f]}\right)^2 |[f]Y'>|[\tilde{f}]\tilde{Y}'>$$
$$- D_{YY}^{[f]} D_{Y'Y}^{[f]} (|[f]Y'>|[\tilde{f}]\tilde{Y}> - |[f]Y>|[\tilde{f}]\tilde{Y}'>).$$
(4.182)

The action of $(i-1, i)$ on $|[f]Y'> |[\tilde{f}]Y'>$ is

$$(i-1,i)|[f]Y'>|[\tilde{f}]\tilde{Y}'> = -\left(D_{Y'Y'}^{[f]}\right)^2 |[f]Y'>|[\tilde{f}]\tilde{Y}'> + \left(D_{YY'}^{[f]}\right)^2 |[f]Y>|[\tilde{f}]\tilde{Y}>$$
$$+ D_{Y'Y'}^{[f]} D_{YY'}^{[f]} (|[f]Y'>|[\tilde{f}]\tilde{Y}> - |[f]Y>|[\tilde{f}]\tilde{Y}'>).$$
(4.183)

Using

$$D_{Y'Y'}^{[f]} = -D_{YY}^{[f]} \quad \text{and} \quad D_{Y'Y}^{[f]} = D_{YY'}^{[f]} \quad \text{for } Y' \neq Y$$

which follows from (4.99b) and the fact that for any transposition $D = D^T$, (see Example 3.2 of Section 3.7) the difference between (4.182) and (4.183) gives

$$(i-1,i)(|[f]Y>|[\tilde{f}]\tilde{Y}> - |[f]Y'>|[\tilde{f}]\tilde{Y}'>)$$
$$= -\left[\left(D_{YY}^{[f]}\right)^2 + \left(D_{YY'}^{[f]}\right)^2\right](|[f]Y>|[\tilde{f}]\tilde{Y}> - |[f]Y'>|[\tilde{f}]\tilde{Y}'>)$$
$$= -(|[f]Y>|[\tilde{f}]\tilde{Y}> - |[f]Y'>|[\tilde{f}]\tilde{Y}'>)$$

i.e. an antisymmetric function. Each transposition brings in a $(-)$ sign and one can write a general antisymmetric function as

$$|[1^n]n> = \left(\frac{1}{d_{[f]}}\right)^{1/2} \sum_Y (-)^{n_Y^f} |[f]Y>|[\tilde{f}]\tilde{Y}> \qquad (4.184)$$

where n_Y^f is the number of transpositions necessary to bring Y to a normal Young tableau (Section 4.3). Hence

$$S([f]Y[\tilde{f}]\tilde{Y}|[1^n]n) = \left(\frac{1}{d_{[f]}}\right)^{1/2} (-)^{n_Y^f}. \qquad (4.185)$$

CG coefficients for the permutation groups S_2–S_6 have been computed by Schindler and Mirman (1978) and Chen and Gao (1981) (see Table 4.13 of Chen (1989) for CG coefficients of S_3–S_5).

*4.9 THE K-MATRIX (ISOSCALAR FACTOR)

Another useful property of the CG coefficients of S_n is their factorizability into a matrix K (Hamermesh 1962, Section 7–14) and a CG coefficient of S_{n-1}. To write down this relation it is necessary to specify the row, say p, of the nth particle, and the row, q, of the $(n-1)$th particle in a Yamanouchi symbol Y. If y is the distribution of the $n-2$ remaining particles one writes (cf. Section 4.4)

$$Y = (p\,q\,y).$$

Then the matrix K is defined by

$$S([f']p'q'y'[f'']p''q''y'' \mid [f]pqy) \\ = K([f']p'[f'']p'' \mid [f]p)S([f'_{p'}]q'y'[f''_{p''}]q''y'' \mid [f_p]qy) \tag{4.186}$$

where f_p is a Young tableau of S_{n-1} derived from $[f]$ by the removal of a box containing the nth particle in the row p.

The factorization property (4.186) is a particular case of a theorem called *Racah's factorization lemma* (Racah 1949), specific to the chain $S_n \supset S_{n-1}$. This lemma applies to any chain of groups which gives a complete labelling of an invariant subspace. In terms of group theory language the matrix K is called an *isoscalar factor*.

A basis vector of S_n is completely defined by the chain (4.96)

$$S_n \supset S_{n-1} \supset S_{n-2} \ldots \supset S_2 \supset S_1.$$

Racah's factorization lemma can be applied to any link in this chain. This implies that any CG coefficient is a product of the isoscalar factors appearing at successive links. The calculation of CG coefficients of S_n simplifies considerably in this way by reducing to the calculation of isoscalar factors.

The isoscalar factors satisfy orthogonality conditions which follow directly from (4.167) and (4.168) and definition (4.186). These are

$$\sum_{p'p''} K([f']p'[f'']p'' \mid [f]p)K([f']p'[f'']p'' \mid [f_1]p_1) = \delta_{ff_1}\delta_{pp_1} \tag{4.187}$$

$$\sum_{fp} K([f']p'[f'']p'' \mid [f]p)K([f']p'_1[f'']p''_1 \mid [f]p) = \delta_{p'p'_1}\delta_{p''p''_1} \tag{4.188}$$

where the sum over fp includes only the partitions which branch from the same f_p when the nth particle is removed from the row p (in Table 4.9 the columns of each square table correspond to such f_p). Hence the relation (4.187) holds for f and f_1 which branch from the same f_p and has no meaning for $f = f_1$ but $p \neq p_1$.

At this stage, it is useful to write (4.99a) and (4.99b) in a single equation as

THE K-MATRIX (ISOSCALAR FACTOR)

$$(i-1,i)|[f]pqy> = \alpha_{pq}^f|[f]pqy> + \beta_{pq}^f|[f]qpy> \qquad (4.189)$$

where, of course, $\beta_{pq}^f = 0$ for (4.99a) and in general $\beta_{pq}^f = \sqrt{1-\left(\alpha_{pq}^f\right)^2}$.

Hamermesh (1962, Section 7 - 14) has derived a recursion relation for the matrix K which in terms of the notation (4.189) takes the following form (see also Harvey 1981a, b):

$$K([f']p'[f'']p''|[f]p)K([f'_{p'}]q'[f''_{p''}]q''|[f_p]q)(\alpha^{f'}_{p'q'}\alpha^{f''}_{p''q''} - \alpha^f_{pq})$$
$$+K([f']p'[f'']q''|[f]p)K([f'_{p'}]q'[f''_{q''}]p''|[f_p]q)\alpha^{f'}_{p'q'}\beta^{f''}_{p''q''}$$
$$+K([f']q'[f'']p''|[f]p)K([f'_{q'}]p'[f''_{p''}]q''|[f_p]q)\beta^{f'}_{p'q'}\alpha^{f''}_{p''q''} \qquad (4.190)$$
$$+K([f']q'[f'']q''|[f]p)K([f'_{q'}]p'[f''_{q''}]p''|[f_p]q)\beta^{f'}_{p'q'}\beta^{f''}_{p''q''}$$
$$=K([f']p'[f'']p''|[f]q)K([f'_{p'}]q'[f''_{p''}]q''|[f_q]p)\beta^f_{pq}.$$

This relation leads to the determination of K appearing in the CG coefficient of S_n once the isoscalar factors associated to CG coefficients of S_{n-1} are known. Some of the symmetry properties of CG, and in particular (4.172), are incorporated in (4.190).

The K-matrices (isoscalar factors) of S_3 related to $[21] \times [21]$ can be obtained directly from the definition (4.186) by using the corresponding CG coefficients of S_3, Table 4.7, and the CG coefficient of S_2:

$$S([2]11[2]11|[2]11) = S([1^2]21[1^2]21|[2]11) = 1 \qquad (4.191)$$
$$S([2]11[1^2]21|[1^2]21) = S([1^2]21[2]11|[1^2]21) = 1 \qquad (4.192)$$

the values of which are obvious. The result is shown in Table 4.9(a). The non-zero K-matrices related to other inner products of S_3 can also be obtained from the definition (4.186) and Table 4.7. They are either 1 or −1.

From the recursion relation (4.190) and the isoscalar factors of S_3, one can now derive the isoscalar factors of S_4. Let us consider as an example the calculation of $K([31]p'[31]p''|[31]2)$ where $f = f' = f''$. First, we take

$$\begin{array}{ccc}
\young(\;\;4,3) & \times \quad \young(\;\;4,3) \quad \longrightarrow & \young(\;\;3,4)
\end{array}$$

$p' = 1 \quad q' = 2 \qquad\qquad p'' = 1 \quad q'' = 2 \qquad\qquad p = 2 \quad q = 1$

$\alpha' = \dfrac{1}{3} \quad \beta' = \dfrac{\sqrt{8}}{3} \qquad \alpha'' = \dfrac{1}{3} \quad \beta'' = \dfrac{\sqrt{8}}{3} \qquad \alpha = -\dfrac{1}{3} \quad \beta = \dfrac{\sqrt{8}}{3}$

130   PERMUTATION GROUP S_n

In this case the recursion relation (4.190) becomes

$$\frac{4}{9}K([31]1[31]1 \mid [31]2)K([21]2[21]2 \mid [3]1)$$

$$+\frac{8}{9}K([31]2[31]2 \mid [31]2)K([3]1[3]1 \mid [3]1)$$

$$=\frac{\sqrt{8}}{3}K([31]1[31]1 \mid [31]1)K([21]2[21]2 \mid [21]2)$$

where only the non-zero terms have been considered. Using Table 4.9(a), one gets

$$\sqrt{2}K([31]1[31]1 \mid [31]2) + 4K([31]2[31]2 \mid [31]2) = 3K([31]1[31]1 \mid [31]1). \quad (4.193)$$

Next we take

$$p' = q' = 1 \qquad p'' = q'' = 1 \qquad p = 2 \quad q = 1$$

$$\alpha' = 1 \quad \beta' = 0 \qquad \alpha'' = 1 \quad \beta'' = 0 \qquad \alpha = -\frac{1}{3} \quad \beta = \frac{\sqrt{8}}{3}$$

for which the relation (4.190) together with Table 4.9(a) give

$$K([31]1[31]1 \mid [31]1) = -\sqrt{2}K([31]1[31]1 \mid [31]2). \quad (4.194)$$

Replacing (4.194) in (4.193) one obtains

$$\sqrt{2}K([31]1[31]1 \mid [31]2) = -K([31]2[31]2 \mid [31]2). \quad (4.195)$$

One can easily see that these are the only non-zero matrix elements of $K([31]p'[31]p''\mid[31]2)$. They have to satisfy the orthogonality relation (4.187) from which we get

$$K([31]2[31]2 \mid [31]2) = \sqrt{\frac{2}{3}}$$

$$K([31]1[31]1 \mid [31]2) = -\sqrt{\frac{1}{3}}. \quad (4.196)$$

In (4.196) a phase convention has been made. The first coefficient has been chosen to be positive. Such a *phase convention* can be generalized. One can choose

$$S([f']Y_n'[f'']Y_n'' \mid [f]Y_n) > 0 \quad (4.197)$$

where Y_n', Y_n'', and Y_n are the normal Young tableaux of $[f']$, $[f'']$ and $[f]$, respectively. This is the case of the first coefficient of (4.196)

$$S([31]2[31]2 \mid [31]2) = K([31]2[31]2 \mid [31]2)S([3]1[3]1 \mid [3]1)$$
$$= K([31]2[31]2 \mid [31]2) > 0.$$

If the coefficient associated to normal Young tableaux is zero then one has to order the Young tableaux and take the first non-vanishing coefficient in that order to be positive. The conventional order is given in Table 4.8 for S_n up to $n = 6$. In the process of finding a non-vanishing coefficient we first vary Y''' and then Y'.

At fixed f, it is enough to impose the constraint (4.197) on the first largest values of f' and f'' in the chain of inequalities of partitions of a given n (see Section 2.5). The others can be seen as conjugate partitions of f' and f'' and have to satisfy the following symmetry properties (Hamermesh 1962, Section 7–14)

$$S([\tilde{f}']\tilde{Y}'[\tilde{f}'']\tilde{Y}'' \mid [f]Y) = (-)^{n^{f'}_{Y'}}(-)^{n^{f''}_{Y''}} S([f']Y'[f'']Y'' \mid [f]Y) \qquad (4.198)$$

$$S([f']Y'[\tilde{f}'']\tilde{Y}'' \mid [\tilde{f}]\tilde{Y}) = (-)^{n^{f''}_{Y''}}(-)^{n^{f}_{Y}} S([f']Y'[f'']Y'' \mid [f]Y) \qquad (4.199)$$

$$S([\tilde{f}']\tilde{Y}'[f'']Y'' \mid [\tilde{f}]\tilde{Y}) = (-)^{n^{f'}_{Y'}}(-)^{n^{f}_{Y}} S([f']Y'[f'']Y'' \mid [f]Y) \qquad (4.200)$$

where the left-hand sides refer to normal Young tableaux, and on the right-hand sides n^f_y is the number of transpositions necessary to bring Y to a normal Young tableau (Table 4.8).

As an example let us consider the group S_4. The inequalities between partitions are

$$[4] > [31] > [22] > [211] > [1^4].$$

Here one requires the following coefficients to be positive:

$S([31]2[31]2 \mid [4]1),$ $S([31]2[31]2 \mid [31]2),$ $S([31]2[31]1 \mid [22]2),$
$S([31]2[31]1 \mid [211]3),$ $S([22]2[22]2 \mid [4]1),$ $S([22]2[22]2 \mid [22]2),$
$S([211]3[211]3 \mid [4]1)$

where, for brevity, only the first digit in the Yamanouchi symbol was indicated, the others being irrelevant. The phases of the others result from the symmetry properties (4.198)–(4.200), the orthogonality relations (4.187)–(4.188) and the recurrence relation (4.190).

The isoscalar factors thus calculated for S_4 are presented in Table 4.9(b). They are grouped under matrices satisfying both orthogonality relations (4.187) and (4.188). Coefficients with phases fixed by (4.197) are marked by an asterisk. Using the isoscalar factors of S_4 one can calculate those of S_5, and so on. The values of some coefficients for S_5 are displayed in Table 4.9(c).[†] The consistency of our phases has been tested through the derivation of a unitary transformation between six-particle basis states written in two different coupling schemes (Stancu 1989).

*4.10 THE $\bar{K}$-MATRIX

The K-matrix is used to assist with the evaluation of one-body matrix elements.

The matrix elements of a two-body operator between two symmetric (or antisymmetric) states ψ_n and φ_n of n particles can be written as:

$$<\psi_n \mid \sum_{i<j}^n V_{ij} \mid \varphi_n> = \frac{n(n-1)}{2} <\psi_n \mid V_{n-1,n} \mid \varphi_n>. \qquad (4.201)$$

[†]Other isoscalar factors of S_5 and some of S_6 can be found in Pepin and Stancu (1996).

Table 4.8 Ordering of basis vectors of irreducible representations of S_n ($n = 3, 4, 5, 6$). The sign below each Young tableau is $(-)^{n_Y^f}$ as in Equation (4.184).

[f]	Young tableaux $[f]Y$ and their phases $(-)^{n_Y^f}$
[21]	1) 12 / 3 / (+) 2) 13 / 2 / (−)
[31]	1) 123 / 4 / (+) 2) 124 / 3 / (−) 3) 134 / 2 / (+)
[22]	1) 12 / 34 / (+) 2) 13 / 24 / (−)
[211]	1) 12 / 3 / 4 / (+) 2) 13 / 2 / 4 / (−) 3) 14 / 2 / 3 / (+)
[41]	1) 1234 / 5 / (+) 2) 1235 / 4 / (−) 3) 1245 / 3 / (+) 4) 1345 / 2 / (−)
[32]	1) 123 / 45 / (+) 2) 124 / 35 / (−) 3) 134 / 25 / (+) 4) 125 / 34 / (+) 5) 135 / 24 / (−)
[311]	1) 123 / 4 / 5 / (+) 2) 124 / 3 / 5 / (−) 3) 134 / 2 / 5 / (+) 4) 125 / 3 / 4 / (+) 5) 135 / 2 / 4 / (−) 6) 145 / 2 / 3 / (+)
[221]	1) 12 / 34 / 5 / (+) 2) 13 / 24 / 5 / (−) 3) 12 / 35 / 4 / (−) 4) 13 / 25 / 4 / (+) 5) 14 / 25 / 3 / (−)
[2111]	1) 12 / 3 / 4 / 5 / (+) 2) 13 / 2 / 4 / 5 / (−) 3) 14 / 2 / 3 / 5 / (+) 4) 15 / 2 / 3 / 4 / (−)

Table 4.8 cont.

[51]   1) 12345   2) 12346   3) 12356   4) 12456   5) 13456
          6   5   4   3   2
         (+)   (–)   (+)   (–)   (+)

[42]   1) 1234   2) 1235   3) 1245   4) 1345   5) 1236   6) 1246
          56   46   36   26   45   35
         (+)   (–)   (+)   (–)   (+)   (–)
       7) 1346   8) 1256   9) 1356
          25   34   24
         (+)   (+)   (–)

[411] 1) 1234   2) 1235   3) 1245   4) 1345   5) 1236   6) 1246
          5   4   3   2   4   3
          6   6   6   6   5   5
         (+)   (–)   (+)   (–)   (+)   (–)
       7) 1346   8) 1256   9) 1356   10) 1456
          2   3   2   2
          5   4   4   3
         (+)   (+)   (–)   (+)

[33]   1) 123   2) 124   3) 134   4) 125   5) 135
          456   356   256   346   246
         (+)   (–)   (+)   (+)   (–)

[321] 1) 123   2) 124   3) 134   4) 125   5) 135   6) 123
          45   35   25   34   24   46
          6   6   6   6   6   5
         (+)   (–)   (+)   (+)   (–)   (–)
       7) 124   8) 134   9) 125   10) 135   11) 145   12) 126
          36   26   36   26   26   34
          5   5   4   4   3   5
         (+)   (–)   (–)   (+)   (–)   (–)
       13) 136   14) 126   15) 136   16) 146
          24   35   25   25
          5   4   4   3
         (+)   (+)   (–)   (+)

[222] 1) 12   2) 13   3) 12   4) 13   5) 14
          34   24   35   25   25
          56   56   46   46   36
         (+)   (–)   (–)   (+)   (–)

Table 4.8 cont.

[3111]	1)	123 4 5 6 (+)	2)	124 3 5 6 (−)	3)	134 2 5 6 (+)	4)	125 3 4 6 (+)	5)	135 2 4 6 (−)	6)	145 2 3 6 (+)
	7)	126 3 4 5 (−)	8)	136 2 4 5 (+)	9)	146 2 3 5 (−)	10)	156 2 3 4 (+)			
[2211]	1)	12 34 5 6 (+)	2)	13 24 5 6 (−)	3)	12 35 4 6 (−)	4)	13 25 4 6 (+)	5)	14 25 3 6 (−)	6)	12 36 4 5 (+)
	7)	13 26 4 5 (−)	8)	14 26 3 5 (+)	9)	15 26 3 4 (−)					
[21111]	1)	12 3 4 5 6 (+)	2)	13 2 4 5 6 (−)	3)	14 2 3 5 6 (+)	4)	15 2 3 4 6 (−)	5)	16 2 3 4 5 (+)		

This means that one has to evaluate matrix elements of an operator acting on the $(n-1)$th and the nth particles only. To do this, one has to expand ψ_n and φ_n in terms of sums of products of states of the first $n-2$ particles and the last two particles. For states having a specific permutation symmetry this can be achieved by using twice the factorization property (4.186) of the CG coefficient of S_n, first to relate the CG coefficient of S_n to the corresponding CG coefficient of S_{n-1} and then the CG coefficient of S_{n-1} to the corresponding CG coefficient of S_{n-2}. This gives

$$S([f']p'q'y'[f'']p''q''y'' \mid [f]pqy)$$
$$= K([f']p'[f'']p'' \mid [f]p)S([f'_{p'}]q'y'[f''_{p''}]q''y'' \mid [f_p]qy)$$
$$= K([f']p'[f'']p'' \mid [f]p)K([f'_{p'}]q'[f''_{p''}]q'' \mid [f_p]q)$$
$$\times S([f'_{p'q'}]y'[f''_{p''q''}]y'' \mid [f_{pq}]y)$$

(4.202)

THE K-MATRIX

Table 4.9 Isoscalar factors (K-matrix) for (a) $S_3 \supset S_2$, (b) $S_4 \supset S_3$ and (c) $S_5 \supset S_4$. Phases chosen according to (4.197) are marked by an asterisk.

(a)

$[f']p'\ [f'']p''$	$[f]p$	$[3]1$	$[21]2$
$[21]2\ [21]2$		$\sqrt{\tfrac{1}{2}}$	$\sqrt{\tfrac{1}{2}}$
$[21]1\ [21]1$		$\sqrt{\tfrac{1}{2}}$	$-\sqrt{\tfrac{1}{2}}$

	$[21]1$	$[1^3]3$
$[21]2\ [21]1$	$-\sqrt{\tfrac{1}{2}}$	$\sqrt{\tfrac{1}{2}}$
$[21]1\ [21]2$	$-\sqrt{\tfrac{1}{2}}$	$-\sqrt{\tfrac{1}{2}}$

(b)

$[f']p'\ [f'']p''$	$[f]p$	$[4]1$	$[31]2$
$[31]2\ [31]2$		$\sqrt{\tfrac{1}{3}}^{*}$	$\sqrt{\tfrac{2}{3}}^{*}$
$[31]1\ [31]1$		$\sqrt{\tfrac{2}{3}}$	$-\sqrt{\tfrac{1}{3}}$

	$[31]1$	$[22]2$	$[211]3$
$[31]2\ [31]1$	$-\sqrt{\tfrac{1}{6}}$	$\sqrt{\tfrac{1}{3}}^{*}$	$\sqrt{\tfrac{1}{2}}^{*}$
$[31]1\ [31]2$	$-\sqrt{\tfrac{1}{6}}$	$\sqrt{\tfrac{1}{3}}$	$-\sqrt{\tfrac{1}{2}}$
$[31]1\ [31]1$	$\sqrt{\tfrac{2}{3}}$	$\sqrt{\tfrac{1}{3}}$	

	$[211]1$
$[31]1\ [31]1$	1

	$[31]2$
$[31]1\ [22]2$	1

	$[211]1$
$[31]1\ [22]2$	1

	$[31]2$
$[31]1\ [211]3$	1

	$[31]1$	$[211]3$
$[31]2\ [22]2$	$\sqrt{\tfrac{1}{2}}$	$-\sqrt{\tfrac{1}{2}}$
$[31]1\ [22]2$	$\sqrt{\tfrac{1}{2}}$	$\sqrt{\tfrac{1}{2}}$

	$[211]1$	$[1^4]4$
$[31]2\ [211]1$	$\sqrt{\tfrac{2}{3}}$	$\sqrt{\tfrac{1}{3}}$
$[31]1\ [211]3$	$\sqrt{\tfrac{1}{3}}$	$-\sqrt{\tfrac{2}{3}}$

	$[31]1$	$[22]2$	$[211]3$
$[31]2\ [211]3$	$-\sqrt{\tfrac{1}{2}}$	$-\sqrt{\tfrac{1}{3}}$	$-\sqrt{\tfrac{1}{6}}$
$[31]1\ [211]3$		$\sqrt{\tfrac{1}{3}}$	$-\sqrt{\tfrac{4}{6}}$
$[31]1\ [211]1$	$\sqrt{\tfrac{1}{2}}$	$-\sqrt{\tfrac{1}{3}}$	$-\sqrt{\tfrac{1}{6}}$

	$[4]1$
$[22]2\ [22]2$	1^{*}

	$[22]2$
$[22]2\ [22]2$	1^{*}

	$[1^4]4$
$[22]2\ [22]2$	1

	$[31]2$
$[22]2\ [211]3$	-1

Table 4.9 *cont.*

	[31]1	[211]3
[22]2 [211]3	$\sqrt{\frac{1}{2}}$	$-\sqrt{\frac{1}{2}}$
[22]2 [211]1	$\sqrt{\frac{1}{2}}$	$\sqrt{\frac{1}{2}}$

	[211]1
[22]2 [211]3	-1

	[4]1	[31]2
[211]3 [211]3	$\sqrt{\frac{2}{3}}^*$	$-\sqrt{\frac{1}{3}}$
[211]1 [211]1	$\sqrt{\frac{1}{3}}$	$\sqrt{\frac{2}{3}}$

	[211]1
[211]3 [211]3	1

	[31]1	[22]2	[211]3
[211]3 [211]3	$-\sqrt{\frac{4}{6}}$	$-\sqrt{\frac{1}{3}}$	
[211]3 [211]1	$\sqrt{\frac{1}{6}}$	$-\sqrt{\frac{1}{3}}$	$\sqrt{\frac{1}{2}}$
[211]1 [211]3	$\sqrt{\frac{1}{6}}$	$-\sqrt{\frac{1}{3}}$	$-\sqrt{\frac{1}{2}}$

(c)

$[f']p'\ [f'']p''$	$[f]p$	[5]1	[41]2
[41]2 [41]2		$\sqrt{\frac{1}{4}}^*$	$\sqrt{\frac{3}{4}}^*$
[41]1 [41]1		$\sqrt{\frac{3}{4}}$	$-\sqrt{\frac{1}{4}}$

	[41]1	[32]2	[311]3
[41]2 [41]1	$-\sqrt{\frac{1}{12}}$	$\sqrt{\frac{5}{12}}^*$	$\sqrt{\frac{1}{2}}^*$
[41]1 [41]2	$-\sqrt{\frac{1}{12}}$	$\sqrt{\frac{5}{12}}$	$-\sqrt{\frac{1}{2}}$
[41]1 [41]1	$\sqrt{\frac{5}{6}}$	$\sqrt{\frac{1}{6}}$	

	[311]1
[41]1 [41]1	1

	[32]1
[41]1 [41]1	1

	[41]2
[41]1 [32]2	1

	[41]1	[32]2	[311]3
[41]2 [32]2	$\sqrt{\frac{5}{15}}$	$\sqrt{\frac{2}{12}}^*$	$\sqrt{\frac{10}{20}}^*$
[41]1 [32]2	$\sqrt{\frac{2}{15}}$	$\sqrt{\frac{5}{12}}$	$-\sqrt{\frac{9}{20}}$
[41]1 [32]1	$\sqrt{\frac{8}{15}}$	$-\sqrt{\frac{5}{12}}$	$-\sqrt{\frac{1}{20}}$

	[32]1	[221]3
[41]2 [32]1	$-\sqrt{\frac{3}{8}}$	$\sqrt{\frac{5}{8}}^*$
[41]1 [32]2	$-\sqrt{\frac{5}{8}}$	$-\sqrt{\frac{3}{8}}$

	[311]1	[221]2
[41]1 [32]2	$-\sqrt{\frac{1}{4}}$	$\sqrt{\frac{3}{4}}$
[41]1 [32]1	$-\sqrt{\frac{3}{4}}$	$-\sqrt{\frac{1}{4}}$

Table 4.9 *cont.*

	$[41]2$
$[41]1\ [311]3$	-1

	$[41]1$	$[32]2$	$[311]3$
$[41]2\ [311]3$	$\sqrt{\frac{1}{3}}$	$\sqrt{\frac{10}{24}}$	$\sqrt{\frac{2}{8}}^{*}$
$[41]1\ [311]3$		$-\sqrt{\frac{9}{24}}$	$\sqrt{\frac{5}{8}}$
$[41]1\ [311]1$	$-\sqrt{\frac{2}{3}}$	$\sqrt{\frac{5}{24}}$	$\sqrt{\frac{1}{8}}$

	$[32]1$	$[221]3$
$[41]1\ [311]3$	$-\sqrt{\frac{1}{16}}$	$\sqrt{\frac{15}{16}}$
$[41]1\ [311]1$	$-\sqrt{\frac{15}{16}}$	$-\sqrt{\frac{1}{16}}$

	$[311]1$	$[221]2$	$[21^3]4$
$[41]2\ [311]1$	$-\sqrt{\frac{2}{8}}$	$\sqrt{\frac{10}{24}}$	$\sqrt{\frac{1}{3}}^{*}$
$[41]1\ [311]3$	$-\sqrt{\frac{1}{8}}$	$\sqrt{\frac{5}{24}}$	$-\sqrt{\frac{2}{3}}$
$[41]1\ [311]1$	$-\sqrt{\frac{5}{8}}$	$-\sqrt{\frac{9}{24}}$	

	$[21^3]1$
$[41]1\ [311]1$	1

	$[32]2$	$[311]3$
$[41]1\ [221]3$	$-\sqrt{\frac{1}{4}}$	$\sqrt{\frac{3}{4}}$
$[41]1\ [221]2$	$-\sqrt{\frac{3}{4}}$	$-\sqrt{\frac{1}{4}}$

	$[32]1$	$[221]3$
$[41]2\ [221]3$	$\sqrt{\frac{5}{8}}$	$-\sqrt{\frac{3}{8}}$
$[41]1\ [221]2$	$-\sqrt{\frac{3}{8}}$	$-\sqrt{\frac{5}{8}}$

	$[311]1$	$[221]2$	$[21^3]4$
$[41]2\ [221]2$	$\sqrt{\frac{10}{20}}$	$\sqrt{\frac{2}{12}}$	$-\sqrt{\frac{5}{15}}$
$[41]1\ [221]3$	$-\sqrt{\frac{1}{20}}$	$-\sqrt{\frac{5}{12}}$	$-\sqrt{\frac{8}{15}}$
$[41]1\ [221]2$	$-\sqrt{\frac{9}{20}}$	$\sqrt{\frac{5}{12}}$	$-\sqrt{\frac{2}{15}}$

	$[21^3]1$
$[41]1\ [221]2$	1

	$[221]3$
$[41]1\ [21^3]4$	-1

	$[311]3$
$[41]1\ [21^3]4$	1

	$[311]1$	$[221]2$	$[21^3]4$
$[41]2\ [21^3]4$	$\sqrt{\frac{1}{2}}$	$-\sqrt{\frac{5}{12}}$	$-\sqrt{\frac{1}{12}}$
$[41]1\ [21^3]4$		$-\sqrt{\frac{2}{12}}$	$\sqrt{\frac{10}{12}}$
$[41]1\ [21^3]1$	$\sqrt{\frac{1}{2}}$	$\sqrt{\frac{5}{12}}$	$\sqrt{\frac{1}{12}}$

Table 4.9 *cont.*

	$[21^3]1$	$[1^5]5$
$[41]2\ [21^3]1$	$\sqrt{\frac{3}{4}}$	$\sqrt{\frac{1}{4}}$
$[41]1\ [21^3]4$	$\sqrt{\frac{1}{4}}$	$-\sqrt{\frac{3}{4}}$

	$[5]1$	$[41]2$
$[32]2\ [32]2$	$\sqrt{\frac{3}{5}}$	$\sqrt{\frac{2}{5}}$
$[32]1\ [32]1$	$\sqrt{\frac{2}{5}}$	$-\sqrt{\frac{3}{5}}$

	$[41]1$	$[32]2$	$[311]3$
$[32]2\ [32]2$	$\sqrt{\frac{1}{3}}$	$\sqrt{\frac{4}{6}}^*$	
$[32]2\ [32]1$	$-\sqrt{\frac{1}{3}}$	$\sqrt{\frac{1}{6}}$	$\sqrt{\frac{1}{2}}^*$
$[32]1\ [32]2$	$-\sqrt{\frac{1}{3}}$	$\sqrt{\frac{1}{6}}$	$-\sqrt{\frac{1}{2}}$

	$[32]1$	$[221]3$
$[32]2\ [32]2$	$\sqrt{\frac{1}{4}}$	$\sqrt{\frac{3}{4}}^*$
$[32]1\ [32]1$	$\sqrt{\frac{3}{4}}$	$-\sqrt{\frac{1}{4}}$

	$[311]1$	$[221]2$	$[21^3]4$
$[32]2\ [32]2$	$\sqrt{\frac{4}{10}}$		$\sqrt{\frac{3}{5}}$
$[32]2\ [32]1$	$-\sqrt{\frac{3}{10}}$	$-\sqrt{\frac{1}{2}}$	$\sqrt{\frac{1}{5}}$
$[32]1\ [32]2$	$\sqrt{\frac{3}{10}}$	$-\sqrt{\frac{1}{2}}$	$-\sqrt{\frac{1}{5}}$

	$[21^3]1$
$[32]1\ [32]1$	1

where S on the left-hand side is a CG coefficient of S_n and on the right-hand side is a CG coefficient of S_{n-2}. The partition $[f'_{p'q'}]$, etc., results from $[f']$ by first removing the nth particle from the row p' and then the $(n-1)$th particle from the row q'. In practical calculations it is useful to have these two particles either in a symmetric or antisymmetric state. This can be achieved by using the diagonalized Young–Yamanouchi–Rutherford representation given by Equations (4.109). Let us first define

$$K_2([f]pq[f']p'q' \mid [f'']p''q'') = K([f']p'[f'']p'' \mid [f]p)K([f'_{p'}]q'[f''_{p''}]q'' \mid [f_p]q) \quad (4.203)$$

and write (4.109a) and (4.109b) in the compact form

$$|[f]\widehat{pq}y\rangle = \gamma^f_{pq}|[f]pqy\rangle + \delta^f_{pq}|[f]qpy\rangle$$
$$= \left[\gamma^f_{pq} + \delta^f_{pq}(pq)\right]|[f]pqy\rangle \quad (4.204)$$

where (pq) is a transposition exchanging the positions of the nth and $(n-1)$th particles and γ and δ are the coefficients of (4.109a) if $\widehat{pq} = \overline{pq}$ and of (4.109b) if $\widehat{pq} = \tilde{pq}$. The compact form (4.204) also contains the cases

$$|[f]\overline{pq}y> = |[f]pqy> \qquad (4.205a)$$

where the particles $n-1$ and n are adjacent in the same row and

$$|[f]\tilde{p}qy> = |[f]qpy> \qquad (4.205b)$$

when the particles $n-1$ and n are adjacent in the same column. In the last two cases obviously $\gamma = 1$ and $\delta = 0$. Then the $\overline{K}$-matrix is defined as (see Harvey 1981a,b):

$$\overline{K}([f]\widehat{pq}[f']\widehat{p'q'}|[f'']\widehat{p''q''})$$
$$= \left(\gamma_{pq}^{f} + \delta_{pq}^{f}(pq)\right)\left(\gamma_{p'q'}^{f'} + \delta_{p'q'}^{f'}(p'q')\right)\left(\gamma_{p''q''}^{f''} + \delta_{p''q''}^{f''}(p''q'')\right) \qquad (4.206)$$
$$\times K_2([f]pq[f']p'q'|[f'']p''q'').$$

Example 4.2 Write $\overline{K}([51]\overline{11}[222]\tilde{23}|[2211]\tilde{24})$ explicitly. The Young tableaux are

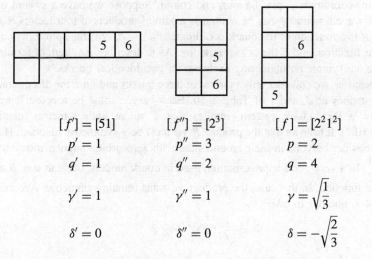

$$[f'] = [51] \qquad [f''] = [2^3] \qquad [f] = [2^2 1^2]$$
$$p' = 1 \qquad p'' = 3 \qquad p = 2$$
$$q' = 1 \qquad q'' = 2 \qquad q = 4$$
$$\gamma' = 1 \qquad \gamma'' = 1 \qquad \gamma = \sqrt{\frac{1}{3}}$$
$$\delta' = 0 \qquad \delta'' = 0 \qquad \delta = -\sqrt{\frac{2}{3}}$$

Then

$$\overline{K}([51]\overline{11}[222]\tilde{23}|[2211]\tilde{24})$$
$$= \sqrt{\frac{1}{3}} K_2([51]11[222]32|[2211]24) - \sqrt{\frac{2}{3}} K_2([51]11[222]32|[2211]42).$$

Another name for a K-matrix is a *one-body fractional parentage coefficient* and for a $\overline{K}$-matrix a *two-body fractional parentage coefficient*. Fractional parentage coefficients can be also defined in relation to CG coefficients of unitary groups (Chen 1989, Section 7.13). Examples of $\overline{K}$-matrices for S_6 can be found in Harvey

(1981a,b), Stancu and Wilets (1988), and Stancu (1989). Applications will be considered in Section 10.4.

4.11 BARYONS AS THREE QUARK-STATES

Nowadays, hadrons can be viewed as composite particles and can be treated as confined systems of quarks and antiquarks interacting via gluon exchange. A meson is a quark–antiquark pair and a baryon is a three-quark system. Systems of more than three particles (quarks, antiquarks) may also exist. This subject is discussed in Chapter 10.

In the standard model, the quarks appear with six distinct flavours: up (u), down (d), strange (s), charm (c), bottom (b), and top (t). Experimental evidence for the t-quark is still sought, although some evidence already exists (see Section 8.8). Some of the quark properties are listed in Table 8.10. The indicated masses are still controversial. The quantities I_3, S, C, B, T are isotopic spin, strangeness, charm, bottomness, and topness. They are discussed in Chapter 8. Moreover, each quark is assumed to appear in three different colours. Hence, the wave function ψ of a multiquark system must contain the following degrees of freedom: space (or momentum) coordinates, spin, flavour, and colour. Suppose we have a system of n quarks. Its wave function ψ can be written as a sum of products of four factors $R \chi \phi C$, where R is a function of the quark coordinates $r_1, \ldots, r_n$, χ is the spin function, ϕ the flavour function, and C the colour function. As ψ describes a system of fermions it must be antisymmetric under any exchange of two identical quarks.

In this section, we consider only systems of three quarks and limit the discussion to the lightest ones u, d, and s of Table 8.10. Each baryon must be a colour singlet which, for a three-quark system, implies that C is an antisymmetric function (Exercise 10.2). It follows that the product $R \chi \phi$ must be a symmetric function. Here, we first consider baryons in their ground state with zero orbital angular momentum and $S = \frac{1}{2}$. In a very good approximation, used in quark models, one can take R as a symmetric function. In this case, the product $\chi \phi$ must remain symmetric. According to Table 4.7, one has to take

$$\Psi = \frac{1}{\sqrt{2}}(\chi^\rho \phi^\rho + \chi^\lambda \phi^\lambda) \qquad (4.207)$$

with ϕ^ρ and ϕ^λ defined in Table 4.1. χ^ρ and χ^λ represent states of spin $S = \frac{1}{2}$ and permutation symmetry ρ and λ, respectively. They can be obtained from $\phi^\rho_{p(n)}$ and $\phi^\lambda_{p(n)}$ by making the replacement

$$u \to \uparrow, \qquad d \to \downarrow$$

where $\uparrow$ and $\downarrow$ are spin $\frac{1}{2}$ single-particle states of projection $S_z = +\frac{1}{2}$ and $S_z = -\frac{1}{2}$, respectively (cf. Equation (4.129)). The three particle states of mixed symmetry and $S = \frac{1}{2}$, $S_z = +\frac{1}{2}$ are

$$\chi^\lambda_+ = -\frac{1}{\sqrt{6}}(\uparrow\downarrow\uparrow + \downarrow\uparrow\uparrow - 2\uparrow\uparrow\downarrow) \qquad (4.208)$$

BARYONS AS THREE-QUARK STATES

$$\chi_+^\rho = \frac{1}{\sqrt{2}}(\uparrow\downarrow\uparrow - \downarrow\uparrow\uparrow) \qquad (4.209)$$

and those having $S = \frac{1}{2}$, $S_z = -\frac{1}{2}$ are

$$\chi_-^\lambda = \frac{1}{\sqrt{6}}(\uparrow\downarrow\downarrow + \downarrow\uparrow\downarrow - 2\downarrow\downarrow\uparrow) \qquad (4.210)$$

$$\chi_-^\rho = \frac{1}{\sqrt{2}}(\uparrow\downarrow\downarrow - \downarrow\uparrow\downarrow). \qquad (4.211)$$

Based on the wave function (4.207) one can derive those properties of baryons which do not depend on R (the colour part C does not enter into the discussion explicitly and thus can be ignored). These properties could be the charge or the magnetic moments. Some examples are given as exercises below.

Another way to get a symmetric $\chi\phi$ function is to take both χ and ϕ symmetric. In this case, the possible ϕ states are listed in Table 4.2 and χ must be an $S = \frac{3}{2}$ state. With the previous notation, the $\chi_{\frac{3}{2}m}$ states take the form

$$\begin{aligned}
\chi_{\frac{3}{2},\frac{3}{2}} &= \uparrow\uparrow\uparrow \\
\chi_{\frac{3}{2},\frac{1}{2}} &= \frac{1}{\sqrt{3}}(\uparrow\uparrow\downarrow + \uparrow\downarrow\uparrow + \downarrow\uparrow\uparrow) \\
\chi_{\frac{3}{2},-\frac{1}{2}} &= \frac{1}{\sqrt{3}}(\uparrow\downarrow\downarrow + \downarrow\uparrow\downarrow + \downarrow\downarrow\uparrow) \\
\chi_{\frac{3}{2},-\frac{3}{2}} &= \downarrow\downarrow\downarrow
\end{aligned} \qquad (4.212)$$

Exercise 4.8 Calculate the proton and neutron charges starting from the wave function (4.207).

Solution The charge operator is defined in general as

$$Q = \sum_{i=1}^{N} q(i) \qquad (1)$$

and according to Table 8.10 one has

$$q|u> = \frac{2}{3}|u> \qquad (2)$$

$$q|d> = -\frac{1}{3}|d>. \qquad (3)$$

For the expectation value of a single-particle operator like (1) between a symmetric (or an antisymmetric) function one can write

$$<\psi|Q|\psi> = N<\psi|q(i)|\psi> \qquad i = 1, 2\ldots, N. \qquad (4)$$

Here we have $N = 3$ and we choose $i = 3$.

142   PERMUTATION GROUP S_n

The wave function of a proton and a neutron in their ground state with $S_z = +\frac{1}{2}$ are

$$\psi_{p\uparrow} = \frac{1}{\sqrt{2}} \left(\phi_p^\rho \chi_+^\rho + \phi_p^\lambda \chi_+^\lambda \right), \quad \psi_{n\uparrow} = \frac{1}{\sqrt{2}} \left(\phi_n^\rho \chi_+^\rho + \phi_n^\lambda \chi_+^\lambda \right) \quad (5)$$

with $\phi_{p(n)}^\rho$ and $\phi_{p(n)}^\lambda$ defined in Table 4.1.
The proton charge is given by

$$\begin{aligned}
\langle \psi_{p\uparrow} | Q | \psi_{p\uparrow} \rangle &= \frac{3}{2} \langle \chi_+^\rho \phi_p^\rho + \chi_+^\lambda \phi_p^\lambda | q(3) | \chi_+^\rho \phi_p^\rho + \chi_+^\lambda \phi_p^\lambda \rangle \\
&= \frac{3}{2} \left[\langle \phi_p^\rho | q(3) | \phi_p^\rho \rangle + \langle \phi_p^\lambda | q(3) | \phi_p^\lambda \rangle \right] \\
&= \frac{3}{2} \left[\frac{1}{2} \langle udu - duu | \frac{2}{3} udu - \frac{2}{3} duu \rangle \right. \\
&\quad \left. + \frac{1}{6} \langle udu + duu - 2\, uud | \frac{2}{3} udu + \frac{2}{3} duu + \frac{2}{3} uud \rangle \right] = 1.
\end{aligned}$$
(6)

In a similar way one can check that the neutron charge is zero.

Exercise 4.9 Calculate the ratio of the magnetic moments μ_p and μ_n of the proton and neutron, respectively.

Solution The z-component of the magnetic moment operator is

$$M_z = \sum_{i=1}^{3} \mu_i\, q(i)\, \sigma_z(i) \quad (1)$$

where, if the quarks are Dirac particles, one has

$$\mu_i = \frac{e\hbar}{2 m_i c} \quad (2)$$

with m_i the mass of the constituent quark i, $q(i)$ the charge operator of the quark i as in the previous exercise, and $\sigma_z(i)$ the z-component of the Pauli matrix.

Here we take $m_u = m_d = m$ and express the results in terms of $\mu_u = \mu_d = \mu$.

The wave functions are given by Equation (5) of Exercise 4.8. Since M_z is a one-body operator also, one can simplify the calculations as in Exercise 4.8. For the proton one has

$$\begin{aligned}
\mu_p &= \langle \psi_{p\uparrow} | M_z | \psi_{p\uparrow} \rangle = 3\mu \langle \psi_{p\uparrow} | q(3) \sigma_z(3) | \psi_{p\uparrow} \rangle \\
&= \frac{3}{2}\mu \left[\langle \phi_p^\rho | q(3) | \phi_p^\rho \rangle \langle \chi_+^\rho | \sigma_z(3) | \chi_+^\rho \rangle + \langle \phi_p^\lambda | q(3) | \phi_p^\lambda \rangle \right. \\
&\quad \left. \langle \chi_+^\lambda | \sigma_z(3) | \chi_+^\lambda \rangle \right] \\
&= \frac{3}{2}\mu \left(\frac{2}{3} \times 1 + 0 \times \left(-\frac{1}{3} \right) \right) = \mu
\end{aligned}$$
(3)

and in a similar way, one gets for the neutron

$$\mu_n = -\frac{2}{3}\mu \tag{4}$$

Hence the ratio is

$$\frac{\mu_p}{\mu_n} = -\frac{3}{2} \tag{5}$$

which is a success of this simple model because the experimental value is −1.46. One can calculate magnetic moments of other baryons of spin $\frac{1}{2}$ by using the same prescription. The derived expressions are shown in Table 4.10, together with the corresponding experimental values. The last row represents the absolute value of the transition magnetic moment $\mu_{\Sigma\Lambda} = <\Sigma^0|M_z|\Lambda^0>$ which is the M1 radiative transition amplitude of

$$\Sigma^0 \to \Lambda^0 + \gamma. \tag{6}$$

For an up-to-date study of the $S = \frac{1}{2}$ baryon magnetic moments, see, for example, Karl 1992a.

The reader is invited to solve Exercise 4.11 located at the end of this chapter.

Table 4.10 Experimental magnetic moments of baryons of spin $\frac{1}{2}$ in units of Bohr magnetons $\mu_N = \frac{e\hbar}{2Mc}$ where M is the nucleon mass (Particle Data Group 1992) and their theoretical expressions in terms of $\bar{\mu}_u = \frac{2}{3}\mu_u$, $\bar{\mu}_d = -\frac{1}{3}\mu_d$, and $\bar{\mu}_s = -\frac{1}{3}\mu_s$ where $\mu_i (i = u, d, s)$ is given by Equation (2) of Exercise 4.9.

Observable	Experimental value	Non-relativistic quark model
μ_p	2.793	$\frac{4}{3}\bar{\mu}_u - \frac{1}{3}\bar{\mu}_d$
μ_n	−1.913	$\frac{4}{3}\bar{\mu}_d - \frac{1}{3}\bar{\mu}_u$
μ_{Λ^0}	−0.613 ± 0.004	$\bar{\mu}_s$
μ_{Σ^+}	2.42 ± 0.05	$\frac{4}{3}\bar{\mu}_u - \frac{1}{3}\bar{\mu}_s$
μ_{Σ^-}	−1.160 ± 0.025	$\frac{4}{3}\bar{\mu}_d - \frac{1}{3}\bar{\mu}_s$
μ_{Ξ^0}	−1.250 ± 0.014	$\frac{4}{3}\bar{\mu}_s - \frac{1}{3}\bar{\mu}_u$
μ_{Ξ^-}	−0.6507 ± 0.0025	$\frac{4}{3}\bar{\mu}_s - \frac{1}{3}\bar{\mu}_d$
$|\mu_{\Lambda\Sigma}|$	1.61 ± 0.08	$\sqrt{\frac{1}{3}}(\bar{\mu}_u - \bar{\mu}_d)$

4.12 BRAID GROUPS AND NEW DEVELOPMENTS

Nowadays interest in symmetric groups has been revived both in mathematics and physics due to the link with braid groups. We shall give below some very elementary ideas about the braid group (for a more detailed description, see, for example, Birman 1974).

One can view the braid group B_n, $n < \infty$, as a generalization of the permutation group. Although the braid group, in contrast to the permutation group S_n, is an infinite dimensional group, its generators σ_i, introduced below, have properties in common with transpositions.

The braid group was first defined by Emil Artin in 1925. There is a homomorphism between B_n and S_n:

$$B_n \to S_n. \tag{4.213}$$

An element of B_n can be depicted by a diagram of n strings joining two sets of n points, each set located on a line, the two lines being parallel, with 'over-crossings' or 'under-crossings' of strings as illustrated in Fig. 4.4 for $n = 3$.

The over- and under-crossings which obviously inspired the name of B_n are essential for the braid group. If neglected, one obtains a trivial case of B_n, the group S_n of permutations of points from one line to the other. If we denote by i and $i+1$ two consecutive points on the top and bottom lines, as in Fig. 4.5, the string starting at i on the top line can reach $i+1$ on the bottom line by (a) under-crossing or (b) over-crossing the string starting at $i+1$ on the top which reaches i on the bottom line.

The corresponding elements of the braid group are denoted by σ_i and σ_i^{-1}, respectively. They both correspond to the same transposition $(i, i+1)$ of S_n:

$$\sigma_i, \sigma_i^{-1} \to (i, i+1). \tag{4.214}$$

The crossings take place out of the plane containing the two parallel lines and for n points one can create $n - 1$ elements $\sigma_1, \sigma_2, \ldots, \sigma_{n-1}$ of the type (4.214). These elements generate the braid group B_n of n strings in $\mathbb{R}^3$ provided they satisfy two relations required by Artin. The first is

$$\sigma_i \sigma_j = \sigma_j \sigma_i \quad \text{for } |i - j| > 1. \tag{4.215}$$

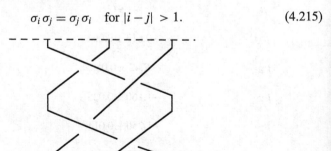

Figure 4.4 An element of the braid group B_3.

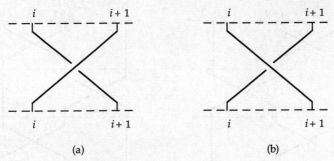

Figure 4.5 (a) A generator σ_i and (b) its inverse σ_i^{-1}.

To understand the meaning of this relation in terms of permutations let us take for illustration $j = i+2$. This gives

$$\sigma_i \sigma_{i+2} = \sigma_{i+2} \sigma_i$$

or, using the correspondence (4.214), one can write

$$(i, i+1)(i+2, i+3) = (i+2, i+3)(i, i+1).$$

This relation suggests that (4.215) is the generalization to B_n of the commutation property of transpositions which have no elements in common. For the symmetric group that was an important property, embodying the cyclic structure of the group.

Artin's second relation is

$$\sigma_i \sigma_{i+1} \sigma_i = \sigma_{i+1} \sigma_i \sigma_{i+1}. \tag{4.216}$$

In terms of transpositions this property becomes

$$(i, i+1)(i+1, i+2)(i, i+1) = (i+1, i+2)(i, i+1)(i+1, i+2).$$

Using (2.18b) on the left-hand side and (4.98) with $v = 2$ on the right-hand side the above relation can be written as

$$(i, i+1, i+2)(i, i+1) = (i, i+2).$$

Due to the fact that a transposition is equal to its inverse one can further write

$$(i, i+1, i+2) = (i, i+2)(i, i+1)$$

which is a particular case of the decomposition (2.18a). Therefore Artin's second property corresponds to the decomposition of elements of B_n in terms of σ_i in two different ways.

The composition of braids can be represented graphically by placing diagrams end to end. As an illustration, the graphical representation of (4.216) is shown in Fig. 4.6.

Artin's second property is typical for the Yang–Baxter equation used recently in statistical physics (Baxter 1982).

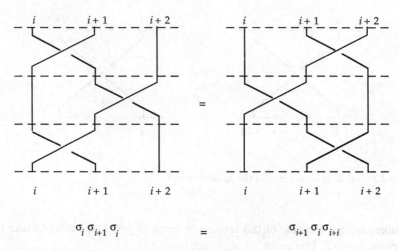

Figure 4.6 Graphical representation of Equation (4.216).

As mentioned above, the braid group is a generalization of the permutation group, and the representation theory of B_n is much more elaborate than that of S_n. However it uses many concepts of the representation theory of S_n, and in particular the Young diagram technique which again plays an essential role (see, for example, Jones 1987; Lawrence 1990).

Braid groups already have a large number of applications. In mathematics they are used in the study of complex functions of hypergeometric type having several variables (Constantinescu and Lüdde 1992) and in physics they are applied to statistical mechanics in connection with the Yang–Baxter equation (Jimbo 1989), to two-dimensional conformal field theory and string theory (Belavin, Polyakov, and Zamolodchikov 1984), and to quantum groups (Fadeev, Reshetikhin, and Takhtajan 1989). For an elementary introduction to quantum groups, see Chapter 5.

SUPPLEMENTARY EXERCISES

4.10 Find the decomposition of the outer product [32] ⊗ [221] of S_5 by using Littlewood's rule.

4.11 Derive the expressions of the magnetic moments given in Table 4.10, column 3, following the prescription of Exercise 4.9. This implies the use of flavour states shown in Table 4.1 and spin states given by Equations (4.208) and (4.209).

(a) Find the values of $\overline{\mu}_u$, $\overline{\mu}_d$ and $\overline{\mu}_s$ which fit the experimental magnetic moments of p, n, and Λ^0. Insert these values in the other expressions of column 3 and compare the results with experiment.

(b) Show that calculating μ_{Λ^0} and μ_{Σ^0} by using the simplified wave functions

$$\psi^\uparrow_{\Sigma^0} = uds\,\chi^\lambda_+ \qquad \psi^\uparrow_{\Lambda^0} = uds\,\chi^\rho_+$$

one obtains the same result as before.

5

LIE GROUPS

Continuous groups were briefly introduced in Chapter 2 as another category of groups used in physical applications. They consist of an infinite number of elements, in contrast to finite groups, and have numerous applications in quantum mechanics and field theories. Although they have some analogies with finite groups (especially the continuous compact groups) they require a distinct treatment of their own. The most familiar example of a continuous group is the rotation group (Chapter 6). Other groups currently used in physics are the special unitary groups SU(n) (Chapter 8) and the Lorentz and Poincaré groups (Chapter 7).

Groups with an infinite number of elements are of two types: discrete and continuous. In the first category the number of elements g_i is countable and one can label them by a subscript i which runs through the integers 1, 2, ..., ∞, because they do not need more than one discrete parameter to label the elements (Hamermesh 1962, Section 8.2). Some examples are:

(1) the sequence 0, 1, –1, 2, –2, ... forms a group where the composition law is addition;
(2) the transformations

$$x' = x + n, \quad n \text{ integer,} \tag{5.1}$$

form a group with addition as a composition law too;
(3) the transformations

$$x' = rx, \quad r \text{ rational}, r \neq 0, \tag{5.2}$$

form a group where the composition law is multiplication.

The second category is formed by the infinite continuous groups where the number of elements is uncountable. To understand them better in relation to discrete groups let us introduce for each group, discrete or continuous, an abstract space, the group manifold, where each point a corresponds to an element g_a of that group

$$a \leftrightarrow g_a.$$

We may say that the multiplication law $g_c = g_a g_b$ defines a function ϕ on the group manifold such that

$$c = \phi(a; b). \tag{5.3}$$

For discrete groups a, b and c take discrete values. For continuous groups ϕ must be continuous and there is an uncountable number of elements in the 'neighbourhood' of the identity element. The most general case is a *topological group* where the group manifold forms a *topological space*. The discussion below is restricted to special cases of continuous groups where the elements can be labelled by a finite set r of continuously varying real parameters. These groups are called *finite continuous groups* or *Lie groups*, from the name of the Norwegian mathematician Sophus Lie. It is impossible to construct a multiplication table for the Lie groups but the structure of the group is determined by a finite set of commutation relations (a Lie algebra) between the so-called generators of the group, the number of which is also equal to r. They are introduced below following closely the approach of Racah 1965.

5.1 INFINITESIMAL TRANSFORMATIONS

One can introduce the Lie groups by starting from a set of n variables x_0^i ($i = 1, 2, \ldots, n$) which may represent the coordinates of a point in a given basis of an n-dimensional space. A basis transformation changes x_0^i into x^i through the set of equations

$$x^i = f^i(x_0^1, x_0^2, \cdots, x_0^n; a^1, a^2, \cdots, a^r) \tag{5.4}$$

where a^ρ ($\rho = 1, 2, \ldots, r$) is a set of real independent parameters and f^i are analytic functions of a^ρ and depend *essentially* on them, or in other words the a^ρ define f^i *uniquely* and *completely*. This means that no two transformations with different parameters are the same for all values of x^0 and that r is the smallest number of parameters needed. The property that the a^ρ are essential parameters is very important and we shall return to it below.

In a shorthand notation one can rewrite (5.4) as

$$x = f(x_0; a) \tag{5.5a}$$

or alternatively

$$x = S_a x_0 \tag{5.5b}$$

where the set of transformations S_a depend on the parameter a and map the point x_0 onto x.

The set of transformations f^i form a group if they satisfy the group axioms (Chapter 2) as follows:

1. Two successive transformations yield a third one which belongs to the same set. Let us take

$$x = f(x_0; a) \quad \text{and} \quad x' = f(x; b). \tag{5.6}$$

It follows that

$$x' = f(x; b) = f(f(x_0; a); b) = f(x_0; c) = f(x_0; \varphi(a; b)) \tag{5.7}$$

i.e. there exists a set of parameters c^ρ defined by

$$c^\rho = \varphi^\rho(a; b). \tag{5.8}$$

So far the requirement (5.8) is the analogue of (5.3) for finite groups. For continuous groups one has in addition the property that φ is an analytic function of a and b, i.e. it possesses derivatives of all orders with respect to a and b.

2. For each transformation there exists a unique *inverse*

$$x_0 = f(x; \bar{a}) \tag{5.9}$$

which also belongs to the set. The uniqueness of $\bar{a}$ is ensured by the condition

$$\left| \frac{\partial f}{\partial x_0} \right| \neq 0 \tag{5.10}$$

i.e. the Jacobian of the transformation f must be different from zero

3. The identity transformation exists and is defined as follows

$$x_0 = f(x; \bar{a}) = f(f(x_0; a); \bar{a}) = f(x_0; \varphi(a; \bar{a})) = f(x_0; a_0). \tag{5.11}$$

For convenience one can take

$$a_0^\rho = 0 \quad (\rho = 1, 2, \ldots, n). \tag{5.12}$$

The basic idea of Sophus Lie was to consider a finite transformation as a succession of infinitesimal transformations. These are transformations in the 'neighbourhood' of the identity element and one can reduce the study of continuous groups to the study of infinitesimal transformations because the structure of this region determines the structure of the whole group.

From the properties introduced above one can see that one can write two equivalent expressions for x:

$$x = f(x_0; a) \tag{5.13a}$$
$$x = f(x; 0). \tag{5.13b}$$

An infinitesimal transformation $x + dx$ of x can thus be obtained in two different ways: by differentiation of (5.13a)

$$x + dx = f(x_0; a + da) \tag{5.14a}$$

or by introducing an infinitesimal parameter δa in (5.13b) such as

$$x + dx = f(x; \delta a). \tag{5.14b}$$

It follows that

$$dx = \left(\frac{\partial f(x_0; b)}{\partial b^\sigma} \right)_{b=a} da^\sigma \tag{5.15a}$$

or
$$dx = \left(\frac{\partial f(x; a)}{\partial a^\sigma}\right)_{a=0} \delta a^\sigma. \tag{5.15b}$$

It is convenient to introduce the notation
$$u^i_\sigma(x) = \left(\frac{\partial f^i(x; a)}{\partial a^\sigma}\right)_{a=0} \tag{5.16}$$

and rewrite (5.15b) as
$$dx^i = u^i_\sigma(x)\delta a^\sigma. \tag{5.17}$$

As in Equation (5.7) one can write
$$x + dx = f(x; \delta a) = f(f(x_0; a); \delta a) = f(x_0; \varphi(a; \delta a)).$$

From the equivalence between (5.14a) and (5.14b) it follows that
$$a + da = \varphi(a; \delta a). \tag{5.18}$$

For $\delta a = 0$ one has
$$x = f(x; 0) = f(x_0; \varphi(a; 0)).$$

Hence
$$a = \varphi(a; 0). \tag{5.19}$$

The relation (5.18) is crucial in the study of infinitesimal transformations. Its study leads to the structure constants and the infinitesimal generators of a group, as shown below.

5.2 STRUCTURE CONSTANTS

This section shows how the structure constants of a Lie group can be introduced and is rather formal.

For an infinitesimal change δa^τ in the parameters a^τ the relation (5.18) can be written as
$$a + da = \varphi(a; \delta a) = \varphi(a; 0) + \left.\frac{\partial \varphi(a; b)}{\partial b^\tau}\right|_{b=0} \delta a^\tau. \tag{5.20}$$

Using (5.19) it follows that
$$da^\rho = \mu^\rho_\tau(a)\delta a^\tau \tag{5.21}$$

where
$$\mu^\rho_\tau(a) = \left.\frac{\partial \varphi(a; b)}{\partial b^\tau}\right|_{b=0} \tag{5.22}$$

152   LIE GROUPS

i.e. da^ρ are linear combinations of δa^τ, and conversely, δa^τ can be written as a linear combination of da^ρ if the matrix μ^ρ_τ is non-singular. Defining λ such that

$$\lambda\mu = 1 \quad \text{or} \quad \lambda^\sigma_\rho \mu^\rho_\tau = \delta^\sigma_\tau \tag{5.23}$$

one can write

$$\delta a^\sigma = \lambda^\sigma_\rho(a) da^\rho. \tag{5.24}$$

With this notation the relation (5.17) can be written as

$$dx^i = u^i_\sigma(x) \lambda^\sigma_\rho(a) da^\rho \tag{5.25}$$

On the other hand (5.15a) can be written as

$$dx^i = \frac{\partial x^i}{\partial a^\rho} da^\rho \tag{5.26}$$

from which it follows that

$$\frac{\partial x^i}{\partial a^\rho} = u^i_\sigma(x) \lambda^\sigma_\rho(a). \tag{5.27}$$

The necessary and sufficient integrability condition (Eisenhart 1933, Smirnov 1964) of the system (5.27) is

$$\frac{\partial^2 x^i}{\partial a^\sigma \partial a^\rho} = \frac{\partial^2 x^i}{\partial a^\rho \partial a^\sigma}. \tag{5.28}$$

This condition leads to the relation

$$\left(u^j_\tau \frac{\partial u^i_\nu}{\partial x^j} - u^j_\nu \frac{\partial u^i_\tau}{\partial x^j} \right) \lambda^\tau_\rho \lambda^\nu_\sigma + u^i_\tau \left(\frac{\partial \lambda^\tau_\sigma}{\partial a^\rho} - \frac{\partial \lambda^\tau_\rho}{\partial a^\sigma} \right) = 0 \tag{5.29}$$

or, using (5.23), one gets

$$\left(u^j_\kappa \frac{\partial u^i_\delta}{\partial x^j} - u^j_\delta \frac{\partial u^i_\kappa}{\partial x^j} \right) = c^\tau_{\kappa\delta}(a) u^i_\tau \tag{5.30}$$

where

$$c^\tau_{\kappa\delta} = \left(\frac{\partial \lambda^\tau_\rho}{\partial a^\sigma} - \frac{\partial \lambda^\tau_\sigma}{\partial a^\rho} \right) \mu^\rho_\kappa \mu^\sigma_\delta \tag{5.31}$$

or alternatively

$$\frac{\partial \lambda^\tau_\nu}{\partial a^\gamma} - \frac{\partial \lambda^\tau_\gamma}{\partial a^\nu} = c^\tau_{\kappa\delta} \lambda^\kappa_\nu \lambda^\delta_\gamma. \tag{5.32}$$

Since u^i_σ are independent of a^τ by definition (5.16), the differentiation of (5.30) gives

$$\frac{\partial c^\tau_{\kappa\delta}}{\partial a^\rho} u^i_\tau = 0. \tag{5.33}$$

The quantities u^i_τ are linearly independent with respect to the index τ. This follows from the property that the parameters a^τ are essential in the transformations (5.4) and is proved in Exercise 5.1 below.

STRUCTURE CONSTANTS

Exercise 5.1 Prove that u^i_τ are linearly independent quantities in both the indices i and τ.

Solution First consider the index τ and suppose a^τ ($\tau = 1, 2, \ldots, r$) are not essential. Then one can find arbitrarily small quantities $\varepsilon^\tau(a)$ such that

$$f^i(x; a) = f^i(x; a + \varepsilon). \tag{1}$$

Expanding the right-hand side and keeping only terms linear in ε^τ one gets

$$\varepsilon^\tau(a) \frac{\partial f^i(x, a)}{\partial a^\tau} = 0. \tag{2}$$

But a^τ are in fact essential parameters which implies that

$$\varepsilon^\tau(a) = 0 \quad \text{for any } a. \tag{3}$$

In the limit $a \to 0$, $\varepsilon^\tau(a) \to \lambda^\tau$, where λ^τ are constant values and (2) becomes

$$\lambda^\tau u^i_\tau = 0, \tag{4}$$

which, together with (3), proves that u^i_τ are independent with respect to the index τ.

To prove linear independence with respect to i we have to prove that all constants λ_i in the linear combination

$$\lambda_i u^i_\tau = 0 \tag{5}$$

are zero. Multiplying (5) by δa^τ and summing over τ one gets

$$\lambda_i \mathrm{d} x^i = 0. \tag{6}$$

By definition $\mathrm{d} x^i$ are linearly independent, from which it follows that

$$\lambda_i = 0 \quad \text{for any } i.$$

If u^i_τ are linearly independent it follows that

$$\frac{\partial c^\tau_{\kappa\delta}}{\partial a^\rho} = 0 \tag{5.34}$$

i.e. $c^\tau_{\kappa\delta}$ are independent of a.

The quantities $c^\tau_{\kappa\sigma}$ are called the *structure constants* of a Lie group and they play an important role in the group properties as will be shown below. The property

$$c^\tau_{\kappa\sigma} = -c^\tau_{\sigma\kappa} \tag{5.35}$$

follows from the relation (5.30).

The parameters of a Lie group are real by definition, and any relation describing the group structure must involve real numbers. For this reason the structure constants must be *real* numbers.

From the structure constants one can construct a symmetric second-rank tensor

$$g_{\rho\tau} = c^\mu_{\rho\lambda} c^\lambda_{\tau\mu} \tag{5.36}$$

known as the *metric tensor* or *Killing form*. Its properties have been used by Cartan to distinguish the semi-simple groups (see later) from other groups.

The tensor $g_{\rho\tau}$ also serves to rise or lower the indices of the structure constants. For example one has

$$c_{\mu\nu\sigma} = c_{\mu\nu}^\lambda g_{\lambda\sigma}. \tag{5.37}$$

One can show that the structure constant of the left hand side of (5.37) is antisymmetric to transpositions of any two indices, which is a generalization of (5.35) (see Exercise 5.2).

5.3 GENERATORS

Let us consider now a function F of the coordinates x^i. An infinitesimal transformation $x^i \rightarrow x^i + dx^i$ changes F infinitesimally by

$$dF = \frac{\partial F}{\partial x^i} dx^i = \delta a^\sigma u_\sigma^i \frac{\partial F}{\partial x^i} = \delta a^\sigma X_\sigma F \tag{5.38}$$

where the operators

$$X_\sigma = u_\sigma^i \frac{\partial}{\partial x^i} \tag{5.39}$$

are called *infinitesimal operators* or *generators* of the group of transformations (5.5). From (5.30) it follows that

$$[X_\kappa, X_\delta] = c_{\kappa\delta}^\tau X_\tau. \tag{5.40}$$

The linear independence of u_τ^i leads to the linear independence of the operators X_τ. They form an r-dimensional vector space. The '*Lie product*' of two basis vectors of this space is defined by their commutator, which gives another basis vector. This means that the set of these basis vectors is closed under the above multiplication law. The r infinitesimal operators span an r-dimensional vector space characterized by the quantities $\Sigma\, a^\tau X_\tau$ where a^τ are real numbers. In this sense they form an algebra, called a *real Lie algebra*. To every Lie group there is a real Lie algebra which is unique. However, in general, several non-isomorphic Lie groups can correspond to the same real Lie algebra.

One should note that the algebra of a Lie group is defined through (5.38) in the neighbourhood of the identity element. But it remains valid for any finite transformation. The transfer of this validity is effected by the group multiplication law. Then, it is natural that the structure constants are independent of the group parameters a^ρ, as shown by equation (5.34).

A real Lie algebra $\mathscr{L}$ of dimensions n (≥ 1) has the following properties:

$$[X_\rho, X_\sigma] \in \mathscr{L} \quad \text{for} \quad X_\rho, X_\sigma \in \mathscr{L} \tag{5.41}$$

$$[\alpha X_\rho + \beta X_\sigma, X_\tau] = \alpha[X_\rho, X_\tau] + \beta[X_\sigma, X_\tau] \tag{5.42}$$

for $X_\rho, X_\sigma, X_\tau \in \mathscr{L}$ and all *real* number α and β;

$$[X_\rho, X_\sigma] = -[X_\sigma, X_\rho] \tag{5.43}$$
$$[[X_\rho, X_\sigma], X_\tau] + [[X_\sigma, X_\tau], X_\rho] + [[X_\tau, X_\rho], X_\sigma] = 0. \tag{5.44}$$

This is the Jacobi identity or the *'associativity'* condition. It can be shown (see, for example, Richtmyer (1981) section 25.3) that it stems from the associativity property of the group, introduced in section 2.1.

A real Lie algebra can be complexified. In this case the quantities $\Sigma a^\tau X_\tau$ involve complex numbers a^τ, hence the linear combinations in (5.42) imply complex numbers. A typical example is the angular momentum ladder operators which represent generators of the rotation group. The general procedure is to obtain a real Lie algebra and take a complex extension of it. It turns out to be more convenient to work with complex Lie algebras. But the structure of a Lie group should be studied on the basis of its real algebra because the passage from complex Lie algebra to a real Lie algebra is more subtle (Cornwell 1984, Chapter 10) and not straightforward. One has to be careful, in general, because the matrices of a real Lie algebra need not themselves to be real. A typical example is su(2) which is formed of the Pauli matrices (Section 8.3).

The use of (5.40) in (5.44) leads to the following relation

$$c^\mu_{\rho\sigma} c^\nu_{\mu\tau} + c^\mu_{\sigma\tau} c^\nu_{\mu\rho} + c^\mu_{\tau\rho} c^\nu_{\mu\sigma} = 0 \tag{5.45}$$

which, besides (5.35), is another important property of the structure constants.

The relations (5.35) and (5.45) result from the assumption that the transformations f^i form a group. Conversely, starting from these relations one can find u's and λ's satisfying (5.30) and (5.32), and determine x^i which satisfy (5.27) and form a group. These findings represent the content of three fundamental theorems proved by Lie, which can be found in Racah (1965).

Exercise 5.2 Prove that the structure constants $c_{\rho\sigma\lambda}$ are:

(1) antisymmetric under transposition of any two indices;
(2) invariant under any circular permutation.

Solution The relation (5.35) proves the antisymmetry under the transposition of the first two indices. Therefore it is enough to prove the antisymmetry at transposing the indices 2 and 3.

From (5.36) and (5.37) it follows that

$$c_{\rho\sigma\lambda} = c^\tau_{\rho\sigma} g_{\tau\lambda} = c^\tau_{\rho\sigma} c^\nu_{\tau\mu} c^\mu_{\lambda\nu}. \tag{1}$$

Applying the property (5.45) to the first and second factors this can be rewritten as

$$c_{\rho\sigma\lambda} = -c^\tau_{\sigma\mu} c^\nu_{\tau\rho} c^\mu_{\lambda\nu} - c^\tau_{\mu\rho} c^\nu_{\tau\sigma} c^\mu_{\lambda\nu} \tag{2}$$

and using (5.35) for the second factor of the first term and the third factor of the second term one gets

$$c_{\rho\sigma\lambda} = c^\tau_{\sigma\mu} c^\nu_{\rho\tau} c^\mu_{\lambda\nu} + c^\tau_{\mu\rho} c^\nu_{\tau\sigma} c^\mu_{\nu\lambda}. \tag{3}$$

One can now permute λ with σ in both the left- and right-hand sides of (3). This introduces

$$c_{\rho\lambda\sigma} = c^\tau_{\lambda\mu} c^\nu_{\rho\tau} c^\mu_{\sigma\nu} + c^\tau_{\mu\rho} c^\nu_{\tau\lambda} c^\mu_{\nu\sigma}. \tag{4}$$

Applying (5.35) to all c's and reordering the terms and factors leads to

$$c_{\rho\lambda\sigma} = -c^\mu_{\sigma\nu} c^\tau_{\rho\mu} c^\nu_{\lambda\tau} - c^\nu_{\tau\rho} c^\mu_{\nu\sigma} c^\tau_{\mu\lambda}. \tag{5}$$

Making a circular permutation of the repeated indices $\mu \to \tau \to \nu$ in the first term and $\tau \to \mu \to \nu$ in the second term, one gets

$$c_{\rho\lambda\sigma} = -c^\tau_{\sigma\mu} c^\nu_{\rho\tau} c^\mu_{\lambda\nu} - c^\tau_{\mu\rho} c^\nu_{\tau\sigma} c^\mu_{\nu\lambda} = -c_{\rho\sigma\lambda} \tag{6}$$

which proves the antisymmetry at permuting λ with σ. The property (2) is an immediate consequence of (1) because any circular permutation of three indices can be written as a product of two adjacent transpositions according to the relation (2.18b).

Due to the property (2) of the structure constants and the symmetry property of $g_{\rho\tau}$ one can write

$$c_{\mu\nu\sigma} = c_{\nu\sigma\mu} = c^\tau_{\nu\sigma} g_{\tau\mu} = g_{\mu\tau} c^\tau_{\nu\sigma} \tag{7}$$

which is another way of writing (5.37).

Using (5.38) one can write

$$F(x + dx) = F(x) + \frac{\partial F}{\partial x^i} dx^i = S_{\delta a} F \tag{5.46}$$

where the operator

$$S_{\delta a} = 1 + \delta a^\sigma X_\sigma \tag{5.47}$$

effects the infinitesimal change $F \to F + dF$ induced by the infinitesimal parameters δa^σ. Two successive infinitesimal transformations $S_{\delta a} S_{\delta b}$ acting on F give another infinitesimal transformation

$$S_{\delta a} S_{\delta b} = (1 + \delta a^\sigma X_\sigma)(1 + \delta b^\rho X_\rho) = 1 + \delta a^\sigma X_\sigma + \delta b^\rho X_\rho \tag{5.48}$$

because only the first order infinitesimals have to be retained. The relation (5.48) shows that the multiplication of two elements of the group corresponds to the addition of their parameters for infinitesimal transformations.

It is useful to generate finite transformations also. Let us consider first a one parameter Lie group (a group of order one) and write an infinitesimal change δa of its parameter a as

$$\delta a = \frac{a}{N} \tag{5.49}$$

with N an arbitrarily large number. Applying the transformation (5.47) N times one gets

$$S_a = \left(1 + \frac{a}{N}X\right)^N$$

which, in the limit $N \to \infty$, becomes

$$S_a = e^{aX}. \tag{5.50}$$

This can be generalized to a group of order r by writing

$$S_a = e^{a^\rho X_\rho} \tag{5.51}$$

where a^ρ corresponds to a special choice, called 'canonical' of group parameters. It is always valid for connected and compact Lie groups as defined in Section 5.7 (see, for example, Cornwell 1984, Chapter 10, Section 5).

The inverse of the finite transformation (5.51) is

$$S_a^{-1} = e^{-a^\rho X_\rho} \tag{5.52}$$

which is a transformation with parameters of opposite signs to those of (5.51).

In general, the infinitesimal generators X_ρ do not commute and hence, in any particular application of the transformation (5.51) or its inverse (5.52), care must be taken of the order of the operators. An example is the rotation group, Chapter 6.

The operator (5.47) can be viewed as a first-order Taylor expansion of (5.51) with $a^\rho \to \delta a^\rho$. There are situations where the first-order terms vanish and one has to retain second order infinitesimals in the Taylor expansion of (5.51) and (5.52)

$$S_a = 1 + \delta a^\sigma X_\sigma + \frac{1}{2}\delta a^\sigma \delta a^\rho X_\sigma X_\rho \tag{5.53}$$

$$S_a^{-1} = 1 - \delta a^\sigma X_\sigma + \frac{1}{2}\delta a^\sigma \delta a^\rho X_\sigma X_\rho. \tag{5.54}$$

One of Lie's fundamental theorems ensures that one never needs to go beyond the second-order infinitesimals because they are related to commutators and these define the structure of the group. A typical example which will be used below is the product $S_a S_b S_a^{-1} S_b^{-1}$. A straightforward calculation based on (5.53) and (5.54) gives

$$S_a S_b S_a^{-1} S_b^{-1} = 1 + \delta a^\sigma \delta b^\rho [X_\sigma, X_\rho]. \tag{5.55}$$

In physical applications of Lie groups the generators X_σ (or linear combinations of them) correspond to observables. The invariance of a Hamiltonian under a group of transformations means

$$[H, S_a] = 0 \tag{5.56}$$

which, using (5.47) or (5.51), leads to

$$[H, X_\sigma] = 0 \tag{5.57}$$

inasmuch as a^σ or δa^σ are linearly independent. Hence the invariance of a Hamiltonian under transformations forming a Lie group means that H commutes with all the generators of that group (examples will be encountered in Chapters 6–8).

Exercise 5.3 Consider the group of linear transformations

$$x' = ax + b \tag{1}$$

depending on the parameters a and b. Find its generators and Lie algebra.

Solution The identity transformation is $x' = x$. An infinitesimal transformation around unity gives

$$x' = (1 + \delta a)x + \delta b \tag{2}$$

hence

$$dx = \delta a\, x + \delta b. \tag{3}$$

Under this transformation $F(x)$ changes infinitesimally to

$$F(x + dx) = F(x) + dx\frac{dF}{dx} = \left[1 + (\delta a\, x + \delta b)\frac{d}{dx}\right]F.$$

Hence the two generators are

$$X_a = x\frac{d}{dx}, \quad X_b = \frac{d}{dx}. \tag{4}$$

The Lie algebra is

$$[X_a, X_b] = -X_b. \tag{5}$$

5.4 SIMPLE AND SEMI-SIMPLE GROUPS

We now make use of the relation (5.55) in order to find some properties of the structure constants of simple and semi-simple groups introduced in Section 2.7.

Apart from that, one can also see that (5.55) has an immediate consequence for an abelian group for which

$$S_a S_b = S_b S_a$$

or

$$S_a S_b S_a^{-1} S_b^{-1} = 1.$$

Using (5.55) one gets

$$1 + \delta a^\rho \delta b^\sigma [X_\rho, X_\sigma] = 1. \tag{5.58}$$

SIMPLE AND SEMI-SIMPLE GROUPS

Hence

$$[X_\rho, X_\sigma] = 0 \tag{5.59}$$

and consequently all structure constants vanish for an abelian group

$$c_{\rho\sigma}^\tau = 0 \quad \text{for all } \rho, \sigma \text{ and } \tau. \tag{5.60}$$

The algebra of an abelian group is said to be abelian or commutative.

In Section 2.2 we saw that a subgroup is a subset S' of elements of a group S which forms itself a group under the same law of composition as S. Let us take two elements S'_a and S'_b of S'. Then

$$S'_a S'_b \in S'$$

and also

$$S'_a S'_b (S'_a)^{-1} (S'_b)^{-1} = S'_c \in S'. \tag{5.61}$$

Using (5.55) in the left-hand side and (5.47) in the right-hand side of (5.61), one has

$$1 + \delta a^\rho \delta b^\sigma [X_\rho, X_\sigma] = 1 + \delta c^\tau X_\tau. \tag{5.62}$$

From (5.40), one gets

$$\sum_{\tau=1}^r \delta a^\rho \delta b^\sigma c_{\rho\sigma}^\tau X_\tau = \sum_{\tau=1}^p \delta c^\tau X_\tau$$

where r is the order of S and p the order of S'. As $r > p$, one can divide the sum in the left-hand side into two distinct contributions and write

$$\sum_{\tau=1}^p (\delta a^\rho \delta b^\sigma c_{\rho\sigma}^\tau - \delta c^\tau) X_\tau + \sum_{\tau=p+1}^r \delta a^\rho \delta b^\sigma c_{\rho\sigma}^\tau X_\tau = 0. \tag{5.63}$$

X_τ are linearly independent quantities, from which it follows that

$$\sum_{\rho,\sigma=1}^p \delta a^\rho \delta b^\sigma c_{\rho\sigma}^\tau = 0 \quad \text{for } \tau > p \tag{5.64}$$

or

$$c_{\rho\sigma}^\tau = 0 \quad \text{for } \tau > p \text{ and } \rho, \sigma \leq p. \tag{5.65}$$

By definition (Section 2.6) an invariant subgroup H is a subgroup of S if it contains all conjugates of its elements (or equivalently a certain number of full classes of S). Let us denote by p and r the order of H and S, respectively ($p < r$), and suppose $h_1 \in H$ and $s \in S$. Then

$$sh_1 s^{-1} \in H$$

and also

$$sh_1 s^{-1} h_1^{-1} = h_2 \in H. \tag{5.66}$$

Using (5.55) in the left-hand side and (5.47) in the right-hand side one writes

$$1 + \delta s^\rho \delta h_1^\sigma [X_\rho, X_\sigma] = 1 + \delta h_2^\tau X_\tau$$

160   LIE GROUPS

or, based on (5.40), one gets

$$\sum_{\rho=1}^{r}\sum_{\sigma=1}^{p}\sum_{\tau=1}^{r} \delta s^\rho \delta h_1^\sigma c_{\rho\sigma}^\tau X_\tau = \sum_{\tau=1}^{p} \delta h_2^\tau X_\tau \qquad (5.67)$$

where the sums over repeated indices have been written explicitly. As in the previous case the sum over τ in the left-hand side can be split into two parts:

$$\sum_{\tau=1}^{p}\left(\sum_{\rho=1}^{r}\sum_{\sigma=1}^{p} \delta s^\rho \delta h_1^\sigma c_{\rho\sigma}^\tau - \delta h_2^\tau\right) X_\tau + \sum_{\tau=p+1}^{r}\sum_{\rho=1}^{r}\sum_{\sigma=1}^{p} \delta s^\rho \delta h_1^\sigma c_{\rho\sigma}^\tau X_\tau = 0. \qquad (5.68)$$

From the linear independence of the generators it follows that

$$\sum_{\rho=1}^{r}\sum_{\sigma=1}^{p} c_{\rho\sigma}^\tau \delta s^\rho \delta h_1^\sigma = 0 \quad \text{for } \tau > p \qquad (5.69)$$

and since δs^ρ and δh_1^σ are independent variations one finally gets

$$c_{\rho\sigma}^\tau = 0 \quad \text{for } \tau > p, \sigma \leq p \text{ and any } \rho. \qquad (5.70)$$

According to Section 2.7, a group is simple if it has no proper invariant subgroup, and semi-simple if it has no abelian invariant subgroup. A simple group is also semi-simple and many continuous groups of physical interest are semi-simple. Therefore it is important to find a way to identify them from the properties (5.60) and (5.70) of the structure constants. At this stage the second-rank tensor (5.36), built from structure constants, comes into play.

Suppose that a group possesses an abelian invariant subgroup. If the indices of the generators of the abelian invariant subgroup are denoted by $\bar\rho, \bar\sigma, \bar\lambda,\ldots$ then according to (5.36) one can write

$$g_{\rho\bar\sigma} = c_{\rho\lambda}^\mu c_{\bar\sigma\mu}^\lambda = -c_{\rho\lambda}^\mu c_{\mu\bar\sigma}^\lambda$$

and using the condition (5.70) for the second factor in the above expression one gets

$$g_{\rho\bar\sigma} = -c_{\rho\lambda}^\mu c_{\mu\bar\sigma}^{\bar\lambda}$$

and using again (5.70) for the first factor, the tensor element $g_{\rho\bar\sigma}$ becomes

$$g_{\rho\bar\sigma} = -c_{\rho\lambda}^{\bar\mu} c_{\bar\mu\bar\sigma}^{\bar\lambda}$$
$$= c_{\rho\lambda}^{\bar\mu} c_{\bar\sigma\bar\mu}^{\bar\lambda}$$

The second factor above is a structure constant of the invariant abelian subgroup and based on (5.60) one obtains

$$g_{\rho\bar\sigma} = 0. \qquad (5.71)$$

This implies

$$\det|g_{\rho\sigma}| = 0 \qquad (5.72)$$

SIMPLE AND SEMI-SIMPLE GROUPS

which is the necessary condition for a group not to be semi-simple. Its contrary

$$\det |g_{\rho\sigma}| \neq 0 \tag{5.73}$$

is the sufficient condition for a group to be semi-simple. That (5.73) is both a necessary and sufficient condition has been proved by Cartan.

The condition (5.73) means that the matrix $g_{\rho\sigma}$ has an inverse $g^{\rho\sigma'}$ which satisfies

$$g_{\rho\sigma} g^{\rho\sigma'} = \delta_\sigma^{\sigma'}. \tag{5.74}$$

Chapters 6–8 contain examples of semi-simple groups which by definition satisfy (5.73). To gain more familiarity with Lie groups we show in Exercise 5.4 below a simple example where (5.73) is not satisfied.

Exercise 5.4 Show that the Euclidean group in two dimensions E_2 defined by the transformation

$$x' = x\cos\theta + y\sin\theta - \rho_1 \tag{1a}$$
$$y' = -x\sin\theta + y\cos\theta - \rho_2 \tag{1b}$$

is not semi-simple.

Solution The transformation (1) represents geometrically two successive operations:

(1) a translation of the origin of the coordinate system by $\boldsymbol{\rho} = (\rho_1, \rho_2)$, in the new system, the vector $\boldsymbol{r} = (x, y)$ has the components

$$x' = x - \rho_1, \qquad y' = y - \rho_2;$$

(2) a rotation through an angle θ about the z-axis of the translated system.

Note that in Example 3.1, instead of the system of coordinates, the vector $\boldsymbol{r}$ has been rotated. To rotate the system amounts to transposing the rotation matrix resulting from (1). In general, the Euclidean group E_2 or E_3 in two or three dimensions, respectively, contains translations and rotations, i.e. operations which can be applied to a rigid body (see Section 6.3).

Consider an infinitesimal rotation

$$\cos\theta \approx 1, \qquad \sin\theta \approx \theta \tag{2}$$

Then

$$x' - x = \theta y - \rho_1 \tag{3a}$$
$$y' - y = -\theta x - \rho_2. \tag{3b}$$

This is a generalization of Exercise 5.3. With $x^1 = x$, $x^2 = y$ a scalar function $F(x^i)$ changes infinitesimally into

$$F(x^i + dx^i) = F(x^i) + dx\frac{\partial F}{\partial x} + dy\frac{\partial F}{\partial y}$$

$$= \left[1 + (\theta y - \rho_1)\frac{\partial}{\partial x} + (-\theta x - \rho_2)\frac{\partial}{\partial y}\right]F.$$

To each infinitesimal parameter, here θ, ρ_1 and ρ_2, there corresponds a generator, cf. (5.38). These are

$$X_1 = y\frac{\partial}{\partial x} - x\frac{\partial}{\partial y}, \qquad X_2 = -\frac{\partial}{\partial x}, \qquad X_3 = -\frac{\partial}{\partial y}. \tag{4}$$

Their algebra is

$$[X_1, X_2] = X_3, \qquad [X_1, X_3] = -X_2, \qquad [X_2, X_3] = 0. \tag{5}$$

Hence

$$c_{12}^3 = 1, \qquad c_{13}^2 = -1, \qquad c_{23}^1 = 0 \tag{6}$$

and

$$g_{\rho\sigma} = \begin{pmatrix} -2 & 0 & 0 \\ 0 & 0 & 0 \\ 0 & 0 & 0 \end{pmatrix} \tag{7}$$

which shows that the group is not semi-simple because

$$\det|g_{\rho\sigma}| = 0. \tag{8}$$

5.5 SIMPLE AND SEMI-SIMPLE LIE ALGEBRAS

The concepts of simple and semi-simple Lie groups can also be introduced in terms of properties of their algebras. For this purpose one needs to define a subalgebra. In analogy to subgroups a subalgebra $\mathscr{L}'$ of a Lie algebra is a subset of elements of $\mathscr{L}$ which by themselves form an algebra with the same commutator as that of $\mathscr{L}$.

$\mathscr{L}'$ is said to be a proper subalgebra of $\mathscr{L}$ if at least one element of $\mathscr{L}$ is not contained in $\mathscr{L}'$. The dimension of $\mathscr{L}'$ is accordingly smaller than that of $\mathscr{L}$. (Any algebra and subalgebra has at least one element.)

A subalgebra $\mathscr{L}'$ of $\mathscr{L}$ is said to form an *ideal* or an *invariant subalgebra* of $\mathscr{L}$ if the commutator

$$[X_\rho, X_\sigma] = c_{\rho\sigma}^\tau X_\tau \in \mathscr{L}' \tag{5.75}$$

for all $X_\rho \in \mathscr{L}'$ and $X_\sigma \in \mathscr{L}$. This definition is equivalent to Equation (5.70). If the algebra $\mathscr{L}$ contains members that are not in the ideal then the ideal is called a *proper ideal*.

For real Lie algebras there are some theorems which show the connection between concepts introduced for groups and their analogues for the corresponding algebras. One important example is the following (see Cornwell 1984, Chapter 11, Section 2):

Theorem If G and G' are linear Lie groups, $\mathscr{L}$ and $\mathscr{L}'$ their corresponding Lie algebras, and G' is a subgroup of G, then $\mathscr{L}'$ is a subalgebra of $\mathscr{L}$. If G' is an invariant subgroup of G, then $\mathscr{L}'$ is an ideal of $\mathscr{L}$.

The concepts of simple and semi-simple algebras are introduced by the following definitions:

(1) A Lie algebra $\mathscr{L}$ is said to be simple if it is not abelian and does not possess a proper invariant Lie subalgebra.
(2) A Lie algebra $\mathscr{L}$ is said to be semi-simple if it does not possess an abelian invariant subalgebra.

As for groups, if an algebra is simple it is also semi-simple and the converse is not true.

If $\mathscr{L}$ is abelian then it is neither simple nor semi-simple. As all one-dimensional Lie algebras are abelian, simple and semi-simple Lie algebras must have dimensions greater than one.

In the above context simple and semi-simple groups can be defined as follows:

(1) A Lie group is said to be simple if and only if its real Lie algebra is simple.
(2) A Lie group is semi-simple if and only if its real Lie algebra is semi-simple.

The semi-simple Lie algebras are very important both for the mathematical study of Lie groups (see Section 5.8) and also for their physical applications.

In section 5.3–5.5 we have discussed the connection between Lie groups and (finite dimensional) Lie algebras. Infinite dimensional algebras do exist. Two particular classes, which have generated much interest during the last two decades, are the Kac–Moody algebra and the Virasoro algebra. They have applications in several areas of mathematics and physics (Cornwell 1989, Kac 1990, Bailin and Love 1994).

5.6 EXAMPLES OF LIE GROUPS

Before proceeding further with the properties of Lie groups it is useful to give some examples. The elements of the groups we are going to consider here and in the subsequent chapters are square $n \times n$ matrices. The reason is that any Lie group that is important in physics is a *linear* Lie group and its representations are matrices (Cornwell 1984, Chapter 3, Section 1). To form a group these matrices have to be non-singular. The law of composition of the group is matrix multiplication which has the property of being associative. The elements of a matrix can be real ($\mathbb{R}$) or complex ($\mathbb{C}$). In physical applications these matrices, denoted generally by a, result from a transformation of a vector space $x = (x_1, x_2, \ldots, x_n)$ into another vector space $x' = (x'_1, x'_2, \ldots, x'_n)$

$$x' = ax. \quad (5.76)$$

A group of transformations can be defined in terms of the properties of a. A list of relevant properties of matrices is given in Table 5.1. Alternatively, a group can also be defined by the property of leaving invariant some functional form of x as indicated below.

Table 5.1 Matrix properties relevant to the definition of various continuous groups. The inverse, transpose, complex conjugate, and Hermitian conjugate are denoted by a^{-1}, a^T, a^*, and a^+, respectively.

Name	Property
symmetric	$a = a^T$
skew symmetric	$a = -a^T$
orthogonal	$a^{-1} = a^T$
real	$a = a^*$
imaginary	$a = -a^*$
Hermitian	$a = a^+$
skew hermitian	$a = -a^+$
unitary	$a^{-1} = a^+$

The n^2 (real or complex) matrix elements are parameters which vary continuously. The constraints imposed by the properties listed in Table 5.1 usually restrict the number of independent parameters.

In Table 5.2, we list some of the most important groups, the elements of which are matrices of degree n. The first column indicates the name and the commonly used notation for a specific group. The second column represents the definition of the group through the properties of a. Column 3 gives the group order. We give some further details below.

The general linear group GL(n, $\mathbb{C}$) of complex matrices of degree n is the largest linear matrix group. The other groups listed below are subgroups of this group. The order of the GL(n, $\mathbb{C}$) group is given by twice the number, n^2, of matrix elements because each element is complex. On the other hand, GL(n, $\mathbb{R}$) the elements of which are real matrices, is of order n^2. It follows that

$$\text{GL}(n, \mathbb{C}) \supset \text{GL}(n, \mathbb{R}). \quad (5.77)$$

Table 5.2 Lie groups.

Name	Definition	Order	Remark
Complex general linear $GL(n,C)$	$\det a \neq 0$	$2n^2$	$GL(n, C) \supset GL(n,R)$
Real general linear $GL(n,R)$ or $GL(n)$	$a = a^*$ $\det a \neq 0$	n^2	
Complex special linear $SL(n,C)$	$\det a = 1$	$2n^2 - 2$	$SL(n, C) \supset SL(n, R)$
Real special linear $SL(n,R)$ or $SL(n)$	$a = a^*$ $\det a = 1$	$n^2 - 1$	
Unitary $U(n)$ or U_n	$a^{-1} = a^+$ $\|\det a\| = 1$	n^2	Isomorphic with $GL(n,R)$
Special unitary $SU(n)$ or SU_n	$a^{-1} = a^+$ $\det a = 1$	$n^2 - 1$	Isomorphic with $SL(n,R)$
Complex orthogonal $O(n,C)$	$a^{-1} = a^T$ $\det a = \pm 1$	$n(n-1)$	$O(n, C) \supset O(n, R)$
Real orthogonal $O(n,R)$ or $O(n)$ or O_n	$a = a^*$ $a^{-1} = a^T$ $\det a = \pm 1$	$\dfrac{n(n-1)}{2}$	Isomorphic to proper ($\det a = 1$) or improper ($\det a = -1$) rotation group
Real special orthogonal $SO(n,R)$ or $SO(n)$ or SO_n	$a = a^*$ $a^{-1} = a^T$ $\det a = 1$	$\dfrac{n(n-1)}{2}$	Isomorphic to proper rotation group in n dimensions
$O(n,m)$	$a^* = a$ $a^T g a = g$		$g = \begin{pmatrix} I_n & 0 \\ 0 & -I_m \end{pmatrix}$ $O(3,1)$ is isomorphic to the homogeneous Lorentz group
Complex symplectic $Sp(2n,C)$	$a^T g a = g$	$4n^2 + 2n$	Symplectic groups are only for even dimensional space and are unimodular ($\det a = 1$) $g^T = -g$
Real symplectic $Sp(2n,R)$	$a^T g a = g$ $a = a^*$	$2n^2 + n$	
Unitary symplectic $Sp(2n)$	$a^T g a = g$ $a^{-1} = a^+$	$2n^2 + n$	$g = \begin{pmatrix} 0 & I_n \\ -I_n & 0 \end{pmatrix}$ $SU(2n) \supset Sp(2n)$

I_n is a unit matrix of dimension n.

Imposing the restriction det $a = 1$ (for a unimodular group) one obtains the special linear complex, SL(n, $\mathbb{C}$), or real, SL(n, $\mathbb{R}$), group. In the first case det $a = 1$ imposes two constraints and in the second case one constraint, which give $2n^2 - 2$ for the order of SL(n,$\mathbb{C}$) and $n^2 - 1$ for the order of SL(n,$\mathbb{R}$). It follows that

$$\mathrm{GL}(n, \mathbb{C}) \supset \mathrm{SL}(n, \mathbb{C}) \supset \mathrm{SL}(n, \mathbb{R}). \tag{5.78}$$

The elements of a unitary group U(n) are unitary matrices. The unitarity ensures that the linear transformation (5.76) leaves invariant the quadratic Hermitian form

$$\sum_{i=1}^{n} x_i x_i^*. \tag{5.79}$$

The unitarity condition

$$a^+ a = a a^+ = 1 \tag{5.80}$$

stipulates that

$$\det a = \mathrm{e}^{i\varphi} \tag{5.81}$$

and imposes n constraints from the diagonal elements and $n(n-1)$ from the off-diagonal matrix elements, leaving

$$2n^2 - n - n(n-1) = n^2 \tag{5.82}$$

independent real parameters which is the order r of U(n). For special unitary groups, SU(n), the extra restriction det $a = 1$ ($\varphi = 0$ in Equation (5.81)) reduces the order to $r = n^2 - 1$. The particular example of SU(2), which has three independent parameters, was discussed in Exercise 2.3. The unitary group will be presented in Chapter 8.

The complex linear transformations (5.76) which leave invariant the complex quadratic form

$$\sum_{i=1}^{n} (x_i)^2 \tag{5.83}$$

have the property

$$a^{\mathrm{T}} a = a a^{\mathrm{T}} = 1 \tag{5.84}$$

and form the complex linear orthogonal group O(n, $\mathbb{C}$). The orthogonality condition (5.84) imposes $2n$ constraints from the diagonal elements and n(n - 1) from the off-diagonal elements (real and imaginary parts separately) which leads to the order of O(n, $\mathbb{C}$) being given by

$$r = 2n^2 - 2n - n(n-1) = n(n-1). \tag{5.85}$$

Restricting the matrices a to be real, one obtains the real orthogonal group O(n, $\mathbb{R}$) of order

$$r = n^2 - n - \frac{n(n-1)}{2} = \frac{n(n-1)}{2}. \tag{5.86}$$

The orthogonality condition (5.84) also gives

$$\det a = \pm 1 \tag{5.87}$$

and the elements of O(n, $\mathbb{C}$) or O(n, $\mathbb{R}$) can be divided into two sets, one associated with det $a = +1$ and another one associated with det $a = -1$, which are disconnected (see next section). The set with det $a = +1$ forms the subgroup SO(n, $\mathbb{C}$) for $a \neq a^*$ and SO(n, $\mathbb{R}$) for $a = a^*$. They are still of order $n(n-1)$ and $\frac{n(n-1)}{2}$, respectively. These groups represent 'proper' rotations. The set associated with det $a = -1$ represent 'improper' rotations because it includes inversions (or reflections) denoted by the operator I. The inversions are responsible for det $a = -1$. So one can write in this case, for example,

$$O(n) = SO(n) \times I \tag{5.88}$$

where O(n) is an alternative notation for O(n, $\mathbb{R}$). The group O(n) will be presented in Chapter 6.

A generalization of the group O(n) is the group O(n, m) of real linear transformations which leave invariant the quantity

$$\sum_{i=1}^{n} x_i^2 - \sum_{j=n+1}^{n+m} (x_j)^2. \tag{5.89}$$

The matrix a leaving (5.89) invariant must obey

$$a^T g a = g \tag{5.90}$$

with

$$g = \begin{pmatrix} I_n & 0 \\ 0 & -I_m \end{pmatrix} \tag{5.91}$$

where I_n is a unit $n \times n$ matrix. By generalization of (5.86) one finds that the order of O(n, m) is

$$r = \frac{1}{2}(n+m)(n+m-1). \tag{5.92}$$

The Minkowski space x^1, x^2, x^3, i x^0 where the squared distance from origin is

$$s^2 = (x^0)^2 - (x^1)^2 - (x^2)^2 - (x^3)^2 \tag{5.93}$$

provides a particular case of (5.89). The group which leaves (5.93) invariant is the homogeneous Lorentz group which is isomorphic to O(3, 1) and can thus be introduced as a particular example of O(n, m), as shown in Chapter 7.

The symplectic group is formed by $2n \times 2n$ matrices which leave invariant the skew symmetric bilinear form

$$\sum_{k=1}^{n}(x_k y_{-k} - x_{-k} y_k). \tag{5.94}$$

The invariance of (5.94) implies

$$a^T g a = g \tag{5.95}$$

where g is now the skew symmetric matrix

$$g = \begin{pmatrix} 0 & I_n \\ -I_n & 0 \end{pmatrix}. \tag{5.96}$$

One sees that a symplectic transformation can be defined in an even dimensional space only. There are three types of symplectic groups

    a real   Sp($2n$, $\mathbb{R}$)
    a complex   Sp($2n$, $\mathbb{C}$)
    a unitary   Sp($2n$).

For Sp($2n$, $\mathbb{R}$) the number of constraints resulting from the diagonal elements of (5.95) is equal to n and there are $4C_n^2$ constraints from the off-diagonal elements, so that its order is

$$r = (2n)^2 - (2n^2 - n) = n(2n + 1). \tag{5.97}$$

For Sp($2n$, $\mathbb{C}$) both the number of real parameters in the matrix elements and the number of constraints is twice as large so that its order is

$$r = 2n(2n + 1). \tag{5.98}$$

For Sp($2n$) the unitarity condition (5.80) imposes $2n + C_{2n}^2$ extra constraints with respect to Sp($2n$, $\mathbb{C}$) so that its order, using (5.98), becomes

$$r = 2n(2n + 1) - 2n - n(2n - 1) = 2n^2 + n. \tag{5.99}$$

All symplectic groups are unimodular (see Hamermesh 1962, Section 10–8). They are related to other groups as follows

    GL($2n$, $\mathbb{C}$) $\supset$ Sp($2n$, $\mathbb{C}$) $\supset$ Sp($2n$, $\mathbb{R}$)
    Sp($2n$, $\mathbb{C}$) $\supset$ Sp($2n$)
    SU($2n$) $\supset$ Sp($2n$).

The unitary simplectic group has a wide range of applications in physics, as for example in the classification of states in jj-coupling (Hamermesh 1962, Section 11-9; Wybourne 1970). Other groups, such as Sp($2n$, $\mathbb{R}$), have been used in spectrum-generating algebra methods, for example in the derivation of a baryon resonance mass formula (Bowler *et al.* 1981).

The treatment of the symplectic group is beyond the scope of this book and we refer the reader to Hamermesh (1962), Chapter 10 and Wybourne (1970), Chapter 8.

The algebras of the groups listed in Table 5.2 are denoted by the same names but in lower case, for example the algebra of SU(n) is denoted by su(n).

5.7 COMPACTNESS

Infinitesimal transformations parametrize the group elements in the neighbourhood of the identity element and they reflect *local* properties of the group. Although most information concerning the structure of a Lie group comes from the study of its local properties, embedded in its real Lie algebra, there are *global* properties (properties of the whole group) which are also important. A particular example is the compactness.

It is rather complicated, to define a compact set in a general topological space, but for Lie groups, a simple definition is possible and allows us to distinguish between a compact and a non-compact Lie group. Compactness of Lie groups is defined as in a finite dimensional Euclidean space but it refers to the space of group parameters. There is a theorem, known as Heine–Borel theorem, which states that a subset of points in a finite dimensional Euclidean space is *compact* if and only if it is *closed* and *bounded*.

A set is bounded if it can be contained in a finite sphere of the space. Therefore in a Euclidean space any region of finite extension is compact and a region extending to infinity is not compact.

A set of points in an interval $[a, b]$ is closed if and only if both ends of the interval can be attained. This definition refers to *connected* sets only. In group theory language a connected group means that, given an arbitrary element g, one can reach the identity element by a continuous variation of the r parameters of the group. For example, the orthogonal group SO(n) of matrices of determinant $+1$ is connected but the more general group O(n) of matrices of det $a = 1$ is not connected because one cannot pass continuously from det $a = +1$ to det $a = -1$.

Alternatively O(n) has two connected components, one formed of elements with det $a = 1$, the other with det $a = -1$, and they cannot be connected to each other. In general, one says that a Lie group is connected if it possesses only one connected component.

In conclusion a Lie group is said to be compact if its parameters a^1, a^2, ..., a^r range over closed finite intervals $\alpha^\rho \leq a^\rho \leq \beta^\rho$, $\rho = 1, 2, \ldots, r$.

Many groups of physical interest are compact. For example, for U(n) the unitarity condition requires

$$\sum_{j=1}^{n} |a_{ij}|^2 = 1 \quad \text{for all } i$$

which implies that $|a_{ij}| \leq 1$ for all i and j, i.e. the parameters of U(n) are restricted to vary over a closed finite interval. Thus the group U(n) is compact. The same type of arguments hold for SU(n) or O(n) which are subgroups of U(n).

The groups which are not compact fail to be compact because their set of parameters has infinite range and is therefore unbounded. Examples are the group of translations (or any other group which has the translation group as a subgroup, like the Euclidean group or the Poincaré group) and the proper Lorentz group (or the Poincaré group because it contains the proper Lorentz group as a subgroup).

The distinction between compact and non-compact Lie groups is extremely important because their representation theory is very different. For compact Lie groups the theory bears a strong similarity to that of finite groups. An important result, mentioned in Section 3.10, is that every finite dimensional representation of a compact Lie group is equivalent to a unitary representation and therefore is fully reducible. For a non-compact group the finite dimensional representations are no longer unitary and all unitary representations are infinite dimensional.

The Lie algebra of a compact (non-compact) Lie group is also said to be compact (non-compact). There is a theorem which states that a necessary and sufficient condition for a semi-simple Lie algebra to be compact is that the matrix (5.36) be negative definite. This theorem is not proved, but is illustrated by Exercise 5.5 later in this chapter.

5.8 DIRECT AND SEMI-DIRECT SUMS OF LIE ALGEBRAS

A real or complex Lie algebra $\mathscr{L}$ is a *direct sum* of two Lie algebras $\mathscr{L}_1$ and $\mathscr{L}_2$,

$$\mathscr{L} = \mathscr{L}_1 \oplus \mathscr{L}_2,$$

if the vector space of $\mathscr{L}$ is the direct sum of the vector spaces of $\mathscr{L}_1$ and $\mathscr{L}_2$ and if any element of $\mathscr{L}_1$ commutes with any element of $\mathscr{L}_2$.

If $\mathscr{L}_1$ and $\mathscr{L}_2$ are the Lie algebras of the Lie groups H_1 and H_2 then the Lie algebra of the direct product $H_1 \times H_2$ is isomorphic to $\mathscr{L}_1 \oplus \mathscr{L}_2$. An example is the group SO(4) = SO(3) × SO(3). Its algebra is isomorphic to so(3) × so(3).

The representations of a direct product group $H_1 \times H_2$ are given by the direct product of representations (Section 3.9) of H_1 and H_2. The algebraic analogue of this result is the following. Let D_1 and D_2 be representations of dimensions n_1 and n_2 of two Lie algebras $\mathscr{L}_1$ and $\mathscr{L}_2$, respectively. Then the matrices

$$D = D_1 \otimes I_{n_2} + D_2 \otimes I_{n_1}$$

where I_n is a unit $n \times n$ matrix, provide an $n_1 n_2$-dimensional representation of $\mathscr{L}_1 \oplus \mathscr{L}_2$

The algebra of a direct product group is always isomorphic to a direct sum algebra. Conversely, if a Lie group has an algebra isomorphic to a direct sum it does not always mean that the group is a direct product. This is the case for U(n) with $n \geq 2$. The algebra of U(n) is isomorphic to u(1) + su(n) where u(1) $\sim I_n$ (Chapter 8). But U(n) is not the direct product of U(1) and SU(n) because these two groups have two elements in common I_n and $-I_n$ and this is at variance with the definition of a direct product group which requires only the identity element to be in common. In such cases the difficulty in representation theory is overcome by use of the so-called *universal covering group* which reduces the group representation problem to that of SU(n). For details of the universal covering group, see Hamermesh (1962), Section 8-14 or Cornwell (1984), Chapter II, Section 7.

An algebra $\mathscr{L}$ is the semi-direct sum of two algebras $\mathscr{L}_1$ and $\mathscr{L}_2$

$$\mathscr{L} = \mathscr{L}_1 \oplus_S \mathscr{L}_2$$

if

$$[\mathscr{L}_1, \mathscr{L}_1] \subset \mathscr{L}_1, \quad [\mathscr{L}_2, \mathscr{L}_2] \subset \mathscr{L}_2 \quad \text{and} \quad [\mathscr{L}_1, \mathscr{L}_2] \subset \mathscr{L}_1.$$

The algebras of semi-direct product groups (Section 2.7) provide examples of semi-direct sums. Well-known cases are the Euclidean groups E_2, E_3, and the Poincaré group. Let us illustrate the discussion with the group E_2 of Exercise 5.4. In the notation of that exercise $\mathscr{L}_1$ is formed from the generators X_2 and X_3, i.e. the generators of the translation group in two dimensions and $\mathscr{L}_2$ is just X_1, the generator of the rotation group in two dimensions.

A powerful technique in constructing the representations of semi-direct product groups is the method of *induced* representations (see, for example, Cornwell 1984, Chapter 5, Section 7). By using this method one can construct representations of a group in terms of the representations of its subgroups.

5.9 CLASSIFICATION OF SEMI-SIMPLE GROUPS

In Sections 5.1–5.3 we saw that the study of Lie groups reduces to the study of infinitesimal transformations which generate a Lie algebra for each group.

For semi-simple groups the study is further simplified by the classification of semi-simple groups introduced by Cartan. This classification gives few general categories of groups which have properties in common. From these properties one can recognize the group responsible for a specific physical symmetry. We recall that most groups of importance in physics are semi-simple. For groups which have physical applications but are not semi-simple the problem can, in general, be reduced to the study of semi-simple groups. One example is $U(n)$ which is not semi-simple because $U(1)$ is an invariant abelian subgroup of $U(n)$. But, as mentioned in the previous section, the study of $U(n)$ can be reduced to the study of $SU(n)$, which is semi-simple.

Other non-semi-simple groups may be described as semi-direct products of two groups (Section 2.7) in which one is semi-simple and the other is a trivial case of representation theory, so attention has to be concentrated on the semi-simple group. Examples are the Euclidean groups $E_2 = T_2 \wedge R_2$, $E_3 = T_3 \wedge R_3$ and the Poincaré group $P = T_4 \wedge L$ where the translation groups T_2, T_3, and T_4 in two, three, and four dimensions are abelian and thus trivial in representation theory. The other factors of the semi-direct products, R_2, R_3 or the Lorentz group L are semi-simple.

Standard form of a semi-simple Lie algebra

The starting point of Cartan's classification is to obtain a standard coordinate system for the infinitesimal operators. As discussed in Section 5.3 these operators form an

r-dimensional linear vector space. We shall search for a basis in this space which gives the maximum number of generators which commute among themselves. The form taken by the commutators in this basis leads to the so-called *standard form* of a Lie algebra.

Following Cartan this amounts to solving the following eigenvalue problem

$$[A, X] = \rho X \tag{5.100}$$

where A is an *arbitrarily fixed* linear combination

$$A = a^\mu X_\mu \tag{5.101}$$

and X is an eigenvector corresponding to the eigenvalue ρ to be found

$$X = x^\mu X_\mu. \tag{5.102}$$

Using the definitions of A and X and the commutation relation (5.40) one rewrites (5.100) as

$$a^\mu x^\nu c_{\mu\nu}^\tau X_\tau = \rho x^\tau X_\tau. \tag{5.103}$$

Since X_τ are linearly independent one gets the following system of equations for x^ν

$$(a^\mu c_{\mu\nu}^\tau - \rho \delta_\nu^\tau) x^\nu = 0. \tag{5.104}$$

The existence of a non-trivial solution requires

$$\det |a^\mu c_{\mu\nu}^\tau - \rho \delta_\nu^\tau| = 0. \tag{5.105}$$

The secular equation (5.105) has r solutions, called *roots*. Some of them may be degenerate. Generally, r linearly independent eigenvectors may not exist if the secular equation has degenerate solutions. But for semi-simple groups Cartan showed that if A is chosen in such a way that the secular equation (5.105) has the maximum number of different solutions, in that case only $\rho = 0$ is degenerate. Moreover if $\rho = 0$ is ℓ times degenerate there are ℓ linearly independent eigenvectors corresponding to the eigenvalue $\rho = 0$, which commute among themselves. They are usually denoted by $H_1, H_2, ..., H_\ell$ and this subset of generators is called a *Cartan subalgebra*. The value of ℓ gives the *rank* of the semi-simple group. One has $\ell \geq 1$ because $[A, A] = 0$ gives $\ell = 1$ as the smallest ℓ value.

The eigenvectors corresponding to the non-vanishing $r - \ell$ roots are usually denoted by E_α. Hence the eigenvalue equations are

$$[A, H_i] = 0 \quad (i = 1, 2, \ldots, \ell) \tag{5.106}$$
$$[A, E_\alpha] = \alpha E_\alpha \quad (\alpha = 1, 2, \ldots, r - \ell) \tag{5.107}$$

where Latin subscripts are used for vectors of type H and greek subscripts for vectors of type E.

Note that $[A, A] = 0$ implies that A corresponds to $\rho = 0$ therefore it must be a combination of H_i, so we write

$$A = \lambda^i H_i \tag{5.108}$$

where summation over repeated indices is understood.

CLASSIFICATION OF SEMI-SIMPLE GROUPS

The commutator $[H_i, E_\alpha]$

One starts from the Jacobi identity

$$[A, [H_i, E_\alpha]] + [H_i, [E_\alpha, A]] + [E_\alpha, [A, H_i]] = 0. \tag{5.109}$$

Using (5.106) and (5.107) one gets

$$[A, [H_i, E_\alpha]] = \alpha[H_i, E_\alpha] \tag{5.110}$$

which shows that $[H_i, E_\alpha]$ is a vector of type E_α

$$[H_i, E_\alpha] = \alpha_i E_\alpha \tag{5.111}$$

which contains the structure constant

$$c_{i\alpha}^\tau = \alpha_i \delta_\alpha^\tau. \tag{5.112}$$

From (5.107), (5.108), and (5.111) one establishes that

$$\alpha = \lambda^i \alpha_i \tag{5.113}$$

from which one can interpret a root as a vector of covariant components α_i in an l-dimensional space.

The commutator $[E_\alpha, E_\beta]$

Here one starts from the Jacobi identity

$$[A, [E_\alpha, E_\beta]] + [E_\alpha, [E_\beta, A]] + [E_\beta, [A, E_\alpha]] = 0. \tag{5.114}$$

The relation (5.107) implies

$$[A, [E_\alpha, E_\beta]] = (\alpha + \beta)[E_\alpha, E_\beta] \tag{5.115}$$

which means that $[E_\alpha, E_\beta]$ is an eigenvector associated to $\alpha + \beta$ if the sum of α and β is also a root. If it is not

$$[E_\alpha, E_\beta] = 0 \tag{5.116}$$

or

$$c_{\alpha\beta}^\tau = 0 \quad \text{for } \tau \neq \alpha + \beta. \tag{5.117}$$

If $\alpha + \beta$ is a root there are two possibilities:

(1) $\alpha + \beta \neq 0$ in which case one writes

$$[E_\alpha, E_\beta] = N_{\alpha\beta} E_{\alpha+\beta} \tag{5.118}$$

where

$$N_{\alpha\beta} = c_{\alpha\beta}^{\alpha+\beta}; \tag{5.119}$$

(2) $\alpha + \beta = 0$ or $\beta = -\alpha$ in which case $[E_\alpha, E_{-\alpha}]$ must be a linear combination of H_i

$$[E_\alpha, E_{-\alpha}] = \alpha^i H_i \tag{5.120}$$

with the structure constant

$$c^i_{\alpha-\alpha} = \alpha^i. \tag{5.121}$$

The quantities α^i can be related to α_i as shown below. Equations (5.37) and (5.74) give

$$c^\tau_{\nu\sigma} = g^{\mu\tau} c_{\nu\sigma\mu}. \tag{5.122}$$

Taking $\nu = \alpha$, $\sigma = -\alpha$, and $\tau = i$ (5.122) becomes

$$c^i_{\alpha-\alpha} = g^{ki} c_{\alpha-\alpha k} \tag{5.123}$$

because $g^{k\alpha} = 0$ (see equation (5.128)). The symmetry of the tensor g^{ik}, the invariance of the structure constants under circular permutation of their indices (Exercise 5.2), and definition (5.37) lead to

$$\begin{aligned} c^i_{\alpha-\alpha} &= g^{ik} c_{k\alpha-\alpha} = g^{ik} c^\lambda_{k\alpha} g_{\lambda-\alpha} \\ &= g^{ik} c^\alpha_{k\alpha} g_{\alpha-\alpha}. \end{aligned} \tag{5.124}$$

The last equality holds due to (5.112). Moreover one can normalize E_α to give $g_{\alpha-\alpha} = 1$ (see for example Exercise 5.5). In this case, using (5.121) and (5.112) one obtains

$$\alpha^i = g^{ik} \alpha_k. \tag{5.125}$$

The quantities α^i are interpreted as the contravariant components of the root α.

All the commutation relations derived above give the standard form of a semi-simple Lie algebra, also referred to as the Cartan–Weyl basis. Here it is again:

$$[H_i, H_k] = 0 \quad (i, k = 1, 2, \ldots, \ell) \tag{5.126a}$$
$$[H_i, E_\alpha] = \alpha_i E_\alpha \tag{5.126b}$$
$$[E_\alpha, E_\beta] = N_{\alpha\beta} E_{\alpha+\beta} \quad (\alpha + \beta \neq 0) \tag{5.126c}$$
$$[E_\alpha, E_{-\alpha}] = \alpha^i H_i. \tag{5.126d}$$

The Cartan–Weyl basis is often used in physics. However there are other bases which have other advantages, as for example the Chevalley basis (Wybourne 1974, Chapter 8).

Properties of the metric tensor

It has already been mentioned in Section 5.4 that Cartan's criterion for a group to be semi-simple is to have

$$\det |g_{\rho\sigma}| \neq 0.$$

Let us see how this appears in the Cartan–Weyl basis. It is convenient to order the basis as

$$1, 2, \ldots, l, \alpha, -\alpha, \beta, -\beta, \ldots, \text{etc.}$$

Then, using (5.112) one has

$$g_{ik} = \sum_{\alpha,\beta} c_{i\alpha}^{\beta} c_{k\beta}^{\alpha} = \sum_{\alpha,\beta} \alpha_i \beta_k \delta_{\alpha\beta} = \sum_{\alpha} \alpha_i \alpha_k \tag{5.127}$$

and

$$g_{i\alpha} = c_{i\beta}^{\lambda} c_{\alpha\lambda}^{\beta} = \beta_i \delta_{\beta}^{\lambda} c_{\alpha\lambda}^{\beta} = \beta_i c_{\alpha\beta}^{\beta} = 0, \tag{5.128}$$

because $\alpha \neq 0$; and

$$\begin{aligned} g_{\alpha\beta} &= c_{\alpha i}^{\tau} c_{\beta\tau}^{i} + c_{\alpha-\alpha}^{i} c_{\beta i}^{-\alpha} + \sum_{\tau \neq 0} c_{\alpha\tau}^{\alpha+\tau} c_{\beta\,\alpha+\tau}^{\tau} \\ &= c_{i\alpha}^{\alpha} c_{\alpha\beta}^{i} + c_{\alpha-\alpha}^{i} c_{\beta i}^{-\alpha} + \sum_{\tau \neq 0} c_{\alpha\tau}^{\alpha+\tau} c_{\beta\,\alpha+\tau}^{\tau}. \end{aligned} \tag{5.129}$$

Both (5.112) and (5.117) show that one must have $\beta = -\alpha$ in order to obtain a non-vanishing value for $g_{\alpha\beta}$. As already mentioned, one can normalize E_α to give (see Exercise 5.5)

$$g_{\alpha-\alpha} = 1. \tag{5.130}$$

The tensor $g_{\rho\sigma}$ can therefore be written in the form

$$g_{\rho\sigma} = \begin{pmatrix} g_{ik} & & & 0 \\ & \begin{matrix} 0 & 1 \\ 1 & 0 \end{matrix} & 0 & 0 \\ & 0 & \begin{matrix} 0 & 1 \\ 1 & 0 \end{matrix} & 0 \\ 0 & & & \ddots \end{pmatrix}. \tag{5.131}$$

It follows that (5.73) is fulfilled if

$$\det |g_{ik}| \neq 0, \tag{5.132}$$

a condition which is satisfied by (5.127) when the sum extends over the entire ℓ-dimensional space.

Exercise 5.5 Find the normalization which satisfies (5.130) for the rotation group R_3.

Solution The generators of the rotation group (Section 6.1) are the components of the angular momentum, which satisfy the algebra

$$[L_a, L_b] = \mathrm{i}\, \varepsilon_{abc} L_c. \tag{1}$$

If one defines the operators H_i and E_α as

$$H_0 = L_3 \tag{2}$$

$$E_1 = \frac{1}{2}(L_1 + i L_2) \tag{3}$$

$$E_{-1} = \frac{1}{2}(L_1 - i L_2) \tag{4}$$

one finds the standard form

$$[H_0, E_{\pm 1}] = \pm E_{\pm 1} \tag{5}$$

$$[E_1, E_{-1}] = \frac{1}{2} H_0. \tag{6}$$

The group is of rank $\ell = 1$ and its algebra has two non-vanishing roots $\alpha = 1$ and $\alpha = -1$. The non-vanishing structure constants are

$$c^1_{0\,1} = 1, \qquad c^{-1}_{0\,-1} = -1, \qquad c^0_{1\,-1} = \frac{1}{2} \tag{7}$$

which give

$$g_{00} = c^1_{0\,1} c^1_{0\,1} + c^{-1}_{0\,-1} c^{-1}_{0\,-1} = 2 \tag{8}$$

$$g_{1\,-1} = c^1_{1\,0} c^0_{-1\,1} + c^0_{1\,-1} c^{-1}_{-1\,0} = 1 \tag{9}$$

i.e. the desired answer. Note that

$$\det |g_{\rho\sigma}| = -2 \tag{10}$$

i.e. the determinant is negative, which is a characteristic of compact groups as mentioned in Section 5.7.

Properties of the root vectors

The scalar product of two roots α and β is defined as

$$(\alpha, \beta) = \alpha^i \beta_i. \tag{5.133}$$

The classification of semi-simple groups given by Cartan is based on the properties of roots. These properties limit the number of roots to a few, as shown below. A list of properties can be formulated as follows.

1. If α is a root then $-\alpha$ is also a root. This property is expressed by (5.129) and the discussion following it, and is crucial for the semi-simple character of the group because $g_{\alpha\tau} = 0$ for $\tau \neq -\alpha$. Then if $-\alpha$ did not exist as a root the row of α in $g_{\alpha\tau}$ would vanish in $|g_{\alpha\tau}|$ and the Cartan criterion would be violated.

2. If α and β are root vectors then

$$2\frac{(\alpha, \beta)}{(\alpha, \alpha)} \tag{5.134}$$

is an integer.

3. If α and β are roots then so is

$$\beta - \alpha \frac{2(\alpha, \beta)}{(\alpha, \alpha)}. \qquad (5.135)$$

Properties 2 and 3 form the content of a *theorem* proved by Racah (1965) and here we follow this reference very closely. This theorem states that if α and γ are roots and $\alpha + \gamma$ is not a root, then there is a string of roots

$$\gamma, \gamma - \alpha, \ldots, \gamma - g\alpha \qquad (5.136)$$

with

$$g = \frac{2(\alpha, \gamma)}{(\alpha, \alpha)} = \text{integer}. \qquad (5.137)$$

In other words γ and $\gamma - g\alpha$ are the upper and lower limits of the string.

Since $\alpha + \gamma$ is not a root one has $[E_\alpha, E_\gamma] = 0$ and according to (5.126) one can write

$$\begin{aligned}
[E_{-\alpha}, E_\gamma] &= N_{-\alpha\gamma} E_{\gamma-\alpha} = E'_{\gamma-\alpha} \\
[E_{-\alpha}, E'_{\gamma-\alpha}] &= E'_{\gamma-2\alpha} \\
&\cdots\cdots\cdots\cdots\cdots\cdots \\
[E_{-\alpha}, E'_{\gamma-j\alpha}] &= E'_{\gamma-(j+1)\alpha} \\
&\cdots\cdots\cdots\cdots\cdots\cdots \\
[E_{-\alpha}, E'_{\gamma-g\alpha}] &= E'_{\gamma-(g+1)\alpha} = 0
\end{aligned} \qquad (5.138)$$

where the last commutator indicates that the process of subtracting α from γ must terminate after g steps. But the operator $E'_{\gamma-j\alpha}$ with $j < g$ can also be obtained from

$$[E_\alpha, E'_{\gamma-(j+1)\alpha}] = \mu_{j+1} E'_{\gamma-j\alpha}. \qquad (5.139)$$

To derive the constants μ_{j+1} we replace $E'_{\gamma-(j+1)\alpha}$ in (5.139) by using (5.138) and the Jacobi identity

$$\begin{aligned}
\mu_{j+1} E'_{\gamma-j\alpha} &= [E_\alpha, [E_{-\alpha}, E'_{\gamma-j\alpha}]] \\
&= -[E_{-\alpha}, [E'_{\gamma-j\alpha}, E_\alpha]] - [E'_{\gamma-j\alpha}, [E_\alpha, E_{-\alpha}]] \\
&= \mu_j E'_{\gamma-j\alpha} + \Sigma \, \alpha^i [H_i, E'_{\gamma-j\alpha}] \\
&= \{\mu_j + (\alpha, \gamma) - j(\alpha, \alpha)\} E'_{\gamma-j\alpha}
\end{aligned}$$

which leads to the recurrence relation

$$\mu_{j+1} = \mu_j + (\alpha, \gamma) - j(\alpha, \alpha). \qquad (5.140)$$

The quantity μ_0 is not defined by (5.140) because $\alpha + \gamma$ is not a root, so one can set $\mu_0 = 0$. By iteration one obtains

$$\mu_{j+1} = (j+1)(\alpha, \gamma) - \frac{j(j+1)}{2}(\alpha, \alpha).$$

or
$$\mu_j = j(\alpha, \gamma) - \frac{j(j-1)}{2}(\alpha, \alpha). \tag{5.141}$$

As indicated in (5.138) the chain terminates at $j = g$, and (5.139) still holds provided $\mu_{g+1} = 0$ which stipulates

$$(\alpha, \gamma) = \frac{g}{2}(\alpha, \alpha) \tag{5.142}$$

hence

$$\mu_j = \frac{j(g-j+1)}{2}(\alpha, \alpha) \tag{5.143}$$

with j and g non-negative integers and $(\alpha, \alpha) \neq 0$.

Let us show that for semi-simple groups one must always have $(\alpha, \alpha) \neq 0$. If (α, α) were zero, according to (5.142) α would be orthogonal to every root. Then one would have

$$\alpha^1 \alpha_1 + \alpha^2 \alpha_2 + \cdots + \alpha^\ell \alpha_\ell = 0$$
$$\alpha^1 \beta_1 + \alpha^2 \beta_2 + \cdots + \alpha^\ell \beta_\ell = 0$$
$$\cdots\cdots\cdots\cdots\cdots\cdots\cdots\cdots\cdots$$
$$\alpha^1 \gamma_1 + \alpha^2 \gamma_2 + \cdots + \alpha^\ell \gamma_\ell = 0$$

which is a system of $r - \ell$ equations where the $r - \ell$ non-vanishing roots have been named $\alpha, \beta, \ldots, \gamma$. Multiplying the first equation by α_i, the second by β_i and the last by $\gamma_i (i = 1, 2, \ldots, \ell)$ and adding them together for fixed i, by (5.127) one obtains the following system of ℓ equations for the ℓ unknown quantities $\alpha^1, \alpha^2, \ldots, \alpha^\ell$:

$$\alpha^1 g_{11} + \alpha^2 g_{12} + \cdots \alpha^\ell g_{1\ell} = 0$$
$$\alpha^1 g_{21} + \alpha^2 g_{22} + \ldots \alpha^\ell g_{2\ell} = 0$$
$$\cdots\cdots\cdots\cdots\cdots\cdots\cdots\cdots\cdots$$
$$\alpha^1 g_{\ell 1} + \alpha^2 g_{\ell 2} + \cdots \alpha^\ell g_{\ell\ell} = 0.$$

A non-trivial solution exists only if $\det |g_{ik}| = 0$, but this violates Cartan's criterion. Hence, if α is a non-vanishing solution, one must have $(\alpha, \alpha) \neq 0$. This allows us to rewrite (5.142) as (5.137) and to finally obtain the desired string (5.136).

The geometrical interpretation of this result is that, starting with a root γ a new root is obtained by reflecting the root γ into the hyperplane orthogonal to α, as schematically illustrated below.

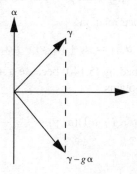

CLASSIFICATION OF SEMI-SIMPLE GROUPS

The reflection of γ produces the root $\gamma - g\alpha$ because the projection of $\gamma - g\alpha$ along the direction of α is

$$(\gamma, \alpha) - g(\alpha, \alpha) = -(\gamma, \alpha)$$

due to (5.137).

The reflections by which roots go to roots are called Weyl reflections. They form a group, called the Weyl group, which is of finite order. The string of roots (5.136) is invariant under a Weyl reflection.

We can now set $\gamma = \beta + j\alpha$ with $j \geq 0$ (integer) in order to understand property 3. The string (5.136) becomes

$$\beta + j\alpha, \ \beta + (j-1)\alpha, \ \cdots, \ \beta, \ \cdots, \ \beta - k\alpha \qquad (5.144)$$

with $k = g - j =$ integer. The value of g of (5.137) becomes

$$g = \frac{2(\alpha, \beta) + 2j(\alpha, \alpha)}{(\alpha, \alpha)} = 2\frac{(\alpha, \beta)}{(\alpha, \alpha)} + 2j \qquad (5.145)$$

hence $\dfrac{2(\alpha, \beta)}{(\alpha, \alpha)}$ is also an integer and this proves property 2. Since the string ends at $\beta - k\alpha$ where

$$k = g - j = 2\frac{(\alpha, \beta)}{(\alpha, \alpha)} + j \qquad (5.146)$$

it must contain the root $\beta - 2\dfrac{(\alpha, \beta)}{(\alpha, \alpha)}\alpha$ because obviously $2\dfrac{(\alpha, \beta)}{(\alpha, \alpha)} < k$. This is precisely the meaning of property 3.

Consequence 1

If α is a root, 2α cannot be a root since $[E_\alpha, E_\alpha] = 0$. Furthermore no multiple $k\alpha$ with $k > 2$ is a root because the string (5.136) with $\gamma = \alpha$ terminates at α on the left. Then, the whole string (5.136) with $\gamma = \alpha$ will be

$$\alpha, \ 0, \ -\alpha \qquad (5.147)$$

because $g = 2$.

Consequence 2

Taking $\beta \neq \pm \alpha$ one can build an α-string containing β in the form (5.144). One can prove that such a string consists of at most four roots or in other words the only possible values of $2\dfrac{(\alpha, \beta)}{(\alpha, \alpha)}$ are

$$2\frac{(\alpha, \beta)}{(\alpha, \alpha)} = 0, \ \pm 1, \ \pm 2, \ \pm 3 \qquad (5.148)$$

and the values of $2\dfrac{(\alpha, \beta)}{(\alpha, \alpha)}$ are called *Cartan integers*. Let us assume that the α-string has five roots. It is convenient to label them as

$$\beta + 2\alpha, \ \beta + \alpha, \ \beta, \ \beta - \alpha, \ \beta - 2\alpha. \qquad (5.149)$$

From consequence 1 it follows that 2α and $2(\beta + \alpha)$ are not roots if α and $\beta + \alpha$ are roots. These can be written as

$$2\alpha = (\beta + 2\alpha) - \beta \quad \text{and} \quad 2(\beta + \alpha) = (\beta + 2\alpha) + \beta$$

which shows that the β-string containing $\beta + 2\alpha$ consists of a single member, $\beta + 2\alpha$ or equivalently $(\beta + 2\alpha, \beta) = 0$. Similarly neither $-2\alpha = \beta - 2\alpha - \beta$ nor $2(\beta - \alpha) = \beta - 2\alpha + \beta$ are roots, so that $(\beta - 2\alpha, \beta) = 0$ holds too. Adding the two vanishing scalar products one obtains $(\beta, \beta) = 0$ or $\beta = 0$, which contradicts the initial assumption. Consequently one abandons either the root $\beta + 2\alpha$ or the root $\beta - 2\alpha$ and one remains with only four roots, which imposes the constraint

$$j + k \leq 3 \tag{5.150}$$

in (5.144). On the other hand, (5.146) yields

$$2\frac{(\alpha, \beta)}{(\alpha, \alpha)} = k - j \tag{5.151}$$

which, together with (5.150), leads to (5.148).

Normalization

The structure constants $N_{\alpha\beta}$ defined by (5.119) give the relative magnitude of the operators E.

A by-product of the theorem proved in the previous subsection is that $N_{\alpha\beta}$ does not vanish if the sum $\alpha + \beta$ of two roots is also a root. From (5.138), (5.139), (5.143), and (5.146), one can establish that

$$N_{-\alpha, \alpha+\beta} N_{\alpha\beta} = \mu_j = \frac{j(k+1)}{2}(\alpha, \alpha). \tag{5.152}$$

For $j \geq 1$ the right-hand side is non-zero, thus $N_{\alpha\beta} \neq 0$.

Some useful relations can be established between the constants $N_{\alpha\beta}$. From (5.118) one has

$$N_{\alpha\beta} = -N_{\beta\alpha}. \tag{5.153}$$

Next, one can choose[†] $H_i = H_i^+$ and $E_\alpha^+ = E_{-\alpha}$. As the structure constants are real one obtains from (5.126c)

$$N_{\alpha\beta} = -N_{-\alpha, -\beta}. \tag{5.154}$$

Furthermore, one can start from the Jacobi identity

$$[E_\alpha, [E_{-\alpha-\beta}, E_\beta]] + [E_{-\alpha-\beta}, [E_\beta, E_\alpha]] + [E_\beta, [E_\alpha, E_{-\alpha-\beta}]] = 0.$$

[†] Another possible choice is $E_\alpha^+ = -E_\alpha$. This gives $N_{\alpha\beta} = N_{-\alpha, -\beta}$. The hermiticity of H_i can be accepted for physical reasons. It corresponds to observables; see the following chapters.

CLASSIFICATION OF SEMI-SIMPLE GROUPS

First using (5.126c) one gets

$$N_{-\alpha-\beta,\beta}[E_\alpha, E_{-\alpha}] - N_{\alpha\beta}[E_{-\alpha-\beta}, E_{\alpha+\beta}] + N_{\alpha,-\alpha-\beta}[E_\beta, E_{-\beta}] = 0$$

and by (5.126d) and the linear independence of H_i one obtains

$$(N_{\alpha\beta} + N_{-\alpha-\beta,\beta})\alpha^i + (N_{\alpha\beta} + N_{\alpha,-\alpha-\beta})\beta^i = 0$$

which is valid for any α^i and β^i. Hence

$$N_{\alpha\beta} = N_{\beta,-\alpha-\beta} = N_{-\alpha-\beta,\alpha}. \tag{5.155}$$

It is enough to choose the phase of $N_{\alpha\beta}$. The other structure constants will be determined through (5.154) and (5.155). As an example, see SU(3) in Chapter 8.

Root diagrams

We have seen that each non-vanishing root can be interpreted as a vector of α_i covariant components in an ℓ-dimensional space. The graphical representation of root vectors is called a *root diagram*. The complete classification of simple Lie groups found algebraically by Cartan can also be obtained by using the diagram method. This is due to Van der Waerden, who showed that to every diagram there corresponds only one infinitesimal Lie group. The classification of semi-simple groups reduces in the end to the classification of simple groups because, as Cartan showed, any semi-simple group is a direct product of simple groups.

Let us consider two roots, α and β, and denote

$$2\frac{(\alpha, \beta)}{(\alpha, \alpha)} = m, \quad 2\frac{(\alpha, \beta)}{(\beta, \beta)} = n \tag{5.156a}$$

where m and n are integers, as shown earlier in this section. The angle ϕ between the two roots is given by

$$\cos^2 \phi = \frac{(\alpha, \beta)^2}{(\alpha, \alpha)(\beta, \beta)} = \frac{mn}{4}. \tag{5.156b}$$

From the Schwartz inequality $(\alpha, \beta)^2 \leq (\alpha, \alpha)(\beta, \beta)$ it follows that

$$\frac{mn}{4} \leq 1. \tag{5.157}$$

The sets $m = 4, n = 1$ and $m = 1, n = 4$ are excluded by (5.148). The remaining cases are shown in Table 5.3.

Let us first consider groups of rank $\ell = 1$. The root diagram is one-dimensional, and according to the discussion above, there are just two non-vanishing roots $\pm\alpha$. Hence there is only one possible diagram, shown in Fig. 5.1. This corresponds to $\phi = 0$ ($\beta = \alpha$) and it represents, for example, the algebra so(3) (see Exercise 5.5).

Figure 5.1 Root diagram of the Lie algebra A_1, B_1 or C_1.

Table 5.3 The angle ϕ between two consecutive non-vanishing roots α and β and the ratio of their lengths.

m	n	ϕ	$\dfrac{|\beta|}{|\alpha|}$
0	0	90°	undetermined
1	1	60°	1
2	1	45°	$\sqrt{2}$
1	2	45°	$\dfrac{1}{\sqrt{2}}$
2	2	0°	1
3	1	30°	$\sqrt{3}$
1	3	30°	$\dfrac{1}{\sqrt{3}}$

The other cases of Table 5.3 can be illustrated for groups of rank $\ell = 2$. First note that $m = n = 0$ ($\phi = 90$) does not represent a simple group but a direct product of simple groups. The corresponding diagram is shown in Fig. 5.2.

One can view this as a diagram with two orthogonal parts, each corresponding to an invariant subalgebra. Let us fix α such that $\alpha = \alpha_1$ and β such that $\beta = \beta_2$. Then E_α, $E_{-\alpha}$ and H_1 form one invariant subalgebra, $\mathscr{L}_1$, and E_β, $E_{-\beta}$, and H_2 another one, $\mathscr{L}_2$. Then

$$[E_\alpha, E_\beta] = 0 \quad \text{because } \alpha + \beta \text{ is not a root}$$
$$[H_1, E_\beta] = 0 \quad \text{because } \beta_1 = 0 \tag{5.158}$$
$$[H_2, E_\alpha] = 0 \quad \text{because } \alpha_2 = 0.$$

The diagram in Fig 5.2 represents the algebra $\mathscr{L}$ of SO(4) which is the direct sum $\mathscr{L} = \mathscr{L}_1 \oplus \mathscr{L}_2$ of two so(3) algebras. In the notation of Cartan's thesis, Fig. 5.2 is a diagram of type D_2.

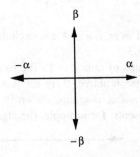

Figure 5.2 Root diagram of the Lie algebra D_2.

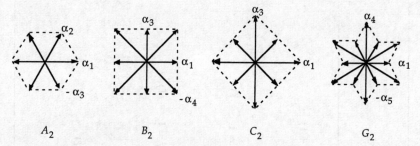

Figure 5.3 Root diagrams of the Lie algebras A_2, B_2, C_2, and G_2.

The interesting cases for $\ell = 2$ remain: $\phi = 30°, 45°$, and $60°$. The corresponding diagrams are shown in Fig. 5.3. The diagrams are labelled by the letters used by Cartan in his thesis. The diagrams A_2, B_2, C_2 and G_2 are associated with the roots of su(3), so(5), and sp(4) algebras and of the Lie algebra of an exceptional group called G_2, respectively. The diagram C_2, isomorphic to B_2, differs from C_2 by a rotation of $45°$.

In general, according to Cartan, there are four infinite sets of simple (complex) Lie algebras having root diagrams denoted by A_ℓ, B_ℓ, C_ℓ, and D_ℓ characterized by $\phi = 60°, 45°, 45°$, and $90°$, respectively. They are called *classical* Lie algebras and their realizations in terms of matrices give the linear Lie groups indicated below:

(1) for A_ℓ, $\ell = 1, 2, 3, \cdots$ $SU(\ell+1)$ or $SL(\ell+1)$;
(2) for B_ℓ, $\ell = 1, 2, 3, \cdots$ $SO(2\ell+1)$;
(3) for C_ℓ, $\ell = 1, 2, 3, \cdots$ $Sp(2\ell)$;
(4) for D_ℓ, $\ell = 3, 4, 5, \cdots$ $SO(2\ell)$.

Apart from that there are another five possibilities corresponding to Cartan's *exceptional* simple Lie groups E_6, E_7, E_8, F_4, and G_2.

For any ℓ the roots can be obtained by starting from a set e_i of mutually orthogonal unit vectors and proceeding as indicated below in detail. The result is summarized in the last column of Table 5.4. The case $\ell = 1$ is a limiting case and is excluded from the discussion. The diagrams A_1, B_1 and C_1 are identical (Fig. 5.1).

A_ℓ. The roots can be obtained from $\ell+1$ unit vectors by forming $2C^2_{\ell+1} = \ell(\ell+1)$ differences of type $e_i - e_k$. The group order is $\ell(\ell+1) + \ell = \ell^2 + 2\ell$, consistent with the order of $SU(n)$ in Table 5.2 if $n = \ell+1$. In particular, for A_2 the unit vectors are $e_1 = (1, 0, 0)$, $e_2 = (0, 1, 0)$, $e_3 = (0, 0, 1)$, and the roots are

$$\begin{aligned} \alpha_1 &= e_1 - e_2 = (1, -1, 0) & \alpha_4 &= -\alpha_1 \\ \alpha_2 &= e_2 - e_3 = (0, 1, -1) & \alpha_5 &= -\alpha_2 \\ \alpha_3 &= e_3 - e_1 = (-1, 0, 1) & \alpha_6 &= -\alpha_3. \end{aligned} \qquad (5.159)$$

Table 5.4 Dynkin diagrams and roots of the classical Lie groups.

Cartan's notation	Group	Group order	Dynkin diagram	Roots
A_ℓ	$SU(\ell+1)$	$\ell(\ell+2)$	$\underset{\alpha_1}{\underset{e_1-e_2}{\circ}} - \underset{\alpha_2}{\underset{e_2-e_3}{\circ}} - \cdots - \underset{\alpha_{\ell-1}}{\underset{e_{\ell-1}-e_\ell}{\circ}} - \underset{\alpha_\ell}{\underset{e_\ell-e_{\ell+1}}{\circ}}$	$e_i - e_j \ (i,j=1,\cdots,\ell+1)$
B_ℓ	$SO(2\ell+1)$	$\ell(2\ell+1)$	$\underset{\alpha_1}{\underset{e_1-e_2}{\circ}} - \underset{\alpha_2}{\underset{e_2-e_3}{\circ}} - \cdots - \underset{\alpha_{\ell-1}}{\underset{e_{\ell-1}-e_\ell}{\circ}} = \underset{\alpha_\ell}{\underset{e_\ell}{\bullet}}$	$\pm 2e_i$ and $\pm e_i \pm e_j \ (i,j=1,\cdots,\ell)$
C_ℓ	$Sp(2\ell)$	$\ell(2\ell+1)$	$\underset{\alpha_1}{\underset{e_1-e_2}{\bullet}} - \underset{\alpha_2}{\underset{e_2-e_3}{\bullet}} - \cdots - \underset{\alpha_{\ell-1}}{\underset{e_{\ell-1}-e_\ell}{\bullet}} = \underset{\alpha_\ell}{\underset{2e_\ell}{\circ}}$	$\pm 2e_i$ and $\pm e_i \pm e_j \ (i,j=1,\cdots,\ell)$
D_ℓ	$SO(2\ell)$ $\ell \geq 3$	$\ell(2\ell-1)$	$\underset{\alpha_1}{\underset{e_1-e_2}{\circ}} - \underset{\alpha_2}{\underset{e_2-e_3}{\circ}} - \cdots - \underset{\alpha_{\ell-2}}{\underset{e_{\ell-2}-e_{\ell-1}}{\circ}} \diagdown^{\underset{\alpha_{\ell-1}}{\underset{e_{\ell-1}-e_\ell}{\circ}}}_{\underset{\alpha_\ell}{\underset{e_{\ell-1}+e_\ell}{\circ}}}$	$\pm e_i \pm e_j \ (i,j=1,\cdots,\ell)$

One has, for example,
$$\cos^2 \phi = \frac{(e_1 - e_2, e_2 - e_3)^2}{(e_1 - e_2, e_1 - e_2)(e_2 - e_3, e_2 - e_3)} = \frac{1}{4}$$
hence $\phi = 60°$. The vectors α_i lie in a plane because $\alpha_1 + \alpha_2 + \alpha_3 = 0$, and together with $-\alpha_1$, $-\alpha_2$ and $-\alpha_3$, form the diagram A_2 of Fig. 5.3.

B_ℓ. The root vectors can simply be constructed as $\pm e_i$ and $\pm e_i \pm e_k$ ($i, k = 1, 2, \ldots, \ell$). There are in all $2\ell + 4C_\ell^2 = 2\ell^2$ such vectors and the group order is $\ell + 2\ell^2 = \ell(2\ell + 1)$ consistent with the order of SO(n) in Table 5.2 if $n = 2\ell + 1$. In particular for B_2 the roots can be labelled as

$$\begin{aligned}
\alpha_1 &= e_1 = (1, 0) & \alpha_5 &= -\alpha_1 = (-1, 0) \\
\alpha_2 &= e_1 + e_2 = (1, 1) & \alpha_6 &= -\alpha_2 = (-1, -1) \\
\alpha_3 &= e_2 = (0, 1) & \alpha_7 &= -\alpha_3 = (0, -1) \\
\alpha_4 &= -e_1 + e_2 = (-1, 1) & \alpha_8 &= -\alpha_4 = (1, -1).
\end{aligned} \quad (5.160)$$

C_ℓ. The procedure is the same as for B_ℓ but the realization of this algebra is the group Sp(2ℓ). For $\ell = 2$, C_2 differs from B_2 by a rotation of $45°$, as shown in Fig. 5.3, which leads to a relabelling of the roots (5.160). One says that the diagrams B_2 and C_2 are isomorphic, or alternatively, the algebras so(5) and sp(4) are isomorphic.

D_ℓ. The non-vanishing root vectors are described by $\pm e_i \pm e_k$ ($i, k = 1, 2, \ldots, \ell$) and the group order is $\ell + 4C_\ell^2 = \ell(2\ell - 1)$, i.e. the order of SO($2\ell$). For $\ell = 2$ the group is not simple, as discussed above. For $\ell = 3$ one can show that the diagram D_3 is isomorphic to A_3. In the e_i ($i = 1, \ldots, 4$) vector space of A_3 one makes the transformation

$$e'_1 = \frac{1}{2}(e_1 + e_2 - e_3 - e_4)$$

$$e'_2 = \frac{1}{2}(e_1 - e_2 + e_3 - e_4)$$

$$e'_3 = \frac{1}{2}(e_1 - e_2 - e_3 + e_4)$$

$$e'_4 = \frac{1}{2}(e_1 + e_2 + e_3 + e_4)$$

under which the roots of A_3 become

$$\begin{aligned}
\pm(e'_1 - e'_2) &= \pm(e_2 - e_3), & \pm(e'_2 - e'_3) &= \pm(e_3 - e_4), \\
\pm(e'_1 - e'_3) &= \pm(e_2 - e_4), & \pm(e'_2 - e'_4) &= \mp(e_2 + e_4), \\
\pm(e'_1 - e'_4) &= \mp(e_3 + e_4), & \pm(e'_3 - e'_4) &= \mp(e_2 + e_3),
\end{aligned} \quad (5.161)$$

i.e. are the same as those of D_3. This implies that the algebras su(4) and so(6) are isomorphic. For $\ell \geq 4$ the D_ℓ diagrams correspond to non-isomorphic Lie algebras.

In all the cases discussed above, an isomorphism between two algebras implies a homomorphic mapping between the corresponding linear Lie groups. For example SU(2) → SO(3) and SU(4) → SO(6).

The root diagrams of the exceptional Lie algebras G_2, F_4, E_6, E_7, and E_8 can also be obtained from the set of unit vectors e_i (see Racah 1965). The result is summarized in the last column of Table 5.5. The exceptional groups E_6, E_7, E_8, and F_4 have roots in common with A_5, A_7, D_8, and B_4, respectively, plus other roots as indicated in the table. The order is easy to calculate. Let us take for example E_6. There are 30 non-vanishing roots in common with A_5 plus 42 roots as given by the last column of Table 5.5, plus six zero-roots (the rank is equal to the number of simple roots, introduced in the next section) which give 78 for the group order.

Dynkin diagrams

For groups of rank $\ell > 2$, diagrams as shown in Fig. 5.3 can no longer be drawn. However, such groups can be represented by other two-dimensional diagrams which contain the essential information about the root vectors. These diagrams were invented by Dynkin (1947) and are valid for algebras of any rank.

The key to Dynkin diagrams is that one can find all roots from a minimum system of simple roots, to be defined below.

First we introduce the notion of *positive root*. A root is said to be positive with respect to a chosen basis if the first non-vanishing component is positive. The notation for a positive root α is '$\alpha > 0$'. Examples of positive roots are α_1, α_2, and $-\alpha_3$ for A_2 and α_1, α_2, α_3, and $-\alpha_4$ for B_2 in Fig. 5.3. In general half of the non-vanishing roots are positive.

A non-vanishing root is *simple* if it is positive but cannot be expressed as a sum of two positive roots. In the case of A_2 only the roots α_2 and $-\alpha_3$ are simple. For the diagram B_2, α_3, and $-\alpha_4$ are simple.

The simple roots have the following important properties.

(1) The number of simple roots is equal to the rank l of the group (for a proof, see for example Cornwell 1984, Appendix E, Section 9).

(2) If α and β are two simple roots and $\alpha \neq \beta$ then

   (a) their difference is not a root. This property can be easily understood. If the difference $\alpha - \beta$ is positive then writing $\alpha = (\alpha - \beta) + \beta$ implies that α cannot be simple. If $\beta - \alpha$ is positive, as $\beta = (\beta - \alpha) + \alpha$, it follows that β cannot be simple either. In both cases there is a conflict with the initial assumption.

   (b) $(\alpha, \beta) \leq 0$ (for proof see, for example, Cornwell 1984, Chapter 13, Section 7). Combined with (5.148) this gives

$$2\frac{(\alpha, \beta)}{(\alpha, \alpha)} = 0, -1, -2, -3 \qquad (5.162)$$

Group	Group order	Dynkin diagram	Roots
G_2	14	●═○ α_2 α_1	$e_i - e_j$ ($i,j = 1, 2, 3; i \neq j$) $\pm 2e_i \mp e_j \mp e_k$ ($i,j,k = 1, 2, 3; i \neq j \neq k$)
F_4	52	○—●═●—○ α_1 α_2 α_3 α_4	As for B_4 plus 16 roots of type $\frac{1}{2}(\pm e_1 \pm e_2 \pm e_3 \pm e_4)$
E_6	78	○—○—○—○—○ with α_6 branch α_1 α_2 α_3 α_4 α_5, α_6	As for A_5 plus $\pm\sqrt{2}e_7$ and 40 roots of type $\frac{1}{2}(\pm e_1 \pm e_2 \pm e_3 \pm e_4 \pm e_5 \pm e_6) \pm e_7/\sqrt{2}$ (an arbitrary choice of three + and three − signs in bracket)
E_7	133	○—○—○—○—○—○ with α_7 branch α_1 α_2 α_3 α_4 α_5 α_6, α_7	As for A_7 plus 70 roots of type $\frac{1}{2}(\pm e_1 \pm e_2 \pm e_3 \pm e_4 \pm e_5 \pm e_6 \pm e_7 \pm e_8)$ (an arbitrary choice of four + and four − signs in bracket)
E_8	248	○—○—○—○—○—○—○ with α_8 branch α_1 α_2 α_3 α_4 α_5 α_6 α_7, α_8	As for D_8 plus 128 roots of type $\frac{1}{2}(\pm e_1 \pm e_2 \pm e_3 \pm e_4 \pm e_5 \pm e_6 \pm e_7 \pm e_8)$ (with an even number of + signs)

from which it follows that the angle ϕ between two simple roots is 90°, 120°, 135°, or 150°.

In a Dynkin diagram each simple root is represented by a circle and two circles are joined by $n\,m$ lines, where the product $n\,m$ is defined as in (5.156b). The four possibilities are shown in Table 5.6 where $(\alpha, \alpha) \leq (\beta, \beta)$.

Table 5.6 Simple roots and Dynkin's rule.

$n\,m$	ϕ	$\dfrac{(\beta, \beta)}{(\alpha, \alpha)}$	Dynkin diagram	Algebra
0	90°		○ ○	D_2
1	120°	1	○—○	A_2
2	135°	2	○=○	B_2, C_2
3	150°	3	○≡○	G_2

Note that circles corresponding to orthogonal roots are unconnected. The convention is that circles corresponding to the shortest roots are filled and those corresponding to the longest ones are left empty. For example, in the notation of Fig. 5.3, the Dynkin diagrams for A_2, B_2, and G_2 are

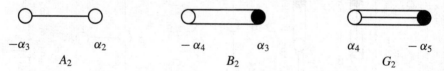

The Dynkin diagrams of the classical and exceptional Lie algebras are given in Tables 5.4 and 5.5, respectively.

In drawing Dynkin diagrams the following rules must be observed (for a proof see, for example, Elliott and Dawber 1979, Section 20.4).

(1) There are only connected diagrams. By construction, lack of connecting lines means that roots in one part are orthogonal to roots in another part of the diagram and this implies that the group is not simple, as for D_2.
(2) There can be no loops.
(3) There are at most three lines emerging from a circle. The three-line case appears

in all diagrams except for A_ℓ. This a rule excludes pieces of type

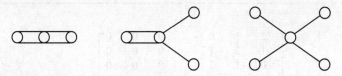

By its extension one can show that a Dynkin diagram can have at most one double line ⊙═⊙ or a junction ─⊂ but never both.

(4) A string with one junction and p, q, and r roots in each of the three legs must obey the constraint

$$p+q+r+1=\ell \qquad (5.163)$$

where ℓ is the rank of the group. This rule follows from the property that the number of simple roots is equal to the rank of the group. The diagrams D_ℓ, E_6, E_7, and E_8 obey this rule. Note that the angle between the legs of a junction has no significance.

From the Dynkin diagram one can construct the *Cartan matrix* A of each algebra. This matrix plays a crucial role in determining all non-vanishing roots from the simple roots.

The Cartan matrix is an $\ell \times \ell$ matrix, the elements of which are defined in terms of the simple roots $\alpha_1, \alpha_2, \ldots \alpha_\ell$ as

$$A_{ik} = \frac{2(\alpha_i, \alpha_k)}{(\alpha_i, \alpha_i)}. \qquad (5.164)$$

Obviously $A_{ii} = 2$ for all $i = 1, 2, \ldots, \ell$ and from (5.162) it follows that for the non-diagonal elements the only possible values are 0, -1, -2, or -3. If $A_{ik} = 0$ ($i \neq k$) it follows that $A_{ki} = 0$ also, because $A_{ik} \sim (\alpha_i, \alpha_k)$, $A_{ki} \sim (\alpha_k, \alpha_i) = (\alpha_i, \alpha_k)$. For the other non-zero off-diagonal matrix elements the equality $A_{ik} = A_{ki}$ does not necessarily hold, so that the Cartan matrices are not always symmetric. Using Tables 5.4 and 5.5 the Cartan matrices can be readily obtained. They are exhibited in Table 5.7 for both the classical and exceptional Lie groups (see also Exercise 5.6).

Exercise 5.6 Find the Cartan matrix of D_3.

Solution From Table 5.4 we see that the Dynkin diagram of D_3 is

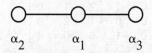

which looks like that for A_3 but with α_1 and α_2 interchanged. The matrix elements (5.164) should be rewritten as

$$A_{ik} = 2\frac{|\alpha_k|}{|\alpha_i|}\cos\phi.$$

Here $|\alpha_1| = |\alpha_2| = |\alpha_3|$. The roots α_2 and α_3 are orthogonal so that $A_{23} = A_{32} = 0$.

Table 5.7 Cartan matrices.

$$A_\ell : \quad A = \begin{bmatrix} 2 & -1 & 0 & \cdots & 0 & 0 & 0 \\ -1 & 2 & -1 & \cdots & 0 & 0 & 0 \\ 0 & -1 & 2 & \cdots & 0 & 0 & 0 \\ \vdots & \vdots & \vdots & & \vdots & \vdots & \vdots \\ 0 & 0 & 0 & \cdots & 2 & -1 & 0 \\ 0 & 0 & 0 & \cdots & -1 & 2 & -1 \\ 0 & 0 & 0 & \cdots & 0 & -1 & 2 \end{bmatrix}$$

$$B_\ell : \quad A = \begin{bmatrix} 2 & -1 & 0 & \cdots & 0 & 0 & 0 \\ -1 & 2 & -1 & \cdots & 0 & 0 & 0 \\ 0 & -1 & 2 & \cdots & 0 & 0 & 0 \\ \vdots & \vdots & \vdots & & \vdots & \vdots & \vdots \\ 0 & 0 & 0 & \cdots & 2 & -1 & 0 \\ 0 & 0 & 0 & \cdots & -1 & 2 & -1 \\ 0 & 0 & 0 & \cdots & 0 & -2 & 2 \end{bmatrix}$$

$$C_\ell : \quad A = \begin{bmatrix} 2 & -1 & 0 & \cdots & 0 & 0 & 0 \\ -1 & 2 & -1 & \cdots & 0 & 0 & 0 \\ 0 & -1 & 2 & \cdots & 0 & 0 & 0 \\ \vdots & \vdots & \vdots & & \vdots & \vdots & \vdots \\ 0 & 0 & 0 & \cdots & 2 & -1 & 0 \\ 0 & 0 & 0 & \cdots & -1 & 2 & -2 \\ 0 & 0 & 0 & \cdots & 0 & -1 & 2 \end{bmatrix}$$

$$D_\ell(\ell \geq 3): \quad A = \begin{bmatrix} 2 & -1 & 0 & \cdots & 0 & 0 & 0 & 0 \\ -1 & 2 & -1 & \cdots & 0 & 0 & 0 & 0 \\ 0 & -1 & 2 & \cdots & 0 & 0 & 0 & 0 \\ \vdots & \vdots & \vdots & & \vdots & \vdots & \vdots & \vdots \\ 0 & 0 & 0 & \cdots & 2 & -1 & 0 & 0 \\ 0 & 0 & 0 & \cdots & -1 & 2 & -1 & -1 \\ 0 & 0 & 0 & \cdots & 0 & -1 & 2 & 0 \\ 0 & 0 & 0 & \cdots & 0 & -1 & 0 & 2 \end{bmatrix}$$

$$E_6 : \quad A = \begin{bmatrix} 2 & -1 & 0 & 0 & 0 & 0 \\ -1 & 2 & -1 & 0 & 0 & 0 \\ 0 & -1 & 2 & -1 & 0 & -1 \\ 0 & 0 & -1 & 2 & -1 & 0 \\ 0 & 0 & 0 & -1 & 2 & 0 \\ 0 & 0 & -1 & 0 & 0 & 2 \end{bmatrix}$$

$$E_7 : \quad A = \begin{bmatrix} 2 & -1 & 0 & 0 & 0 & 0 & 0 \\ -1 & 2 & -1 & 0 & 0 & 0 & 0 \\ 0 & -1 & 2 & -1 & 0 & 0 & -1 \\ 0 & 0 & -1 & 2 & -1 & 0 & 0 \\ 0 & 0 & 0 & -1 & 2 & -1 & 0 \\ 0 & 0 & 0 & 0 & -1 & 2 & 0 \\ 0 & 0 & -1 & 0 & 0 & 0 & 2 \end{bmatrix}$$

CLASSIFICATION OF SEMI-SIMPLE GROUPS

Table 5.7 *cont.*

$$E_8: \quad A = \begin{bmatrix} 2 & -1 & 0 & 0 & 0 & 0 & 0 & 0 \\ -1 & 2 & -1 & 0 & 0 & 0 & 0 & 0 \\ 0 & -1 & 2 & -1 & 0 & 0 & 0 & -1 \\ 0 & 0 & -1 & 2 & -1 & 0 & 0 & 0 \\ 0 & 0 & 0 & -1 & 2 & -1 & 0 & 0 \\ 0 & 0 & 0 & 0 & -1 & 2 & -1 & 0 \\ 0 & 0 & 0 & 0 & 0 & -1 & 2 & 0 \\ 0 & 0 & -1 & 0 & 0 & 0 & 0 & 2 \end{bmatrix}$$

$$F_4: \quad A = \begin{bmatrix} 2 & -1 & 0 & 0 \\ -1 & 2 & -1 & 0 \\ 0 & -2 & 2 & -1 \\ 0 & 0 & -1 & 2 \end{bmatrix}$$

$$G_2: \quad A = \begin{bmatrix} 2 & -1 \\ -3 & 2 \end{bmatrix}$$

In Table 7.3 of Wybourne (1974), the entries for B_ℓ and F_4 are incorrect.

According to Table 5.6 the angle of (α_1, α_2) and of (α_1, α_3) is $120°$. So the matrix is

$$A = \begin{vmatrix} 2 & -1 & -1 \\ -1 & 2 & 0 \\ -1 & 0 & 2 \end{vmatrix}.$$

The Cartan matrix can be used to construct the complete system of non-vanishing roots. It is enough to construct the positive roots. Any positive root $\beta > 0$ can be written as a linear combination of simple roots (Cornwell 1984, Chapter 13, Section 7)

$$\beta = \sum_{j=1}^{\ell} k_j \alpha_j \tag{5.165}$$

where the coefficients k_j are non-negative integers. The quantity $\sum_{j=1}^{\ell} k_j$ is called the level of the root β. The positive roots of level 1 are the simple roots.

The procedure is iterative. One supposes that all positive roots of level k are known and one constructs the roots of level $k + 1$. Suppose β is a root of level k. Then one must find those α_i for which $\beta + \alpha_i$ is a root. But $\beta + \alpha_i$ is a root if the α_i-string containing β is of type (5.144), i.e.

$$\beta - p\alpha_i, \cdots, \beta, \cdots, \beta + q\alpha_i \tag{5.166}$$

with $q > 0$. The number p is known because all roots of level k are supposed to be known and it remains to determine q by (5.146), where $\alpha \to \alpha_i, j \to q$ and $k \to p$. Using (5.164) and (5.165) one can write

$$q = p - \frac{2(\beta, \alpha_i)}{(\alpha_i, \alpha_i)} = p - \sum_{j=1}^{\ell} k_j A_{ij} \qquad (5.167)$$

which shows how q can be found from the Cartan matrix of a given algebra at each level k.

Example 5.1 The roots of A_3. The Cartan matrix is

$$A = \begin{vmatrix} 2 & -1 & 0 \\ -1 & 2 & -1 \\ 0 & -1 & 2 \end{vmatrix}$$

and the roots of level 1 are α_1, α_2, α_3.

The α_1-string containing α_1 is $-\alpha_1$, 0, α_1. The α_2-string containing α_1 is α_1, $\alpha_1 + \alpha_2$ because $\alpha_1 - \alpha_2$ is not a root, which implies $p = 0$ in (5.167) and using the Cartan matrix one gets

$$q = -2 \frac{(\alpha_1, \alpha_2)}{(\alpha_2, \alpha_2)} = 1$$

which indicates that $\alpha_1 + 2\alpha_2$ is not a root. The α_3-string containing α_1 consists of α_1 only, because $p = 0$ and $A_{31} = 0$. In a similar way the α_1, α_2 and α_3 strings containing α_2 or α_3 can be built. The result is that the only roots of level 2 are $\alpha_1 + \alpha_2$ and $\alpha_2 + \alpha_3$.

Next we build roots of level 3 based on those of level 2. The only new root which appears can be obtained for example from the α_3-string containing $\alpha_1 + \alpha_2$. In this string $\alpha_1 + \alpha_2 - \alpha_3$ is not a positive root because of (5.165), so that $p = 0$ again. Then

$$q = -2 \frac{(\alpha_1 + \alpha_2, \alpha_3)}{(\alpha_3, \alpha_3)} = -A_{31} - A_{32} = 1$$

which shows that $\alpha_1 + \alpha_2 + \alpha_3$ is a root. No roots of level 4 appear from those of level 3 so the process terminates, giving six positive roots α_1, α_2, α_3, $\alpha_1 + \alpha_2$, $\alpha_2 + \alpha_3$, $\alpha_1 + \alpha_2 + \alpha_3$. The six negative roots are the negative of these. The group is of rank 3 and in all there are 15 roots which is just the order of A_3.

5.10 REPRESENTATIONS OF SEMI-SIMPLE GROUPS

The theory of representations of semi-simple groups is a vast subject and here we shall only present the basic concepts and try to give an idea of how to construct representations in general. In the subsequent chapters we shall discuss applications to the physically important cases of R_3, the Lorentz group L, and SU(3).

The definition of a representation of a Lie group is the same as that of a finite group and the basic properties are given in Chapter 3. The closest to finite groups are the compact Lie groups. The reason is that a sum over the group elements—appearing in various theorems—can be replaced by an invariant integral (Appendix C). For example, the orthogonality theorem can be generalized to Lie groups in this way (Section 3.10). An important outcome is that every representation of a compact Lie group is equivalent to a unitary representation and is therefore fully reducible (Section 3.8). The full reducibility of representations of semi-simple groups has been proved by Weyl and is known as one of Weyl's theorems. By imposing some particular restrictions on the infinitesimal parameters of the transformation (5.47) Weyl showed that the invariant integral converges and from there the full reducibility follows.

The problem of finding the representations of a Lie group can be reduced to that of finding the representations of the corresponding algebra. The basic concepts for describing the representations of a Lie algebra are the *weights* and the *invariant operators*, also called Casimir operators. These are discussed later in the section. The invariant operators provide the indices of a representation, while the weights give a complete labelling of the vectors of the invariant space of a representation. For example, for the rotation group R_3, the Casimir operator is J^2 and any representation of R_3 carries the index j related to the eigenvalue $j(j+1)$ of J^2. The weights are the eigenvalues m of J_z and any vector space of a given representation is completely labelled by $|jm>$. The value $m=j$ is called the highest weight. For groups of rank $\ell > 1$ there are several invariant operators which commute among themselves. There is a highest weight for each irreducible representation and its components determine the eigenvalues of the commuting invariants.

For each Lie group an important role is played by its *fundamental* representation. This is just the linear transformation (5.76) which generates the $n \times n$ matrix denoted generally by a. If the group has the order r, the matrices obtained from acting with the r generators in the same vector space which produced the matrix a form the fundamental representation of the corresponding Lie algebra. Starting from the fundamental representation one can construct representations of larger dimensions by considering direct products of representations. This is achieved in a systematic way by the tensor method which was briefly introduced in Section 4.5 and will be more thoroughly developed in Section 5.11. It will be shown how the Young diagrams offer a powerful technique in using the tensor method.

Weights

The weight of a representation can be defined starting from the Cartan–Weyl basis (5.126) for example. This basis contains the mutually commuting operators H_i $(i = 1, 2, \ldots, \ell)$ and $r - \ell$ operators of type E_α. If ϕ_m is a vector of a representation space and satisfies the eigenvalue equations

$$H_i \phi_m = m_i \phi_m \quad (i = 1, 2, \cdots, \ell) \tag{5.168}$$

the set of eigenvalues $m_1, m_2, \ldots, m_\ell$ form the covariant components of a vector which is called the *weight* of ϕ. Thus the weight space is ℓ-dimensional. A weight is called *simple* if it has no degeneracy.

The weights have various properties and some of them are similar to the properties of roots. As for roots, a weight is said to be *positive* with respect to a chosen basis if its first non-vanishing component is positive. A weight m is *higher* than another one m' if the difference $m - m'$ is positive. Among all the weights of a representation one always finds one which is the *highest weight*. Of course there is an arbitrariness in this definition because the order of the operators H_i in (5.168) is arbitrary but if an order is chosen it must remain the same for all representations of a given group.

We give below a list of properties of weights without proof. (For a proof, see Racah 1965).

1. Every representation has at least one weight.

2. Eigenvectors belonging to different weights are linearly independent.

3. If ϕ is an eigenvector of weight m then $H_i \phi$ and $E_\alpha \phi$ have the weights m and $m + \alpha$, respectively. This is obvious from the relations (5.126) and (5.168) because

$$H_i H_i \phi = m_i H_i \phi \tag{5.169}$$

and

$$H_i E_\alpha \phi = E_\alpha H_i \phi + \alpha_i E_\alpha \phi = (m_i + \alpha_i) E_\alpha \phi. \tag{5.170}$$

Next, if we consider the vector $E_\beta E_\alpha \phi$ and again use (5.126), (5.168), and also (5.170) we get

$$H_i E_\beta E_\alpha \phi = E_\beta H_i E_\alpha \phi + \beta_i E_\beta E_\alpha \phi = (m_i + \alpha_i + \beta_i) E_\beta E_\alpha \phi. \tag{5.171}$$

In this way one can obtain the weight of all possible vectors ... $E_\alpha E_\beta E_\gamma \phi$.

4. If a representation is irreducible the operators H_i may simultaneously be expressed in diagonal form. Based on this argument, the space vectors of an irreducible representation will be labelled with indices related to the eigenvalues of all H_i of a given algebra. This is the first step towards labelling the space vectors of an irreducible representation.

5. For any weight m and a root α the ratio $\dfrac{2(m, \alpha)}{(\alpha, \alpha)}$ is an integer and $m - \dfrac{2(m, \alpha)}{(\alpha, \alpha)} \alpha$ is a weight. This is a property similar to that of roots and it limits the string of weights as its analogue (Section 5.9) did for roots. The proof is similar. The only difference appears in the fact that weights can have a multiplicity larger than 1 while non-vanishing roots are non-degenerate, so that some special care must be taken. The weight $m - \dfrac{2(m, \alpha)}{(\alpha, \alpha)} \alpha$ can be obtained from m by a Weyl reflection as for roots and the weights obtained from one another by a Weyl reflection are called *equivalent*. A weight is called *dominant* if it is higher than its equivalents.

6. If a representation is irreducible its highest weight is simple.

7. Two irreducible representations are equivalent if their highest weights are equal. This means one can distinguish between non-equivalent irreps by using the highest weights as a label.

REPRESENTATION OF SEMI-SIMPLE GROUPS

Invariant operators

We have mentioned that for the group R_3 the highest weight of an irrep is j. But this is also connected with the eigenvalue of J^2 which is an operator which commutes with all the generators J_x, J_y, J_z of R_3, hence with any rotation in three dimensions. J^2 is a particular case of an invariant operator and its generalization to any semi-simple group was given by Casimir as

$$G^2 = g^{\rho\sigma} X_\rho X_\sigma = g_{\rho\sigma} X^\rho X^\sigma \tag{5.172}$$

which is quadratic in the generators X^σ and is called the *Casimir operator*. That G^2 commutes with every generator can easily be shown using definition (5.37) of the structure constants and their symmetry properties (Exercise 5.2). One has

$$
\begin{aligned}
[G^2, X_\rho] &= [g_{\alpha_1 \alpha_2} X^{\alpha_1} X^{\alpha_2}, X_\rho] \\
&= g_{\alpha_1 \alpha_2}(X^{\alpha_1}[X^{\alpha_2}, X_\rho] + [X^{\alpha_1}, X_\rho]X^{\alpha_2}) \\
&= g_{\alpha_1 \alpha_2}(c^{\alpha_2}{}_{\rho\tau} X^{\alpha_1} X^\tau + c^{\alpha_1}{}_{\rho\tau} X^\tau X^{\alpha_2}) \\
&= c_{\alpha_1 \rho\tau} X^{\alpha_1} X^\tau + c_{\alpha_2 \rho\tau} X^\tau X^{\alpha_2} \\
&= c_{\alpha_1 \rho\tau} X^{\alpha_1} X^\tau + c_{\rho\tau \alpha_1} X^\tau X^{\alpha_1} \\
&= c_{\alpha_1 \rho\tau} X^{\alpha_1} X^\tau + c_{\rho \alpha_1 \tau} X^{\alpha_1} X^\tau = 0
\end{aligned}
$$

where in the penultimate row the invariance under a circular permutation of indices and in the last row the antisymmetry under a permutation of two indices has been used.

The tensor $g^{\rho\sigma}$ can also be written in a form similar to (5.131) for $g_{\rho\sigma}$. This gives

$$G^2 = g^{ik} H_i H_k + \sum_\alpha E_\alpha E_{-\alpha}. \tag{5.173}$$

We are interested in acting with G^2 on the vector of highest weight. Let us call it $\phi_{\max}$. The eigenvalue equations (5.168) become

$$H_i \phi_{\max} = L_i \phi_{\max}. \tag{5.174}$$

If $\phi_{\max}$ is the vector of highest weight and α is a positive root we conclude from (5.170) that

$$E_\alpha \phi_{\max} = 0 \tag{5.175}$$

because $L_i + \alpha_i$ cannot be a weight. Then the sum in (5.173) can be restricted to positive roots α^+ only

$$
\begin{aligned}
G^2 \phi_{\max} &= g^{ik} H_i H_k \phi_{\max} + \sum_{\alpha^+} E_\alpha E_{-\alpha} \phi_{\max} \\
&= g^{ik} H_i H_k \phi_{\max} + \sum_{\alpha^+} [E_\alpha, E_{-\alpha}] \phi_{\max}.
\end{aligned}
$$

Using (5.126d) and (5.174) one obtains

$$G^2 \phi_{\max} = (g^{ik} L_i L_k + \sum_{\alpha^+} \alpha^i L_i) \phi_{\max} \tag{5.176}$$

which shows that the eigenvalue of G^2 is determined by the components $L_1, L_2, \ldots, L_\ell$ of the maximum weight for each group, specified by its roots α^i. It does not work the other way round because one cannot determine the ℓ components of the highest weight from only one eigenvalue. A solution was proposed by Racah who generalized the definition of Casimir by introducing the invariant operators

$$I_n = c^{\beta_2}_{\alpha_1\beta_1} c^{\beta_3}_{\alpha_2\beta_2} \cdots c^{\beta_1}_{\alpha_n\beta_n} X^{\alpha_1} X^{\alpha_2} \cdots X^{\alpha_n} \tag{5.177}$$

with $n \geq 2$. Although introduced by Racah, these invariants are often improperly called Casimir operators. Racah (1950) showed that, for a semi-simple Lie group of rank ℓ one can construct ℓ independent invariants, and their eigenvalues completely specify the irreducible representations. In Exercise 5.7 the formula (5.177) is applied to SO(3) to show that only one independent invariant can be derived. In general the order of independent invariants of the classical Lie groups is

$$A_\ell : \quad I_2, I_3, \cdots, I_{\ell+1}$$
$$B_\ell : \quad I_2, I_4, \cdots, I_{2\ell}$$
$$C_\ell : \quad I_2, I_4, \cdots, I_{2\ell}$$
$$D_\ell : \quad I_2, I_4, \cdots, I_{2\ell}.$$

The role of the invariant operators in labelling irreps can be understood through Schur's lemma 2 (Section 3.10). By definition an I_n commutes with all the generators, therefore it commutes with any finite dimensional irreducible representation, hence it is a multiple of the unit operator, $I_n = \lambda_n I$ where I is the unit matrix. All the I_n can be diagonalized simultaneously and the ℓ eigenvalues λ_n associated with an invariant subspace can be used to label an irreducible representation, provided non-equivalent irreps belong to different eigenvalues. Such condition can be expressed as follows. If $L = (L_1, L_2, \ldots, L_\ell)$ is the highest weight of a given irrep one must obtain a system of ℓ equations of the form

$$\lambda_i = \varphi_i(L_1, L_2, \cdots, L_\ell) \quad i = 1, 2, \ldots, \ell \tag{5.178}$$

which has a unique solution. There are some non-equivalent representations for which the operators (5.177) do not lead to a unique solution. But Racah (1965) proved that for every semi-simple group there exists ℓ alternative invariants whose eigenvalues λ_i are in one-to-one correspondence with the irreducible representations. As an example, details about the invariants of A_2 (SU(3)) are given in Chapter 8.

In the next section we shall see that an alternative and very convenient way to uniquely label the irreps of a Lie group is achieved through the use of Young diagrams. This method is not restricted to semi-simple groups.

Exercise 5.7 Calculate the invariants I_2 and I_3 for the group SO(3) (isomorphic to R_3) using Exercise 5.5.

Solution Exercise 5.5 gives the algebra of SO(3) and the metric tensor $g_{\rho\sigma}$ from which one can immediately derive

REPRESENTATION OF SEMI-SIMPLE GROUPS

$$g^{\rho\sigma} = \begin{pmatrix} \frac{1}{2} & 0 & 0 \\ 0 & 0 & 1 \\ 0 & 1 & 0 \end{pmatrix}. \tag{1}$$

Using the notation of Exercise 5.5 one can write

$$\begin{aligned} X^1 &= g^{1\,-1}X_{-1} = E_{-1} \\ X^{-1} &= g^{-1\,1}X_1 = E_1 \\ X^0 &= g^{0\,0}X_0 = \frac{1}{2}H_0 \end{aligned} \tag{2}$$

According to (5.177) the invariant I_2 is

$$\begin{aligned} I_2 &= c_{\alpha_1\beta_1}^{\beta_2} c_{\alpha_2\beta_2}^{\beta_1} X^{\alpha_1} X^{\alpha_2} = g_{\alpha_1\alpha_2} X^{\alpha_1} X^{\alpha_2} \\ &= g_{0\,0}X^0X^0 + g_{1\,-1}X^1X^{-1} + g_{-1\,1}X^{-1}X^1 \\ &= \frac{1}{2}H_0^2 + E_1 E_{-1} + E_{-1} E_1 \\ &= \frac{1}{2}(L_1^2 + L_2^2 + L_3^2) = \frac{1}{2}L^2. \end{aligned} \tag{3}$$

As expected, I_2 is the Casimir operator (5.172) of SO(3) and is proportional to L^2.

According to (5.177) the invariant operator I_3 is

$$\begin{aligned} I_3 &= c_{\alpha_1\beta_1}^{\beta_2} c_{\alpha_2\beta_2}^{\beta_3} c_{\alpha_3\beta_3}^{\beta_1} X^{\alpha_1} X^{\alpha_2} X^{\alpha_3} \\ &= (c_{0\,1}^1 c_{0\,1}^1 c_{0\,1}^1 + c_{0\,-1}^{-1} c_{0\,-1}^{-1} c_{0\,-1}^{-1}) X^0 X^0 X^0 \\ &\quad + c_{1\,0}^1 c_{0\,1}^0 c_{-1\,1}^0 X^1 X^0 X^{-1} + c_{1\,-1}^0 c_{-1\,0}^{-1} c_{0\,-1}^0 X^1 X^{-1} X^0 \\ &\quad + c_{0\,-1}^{-1} c_{1\,-1}^0 c_{-1\,0}^{-1} X^0 X^1 X^{-1} + c_{-1\,0}^{-1} c_{0\,-1}^0 c_{1\,-1}^0 X^{-1} X^0 X^1 \\ &\quad + c_{-1\,1}^0 c_{1\,0}^1 c_{0\,1}^1 X^{-1} X^1 X^0 + c_{0\,1}^1 c_{-1\,1}^0 c_{1\,0}^1 X^0 X^{-1} X^1 \\ &= \frac{1}{2}[X^1, X^0] X^{-1} + \frac{1}{2}[X^{-1}, X^1] X^0 + \frac{1}{2}[X^0, X^{-1}] X^1 \\ &= \frac{1}{4}[E_{-1}, H_0] E_1 + \frac{1}{4}[E_1, E_{-1}] H_0 + \frac{1}{4}[H_0, E_1] E_{-1}. \end{aligned} \tag{4}$$

The commutators, as given in Exercise 5.5, lead to

$$I_3 = \frac{1}{8}(L_1^2 + L_2^2 + L_3^2) \tag{5}$$

which shows that I_3 is also proportional to L^2. Any other I_n with $n > 3$ will give the same result. This suggests that for SO(3), which is a group of rank $\ell = 1$, there is only one independent invariant operator.

Exercise 5.8 Show that the general linear group is not semi-simple.

Solution Let us derive its generators according to Section 5.3. Consider the infinitesimal transformation

$$x'^i = (\delta^i_j + \varepsilon^i_j)x^j \tag{1}$$

which induces the following change in the scalar function F

$$F(x'^i) = F(x^i + \varepsilon^i_j x^j) = F(x^i) + \varepsilon^i_j x^j \frac{\partial F}{\partial x^i}$$
$$= (1 + \varepsilon^i_j x^j \frac{\partial}{\partial x^i})F.$$

Hence the n^2 generators are

$$X_{ij} = x^i \frac{\partial}{\partial x^j} \quad (i,j = 1, 2, \cdots, n). \tag{2}$$

The Lie algebra of the general linear group turns to be

$$[X_{ik}, X_{mn}] = X_{in}\delta_{km} - X_{mk}\delta_{in}. \tag{3}$$

Define the operator

$$A = \sum_j X_{jj}. \tag{4}$$

Using (3) one finds that A commutes with all the generators

$$[A, X_{mn}] = \sum_j [X_{jj}, X_{mn}] = \sum_j (X_{jn}\delta_{jm} - X_{mj}\delta_{nj}) = 0. \tag{5}$$

Actually the operator A generates an abelian subgroup, the dilatation group, which is an invariant subgroup as indicated by (5). Therefore the general linear group is not semi-simple.

The dilatation group is defined by the transformation

$$x'^i = x^i + k x^i \tag{6}$$

and its generator is obtained as above:

$$F(x'^i) = F(x^i + k x^i) = (1 + k \sum x^i \frac{\partial}{\partial x^i})F = (1 + kA)F. \tag{7}$$

5.11 THE TENSOR METHOD (REVISITED)

In Section 4.5 we saw that, starting from a vector of n components which transform among themselves under transformations belonging to a linear group G, one can construct a tensor T of rank k which constitutes the basis of an n^k-dimensional representation of G. Under the action of an element $a \in$ G the tensor components transform as

$$T'_{i_1 i_2 \cdots i_k} = a_{i_1 i'_1} a_{i_2 i'_2} \cdots a_{i_k i'_k} T_{i'_1 i'_2 \cdots i'_k} \tag{5.179}$$

where each index i or i' can take values from 1 to n. The $n^k \times n^k$ matrix can be reduced with respect to both G and the symmetric group S_k due to the commutation between any element of G with any element of S_k. This reduction procedure is valid for GL(n, $\mathbb{C}$) but also for its subgroups GL(n, $\mathbb{R}$), SL(n, $\mathbb{C}$), SL(n, $\mathbb{R}$), U(n), and SU(n). For some other subgroups further reductions can be made with an operation called *contraction* which also commutes with the elements of the subgroup. For reduction with respect to O(n) and Sp(n) see, for example, Hamermesh (1962), Chapter 10. We shall present the contraction procedure only in Chapter 8 for SU(3).

Irreducible tensors with respect to GL(n, $\mathbb{C}$)

As shown in Section 4.5 the tensors of rank k with a particular permutation symmetry transform among themselves and thus constitute an invariant subspace of an irreducible representation of S_k. According to Section 4.3 there is a one-to-one correspondence between an irrep of S_k and a standard Young diagram with k boxes, and to each component of the tensor forming an invariant subspace one can thus associate a Young tableau obtained from filling the Young diagram with figures in increasing order from left to right and top to bottom in all possible ways.

The first question which arises is which Young diagrams are compatible with a specific n and k. In order to answer this question, in (4.113) it is convenient to make the identification $x_{i_r} \leftrightarrow \alpha_r$ where α_r designates a single particle state as in (4.15). Then the discussion can be made in terms of Weyl tableaux (Section 4.3). One property of a Weyl tableau is that the same particle state index cannot appear twice in a column because a column represents an antisymmetric state. Therefore, if $k > n$, only diagrams with a maximum of n rows are allowed. In general the partition $[f_1, f_2, \ldots, f_n]$ of a given Young diagram must satisfy

$$f_1 + f_2 + \cdots + f_n = k, \qquad f_1 \geq f_2 \geq \cdots \geq f_n \geq 0. \qquad (5.180)$$

If the diagram contains less than n rows, some of the f's are zero and can be omitted as in (4.2b).

One should also add that there exists a non-vanishing tensor for any possible partition (5.180), which means that every Young diagram with n rows or less can be realized for the general linear group.

The discussion given in Section 4.5 for the case $k = n = 3$ is not restricted to SU(3) only. It remains valid for GL(3, $\mathbb{C}$) as well, because no use has been made of the fact that $\det a = 1$ and $a^+ = a^{-1}$. Here we look at $k = 3$ and an arbitrary n and try to determine the dimension of the irreducible representations with the help of outer products (Section 4.6). Following the discussion in Section 4.5 the product of a symmetric rank 2 tensor and a vector x gives the outer product

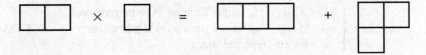

The number of independent components on the left-hand side is given by the dimension of the representation ☐☐, which is equal to $\frac{1}{2}n(n+1)$ (see Section 4.5) times n, i.e. the number of components of the vector x. For the symmetric rank 3 tensor one can generalize the discussion given for $k = n = 3$. There are n components with equal indices $i_1 = i_2 = i_3$ plus $2C_n^2$ components with two identical indices $i_1 = i_2 \neq i_3$, etc., plus C_n^3 components with $i_1 \neq i_2 \neq i_3$ which give a total of C_{n+2}^3 independent components. It follows that the number of independent components of the tensor represented by ☐☐☐ is

$$\frac{n^2(n+1)}{2} - \frac{n(n+1)(n+2)}{6} = \frac{n(n^2-1)}{3}. \tag{5.181}$$

For $n = 3$ this gives eight components, consistent with the case $k = n = 3$ described in Section 4.5, which proves that the discussion there is not necessarily restricted to SU(3). The group SU(3) was used for its relevance in physics and for calculating the dimensions of irreducible representations with a standard well-known formula. The general formula for the dimension of an irreducible representation of $GL(n, \mathbb{C})$ or $GL(n, \mathbb{R})$ can be found, for example, in Hamermesh (1962) (Equation 10–25).

For totally symmetric representations, one has $f_1 = k$ and the dimension $d_{[k,0^{n-1}]}$ of the carrier tensor space is given by the number of ways one can choose k objects among n objects allowing repetition (the counting problem of a Bose–Einstein gas). Then

$$d_{[k,0^{n-1}]} = \frac{(n+k-1)!}{k!(n-1)!}. \tag{5.182}$$

For the particular case $k = n$ the completely antisymmetric tensor has only one component (a Slater determinant) which can be obtained by applying the antisymmetrizer (4.5) to $x_1(1)x_2(2)\ldots x_n(n)$

$$(T^A)_{12\cdots n} = \frac{1}{n!}\sum_P \delta_P P x_1(1)x_2(2)\cdots x_n(n). \tag{5.183}$$

The transformed tensor reads

$$\begin{aligned}(T^A)'_{12\cdots n} &= \sum_{i_1 i_2 \cdots i_n} a_{1 i_1} a_{2 i_2} \cdots a_{n i_n} \frac{1}{n!}\sum_P \delta_P P x_{i_1}(1)x_{i_2}(2)\cdots x_{i_n}(n) \\ &= \sum_{i_1 i_2 \cdots i_n} a_{1 i_1} a_{2 i_2} \cdots a_{n i_n} \varepsilon_{i_1 i_2 \cdots i_n} \frac{1}{n!}\sum_P \delta_P P x_1(1)x_2(2)\cdots x_n(n) \\ &= \det a\, (T^A)_{12\cdots n}\end{aligned} \tag{5.184}$$

where

$$\varepsilon_{i_1 i_2 \cdots i_n} = \begin{cases} +1 & \text{if } i_1 i_2 \cdots i_n \text{ results from an even permutation} \\ -1 & \text{if } i_1 i_2 \cdots i_n \text{ results from an odd permutation} \\ 0 & \text{for any repeated index.} \end{cases} \tag{5.185}$$

THE TENSOR METHODS 201

It follows that for unimodular groups, det $a = 1$, the antisymmetric tensor is an invariant. The tensor $\varepsilon_{i_1 i_2 \ldots i_n}$, called unit antisymmetric tensor, is itself an invariant in that case. We shall return to it in relation to SU(3).

Exercise 5.9 By using outer products find the dimension of the irreducible representation of GL(4) generated by a rank 4 tensor.

Solution This exercise can be viewed as an extension of the rank 3 tensor case discussed in Section 4.5. Here we start from the vector

$$x = \begin{pmatrix} u \\ d \\ s \\ c \end{pmatrix}.$$

One can combine the components of this vector in several ways to obtain a rank 4 tensor. The distinct possibilities are

(1) $uuuu, dddd, ssss, cccc$ 4

(2) $uuud, uddd, uuus, usss, uuuc, uccc,$ 12
     $ddds, dsss, sssc, sccc, dddc, dccc$

(3) $uudd, uuss, uucc, ddss, ddcc, sscc$ 6

(4) $uuds, uudc, uusc, udds, uddc, ddsc,$ 12
     $udcc, uscc, dscc, udss, ussc, dssc$

(5) $udsc$ 1

Note that, in agreement with the definition of a Weyl tableau, in each combination the components of x have been selected so as to respect an order which we chose to be

$$u < d < s < c.$$

The total number of possibilities is 35. This is the dimension of the representation [4] consistent with Equation (5.182). Next, we use the outer product

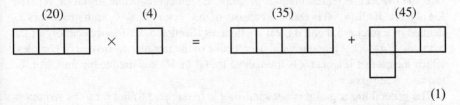

$$\tag{1}$$

The number of independent components on the left hand side is given by the product of dimensions of the representations [3] and [1]. According to the discussion preceding Equation (5.181) this is

$$\frac{n(n+1)(n+2)}{6} \times n = 20 \times 4. \tag{2}$$

The dimension of the tensor space spanning the representation [31] of GL(4) is therefore $20 \times 4 - 35 = 45$. On the other hand, the antisymmetric rank 3 tensor in GL(4) has $C_4^3 = 4$ components and the rank 4 antisymmetric tensor has one component. From the outer product

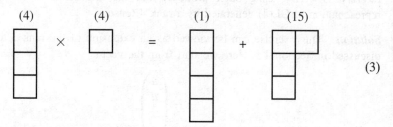

(3)

it turns out that the [211] representation of GL(4) has the dimension $4 \times 4 - 1 = 15$. Finally in the outer product

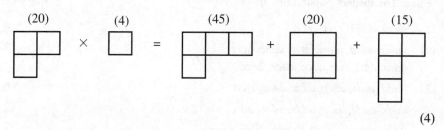

(4)

the number of independent components is 20×4 where 20 is the number coming out from (5.181) with $n = 4$, so that the dimension of [22] is $20 \times 4 - 45 - 15 = 20$.

One can check that the dimensions of the irreps [f] of GL(4) found above are the same as those of [f] of SU(4) from Table 8.7. This remark remains valid for any GL(n) and SU(n) in agreement with the discussion given below.

Irreducible representations of SL(n), U(n) and SU(n)

First the irreducible representations of GL(n, $\mathbb{C}$) remain reducible for GL(n, $\mathbb{R}$). The Lie algebra of GL(n, $\mathbb{R}$) is real and consists of the vectors $\Sigma\, a_{ij} X_{ij}$ with generators X_{ij} defined in Exercise 5.8 and a_{ij} real coefficients (Section 5.3). The reducibility of the matrices $\Sigma\, a_{ij} D(X_{ij})$ remains valid irrespective of the fact that a_{ij} are real or complex, which means that if a matrix is irreducible for GL(n, $\mathbb{R}$) it is irreducible for GL(n, $\mathbb{C}$) too and vice versa.

The general linear group is not semi-simple (Exercise 5.8) but it can be written as the direct product of two of its subgroups: the special linear group and the dilatation group which is abelian. For this reason the general linear group and the special linear group have many properties in common and in particular the irreducible representations of GL(n, $\mathbb{C}$) remain irreducible for SL(n, $\mathbb{C}$). The argument is rather simple. For unimodular groups, the condition det $a = 1$ imposed on the infinitesimal transformation (5.47) implies

$$\operatorname{tr} D(X_{ij}) = 0 \tag{5.186}$$

which has no effect on the reducibility. Hence reducibility is the same for SL(n, $\mathbb{C}$) or GL(n, $\mathbb{C}$). The statement remains valid if $\mathbb{C} \to \mathbb{R}$.

By extending these arguments one can show that an irreducible representation of GL(n, $\mathbb{C}$) remains irreducible for U(n) and SU(n). In conclusion an irrep of GL(n, $\mathbb{C}$) does not decompose if we restrict to GL(n, $\mathbb{R}$), SL(n, $\mathbb{C}$), SL(n, $\mathbb{R}$), U(n), or SU(n), and it can be illustrated by the same Young diagram for all these groups. However, each time we restrict to a unimodular subgroup of GL(n, $\mathbb{C}$), an equivalence appears between the representations associated with the partitions $[f_1, f_2, \ldots, f_n]$ and $[f_1 + s, f_2 + s, \ldots, f_n + s]$, where s is the number of columns of length n. For example, for SU(3) the representations

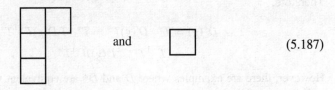

$$\tag{5.187}$$

are equivalent.

The contragradient representation

For a given matrix representation defined by the mapping $g \to D(g)$ the *contragradient* representation is defined by

$$g \to D^{\mathrm{T}}(g^{-1}) = (D^{\mathrm{T}})^{-1}(g).$$

If D is unitary, so is the contragradient representation, and in this case $(D^{\mathrm{T}})^{-1} = D^*$.

The representation D^* which is obtained from D by complex conjugation is called the *conjugate* or *complex conjugate* representation. Thus, in the case of unitary representations, the contragradient and the conjugate representations are identical. (For SU(2) and SU(3), see Chapter 8).

It is easy to define the contragradient representation in terms of Young diagrams. For a Young diagram given by the partition $[f_1, f_2, \ldots, f_n]$, it can be shown that the corresponding contragradient representation is described by the partition $[f_1 - f_n, f_1 - f_{n-1}, \ldots, f_1 - f_2]$. This partition is such that it completes a rectangle with n rows and f_1 columns. For example, the representation [6541] of SU(4) and its contragradient representation [521] are

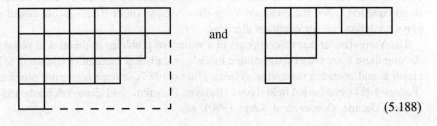

$$\tag{5.188}$$

A general proof for the Young diagram of a conjugate representation of SU(n) is given in Section 18.9 of Elliott and Dawber (1979).

If all matrices of a representation are real, then the representation is *real*. If D is equivalent to a real representation D', then D is equivalent to D^*. This is easy to see because the equivalence of D' and D implies

$$D'(g) = T D(g) T^{-1} \quad \text{for all } g \in G$$

where T is a nonsingular matrix (see section 3.4).

Taking the complex conjugate, one has

$$D'(g) = T^* D^*(g) T^{*-1}.$$

Therefore,

$$D^*(g) = T^{*-1} D'(g) T^* = T^{*-1} T D(g) T^{-1} T^*$$
$$= (T^{-1} T^*)^{-1} D(g) T^{-1} T^*.$$

However, there are examples where D and D^* are equivalent without being real.

If D is equivalent to a real representation, it is called *potentially real* or *real*. For example, all irreps of the rotation group R_3 are potentially real. If D is equivalent to D^* but D is not equivalent to a real representation, it is called *pseudoreal*. Examples can be found in Cornwell (1984), Chapter 15, Section 4 and Appendix F.

Theorem An irrep D is potentially real or pseudoreal if and only if its characters are real. For a proof, see Cornwell 1984, Chapter 5, Section 8.

For a general discussion on real representations, the reader is also referred to Hamermesh (1962), Section 5.5.

5.12 QUANTUM GROUPS

Quantum groups or quantum algebras constitute a new and growing field of extensions to group theory, which made its appearance in the 1980s.

The appropriate name is quantum algebra, which is an abbreviation of quantized universal enveloping (QUE) algebra. Quantum algebras are generalizations of the Lie algebras developed in Chapter 5. In the same way that the Jacobi identity (Section 5.3) plays the role of an 'associativity' condition necessary to identify a Lie algebra, so an equation called the quantum Yang–Baxter equation or Braid–Jacobi equation, plays a similar role for quantum algebras.

The Yang–Baxter equation appears in a variety of problems in theoretical physics. Among these there are exactly soluble models in statistical mechanics (Baxter 1982), classical and quantum integrable systems (Jimbo 1985), inverse scattering problems (Fadeev 1984), conformal field theory (Belavin, Polyakov, and Zamolodchikov 1984; Alvarez-Gaumé, Gomez, and Sierra 1990), etc.

A QUE algebra is described as a q-deformation of a Lie algebra in terms of a parameter $q = e^s$ with s real or $q = e^{is}$ with s real. When the deformation parameter q is set equal to one, the usual Lie algebra is obtained. Mathematically, the deformed algebras have been identified (Drinfeld 1986) as Hopf algebras. There is also a strong connection between quantum algebras and braid groups introduced in Section 4.12.

As mentioned above, quantum groups were initially discovered as solutions of the quantum Yang–Baxter equation and now they are being applied in a large variety of branches of physics. Apart from applications of a more fundamental nature, such as those mentioned above, they are being related to models of molecular physics, quantum optics, nuclear physics, condensed matter physics, and elementary particles. Roughly, there are two kinds of tendencies in these applications. One is to identify existing models with a deformed algebra (see, for example, Bonatsos and Daskaloyannis 1993). Another is to start from models which obey a Lie algebra structure, introduce a quantum deformation, and then study how the physical properties of the system change as a function of this deformation (for example, Bonatsos, Brito, Menezes, Providência and da Providência 1993a).

The interest in searching for applications was especially reinforced by the introduction, independently by Biedenharn (1989) and Macfarlane (1989), of the q-deformed harmonic oscillators. This is a realization of the quantum group $SU_q(2)$ introduced formerly by Sklyanin (1980) and Kulish and Reshetikin (1981). The defining algebraic relations for the quantum group $SU_q(n)$ corresponding to a deformation of the classical simple Lie algebra A_{n-1} (Section 5.9) were given by Jimbo (1985). A q-deformed boson realization of the quantum group $SU_q(n)$ was proposed by Sun and Fu (1989).

We illustrate below some features of quantum algebras by giving a brief description of the q-boson realization of $SU_q(2)$ as proposed by Biedenharn (1989).

The quantum group $SU_q(2)$ is generated by the operators J_+^q, J_-^q, and J_z^q which obey the commutation relations

$$[J_z^q, J_\pm^q] = \pm J_\pm^q \qquad (5.189)$$

$$[J_+^q, J_-^q] = \frac{q^{J_z^q} - q^{-J_z^q}}{q^{1/2} - q^{-1/2}} \qquad (5.190)$$

where q is a real number. The above relation no longer defines a Lie algebra but a deformation of the universal enveloping algebra of the Lie algebra su(2). In the limit $q \to 1$ it contracts to su(2).

Biedenharn starts from the Jordan–Schwinger approach of the theory of angular momentum. In this method one writes the SU(2) generators J_+, J_-, and J_z in terms of a pair a_1 and a_2 of independent commuting boson operators

$$[a_i, a_j^+] = \delta_{ij} \qquad (5.191)$$

in the following form

$$J_+ = a_1^+ a_2, \qquad J_- = a_2^+ a_1, \qquad J_z = \frac{1}{2}(a_1^+ a_1 - a_2^+ a_2). \qquad (5.192)$$

In quantum algebra one defines a q-analogue to the harmonic oscillator by introducing a q-creation operator a^{q^+}, its Hermitian conjugate the q-destruction operator a^q, and the q-boson vacuum ket $|0>_q$ defined by

$$a^q|0>_q = 0. \tag{5.193}$$

Then one *postulates* the algebraic relation

$$a^q a^{q^+} - q^{1/2} a^{q^+} a^q = q^{-N^q/2} \tag{5.194}$$

where N^q is the number operator defined to satisfy the commutation relations

$$[N^q, a^{q^+}] = a^{q^+} \tag{5.195}$$
$$[N^q, a^q] = -a^q \tag{5.196}$$

as for the harmonic oscillator. Then the state vectors $|n>_q$ are constructed in the standard way used for the harmonic oscillator (Messiah 1964, Chapter XII). By definition one takes

$$|n>_q = \mathcal{N}_q (a^{q^+})^n |0>_q. \tag{5.197}$$

To find the normalization constant $\mathcal{N}_q$ one has to iterate equation (5.194). First one multiplies it on the right by a^{q^+}

$$a^q a^{q^+} a^{q^+} - q^{1/2} a^{q^+} a^q a^{q^+} = q^{-N^q/2} a^{q^+}. \tag{5.198}$$

In the left hand side one uses it again in the second term for commuting a^q with a^{q^+}. This gives

$$a^q a^{q^+} a^{q^+} - q a^{q^+} a^{q^+} a^q = q^{1/2} a^{q^+} q^{-N^q/2} + q^{-N^q/2} a^{q^+}. \tag{5.199}$$

The second term on the right hand side can be rewritten in a different way if one takes $q = e^s$. Then by expanding the exponential and using (5.195) one gets

$$\begin{aligned} q^{-N^q/2} a^{q^+} &= \sum_{n=0}^{\infty} \frac{(-)^n s^n (N^q)^n}{2^n n!} a^{q^+} \\ &= a^{q^+} \sum_{n=0}^{\infty} \frac{(-)^n s^n (N^q+1)^n}{2^n n!} = q^{-1/2} a^{q^+} q^{-N^q/2}. \end{aligned} \tag{5.200}$$

With this, (5.199) becomes

$$a^q (a^{q^+})^2 - q(a^{q^+})^2 a^q = (q^{1/2} + q^{-1/2}) a^{q^+} q^{-N^q/2}. \tag{5.201}$$

By repeating the procedure $n-1$ times one obtains

$$a^q (a^{q^+})^n - q^{n/2} (a^{q^+})^n a^q = [n]_q (a^{q^+})^{n-1} q^{-N^q/2} \tag{5.202}$$

where

$$[n]_q = q^{(n-1)/2} + q^{(n-3)/2} + \cdots + q^{-(n-1)/2} = \frac{q^{n/2} - q^{-n/2}}{q^{1/2} - q^{-1/2}}. \tag{5.203}$$

From (5.202) one can determine the norm $\mathcal{N}_q$ and the normalized state (5.197) becomes

$$|n>_q = ([n]_q!)^{-1/2}(a^{q^+})^n|0>_q. \tag{5.204}$$

From the symmetry of (5.203) with respect to the transformation $q \to q^{-1}$ one can see that the same result would have been obtained starting from a relation[†] as such (5.194) but with $q \to q^{-1}$, i.e.

$$a^q a^{q^+} - q^{-1/2} a^{q^+} a^q = q^{N^q/2}. \tag{5.205}$$

Now one can define a q-analogue of the Jordan–Schwinger procedure. One introduces a pair of mutually commuting harmonic oscillator systems, a_i^q and $a_i^{q^+}$, with $i = 1, 2$. One defines the q-boson operators for $i = 2$ by (5.194) and for $i = 1$ by (5.205). Then the generators of $SU_q(2)$ are

$$J_+^q = a_1^{q^+} a_2^q \tag{5.206}$$

$$J_-^q = (J_+^q)^+ = a_2^{q^+} a_1^q \tag{5.207}$$

$$J_z^q = \frac{1}{2}(N_1^q - N_2^q). \tag{5.208}$$

The eigenstates $|jm>_q$ of J^2 and J_z are the q-analogues of the angular momentum states, defined as

$$|jm>_q = ([j+m]_q![j-m]_q!)^{-1/2}(a_1^q)^{j+m}(a_2^q)^{j-m}|0>_q \tag{5.209}$$

where $[\ ]_q$ has the same meaning as in (5.203). They satisfy the equations

$$J_\pm|jm>_q = ([j \mp m]_q[j \pm m + 1]_q)^{1/2}|jm \pm 1>_q \tag{5.210}$$

$$J_z|jm>_q = m|jm>_q. \tag{5.211}$$

The operators (5.206)–(5.208) and the states (5.209) can be used to construct finite dimensional $2j + 1$ unitary representations of $SU_q(2)$ for any $j = 0, \frac{1}{2}, 1, \ldots$. In particular, for $j = \frac{1}{2}$ the $SU_q(2)$ generators are the Pauli matrices. However, the usual technique of constructing Clebsch–Gordan coefficients no longer holds because for two distinct systems 1 and 2 the sum $J_i^q = J_i^q(1) + J_i^q(2)$ does not, in general, obey the commutation relation (5.190). A pattern calculus for tensor operators has been proposed by Biedenharn (1990) to solve this problem. Procedures other than the q-boson operators have also been proposed to construct irreps of $SU_q(2)$. Algebraic expressions of q-deformed Clebsch–Gordan coefficients can be found, for example, in Feng Pan (1991) or Smirnov, Tolstoy, and Kharitonov (1993).

In representation theory, one must distinguish between the two cases: $q = e^s$ with s real and $q = e^{is}$ with s real. In the former case, there is a one-to-one correspondence between the deformed algebra and the corresponding Lie algebra. In the latter case, there are more representations for the deformed algebra than for the Lie algebra, except for the situation where q is not a root of unity, when the one-to-one correspondence still holds.

[†] In an alternative formulation one can replace $q^{\frac{1}{2}}$ by q, see e.g. Macfarlane (1989)

SUPPLEMENTARY EXERCISES

5.10 Show that the invariance of the skew symmetric bilinear form of equation (5.94)

$$\sum_{k=1}^{n}(x_k y_{-k} - x_{-k} y_k)$$

under a group of transformations given by (5.76) where a is an $n \times n$ non-singular matrix implies the property

$$a g a^{\mathrm{T}} = g$$

where

$$g = \begin{pmatrix} 0 & I_n \\ -I_n & 0 \end{pmatrix}.$$

5.11
(a) A real Lie algebra has basis elements X_1, X_2, X_3 with commutation relations

$$[X_1, X_2] = X_3, \qquad [X_2, X_3] = -X_1, \qquad [X_3, X_1] = -X_2.$$

Calculate the metric tensor $g_{\mu\nu}$ of (5.36).

(b) Another Lie algebra has basis elements Y_1, Y_2, Y_3 with commutation relations

$$[Y_1, Y_2] = Y_3, \qquad [Y_2, Y_3] = -Y_1, \qquad [Y_3, Y_1] = Y_2.$$

Show by a simple change of basis that (b) has the same structure as (a).

6

THE ORTHOGONAL GROUP

In Section 5.6, the orthogonal group was introduced as the group of transformations which leave invariant the complex quadratic form $\Sigma(x_i)^2$, where x_i are the components of a vector in an n-dimensional space. Under an orthogonal transformation given by the matrix a the vector components become

$$x'_i = a_{ij}x_j. \tag{6.1}$$

This gives

$$\sum_i (x'_i)^2 = \sum_{ijk} a_{ij}a_{ik}x_jx_k. \tag{6.2}$$

The invariance of the quadratic form requires

$$\sum_i a_{ij}a_{ik} = \delta_{jk} \tag{6.3}$$

which is indeed the orthogonality property (5.84). If a is complex, the group is denoted by $O(n,\mathbb{C})$ and if a is real, the group is denoted by $O(n,\mathbb{R})$ or $O(n)$. The subgroup of $O(n)$ with det $a = 1$ is called special orthogonal and is denoted by $SO(n)$.

In this chapter, we shall be concerned mainly with transformations in a real three-dimensional Euclidean space $\mathbb{R}^3$, because many physical applications involve such transformations. The subgroup of transformations with det $a = 1$, $SO(3)$, corresponds to proper rotations. The group $O(3)$ also contains transformations with det $a = -1$ which represent inversions $x \to -x$. These are trivial transformations from the group theory point of view (Chapter 3) so it is enough to study the group $SO(3)$. One says that $SO(3)$ is isomorphic to the rotation group R_3. Hence, we shall also use geometrical arguments in the study of $SO(3)$.

The group $O(4)$ will also be discussed to some extent (Section 6.2). This group is the symmetry group of the hydrogen atom which possesses a *dynamical* symmetry due to its particular form of force law. Also the group $O(3,1)$ will be introduced and its generators will be deduced from those of $O(4)$. The group $O(3,1)$ is isomorphic to the Lorentz group (Chapter 7), which is the analogue in Minkowski space–time of the group of rotations in $\mathbb{R}^3$.

6.1 THE ROTATION GROUP R_3 OR $SO(3)$

The rotation group is the simplest non-trivial example of a semi-simple compact Lie group. As it is often applied in atomic, molecular, nuclear, subnuclear, or solid state physics, one can find its properties in many quantum mechanics books (see, for example, Schiff 1968, Section 27) or in monographs devoted to angular momentum (see, for example, Brink and Satchler 1968). Therefore we present here only some aspects which are relevant from the group theory point of view and refer the reader to the above-mentioned books for more technical aspects. Here, we shall derive the generators of R_3 and the fundamental representation in terms of the Euler angles.

Conventions for rotations

First, we wish to specify a convention which we use for rotations (or translations). A rotation can be viewed in two ways.

1. A rotation may be represented as the action of a linear operator R which rotates x into x' keeping the coordinate system unchanged, that is,

$$x' = Rx \qquad (6.4)$$

or, in terms of components,

$$x'_i = R_{ij} x_j. \qquad (6.5)$$

This is usually called the active point of view and it will be adopted here. It means that the physical system is rotated (or translated).

2. A rotation may be regarded as a coordinate transformation (or rotation of the coordinate system) with a change of basis from e_i to e'_i:

$$e'_i = R e_i \quad \text{or} \quad e'_i = R_{\alpha i} e_\alpha. \qquad (6.6)$$

Then the components of x with respect to the new basis are

$$x'_i = R_{ji} x_j \qquad (6.7)$$

because, due to (6.6), one has

$$x = x_j e_j = x'_i e'_i = x'_i R_{ji} e_j. \qquad (6.8)$$

This is usually called the passive point of view and it is used in most treatments of the Lorentz transformations (Chapter 7). It means that the physical system is fixed and is seen by observers which move with respect to it. Comparison of (6.5) with (6.7) shows that the active and the passive point of view are the inverse of each other. This amounts to a change of sign of the group parameters as in (5.52) when one passes from the active to the passive point of view or vice versa.

In a rotation, a scalar wave function $\psi_\alpha(x)$ transforms into $\psi'_\alpha(x)$, such that

$$\psi'_\alpha(x') = \psi_\alpha(x) \qquad (6.9)$$

or, by inverting (6.4)
$$\psi'_\alpha(x') = \psi_\alpha(R^{-1}x'). \tag{6.10}$$
Replacing x' by x one has
$$\psi'_\alpha(x) = \psi_\alpha(R^{-1}x). \tag{6.11}$$
There is a transformation operator S_R which changes ψ_α into ψ'_α, so one can write
$$\psi'_\alpha(x) = S_R \psi_\alpha(x) = \psi_\alpha(R^{-1}x). \tag{6.12}$$
Let us take a Hamiltonian invariant under rotations
$$H(x') = S_R H(x) S_R^{-1}. \tag{6.13}$$
The index α above should be the set $\alpha = (E,m)$ where E is an eigenvalue of H and the quantum number m labels degenerate eigenfunctions. One has
$$H\psi_{E,m} = E\psi_{E,m}. \tag{6.14}$$
In a rotated system, one has
$$H(x') S_R \psi_{E,m} = E S_R \psi_{E,m} \tag{6.15}$$
where $S_R \psi_{E,m}$ is also an eigenfunction of H belonging to the eigenvalue E. Thus it can be written as a linear combination of all eigenfunctions belonging to E:
$$S_R \psi_{E,m}(x) = \sum_{m'} D^E_{m',m} \psi_{E,m'}. \tag{6.16}$$
Note that the summation is carried over the first index of the matrix elements $D^E_{m',m}(R)$ consistent with equation (3.24). The matrix D^E depends on the angles of rotation and is an irreducible representation of the rotation group. The index E of the representation D is related to the angular momentum ℓ. Thus the complete set of degenerate eigenfunctions belonging to a given eigenvalue E forms an invariant subspace of the linear operators S_R. This is consistent with the general remarks made in Section 3.3.

Actually $\psi_{E,m} \sim Y_{\ell m}(\Omega)$ or in other words the spherical harmonics $Y_{\ell m}$ for a fixed ℓ mix among themselves under a rotation. The case $\ell = 1$ will be presented in detail below.

Geometrical considerations

A rotation is a transformation of coordinates which leaves invariant lengths of vectors and angles between them. Let us take a Cartesian system of axes Ox, Oy, Oz and rotate it about a given axis passing through the origin O. This produces another Cartesian system of axes Ox', Oy', Oz'. In the fixed system, the unit vectors are

$$e_1 = \begin{pmatrix} 1 \\ 0 \\ 0 \end{pmatrix}, \quad e_2 = \begin{pmatrix} 0 \\ 1 \\ 0 \end{pmatrix}, \quad e_3 = \begin{pmatrix} 0 \\ 0 \\ 1 \end{pmatrix}. \tag{6.17}$$

According to (6.6), the rotated unit vectors become

$$e'_1 = \begin{pmatrix} R_{11} \\ R_{21} \\ R_{31} \end{pmatrix}, \quad e'_2 = \begin{pmatrix} R_{12} \\ R_{22} \\ R_{32} \end{pmatrix}, \quad e'_3 = \begin{pmatrix} R_{13} \\ R_{23} \\ R_{33} \end{pmatrix}. \quad (6.18)$$

Their lengths remain invariant by definition

$$R_{11}^2 + R_{21}^2 + R_{31}^2 = 1$$
$$R_{12}^2 + R_{22}^2 + R_{32}^2 = 1$$
$$R_{13}^2 + R_{23}^2 + R_{33}^2 = 1$$

and their orthogonality is preserved

$$R_{11}R_{12} + R_{21}R_{22} + R_{31}R_{32} = 0, \quad \text{etc.}$$

Putting these relations together we obtain the orthogonality property (6.3) for a matrix $a \equiv R$.

Now let us take a vector in the fixed coordinate system

$$x = x_1 \begin{pmatrix} 1 \\ 0 \\ 0 \end{pmatrix} + x_2 \begin{pmatrix} 0 \\ 1 \\ 0 \end{pmatrix} + x_3 \begin{pmatrix} 0 \\ 0 \\ 1 \end{pmatrix}. \quad (6.19)$$

If both the coordinate systems and the vector are rotated, the vector x transformed into x' has the same components in the new basis, so

$$x' = x_1 \begin{pmatrix} R_{11} \\ R_{21} \\ R_{31} \end{pmatrix} + x_2 \begin{pmatrix} R_{12} \\ R_{22} \\ R_{32} \end{pmatrix} + x_3 \begin{pmatrix} R_{13} \\ R_{23} \\ R_{33} \end{pmatrix} \quad (6.20)$$

which implies

$$x'_i = R_{ij}x_j \quad (6.21)$$

i.e. the relation (6.5). Moreover, by writing explicitly the orthogonality relation $e'_1 = e'_2 \times e'_3$ as

$$R_{11} = R_{22}R_{23} - R_{33}R_{32}, \quad \text{etc.} \quad (6.22)$$

one immediately finds that

$$\det R = 1. \quad (6.23)$$

Hence a rotation in a three-dimensional space is an SO(3) transformation. For this reason, an orthogonal transformation with $\det a = 1$ is called a *proper* rotation. It can be applied to a rigid body. One can say that all proper rotations form a group isomorphic to SO(3). A transformation with $\det R = -1$ is called *improper*. The simplest is a spatial inversion $x' = -x$ for which

$$R = \begin{pmatrix} -1 & 0 & 0 \\ 0 & -1 & 0 \\ 0 & 0 & -1 \end{pmatrix}. \tag{6.24}$$

The operation of reflection in a plane, for example, $x \to x$, $y \to y$, $z \to -z$ is also an improper orthogonal transformation. In this case

$$R = \begin{pmatrix} 1 & 0 & 0 \\ 0 & 1 & 0 \\ 0 & 0 & -1 \end{pmatrix}. \tag{6.25}$$

The product of a proper rotation and an inversion gives an improper transformation belonging to O(3). Note that a reflection in a plane can be written as a product of a proper rotation and an inversion. In particular, the transformation (6.25) results from the product of an inversion and a rotation through $\theta = \pi$ about Oz. The matrix corresponding to a rotation about the Oz-axis is given by (3.49) which, enlarged to three dimensions, becomes

$$R_z(\theta) = \begin{pmatrix} e^{-i\theta} & 0 & 0 \\ 0 & e^{i\theta} & 0 \\ 0 & 0 & 1 \end{pmatrix}. \tag{6.26}$$

Thus a rotation through $\theta = \pi$ corresponds to a diagonal matrix with $-1, -1$, and 1 as diagonal elements. Hence (6.25) can be written as

$$\begin{pmatrix} 1 & 0 & 0 \\ 0 & 1 & 0 \\ 0 & 0 & -1 \end{pmatrix} = \begin{pmatrix} -1 & 0 & 0 \\ 0 & -1 & 0 \\ 0 & 0 & -1 \end{pmatrix} \begin{pmatrix} -1 & 0 & 0 \\ 0 & -1 & 0 \\ 0 & 0 & 1 \end{pmatrix}.$$

The group SO(3) has three parameters and for the rotation group R_3 these can be chosen to be the Euler angles α, β and γ, so that an element of R_3 can be written as a product of three successive rotations (see, for example, Brink and Satchler 1968):

(1) A rotation through α about Oz;
(2) A rotation through β about the new Oy' axis;
(3) A rotation through γ about the new Oz'' axis. Thus

$$R(\alpha, \beta, \gamma) = R_{z''}(\gamma) R_{y'}(\beta) R_z(\alpha). \tag{6.27}$$

The explicit form of the rotation operator corresponding to this rotation will be given in the next subsection.

Infinitesimal transformations

In order to find the group generators we have to consider an infinitesimal transformation around the identity element **1**. Let us take

$$R = \mathbf{1} + \rho \tag{6.28}$$

where R, $\mathbf{1}$, and ρ are 3×3 matrices. The orthogonality condition (5.84) gives

$$\rho^T + \rho = 0.$$

Explicitly, one has

$$\rho_{ii} = 0; \qquad \rho_{ij} = -\rho_{ji} \quad (i,j = 1, 2, 3) \tag{6.29}$$

which means that the rotation group R_3 has three independent parameters. That is consistent with the general formula (5.86). Let us choose as independent parameters ρ_{12}, ρ_{23}, and ρ_{31} and introduce the notation

$$\rho_{12} = -\omega_3, \qquad \rho_{23} = -\omega_1, \qquad \rho_{31} = -\omega_2. \tag{6.30}$$

Then, from (6.21) we get

$$\delta x_1 = x_1' - x_1 = -\omega_3 x_2 + \omega_2 x_3$$
$$\delta x_2 = x_2' - x_2 = +\omega_3 x_1 - \omega_1 x_3$$
$$\delta x_3 = x_3' - x_3 = -\omega_2 x_1 + \omega_1 x_2$$

or, briefly

$$\delta \boldsymbol{x} = \boldsymbol{\omega} \times \boldsymbol{x} \tag{6.31}$$

which represents the modification of a vector rotated through an angle $\omega = (\omega_1^2 + \omega_2^2 + \omega_3^2)^{1/2}$ about an axis parallel to $\boldsymbol{\omega} = (\omega_1, \omega_2, \omega_3)$.

The rotation operator S_R is defined by (6.12) where the argument of ψ_α changes from $\boldsymbol{x}$ to

$$R^{-1}\boldsymbol{x} = R^T \boldsymbol{x} = (\mathbf{1} - \rho)\boldsymbol{x} = \boldsymbol{x} - \delta \boldsymbol{x}. \tag{6.32}$$

The infinitesimal transformation (6.12) of ψ_α reads

$$S_R \psi_\alpha(\boldsymbol{x}) = \psi_\alpha(\boldsymbol{x} - \delta \boldsymbol{x}) = \psi_\alpha(\boldsymbol{x}) - \delta x_i \frac{\partial \psi_\alpha}{\partial x_i}$$
$$= \psi_\alpha - (\omega_2 x_3 - \omega_3 x_2) \frac{\partial \psi_\alpha}{\partial x_1} - (\omega_3 x_1 - \omega_1 x_3) \frac{\partial \psi_\alpha}{\partial x_2} \tag{6.33}$$
$$- (\omega_1 x_2 - \omega_2 x_1) \frac{\partial \psi_\alpha}{\partial x_3}.$$

By using the definition (5.39), one obtains the generators of the rotation group as

$$X_1 = x_2 \frac{\partial}{\partial x_3} - x_3 \frac{\partial}{\partial x_2}$$
$$X_2 = x_3 \frac{\partial}{\partial x_1} - x_1 \frac{\partial}{\partial x_3} \tag{6.34}$$
$$X_3 = x_1 \frac{\partial}{\partial x_2} - x_2 \frac{\partial}{\partial x_1}.$$

These are real operators which give rise to a real Lie algebra, but usually one multiplies them by $-i$ (and ω by i in (6.33)) in order to identify them with the three components of the angular momentum

$$J_K = L_K = -iX_K \quad K = 1, 2, 3, \quad \hbar = 1. \tag{6.35}$$

THE ROTATION GROUP

Then, the right-hand side of (6.33) can be written as

$$S_R \psi_a(x) = (1 - i\,\boldsymbol{\omega} \cdot \boldsymbol{J})\psi_a(x) \tag{6.36}$$

which contains the infinitesimal form of the rotation operator. The finite form (5.51) reads

$$S_R = e^{-i\,\boldsymbol{\omega} \cdot \boldsymbol{J}} \tag{6.37}$$

which is a unitary transformation $S_R^+ = S_R^{-1}$ because $\boldsymbol{J}$ is Hermitian.

The algebra of the rotation group is given by the commutation relations of the angular momentum

$$[J_i, J_j] = i\,\varepsilon_{ijk} J_k \quad (\hbar = 1) \tag{6.38}$$

where ε_{ijk} is the Levi–Civita symbol (5.185).

As discussed in Section 5.10, this group has only one invariant operator. This is the Casimir operator (5.172) (see Exercise 5.7)

$$J^2 = J_1^2 + J_2^2 + J_3^2. \tag{6.39}$$

The label of an irreducible representation of R_3 is related to an eigenvalue of (6.39), i.e. to a specific value, j, of the angular momentum. Following Wigner, we denote an irreducible representation of R_3 by D^j (See the next Subsection).

The vector $\boldsymbol{\omega} = \omega\,\boldsymbol{n}$ appearing in (6.31) has geometrical significance. This can easily be seen by writing the displacement (6.31) as

$$|\delta \boldsymbol{x}| = |\boldsymbol{\omega} \times \boldsymbol{x}| = \omega x \sin(\boldsymbol{\omega}, \boldsymbol{x}) = \omega x_\perp.$$

The vector $\delta \boldsymbol{x}$ is perpendicular to the plane defined by the vectors $\boldsymbol{x}$ and $\boldsymbol{n}$ and its length is equal to the product of ω times the radius $x_\perp$ of the circle obtained by rotating $\boldsymbol{x}$ around $\boldsymbol{n}$ through ω (Fig. 6.1). Thus $\boldsymbol{n}$ is the unit vector along the rotation axis and ω is the rotation angle.

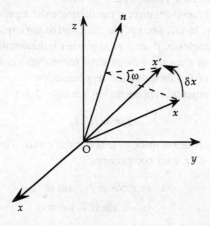

Figure 6.1 Rotation of $\boldsymbol{x}$ through ω about an arbitrary axis $\boldsymbol{n}$.

A rotation about an axis as given by (6.37) can also be viewed as a sequence of three rotations as in Equation (6.27). Accordingly, the rotation operator S_R corresponding to a rotation through Euler angles α, β and γ takes the form

$$S_R(\alpha, \beta, \gamma) = e^{-i\gamma J_{z''}} e^{-i\beta J_{y'}} e^{-i\alpha J_z}. \tag{6.40}$$

One can express $J_{y'}$ and $J_{z''}$ in terms of J_y and J_z and this leads to (Brink and Satchler 1968)

$$S_R(\alpha, \beta, \gamma) = e^{-i\alpha J_z} e^{-i\beta J_y} e^{-i\gamma J_z}, \tag{6.41}$$

which means that the same rotation will be produced by making first a rotation through an angle γ about the original z-axis, followed by a rotation through an angle β still about the original y-axis, and then a rotation through an angle α about the same original z-axis.

Irreducible representations

The irreps of R_3 corresponding to the operator (6.41) are defined by (6.16) as

$$S_R \psi_{jm}(r) = \Sigma D^j_{m'm}(\alpha, \beta, \gamma) \psi_{jm'} \tag{6.42}$$

where the label of an irreducible representation, denoted there by E, is just the angular momentum, as discussed after Equation (6.39).

We shall use the convention of Brink and Satchler (1968) that $S_R(\alpha, \beta, \gamma)$ rotates the physical system counter-clockwise (positive rotation) or, equivalently, rotates the axes clockwise (negative rotation).

Each group has a trivial representation which is called the identical representation. This is a 1×1 matrix. For R_3 this representation has as invariant subspace any scalar, for example the length of a vector to any power.

The first non-trivial representation is the fundamental representation, which in this case is a 3×3 matrix. In this section we shall derive the explicit form of this matrix in terms of the Euler angles α, β, and γ. Any other representation (except the identity representation) is larger than the fundamental representation and can be constructed from direct products of the fundamental representation.

To derive the fundamental representation we take $\psi_\alpha(r) = r$ in (6.12) so that

$$S_R r = R^{-1} r.$$

Let us consider first a rotation through α about the z-axis. Then a vector $r = (x, y, z)$ transforms into $\mathbf{r}' = R^{-1}\mathbf{r}$ with components

$$\begin{aligned} x' &= x \cos \alpha + y \sin \alpha \\ y' &= -x \sin \alpha + y \cos \alpha \\ z' &= z. \end{aligned} \tag{6.43}$$

One can show that the rotation matrix

$$R(\alpha) = \begin{pmatrix} \cos\alpha & \sin\alpha & 0 \\ -\sin\alpha & \cos\alpha & 0 \\ 0 & 0 & 1 \end{pmatrix} \quad (6.44)$$

is reducible. Let us introduce the spherical components of x in terms of spherical harmonics[†] $Y_{\ell m}$

$$x_1 = -\frac{1}{\sqrt{2}}(x+iy) = r\left(\frac{4\pi}{3}\right)^{1/2} Y_{11} \quad (6.45)$$

$$x_0 = z = r\left(\frac{4\pi}{3}\right)^{1/2} Y_{10} \quad (6.46)$$

$$x_{-1} = \frac{1}{\sqrt{2}}(x-iy) = r\left(\frac{4\pi}{3}\right)^{1/2} Y_{1\,-1}. \quad (6.47)$$

The transformation from Cartesian to spherical coordinates is made through the matrix

$$T = \begin{pmatrix} -\frac{1}{\sqrt{2}} & -\frac{i}{\sqrt{2}} & 0 \\ 0 & 0 & 1 \\ \frac{1}{\sqrt{2}} & -\frac{i}{\sqrt{2}} & 0 \end{pmatrix} \quad (6.48)$$

and one finds that

$$T R(\alpha) T^{-1} = \begin{pmatrix} e^{-i\alpha} & 0 & 0 \\ 0 & 1 & 0 \\ 0 & 0 & e^{i\alpha} \end{pmatrix} \quad (6.49)$$

which is reducible.[‡] Then one can write

$$(Tr')_m = \sum_{m'} (T R(\alpha) T^{-1})_{m'm} (Tr)_{m'}. \quad (6.50)$$

Here the $(T r)_m$ are proportional to Y_{1m}. Thus one has

$$Y_{1m}(\theta, \varphi') = e^{-im\alpha} Y_{1m}(\theta, \varphi) = e^{-i\alpha L_z} Y_{1m}(\theta, \varphi) \quad (6.51)$$

where

$$\varphi' = \varphi - \alpha. \quad (6.52)$$

Thus for a rotation through an angle α about the z-axis the matrix elements of D^1 defined by (6.42) are

$$D^1_{m'm}(\alpha, 0, 0) = e^{-im'\alpha} \delta_{m'm}. \quad (6.53)$$

[†] The functions $Y_{\ell m}$ are defined by

$$Y_{\ell m}(\theta, \varphi) = \left[\frac{2\ell+1}{4\pi} \frac{(\ell-m)!}{(\ell+m)!}\right]^{\frac{1}{2}} P_\ell^m(\cos\theta) e^{im\varphi}$$

where P_ℓ^m are associated Legendre functions (see, for example, Jackson (1975), section 3.5.)

[‡] Note that this matrix acts on the column (x_1, x_0, x_{-1}) while (6.26) acts on the column (x_1, x_{-1}, x_0)

This matrix is diagonal, i.e. reducible, which is not surprising because it corresponds to rotations about the z-axis, which form a subgroup R_2 of R_3. (Recall that if a representation is irreducible for a group it can be reducible for a subgroup of that group.)

We next consider a rotation through β about the y-axis. One can easily see that in the same convention as above the rotation matrix is

$$R(\beta) = \begin{pmatrix} \cos\beta & 0 & -\sin\beta \\ 0 & 1 & 0 \\ \sin\beta & 0 & \cos\beta \end{pmatrix}. \tag{6.54}$$

The corresponding similarity transformation with T of (6.48) defines

$$M(\beta) = T R(\beta) T^{-1}$$

the form of which turns out to be

$$M(\beta) = \begin{pmatrix} \frac{1}{2}(1+\cos\beta) & \frac{1}{\sqrt{2}}\sin\beta & \frac{1}{2}(1-\cos\beta) \\ -\frac{1}{\sqrt{2}}\sin\beta & \cos\beta & \frac{1}{\sqrt{2}}\sin\beta \\ \frac{1}{2}(1-\cos\beta) & -\frac{1}{\sqrt{2}}\sin\beta & \frac{1}{2}(1+\cos\beta) \end{pmatrix}. \tag{6.55}$$

This matrix is no longer in diagonal form.

The irreducible representation $d^1(\beta)$ for a rotation through $\alpha = 0$, $\beta \neq 0$, $\gamma = 0$ is a matrix which is the transpose of (6.55) because of the summation convention over the first index in (6.42):

$$d^1(\beta) = M^{\mathrm{T}}(\beta). \tag{6.56}$$

The last step is a rotation through γ about the z-axis again. The result is similar to (6.53):

$$D^1_{m'm}(0, 0, \gamma) = e^{-im\gamma} \delta_{m'm}. \tag{6.57}$$

The fundamental representation of R_3 is the product of three matrices corresponding to successive rotations through α, β and γ. Putting together (6.53), (6.56), and (6.57), one gets

$$D^1_{m'm}(\alpha\beta\gamma) = e^{-im'\alpha} d^1_{m'm}(\beta) e^{-im\gamma}. \tag{6.58}$$

The fundamental representation corresponds to $j = 1$. Similarly one can derive a rotation matrix for any $j = \frac{1}{2}, 1, \frac{3}{2}, \ldots$ in the form

$$D^j_{m'm}(\alpha\beta\gamma) = e^{-im'\alpha} d^j_{m'm}(\beta) e^{-im\gamma} \tag{6.59}$$

where the matrix d^j is generally called the *reduced* rotation matrix. Its matrix elements are real and are given by

$$d^j_{m'm}(\beta) = \sum_t (-)^t \frac{[(j+m')!(j-m')!(j+m)!(j-m)!]^{1/2}}{(j+m'-t)!(j-m-t)!\,t!\,(t+m-m')!} \\ \times \left(\cos\frac{\beta}{2}\right)^{2j+m'-m-2t} \left(\sin\frac{\beta}{2}\right)^{2t+m-m'} \tag{6.60}$$

where the sum is taken over all values of t which lead to non-negative factorials. Properties of $d^j_{m'm}(\beta)$ and $D^j_{m'm}(\alpha\beta\gamma)$ can be found in Appendix V of Brink and Satchler (1968).

The rotation group introduces integer values of j. Based on the homomorphism of SU(2) with R_3 (Chapter 8) one can understand the validity of (6.59) and (6.60) when j is half-integer as well.

A special case of the rotation matrix is

$$D^\ell_{m0}(\alpha\beta\gamma) = \left(\frac{4\pi}{2\ell+1}\right)^{1/2} Y^*_{\ell m}(\beta,\alpha). \tag{6.61}$$

The rotation matrices are unitary because SU(2) is a compact Lie group (Section 3.5). Therefore

$$\sum_{m'} D^j_{mm'}(\alpha\beta\gamma)(D^j_{nm'}(\alpha\beta\gamma))^* = \delta_{mn} \tag{6.62}$$

$$\sum_{m'} (D^j_{m'n}(\alpha\beta\gamma))^* D^j_{m'm}(\alpha\beta\gamma) = \delta_{mn}. \tag{6.63}$$

These relations express the fact that $D^j_{mm'}$ are matrix elements of the unitary transformation (6.41) in an invariant subspace of $2j+1$ orthonormal functions ψ_{jm}.

Besides unitarity the rotation matrices obey another orthogonality relation which represents their orthogonality on the surface of a unit sphere. This gives

$$\frac{1}{8\pi^2}\int_0^{2\pi} d\alpha \int_0^\pi d\beta \sin\beta \int_0^{2\pi} d\gamma \, (D^{j_1}_{m_1\mu_1}(\alpha\beta\gamma))^* D^{j_2}_{m_2\mu_2}(\alpha\beta\gamma)$$
$$= \frac{1}{2j_1+1}\delta_{j_1j_2}\delta_{m_1m_2}\delta_{\mu_1\mu_2}. \tag{6.64}$$

This is a particular example of the orthogonality relation (3.88) with $M=1$ and $n_\mu = 2j_1+1$. The integrand contains the weight function ρ of (3.86),

$$\rho = \frac{1}{8\pi^2}\sin\beta \tag{6.65}$$

resulting from the parametrization of the group in terms of Euler angles

$$0 \le \alpha \le 2\pi, 0 \le \beta \le \pi, 0 \le \gamma \le 2\pi. \tag{6.66}$$

The weight functions for the groups $O(n)$ ($n \ge 2$) were derived by Weyl (1946) (see also Appendix C).

Clebsch–Gordan series

The direct product of two representations gives rise to a representation which is generally reducible. This can be split into a sum of irreps which is called a Clebsch–Gordan (CG) series (Section 3.13). In Section 4.7, we analysed the CG series of the permutation group S_n. For R_3 the problem is analogous and somewhat simpler. The CG series can be obtained through the coupling of angular momenta.

Let us take two irreps D^{j_1} and D^{j_2} corresponding to two angular momenta j_1 and j_2. The addition of these two angular momenta gives an angular momentum j with values between $|j_1 - j_2|$ and $j_1 + j_2$. In terms of group representations this means that the direct product of D^{j_1} and D^{j_2} gives the CG series

$$D^{j_1} \times D^{j_2} = \sum_{j=|j_1-j_2|}^{j_1+j_2} D^j. \tag{6.67}$$

The direct product representation $D^{j_1} \times D^{j_2}$ in the left-hand side of (6.67) is obtained in the direct product space $|j_1 m_1 > |j_2 m_2 >$. These are eigenvectors of J_1^2, J_{1z}, J_2^2 and J_{2z}. They can be written as linear combinations of eigenstates of J^2, J_z, J_1^2 and J_2^2 which correspond to the block diagonal form of the right-hand side of (6.67):

$$|j_1 m_1 > |j_2 m_2 > = \sum_{jm} <jm|j_1 j_2 m_1 m_2> |jm> \tag{6.68}$$

where the sum over j runs from $|j_1 - j_2|$ to $j_1 + j_2$ as in (6.67), and $m = m_1 + m_2$. Note that $|jm>$ is a short hand notation for $|j_1 j_2 jm>$. The coefficients $<jm|j_1 j_2 m_1 m_2>$ of the linear combination (6.68) are called CG coefficients of R_3. The relation (6.68) can be inverted, i.e. one can express a basis vector $|jm>$ of D^j as a linear combination of $|j_1 m_1 > |j_2 m_2 >$:

$$|jm> = \sum_{m_1 m_2} <j_1 j_2 m_1 m_2|jm> |j_1 j_2 m_1 m_2>. \tag{6.69}$$

The properties of R_3 allow CG coefficients to be real so that

$$<jm|j_1 j_2 m_1 m_2> = <j_1 j_2 m_1 m_2|jm>. \tag{6.70}$$

The relations (6.68) and (6.69) represent an orthogonal transformation between two orthogonal bases. In detail this orthogonal transformation can be written

$$\sum_{m_1 m_2} <j_1 j_2 m_1 m_2|jm><j_1 j_2 m_1 m_2|j'm'> = \delta_{jj'} \delta_{mm'} \tag{6.71}$$

$$\sum_{jm} <j_1 j_2 m_1 m_2|jm><j_1 j_2 m_1' m_2'|jm> = \delta_{m_1 m_1'} \delta_{m_2 m_2'}. \tag{6.72}$$

In (6.67) each representation D^j has the dimension $2j + 1$. Suppose each D^j appears only once in the reduction of the direct product. Then, the total number of $|jm>$ states is

$$\sum_{j=|j_1-j_2|}^{j_1+j_2} (2j+1) \tag{6.73}$$

which is equal to $(2j_1 + 1)(2j_2 + 1)$, i.e. the dimension of the direct product space $|j_1 m_1 > |j_2 m_2 >$. This is a confirmation of the fact that D^j appears with multiplicity one in the reduction of $D^{j_1} \times D^{j_2}$.

The CG coefficients have alternative notations. For example

$$<j_1 j_2 m_1 m_2|jm> = C^{j_1\ j_2\ j}_{m_1\ m_2\ m}. \tag{6.74}$$

THE ROTATION GROUP

They are also called vector-addition coefficients. For given j_1 and j_2 they are always restricted by the triangular condition

$$|j_1 - j_2| \leq j \leq j_1 + j_2 \tag{6.75}$$

and they vanish unless

$$m = m_1 + m_2. \tag{6.76}$$

This shows that the sum over m in (6.68) or (6.72) is purely formal, and that for fixed m_1 (or m_2) and m the double sum in (6.69) and (6.71) reduces to a single sum.

It is possible to obtain explicit formulae for the CG coefficients, and a number of notable derivations of the general formula have been given by Wigner (1959), Racah (1942), and Schwinger (1952). (For the relation between CG coefficients, and hypergeometric functions, see Rose (1955), Appendix B.) Tables with analytic expression are available for particular cases. In Table 6.1 we reproduce tables of CG coefficients for $j_2 = \frac{1}{2}$, 1, $\frac{3}{2}$, and $j_1 = j$, arbitrary.

Table 6.1 Clebsch–Gordan coefficients for $j_1 = j$, $j_2 = \frac{1}{2}$, 1, and $\frac{3}{2}$.

$$\left(j\tfrac{1}{2}mm'|JM\right)$$

J	$m' = +\tfrac{1}{2}$	$m' = -\tfrac{1}{2}$
$j+\tfrac{1}{2}$	$\left(\dfrac{j+M+\tfrac{1}{2}}{2j+1}\right)^{1/2}$	$\left(\dfrac{j-M+\tfrac{1}{2}}{2j+1}\right)^{1/2}$
$j-\tfrac{1}{2}$	$-\left(\dfrac{j-M+\tfrac{1}{2}}{2j+1}\right)^{1/2}$	$\left(\dfrac{j+M+\tfrac{1}{2}}{2j+1}\right)^{1/2}$

$$(j\,1\,m\,m'|\,JM)$$

J	$m' = +1$	$m' = 0$	$m' = -1$
$j+1$	$\left\{\dfrac{(j+M)(j+M+1)}{(2j+1)(2j+2)}\right\}^{1/2}$	$\left\{\dfrac{(j-M+1)(j+M+1)}{(2j+1)(j+1)}\right\}^{1/2}$	$\left\{\dfrac{(j-M)(j-M+1)}{(2j+1)(2j+2)}\right\}^{1/2}$
j	$-\left\{\dfrac{(j+M)(j-M+1)}{2j(j+1)}\right\}^{1/2}$	$\dfrac{M}{\{j(j+1)\}^{1/2}}$	$\left\{\dfrac{(j-M)(j+M+1)}{2j(j+1)}\right\}^{1/2}$
$j-1$	$\left\{\dfrac{(j-M)(j-M+1)}{2j(2j+1)}\right\}^{1/2}$	$-\left\{\dfrac{(j-M)(j+M)}{j(2j+1)}\right\}^{1/2}$	$\left\{\dfrac{(j+M+1)(j+M)}{2j(2j+1)}\right\}^{1/2}$

Table 6.1 cont.

$$(j\tfrac{3}{2}mm'|JM)$$

J	$m'=\tfrac{3}{2}$	$m'=\tfrac{1}{2}$
$j+\tfrac{3}{2}$	$\left\{\dfrac{(j+M-\tfrac{1}{2})(j+M+\tfrac{1}{2})(j+M+\tfrac{3}{2})}{(2j+1)(2j+2)(2j+3)}\right\}^{1/2}$	$\left\{\dfrac{3(j+M+\tfrac{1}{2})(j+M+\tfrac{3}{2})(j-M+\tfrac{3}{2})}{(2j+1)(2j+2)(2j+3)}\right\}^{1/2}$
$j+\tfrac{1}{2}$	$-\left\{\dfrac{3(j+M-\tfrac{1}{2})(j+M+\tfrac{1}{2})(j-M+\tfrac{3}{2})}{2j(2j+1)(2j+3)}\right\}^{1/2}$	$-(j-3M+\tfrac{3}{2})\left\{\dfrac{j+M+\tfrac{1}{2}}{2j(2j+1)(2j+3)}\right\}^{1/2}$
$j-\tfrac{1}{2}$	$\left\{\dfrac{3(j+M-\tfrac{1}{2})(j-M+\tfrac{1}{2})(j-M+\tfrac{3}{2})}{(2j-1)(2j+1)(2j+2)}\right\}^{1/2}$	$-(j+3M-\tfrac{1}{2})\left\{\dfrac{j-M+\tfrac{1}{2}}{(2j-1)(2j+1)(2j+2)}\right\}^{1/2}$
$j-\tfrac{3}{2}$	$-\left\{\dfrac{(j-M-\tfrac{1}{2})(j-M+\tfrac{1}{2})(j-M+\tfrac{3}{2})}{2j(2j-1)(2j+1)}\right\}^{1/2}$	$\left\{\dfrac{3(j+M-\tfrac{1}{2})(j-M-\tfrac{1}{2})(j-M+\tfrac{1}{2})}{2j(2j-1)(2j+1)}\right\}^{1/2}$

J	$m'=-\tfrac{1}{2}$	$m'=-\tfrac{3}{2}$
$j+\tfrac{3}{2}$	$\left\{\dfrac{3(j+M+\tfrac{1}{2})(j-M+\tfrac{1}{2})(j-M+\tfrac{3}{2})}{(2j+1)(2j+2)(2j+3)}\right\}^{1/2}$	$\left\{\dfrac{(j-M-\tfrac{1}{2})(j-M+\tfrac{1}{2})(j-M+\tfrac{3}{2})}{(2j+1)(2j+2)(2j+3)}\right\}^{1/2}$
$j+\tfrac{1}{2}$	$(j+3M+\tfrac{3}{2})\left\{\dfrac{j-M+\tfrac{1}{2}}{2j(2j+1)(2j+3)}\right\}^{1/2}$	$\left\{\dfrac{3(j+M+\tfrac{1}{2})(j-M-\tfrac{1}{2})(j-M+\tfrac{1}{2})}{2j(2j+1)(2j+3)}\right\}^{1/2}$
$j-\tfrac{1}{2}$	$-(j-3M-\tfrac{1}{2})\left\{\dfrac{j+M+\tfrac{1}{2}}{(2j-1)(2j+1)(2j+2)}\right\}^{1/2}$	$\left\{\dfrac{3(j+M+\tfrac{1}{2})(j+M+\tfrac{3}{2})(j-M-\tfrac{1}{2})}{(2j-1)(2j+1)(2j+2)}\right\}^{1/2}$
$j-\tfrac{3}{2}$	$-\left\{\dfrac{3(j+M-\tfrac{1}{2})(j+M+\tfrac{1}{2})(j-M-\tfrac{1}{2})}{2j(2j-1)(2j+1)}\right\}^{1/2}$	$\left\{\dfrac{(j+M-\tfrac{1}{2})(j+M+\tfrac{1}{2})(j+M+\tfrac{3}{2})}{2j(2j-1)(2j+1)}\right\}^{1/2}$

The CG coefficients have symmetry properties under the interchange of their indices. In the notation of (6.74) some of these properties are

$$C^{j_1\;j_2\;j}_{m_1\,m_2\,m} = (-)^{j_1+j_2-j}\, C^{j_2\;j_1\;j}_{m_2\,m_1\,m} \tag{6.77}$$

$$C^{j_1\;j_2\;j}_{m_1\,m_2\,m} = (-)^{j_1+j_2-j}\, C^{j_1\;\;j_2\;\;j}_{-m_1\,-m_2\,-m} \tag{6.78}$$

$$C^{j_1\;j_2\;j}_{m_1\,m_2\,m} = (-)^{j_1-m_1}\left(\frac{2j+1}{2j_2+1}\right)^{1/2} C^{j_1\;\;j\;\;j_2}_{m_1\,-m\,-m_2}. \tag{6.79}$$

Wigner modified the CG coefficients in order to simplify their symmetry properties. These symmetrized coefficients, called Wigner or $3-j$ coefficients and denoted by $\begin{pmatrix} j_1 & j_2 & j \\ m_1 & m_2 & m \end{pmatrix}$, are related to (6.74) by

$$C^{j_1\;\;j_2\;\;j}_{m_1\,m_2\,-m} = (-)^{j_1-j_2-m}(2j+1)^{1/2}\begin{pmatrix} j_1 & j_2 & j \\ m_1 & m_2 & m \end{pmatrix}. \tag{6.80}$$

For $3-j$ coefficients the above properties become:

(a) at a circular permutation of columns the $3-j$ coefficient remains unchanged
(b) interchanging any two columns gives

$$(-)^{j_1+j_2+j}\begin{pmatrix} j_1 & j_2 & j \\ m_1 & m_2 & m \end{pmatrix} = \begin{pmatrix} j_2 & j_1 & j \\ m_2 & m_1 & m \end{pmatrix} = \begin{pmatrix} j_1 & j & j_2 \\ m_1 & m & m_2 \end{pmatrix}$$
$$= \begin{pmatrix} j & j_2 & j_1 \\ m & m_2 & m_1 \end{pmatrix}; \qquad (6.81)$$

(c) The same phase $(-)^{j_1+j_2+j}$ appears when $m_i \to -m_i$:

$$\begin{pmatrix} j_1 & j_2 & j \\ m_1 & m_2 & m \end{pmatrix} = (-)^{j_1+j_2+j}\begin{pmatrix} j_1 & j_2 & j \\ -m_1 & -m_2 & -m \end{pmatrix}. \qquad (6.82)$$

Values of $3-j$ coefficients are tabulated in steps of $1/2$ up to any angular momentum less than or equal to eight by Rotenberg *et al.* (1959).

The CG coefficients can help in deriving D^j representations from direct products $D^{j_1} \times D^{j_2}$. The procedure is as follows. On rotation of the coordinate system, the basis state $|jm>$ defined by (6.69) transforms as

$$S_R|jm> = \sum_{m_1 m_2} <j_1 j_2 m_1 m_2|jm> S_R|j_1 m_1> S_R|j_2 m_2> \qquad (6.83)$$

or, according to (6.42), this becomes

$$\sum_{m'} D^j_{m'm}|jm'> = \sum_{\substack{m_1 m_2 \\ \mu_1 \mu_2}} <j_1 j_2 m_1 m_2|jm> D^{j_1}_{\mu_1 m_1} D^{j_2}_{\mu_2 m_2}|j_1 \mu_1> |j_2 \mu_2>. \qquad (6.84)$$

On the left-hand side one can replace $|jm'>$ by its definition (6.69) to get

$$\sum_{\substack{m' \\ m'_1 m'_2}} D^j_{m'm} <j_1 j_2 m'_1 m'_2|jm'> |j_1 m'_1> |j_2 m'_2>$$
$$= \sum_{\substack{m_1 m_2 \\ \mu_1 \mu_2}} <j_1 j_2 m_1 m_2|jm> D^{j_1}_{\mu_1 m_1} D^{j_2}_{\mu_2 m_2}|j_1 \mu_1> |j_2 \mu_2>$$

and, by orthogonality of $|j_1 \mu_1> |j_2 \mu_2>$, one gets

$$\sum_{m'} D^j_{m'm} <j_1 j_2 \mu_1 \mu_2|jm'> = \sum_{m_1 m_2} <j_1 j_2 m_1 m_2|jm> D^{j_1}_{\mu_1 m_1} D^{j_2}_{\mu_2 m_2}. \qquad (6.85)$$

Use of the orthogonality relation (6.71) leads to

$$D^j_{\mu m} = \sum_{\substack{m_1 m_2 \\ \mu_1 \mu_2}} <j_1 j_2 \mu_1 \mu_2 | j \mu><j_1 j_2 m_1 m_2 | j m> D^{j_1}_{\mu_1 m_1} D^{j_2}_{\mu_2 m_2} \qquad (6.86)$$

where the sums are restricted at given μ and m by $\mu_1 + \mu_2 = \mu$ and $m_1 + m_2 = m$.

By using the orthogonality relation (6.72) in (6.85) one can obtain the alternative form

$$D^{j_1}_{\mu_1 m_1} D^{j_2}_{\mu_2 m_2} = \sum_{j \mu m} <j_1 j_2 \mu_1 \mu_2 | j \mu><j_1 j_2 m_1 m_2 | j m> D^j_{\mu m} \qquad (6.87)$$

which represents the reduction of the matrix elements of the direct product matrix $D^{j_1} \times D^{j_2}$.

Exercise 6.1 Construct the reduced rotation matrix $d^1(\beta)$ as defined by (6.56) starting from the direct product of the reduced rotation matrix $d^{1/2}(\beta)$ with itself.

Solution The reduced rotation matrix for $j = \frac{1}{2}$ is given by (8.54) together with (8.47). This is

$$d^{1/2} = \begin{pmatrix} \cos\frac{\beta}{2} & -\sin\frac{\beta}{2} \\ \sin\frac{\beta}{2} & \cos\frac{\beta}{2} \end{pmatrix}. \qquad (1)$$

For $j = 1$ and $j_1 = j_2 = \frac{1}{2}$ the formula (6.86) becomes

$$d^1_{\mu m} = \sum_{\mu_1 m_1} <\frac{1}{2}\frac{1}{2}\mu_1\mu-\mu_1|1\mu><\frac{1}{2}\frac{1}{2}m_1 m-m_1|1m> d^{1/2}_{\mu_1 m_1} d^{1/2}_{\mu-\mu_1, m-m_1} \qquad (2)$$

where the sums over μ_2 and m_2 were eliminated by the constraints $\mu_2 = \mu - \mu_1$ and $m_2 = m - m_1$.

Let us choose the particular matrix element with $\mu = 1$ and $m = 0$. Then in (2) only $\mu_1 = \frac{1}{2}$ is allowed so we get

$$d^1_{10} = <\frac{1}{2}\frac{1}{2}\frac{1}{2}\frac{1}{2}|11> \Big(<\frac{1}{2}\frac{1}{2}\frac{1}{2}-\frac{1}{2}|10> d^{1/2}_{\frac{1}{2}\frac{1}{2}} d^{1/2}_{\frac{1}{2}-\frac{1}{2}} \\ + <\frac{1}{2}\frac{1}{2}-\frac{1}{2}\frac{1}{2}|10> d^{1/2}_{\frac{1}{2}-\frac{1}{2}} d^{1/2}_{\frac{1}{2}\frac{1}{2}} \Big). \qquad (3)$$

The symmetry property (6.78) gives

$$<\frac{1}{2}\frac{1}{2}\frac{1}{2}-\frac{1}{2}|10> = <\frac{1}{2}\frac{1}{2}-\frac{1}{2}\frac{1}{2}|10> \qquad (4)$$

so that

$$d^1_{10} = 2 <\frac{1}{2}\frac{1}{2}\frac{1}{2}\frac{1}{2}|11><\frac{1}{2}\frac{1}{2}\frac{1}{2}-\frac{1}{2}|10> d^{1/2}_{\frac{1}{2}\frac{1}{2}} d^{1/2}_{\frac{1}{2}-\frac{1}{2}}. \qquad (5)$$

THE ROTATION GROUP

From Table 6.1, we get

$$< \frac{1}{2}\frac{1}{2}\frac{1}{2}\frac{1}{2} | 1\,1 > = 1 \qquad (6)$$

$$< \frac{1}{2}\frac{1}{2}\frac{1}{2} - \frac{1}{2} | 1\,0 > = \frac{1}{\sqrt{2}} \qquad (7)$$

and using (1), we obtain

$$d^1_{10} = -\frac{1}{\sqrt{2}} \sin \beta. \qquad (8)$$

By the same procedure the whole matrix (6.56) can be derived.

Spin of a vector particle

In subsection 6.13 the generators, L_i ($i = 1, 2, 3$) in equation (6.35), of the rotation group were obtained from the change under an infinitesimal transformation of a scalar wave function ψ_α. One can also consider the transformation of a vector function

$$V = \begin{pmatrix} V_1 \\ V_2 \\ V_3 \end{pmatrix}. \qquad (6.88)$$

which describes a particle possessing an internal degree of freedom as, for example, an intrinsic angular momentum. Then, when the physical system is rotated, not only does x change into Rx but also the components V_i change, like the components of a vector. The transformation of V is a generalization of (6.12)

$$S_R V(x) = V'(R^{-1}x). \qquad (6.89)$$

In a more appropriate manner, one can write

$$S_R V(x) = R\, V(R^{-1}x). \qquad (6.90)$$

The infinitesimal transformation of a vector is given by (6.31) so we have

$$\begin{aligned} S_R V(x) &= V(R^{-1}x) + \boldsymbol{\omega} \times V(R^{-1}x) \\ &= V(x - \boldsymbol{\omega} \times x) + \boldsymbol{\omega} \times V(x) \\ &= V - i\boldsymbol{\omega} \cdot L\, V(x) + \boldsymbol{\omega} \times V(x) \end{aligned} \qquad (6.91)$$

where (6.36), with $J = L$, has been used in the last row. Furthermore, note that the operator S_R so defined is a sum of three 3×3 matrix operators. The first two terms are proportional to the unit matrix so that they do not affect the components of V. The third term can be recast into the form

$$\boldsymbol{\omega} \times V = -i\boldsymbol{\omega} \cdot S\, V \qquad (6.92)$$

where the components of the vector S are the 3×3 matrices

$$S_x = i\begin{pmatrix} 0 & 0 & 0 \\ 0 & 0 & -1 \\ 0 & 1 & 0 \end{pmatrix}, \quad S_y = i\begin{pmatrix} 0 & 0 & 1 \\ 0 & 0 & 0 \\ -1 & 0 & 0 \end{pmatrix}, \quad S_z = i\begin{pmatrix} 0 & -1 & 0 \\ 1 & 0 & 0 \\ 0 & 0 & 0 \end{pmatrix}. \tag{6.93}$$

Therefore, one can write

$$S_R = 1 - i\,\boldsymbol{\omega} \cdot (\boldsymbol{L} + \boldsymbol{S}) \tag{6.94}$$

so that the generators of the infinitesimal rotation are now the components of the vector

$$\boldsymbol{J} = \boldsymbol{L} + \boldsymbol{S} \tag{6.95}$$

which is the total angular momentum of a particle of angular momentum L and spin S with $S = 1$. A finite transformation has the form (6.37) with J now defined by (6.95).

If the particle has zero angular momentum the infinitesimal change of V can be written explicitly from (6.92) as

$$\delta V_i = -i\,\omega^a S^a_{ij} V_j \tag{6.96}$$

where the components of $\boldsymbol{\omega}$ and S are now denoted by ω^a and S^a ($a = 1, 2, 3$) for convenience. The role of S^a is therefore to rearrange the components of V_i without affecting its x dependence, and one has

$$[L, S] = 0.$$

The infinitesimal change (6.96) has the same structure as the infinitesimal change of a Dirac spinor under a rotation (Section 7.4), the only difference being that S^a are 4×4 matrices for Dirac spinors. We note that the form (6.96) remains valid for any infinitesimal transformation defined in the space of an internal degree of freedom.

Irreducible tensor operators. The Wigner–Eckart theorem

If, under a rotation, a wave function changes from ψ_α to $S_R \psi_\alpha$, the change in the expectation value of an operator Ω is

$$< S_R \psi_\alpha | \Omega | S_R \psi_\beta > = < \psi_\alpha | S_R^+ \Omega S_R | \psi_\beta > \tag{6.97}$$

so one can say that the matrix elements of Ω for rotated states are equal to the matrix elements of the rotated operator

$$\Omega_R = S_R^+ \Omega S_R = S_R^{-1} \Omega S_R \tag{6.98}$$

for the original states. Using (6.37), one finds that an infinitesimal rotation gives

$$\Omega_R = \Omega + i[\boldsymbol{\omega} \cdot \boldsymbol{J}, \Omega]. \tag{6.99}$$

For a scalar operator, one has $\Omega_R = \Omega$ because it commutes with J. For J itself, which is a vector operator, one gets

$$\boldsymbol{J}_R = \boldsymbol{J} + \boldsymbol{\omega} \times \boldsymbol{J}. \tag{6.100}$$

A scalar is a rank-zero and a vector a rank-one operator with respect to rotations. Starting from the components of a vector operator, one can construct higher-rank operators as shown in Sections 4.5 and 5.11. It is convenient to introduce spherical tensors defined as linear combinations of Cartesian tensors. They have the rotation properties of the spherical harmonics

$$S_R Y_{\ell m}(\theta, \varphi) = \sum D^{\ell}_{m'm}(\alpha, \beta, \gamma) Y_{\ell m'}(\theta, \varphi). \tag{6.101}$$

A set of $2J + 1$ operators which transform according to the irreducible representation D^J of the rotation group[†]

$$S_R T_{JM} S_R^+ = \sum_{M'} D^J_{M'M}(\alpha, \beta, \gamma) T_{JM'} \tag{6.102}$$

is called an *irreducible tensor operator* of rank J. For an infinitesimal transformation, the definition (6.102) leads to the following commutation relations (see Brink and Satchler 1968):

$$[J_x \pm i J_y, T_{JM}] = [(J \mp M)(J \pm M + 1)]^{1/2} T_{JM \pm 1} \tag{6.103}$$

$$[J_z, T_{JM}] = M T_{JM}. \tag{6.104}$$

These relations represent an equivalent definition of irreducible tensor operators (6.102) as given by Racah. They can easily be derived by noting that, for an infinitesimal rotation $1 - i \alpha J_k$, one has

$$D^J_{M'M} = \langle JM' | 1 - \alpha J_k | JM \rangle = \delta_{M'M} - i\alpha \langle JM' | J_k | JM \rangle. \tag{6.105}$$

With this rotation the relation (6.102) becomes

$$(1 - i \alpha J_k) T_{JM} (1 + i \alpha J_k) = T_{JM} - i \alpha \sum_{M'} \langle JM' | J_k | JM \rangle T_{JM'}$$

or

$$[J_k, T_{JM}] = \sum_{M'} \langle JM' | J_k | JM \rangle T_{JM'}. \tag{6.106}$$

The non-zero matrix elements of J_z, $J_\pm = J_x \pm J_y$ are

$$\langle jm | J_z | jm \rangle = m \tag{6.107}$$

$$\langle jm \pm 1 | J_\pm | jm \rangle = [(j \mp m)(j \pm m + 1)]^{1/2} \tag{6.108}$$

which lead to (6.103) and (6.104).

The angular momentum is a tensor of rank $J = 1$. Its spherical components are

$$J_\mu = \begin{cases} -\frac{1}{\sqrt{2}}(J_x + i J_y) & \mu = 1 \\ J_z & \mu = 0 \\ \frac{1}{\sqrt{2}}(J_x - i J_y) & \mu = -1. \end{cases} \tag{6.109}$$

[†] A rotation of the quantum system would lead to a transformed operator as in (6.98). The left-hand side of (6.102) is obtained by a rotation of axes or equivalently by an inverse rotation R^{-1} (see the beginning of this section).

Using the CG coefficients

$$<L1\,M+\mu,-\mu|LM> = [L(L+1)]^{-1/2} \begin{cases} [\tfrac{1}{2}(L-M)(L+M+1)]^{1/2} & \mu=1 \\ M & \mu=0 \\ -[\tfrac{1}{2}(L+M)(L-M+1)]^{1/2} & \mu=-1 \end{cases} \quad (6.110)$$

the commutation relations (6.103) and (6.104) can be written in the compact form

$$[J_\mu, T_{LM}] = (-)^\mu <L1\,M+\mu,-\mu|LM> [L(L+1)]^{1/2} T_{L,M+\mu}. \quad (6.111)$$

The close similarity between irreducible tensors and spherical harmonics suggests that irreducible tensors combine like spherical harmonics. Thus, if $T_{j_1 m_1}$ and $T_{j_2 m_2}$ are irreducible tensors of rank j_1 and j_2, respectively the products $T_{j_1 m_1} T_{j_2 m_2}$ form a tensor transforming under the representation $D^{j_1} \times D^{j_2}$. This representation is reducible and its reduction gives the irreducible tensors

$$T_{JM} = \sum_{m_1 m_2} <j_1 j_2 m_1 m_2 | JM> T_{j_1 m_1} T_{j_2 m_2} \quad (6.112)$$

which is the analogue of (6.69) in a tensor operator space.

The tensor operators are important in physics because many operators can be written as tensor operators, for example the electric and magnetic multipole moments (see, for example, Brink and Satchler 1968). It is therefore useful to evaluate their matrix elements with respect to eigenstates of J^2 and J_z.

Let us take a tensor operator T_{JM} acting on a state $|\alpha j m>$ where α is a set of quantum numbers other than j, m, for example a radial quantum number. The resulting state vector $T_{JM}|\alpha j m>$ transforms under the representation $D^J \times D^j = D^j \times D^J$ of the rotation group. This representation is reducible into representations $D^{j'}$ spanned by the states

$$|\alpha' j' m'> = \sum_{Mm} <jJmM|j'm'> T_{JM}|\alpha j m>. \quad (6.113)$$

This is similar to (6.69) but with one of the factors on the right-hand side being an irreducible tensor; the other factor is an angular momentum eigenstate as in (6.69). For this reason the states $|\alpha' j' m'>$ are not generally normalized. Their normalization does not depend on m but could in general depend on J, j, and the form of T_{JM} (see, for example, Schiff 1968, Section 28). By inverting (6.113) with the help of the orthogonality relation (6.72) one gets

$$T_{JM}|\alpha j m> = \sum_{j'm'} <jJmM|j'm'> |\alpha' j' m'>, \quad (6.114)$$

and multiplying on the left with $|\alpha'' j'' m''>$ one gets

$$<\alpha' j' m'|T_{JM}|\alpha j m> = <jJmM|j'm'><\alpha' j' m'|\alpha' j' m'>. \quad (6.115)$$

The normalization factor $<\alpha'j'm'|\alpha'j'm'>$, which is independent of the magnetic quantum numbers, is denoted by

$$<\alpha'j'm'|\alpha'j'm'> = <\alpha'j'||T_J||\alpha j> \tag{6.116}$$

and is called a *reduced matrix element*,[3] so one can rewrite (6.115) as

$$<\alpha'j'm'|T_{JM}|\alpha jm> = <jJmM|j'm'><\alpha'j'||T_J||\alpha j>. \tag{6.117}$$

This is known as Wigner–Eckart theorem (Wigner 1959; Eckart 1930). It states that the matrix element of an irreducible tensor between angular momentum eigenstates depends on the magnetic quantum numbers only through the CG coefficient, which is the geometrical factor taking into account the conservation laws of angular momentum. The reduced matrix element contains the dynamics of the system.

Example 6.1 Using (6.107) – (6.110) one can write the matrix elements of the spherical components J_μ of the angular momentum operator $\boldsymbol{J}$ in the form

$$<j'm'|J_\mu|jm> = \delta_{jj'}[j(j+1)]^{1/2}<j1m\mu|j'm'> \tag{6.118}$$

from which one obtains

$$<j'||\boldsymbol{J}||j> = [j(j+1)]^{1/2}\delta_{jj'}. \tag{6.119}$$

Exercise 6.2 Derive the reduced matrix element $<\ell'||Y_L||\ell>$ obtained in the integration over three spherical harmonics.

Solution A spherical harmonic Y_{LM} is a (multiplicative) irreducible tensor operator. We are interested in the matrix element

$$<\ell'm'|Y_{LM}|\ell m> = \int d\Omega Y^*_{\ell'm'}(\theta,\varphi)Y_{LM}(\theta,\varphi)Y_{\ell m}(\theta,\varphi). \tag{1}$$

Taking $j_1 = L, j_2 = \ell, \mu_1 = M, \mu_2 = m$ and $m_1 = m_2 = 0$ in (6.87) and using (6.61) one obtains

$$Y_{LM}(\theta,\varphi)Y_{\ell m}(\theta,\varphi) = \sum_{kq}\left[\frac{(2L+1)(2\ell+1)}{4\pi(2k+1)}\right]^{1/2} \tag{2}$$
$$<L\ell Mm|kq><L\ell 00|k0> Y_{kq}(\theta,\varphi)$$

where the fact that the CG coefficients are real has been used. From the orthogonality of spherical harmonics one obtains

$$<\ell'm'|Y_{LM}|\ell m> = \left[\frac{(2L+1)(2\ell+1)}{4\pi(2\ell'+1)}\right]^{1/2}<\ell LmM|\ell'm'><\ell L00|\ell'0> \tag{3}$$

---

[3] The definition of the reduced matrix element is not unique. The one used above is that of Rose (1957).

where the symmetry property (6.77) has been used for both CG coefficients. Thus the Wigner–Eckart theorem (6.117) becomes

$$<\ell'm'|Y_{LM}|\ell m> = <\ell L m M|\ell'm'><\ell'||Y_L||\ell> \qquad (4)$$

with

$$<\ell'||Y_L||\ell> = \left[\frac{(2L+1)(2\ell+1)}{4\pi(2\ell'+1)}\right]^{1/2} <\ell L 0 0|\ell' 0>. \qquad (5)$$

Due to the factorization exhibited in (6.117) the theorem is useful in practical calculations when combined with the orthogonality relations (6.71) and (6.72). For example, if $<j'm'|T_{JM}|jm>$ describes a transition from $|jm>$ to $|j'm'>$, and there is no preferred direction, the total probability is proportional to

$$\sum_{m'M} |<j'm'|T_{JM}|jm>|^2 = |<j'||T_J||j>|^2 \sum_{m'M} <jJmM|j'm'>^2$$

$$= |<j'||T_J||j>|^2 \frac{2j'+1}{2j+1} \sum_{m'M} <j'J-m'M|j-m>^2$$

$$= |<j'||T_J||j>|^2 \frac{2j'+1}{2j+1} \qquad (6.120)$$

where, in the second row, symmetry properties (6.78) and (6.79), and in the last row, the orthogonality property (6.71) of CG coefficients, have been used.

6.2 THE GROUP O(4)

In this Section we derive the generators and the Lie algebra of the group O(4) or SO(4). As discussed in Section 5.9, SO(4) is semi-simple but not simple because its algebra is isomorphic to the direct sum of two so(3) algebras. The classical application of SO(4) is in connection with the *dynamical* symmetry of the hydrogen atom which will be discussed in this section. On a similar basis one can derive the generators of O(3,1), which is a particular case of O(n,m) (see Section 5.6). The group O(3,1) is isomorphic to the homogeneous Lorentz group discussed in the next chapter.

The algebra

The procedure used in Section 6.1 for deriving the generators of SO(3) can be extended straightforwardly to a four-dimensional Euclidean space $\mathbb{R}^4$.

We start from the definition (6.3) of an orthogonal transformation, where the matrix a is taken to be real, and consider an infinitesimal transformation of the form

$$a_{ij} = \delta_{ij} + \sigma_{ij}. \qquad (6.121)$$

THE GROUP O(4)

The orthogonality condition (6.3) requires

$$\sigma_{ij} = -\sigma_{ji} \quad (i,j = 1, 2, 3, 4) \tag{6.122}$$

similar to (6.29) for the rotation group R_3.

The six independent parameters of O(4) consistent with (6.122) are denoted by

$$\sigma_{12} = -\omega_3, \sigma_{23} = -\omega_1, \sigma_{31} = -\omega_2 \tag{6.123}$$
$$\sigma_{41} = -\theta_1, \sigma_{42} = -\theta_2, \sigma_{43} = -\theta_3 \tag{6.124}$$

where (6.123) are the same parameters as in (6.30), because O(3) is a subgroup of O(4). With this notation, an infinitesimal transformation takes the form

$$\begin{pmatrix} x'_1 \\ x'_2 \\ x'_3 \\ x'_4 \end{pmatrix} = \begin{pmatrix} 1 & -\omega_3 & \omega_2 & \theta_1 \\ \omega_3 & 1 & -\omega_1 & \theta_2 \\ -\omega_2 & \omega_1 & 1 & \theta_3 \\ -\theta_1 & -\theta_2 & -\theta_3 & 1 \end{pmatrix} \begin{pmatrix} x_1 \\ x_2 \\ x_3 \\ x_4 \end{pmatrix} \tag{6.125}$$

from which the changes in the vector components $\delta x_i = x'_i - x_i$ can be expressed in terms of the parameters ω_i and θ_i ($i = 1, 2, 3$). Thus the infinitesimal change of a scalar function ψ_α reads

$$\begin{aligned} S\psi_\alpha(x) = \psi_\alpha(a^{-1}x) &= \psi_\alpha(x) - \delta x_i \frac{\partial \psi_\alpha}{\partial x_i} \\ &= \left[1 - \omega_1\left(x_2 \frac{\partial}{\partial x_3} - x_3 \frac{\partial}{\partial x_2}\right) - \omega_2\left(x_3 \frac{\partial}{\partial x_1} - x_1 \frac{\partial}{\partial x_3}\right)\right. \\ &\quad - \omega_3\left(x_1 \frac{\partial}{\partial x_2} - x_2 \frac{\partial}{\partial x_1}\right) + \theta_1\left(x_1 \frac{\partial}{\partial x_4} - x_4 \frac{\partial}{\partial x_1}\right) \\ &\quad + \left. \theta_2\left(x_2 \frac{\partial}{\partial x_4} - x_4 \frac{\partial}{\partial x_2}\right) + \theta_3\left(x_3 \frac{\partial}{\partial x_4} - x_4 \frac{\partial}{\partial x_3}\right)\right]\psi_\alpha. \end{aligned} \tag{6.126}$$

To each parameter corresponds a generator defined by the associated bracket. Multiplying each bracket by $-i$, one can define the six generators as

$$\begin{aligned} L_1 &= -i\left(x_2 \frac{\partial}{\partial x_3} - x_3 \frac{\partial}{\partial x_2}\right) & G_1 &= -i\left(x_1 \frac{\partial}{\partial x_4} - x_4 \frac{\partial}{\partial x_1}\right) \\ L_2 &= -i\left(x_3 \frac{\partial}{\partial x_1} - x_1 \frac{\partial}{\partial x_3}\right) & G_2 &= -i\left(x_2 \frac{\partial}{\partial x_4} - x_4 \frac{\partial}{\partial x_2}\right) \\ L_3 &= -i\left(x_1 \frac{\partial}{\partial x_2} - x_2 \frac{\partial}{\partial x_1}\right) & G_3 &= -i\left(x_3 \frac{\partial}{\partial x_4} - x_4 \frac{\partial}{\partial x_3}\right). \end{aligned} \tag{6.127}$$

Then the Lie algebra of O(4) is

$$\begin{aligned} [L_i, L_j] &= i\,\varepsilon_{ijk} L_k \\ [L_i, G_j] &= i\,\varepsilon_{ijk} G_k \\ [G_i, G_j] &= i\,\varepsilon_{ijk} L_k. \end{aligned} \tag{6.128}$$

Now if we put in (6.126)

$$x_4 = i\, x_0 \tag{6.129}$$

with x_0 real and define

$$K_i = i\left(x_i \frac{\partial}{\partial x_0} + x_0 \frac{\partial}{\partial x_i}\right) \tag{6.130}$$

the transformation (6.126) becomes

$$S\psi_\alpha = (1 - i\, \boldsymbol{\omega} \cdot \boldsymbol{L} + i\, \boldsymbol{\beta} \cdot \boldsymbol{K})\psi_\alpha \tag{6.131}$$

where $\omega = (\omega_1, \omega_2, \omega_3)$ and $-i\boldsymbol{\beta} = \boldsymbol{\theta} = (\theta_1, \theta_2, \theta_3)$. The vector $\boldsymbol{\beta}$ can be taken to be real and proportional to the infinitesimal velocity of a uniformly moving system with respect to another such system. In this case (6.131) corresponds to an infinitesimal Lorentz transformation (see the next chapter) where ω_i are the parameters of an infinitesimal rotation and β_i the parameters of a boost. Note that, using K_i instead of G_i, the Lie algebra becomes

$$\begin{aligned} [L_i, L_j] &= i\, \varepsilon_{ijk} L_k \\ [L_i, K_j] &= i\, \varepsilon_{ijk} K_k \\ [K_i, K_j] &= -i\, \varepsilon_{ijk} L_k, \end{aligned} \tag{6.132}$$

which is precisely the algebra of the Lorentz group derived in the next chapter.

Actually, the use of (6.129) is equivalent to the requirement that the scalar

$$x_1^2 + x_2^2 + x_3^2 - x_0^2 \tag{6.133}$$

be an invariant under a group of transformations. This is a particular example of the invariant (5.89) and the associated group is O(3,1). With $x_0 = ct$, (6.133) is the squared distance in Minkowski space. It is therefore not surprising that (6.132) is the same as the Lie algebra of the homogeneous Lorentz group. One says that the homogeneous Lorentz group and O(3,1) are isomorphic.

Exercise 6.3 Find the Lie algebra of the group O(2,1).

Solution One can start with an infinitesimal transformation of type (6.125) but in a three dimensional space

$$\begin{pmatrix} x'_1 \\ x'_2 \\ x'_3 \end{pmatrix} = \begin{pmatrix} 1 & -\omega_3 & \theta_1 \\ \omega_3 & 1 & \theta_2 \\ -\theta_1 & -\theta_2 & 1 \end{pmatrix} \begin{pmatrix} x_1 \\ x_2 \\ x_3 \end{pmatrix}. \tag{1}$$

Then

$$\begin{aligned} \delta x_1 &= x'_1 - x_1 = -\omega_3 x_2 + \theta_1 x_3 \\ \delta x_2 &= x'_2 - x_2 = \omega_3 x_1 + \theta_2 x_3 \\ \delta x_3 &= x'_3 - x_3 = -\theta_1 x_1 - \theta_2 x_2. \end{aligned} \tag{2}$$

Under (1), a scalar function F changes infinitesimally to

$$F(x - \delta x) = \left[1 - \omega_3 \left(x_1 \frac{\partial}{\partial x_2} - x_2 \frac{\partial}{\partial x_1} \right) + \theta_1 \left(x_1 \frac{\partial}{\partial x_3} - x_3 \frac{\partial}{\partial x_1} \right) \right. \\ \left. + \theta_2 \left(x_2 \frac{\partial}{\partial x_3} - x_3 \frac{\partial}{\partial x_2} \right) \right] F. \quad (3)$$

Now we can take $x_3 = ix_0$ and write

$$F(x - \delta x) = (1 - \omega_3 X_3 - i\,\theta_1 X_1 - i\,\theta_2 X_2) F \quad (4)$$

where the generators X_i are real and have the form

$$X_1 = x_1 \frac{\partial}{\partial x_0} + x_0 \frac{\partial}{\partial x_1}$$
$$X_2 = x_2 \frac{\partial}{\partial x_0} + x_0 \frac{\partial}{\partial x_2} \quad (5)$$
$$X_3 = x_1 \frac{\partial}{\partial x_2} - x_2 \frac{\partial}{\partial x_1}.$$

One can find immediately that the Lie algebra is

$$[X_1, X_2] = X_3,\ [X_2, X_3] = -X_1,\ [X_3, X_1] = -X_2. \quad (6)$$

The hydrogen atom. Dynamical symmetries

In Chapters 1 and 3 (Section 3.3), we have shown that symmetry and degeneracy are associated with each other. If a system has a symmetry under some transformations, the generators of the associated group commute with the Hamiltonian and to each degenerate eigenvalue there correspond eigenstates which form an invariant subspace for a specific irreducible representation. If such states form a basis for a reducible representation, it is said that there is an *accidental* degeneracy between the eigenvalues associated with the irreducible parts of this representation and the invariance group being used is not the maximal group of symmetry.

Examples are the hydrogen atom and the isotropic harmonic oscillator. The maximal group of symmetry in each case is not related to geometrical transformations but arises from particular forms of the force law. For that reason these symmetries are called *dynamical* symmetries. In general, it is not easy to infer a dynamical symmetry. For the hydrogen atom or the isotropic harmonic oscillator, dynamical symmetries result from the corresponding classical systems, but in quantum mechanics there are not always classical analogues.

The hydrogen (Kepler) problem or the three dimensional harmonic oscillator can be generalized to N dimensions. Their quantum counterparts can be described by symmetry algebras isomorphic to so($N + 1$) and su(N), respectively (Higgs 1979; Leemon 1979). Then the energy levels of these systems can be determined by purely algebraic means. In this subsection, it will be shown how the energy levels of the hydrogen atom can be found from an algebra isomorphic to so(4), without solving the Schrödinger equation.

There are other known examples of quantum mechanical systems described by Hamiltonians with accidental degeneracies for which a symmetry algebra has been identified as some finite dimensional Lie algebra. As described by Moshinski, Quesne, and Loyola (1990), the determination of these symmetry Lie algebras is more a matter of art than science. In some cases (see, for example, Kibler and Winternitz 1987), however, the usual Lie algebra does not seem to be enough. In those cases quantum algebras, described in Section 5.12 open new possibilities. (For a review, see Bonatsos, Daskaloyannis, and Kokkotas 1993b.)

Here, we discuss the non-relativistic hydrogen atom, the spin being neglected. The Hamiltonian is of the form

$$H = \frac{p^2}{2\mu} - \frac{\kappa}{r} \tag{6.134}$$

with $\kappa = e^2 > 0$. Although the Hamiltonian has rotational symmetry, its eigenvalues E_n depend only on the *total quantum number n* and not on ℓ:

$$E_n = -\frac{\mu e^4}{2\hbar^2 n^2} \quad (n = 1, 2, 3, \ldots). \tag{6.135}$$

E_n is an eigenvalue corresponding to states with $\ell = 0, 1, \ldots, n-1$. The eigenstates of E_n span a space where a representation of O(3) is reducible to

$$D^0 + D^1 + \cdots + D^{n-1}$$

so that the accidental degeneracy is between eigenvalues with $\ell = 0, 1, \ldots, n-1$ each associated to an irrep D^ℓ of O(3). The indication is that O(3) is a subgroup of a larger group of symmetry for the Hamiltonian (6.134). It was found by Fock (1935) that the larger group is O(4). But the Lie algebraic treatment of the hydrogen atom due to Pauli (1926) is much simpler and precedes that of Fock. With his method, Pauli found the energy levels of the hydrogen atom without solving the Schrödinger equation (Schrödinger proposed his equation simultaneously in 1926). The equivalence between Pauli's and Fock's approaches was established by Bargmann (1936).

In classical mechanics, the Hamiltonian of type (6.134) defines the Kepler problem for which a particular solution is an ellipse. The constants of motion are: the total energy, because H is time independent; the angular momentum $L = r \times p$, due to rotational symmetry; and the Runge–Lenz vector which expresses the fact that the orbit is closed, or equivalently, its major axes remain fixed (no precession). The quantum mechanical analogue of the Runge–Lenz vector is

$$M = \frac{1}{2\mu}(p \times L - L \times p) - \frac{\kappa}{r}r. \tag{6.136}$$

One can show that (Greiner and Müller 1989, Chapter 14)

$$[M, H] = 0 \tag{6.137}$$

THE GROUP O(4)

$$L \cdot M = M \cdot L = 0 \tag{6.138}$$

$$M^2 = \frac{2H}{\mu}\left(L^2 + \hbar^2\right) + \kappa^2. \tag{6.139}$$

These relations have been used by Pauli to find the energy levels of the hydrogen atom. One can view the three components of M as generators of some infinitesimal transformations. Together with the components of L, one has six generators. Their algebra is

$$[L_i, L_j] = i\hbar\, \varepsilon_{ijk} L_k \tag{6.140}$$

$$[L_i, M_j] = i\hbar\, \varepsilon_{ijk} M_k \tag{6.141}$$

$$[M_i, M_j] = -2i\,\frac{\hbar}{\mu} H \varepsilon_{ijk} L_k \tag{6.142}$$

with $\hbar \neq 1$.

Now, consider an eigenvalue E of H corresponding to a bound state, so that $E < 0$. In the subspace of eigenvectors associated to E, one can replace H by E and define the vector operator

$$A = \left(-\frac{\mu}{2E}\right)^{1/2} M. \tag{6.143}$$

One can rewrite the algebra (6.140) – (6.142) in terms of A

$$[L_i, L_j] = i\hbar\, \varepsilon_{ijk} L_k \tag{6.144}$$

$$[L_i, A_j] = i\hbar\, \varepsilon_{ijk} A_k \tag{6.145}$$

$$[A_i, A_j] = i\hbar\, \varepsilon_{ijk} L_k, \tag{6.146}$$

which is now a closed algebra and can be recognized as the algebra (6.128) of O(4) by identifying $A_i \to G_i$ and taking $\hbar = 1$. The group O(4) does not represent a geometrical symmetry of H because the fourth coordinate x_4 appearing in (6.127) is fictitious. However, O(4) is the largest symmetry group of the hydrogen atom due to the dependence of type $1/r$ of V in H and its energy levels can be obtained from the algebra (6.144) – (6.146), as shown below.

One can introduce

$$O_i^+ = \frac{1}{2}(L_i + A_i), \qquad O_i^- = \frac{1}{2}(L_i - A_i) \qquad (i = 1, 2, 3) \tag{6.147}$$

and, using (6.144) – (6.146), one can find that

$$[O_i^+, O_j^+] = i\hbar\, \varepsilon_{ijk} O_k^+$$

$$[O_i^-, O_j^-] = i\hbar\, \varepsilon_{ijk} O_k^- \tag{6.148}$$

$$[O_i^+, O_j^-] = 0,$$

so one can see that the set of three operators O_i^+ or O_i^- satisfies the angular momentum commutation relations (6.38) although neither set corresponds to a physical angular momentum. The relations (6.148) are the commutation relations of two so(3) algebras (or two su(2) algebras) which commute with each other. This means that the Lie algebra (6.144)–(6.146) of O(4) or SO(4) is isomorphic to the direct sum so(3) + so(3), in agreement with Section 5.9.

The operators $(O^+)^2$, $(O^-)^2$, O_3^+, O_3^- commute with H and, for a common eigenvector ψ one has

$$(O^+)^2 \psi = \hbar^2 \, a(a+1)\psi \qquad (6.149)$$
$$(O^-)^2 \psi = \hbar^2 \, b(b+1)\psi$$

with a or b taking the values $0, \frac{1}{2}, 1, \ldots$.

The Lie algebra so(4) is of rank $\ell = 2$ so there are two Casimir invariants for SO(4). One can make the choice

$$C_1 = (O^+)^2 + (O^-)^2 = \frac{1}{2}(L^2 + A^2) \qquad (6.150)$$

$$C_2 = (O^+)^2 - (O^-)^2 = L \cdot A. \qquad (6.151)$$

From the definition (6.136) of M and its relation (6.143) with A one can see that

$$L \cdot A = 0 \qquad (6.152)$$

and this implies that

$$a = b. \qquad (6.153)$$

Thus

$$C_1 \psi = \hbar^2 \, 2a(a+1)\psi. \qquad (6.154)$$

Using (6.139) and (6.143), one can rewrite C_1 as

$$C_1 = -\frac{1}{2}\hbar^2 - \frac{\mu}{4E}\kappa^2 \qquad (6.155)$$

so that

$$2a(a+1) = -\frac{1}{2} - \frac{\mu}{4E\hbar^2}\kappa^2. \qquad (6.156)$$

For $\kappa = e^2$ one obtains

$$E = -\frac{\mu \, e^4}{2\hbar^2(2a+1)^2} \qquad (6.157)$$

which is precisely (6.135) with

$$n = 2a + 1 = 1, 2, 3, \cdots. \qquad (6.158)$$

From (6.147) the angular momentum components are

$$L_i = O_i^+ + O_i^-. \qquad (6.159)$$

Due to the fact that O_i^+ and O_i^- commute among themselves, the angular momentum quantum number ℓ is given by the addition law of two angular momenta, so that

$$|a - b| \leq \ell \leq a + b$$

and by (6.153) and (6.158), one obtains

$$\ell = 0, 1, \cdots, n - 1. \tag{6.160}$$

For the other simple dynamical symmetry which is that of the isotropic harmonic oscillator, the reader is referred to Schiff (1968), Section 30.

6.3 THE EUCLIDEAN GROUPS

The Euclidean groups were illustrated in Exercise 5.4 where the algebra of E_2 was derived. Here we proceed to a more systematic study of the Euclidean groups.

As discussed in the introduction, in non-relativistic physics, the laws of nature are invariant under translations and rotations, and the consequence is the conservation of linear momentum and angular momentum. The Euclidean groups combine rotations with translations. These transformations in a two-dimensional space generate the group E_2 and in three-dimensional space the group E_3. The Euclidean groups are similar to the Poincaré group (Section 7.6) but they are simpler and can be used in the study of the Poincaré group.

Here we consider the group E_3. As rotations have been extensively studied in the previous sections, we concentrate for the moment on translations in a three-dimensional space.

Translations

An element $T(a)$ of the translation group in three dimensions, T_3, is defined by

$$T(a)\, x = x + a \tag{6.161}$$

where any vector x is translated by addition of another vector a. The components of a define the group parameters a^i running in the interval

$$-\infty < a^i < \infty \quad (i = 1, 2, 3). \tag{6.162}$$

This indicates that the group parameters vary in a region of infinite extension, thus the group is non-compact. One can note that the group T_3 is a direct product of one-dimensional translations T_1 in x, y and z.

The multiplication law for translations is

$$T(a_1)T(a_2) = T(a_2)T(a_1) = T(a_1 + a_2) \tag{6.163}$$

which expresses the fact that all translations commute, so the group is Abelian. Let us consider an infinitesimal translation with δa small. Then the infinitesimal change of a scalar function is given by an analogue of (6.12)

$$S_T \psi_\alpha(x) = \psi_\alpha(T^{-1}x) = \psi_\alpha(x - \delta a) = \left(1 - \delta a^i \frac{\partial}{\partial x^i}\right)\psi_\alpha. \quad (6.164)$$

The three infinitesimal generators can be defined by

$$P_i = -i\hbar \frac{\partial}{\partial x^i} \quad (6.165)$$

such as to make them correspond to the three components of the momentum operator. From (6.164) one can derive a finite transformation of the form

$$S_T = \exp\left(-\frac{i}{\hbar} a \cdot P\right). \quad (6.166)$$

The irreps of the translation group are one-dimensional because the group is Abelian (Exercise 3.2). They can be easily obtained by acting with (6.166) on

$$\psi_k(r) = \frac{1}{(2\pi)^{3/2}} e^{i k \cdot r}. \quad (6.167)$$

One finds

$$S_T(a)\psi_k(r) = \frac{1}{(2\pi)^{3/2}} e^{-i k \cdot a} \psi_k(r).$$

Thus the irreps are

$$D^k(a) = e^{-i k \cdot a} \quad (6.168)$$

where $k = |k|$.

The group E_3

The group E_3 is defined by the semi-direct product $T_3 \wedge R_3$ so that its elements are products of translations and proper rotations. If $R(\omega)$ denotes a rotation of parameters $\omega_1, \omega_2, \omega_3$ as in Section 6.1 and $T(a)$ is an element of T_3 the general group element of E_3 is written as $T(a) R(\omega)$ and it transforms any vector x into

$$T(a)R(\omega)x = R(\omega) x + a. \quad (6.169)$$

The translations and rotations do not commute so that, reversing the order of the two transformations, one gets

$$R(\omega)T(a) x = R(\omega)(x + a) = R(\omega) x + R(\omega) a. \quad (6.170)$$

But (6.169) can also be written as

$$T(a)R(\omega) x = R(\omega)T(R^{-1}(\omega) a)x \quad (6.171)$$

i.e. as a translation followed by a rotation as in (6.170). The conclusion is that it is enough to consider elements of the form $T(a) R(\omega)$. One can define a transformation operator S which generates the representations of E_3 by a relation analogous to (6.12)

THE EUCLIDEAN GROUPS

$$S(a,\omega)\psi(x) = \psi([T(a)R(\omega)]^{-1}x) = \psi(R^{-1}(\omega)T^{-1}(a)x) = \psi(R^{-1}(x-a)). \quad (6.172)$$

Since E_3 is a semi-direct product of two groups, the labelling of the irreducible representations is not as simple as that of direct product groups, where the elements of the two groups commute as, for example, SO(4) (Section 6.2).

However, due to the fact that T_3 is an abelian subgroup, one can find a convenient procedure, inspired by the method of *induced* representations (Cornwell 1984, Chapter 5, Sect. 7). The irreps have the same multiplication law as the group elements so, due to (6.171), one can write

$$S(a,\omega) = S(a,0)\,S(0,\omega) = S(0,\omega)\,S(R^{-1}(\omega)a,0). \quad (6.173)$$

The procedure we follow next is that of Elliott and Dawber (1979) in their Section 15.1.3. It contains few steps and yields unitary infinite dimensional representations. The first step is to set up the unitary irreps of the invariant subgroup. These are obtained in the basis vectors of type (6.167) denoted here by $|k>$. Each constitutes an invariant subspace for $S(a,0)$

$$S(a,0)|k> = e^{-ia\cdot k}|k>. \quad (6.174)$$

The next step is to show that $S(0,\omega)|k>$ transforms like the representation $D^{|R(\omega)k|}$. Indeed, due to (6.173), one can write

$$\begin{aligned}S(a,0)S(0,\omega)|k> &= S(0,\omega)\,S(R^{-1}(\omega)a,0)|k>\\ &= S(0,\omega)\,e^{-ik\cdot R^{-1}(\omega)a}|k> \quad (6.175)\\ &= e^{-iR(\omega)k\cdot a}\,S(0,\omega)|k>\end{aligned}$$

where, in the last row, the invariance of the scalar product under rotations $k\cdot R^{-1}a = Rk\cdot a$ has been used. The meaning of (6.175) is that if the representation space of $S(a,0)$ contains the vector $|k>$ it must also contain all vectors $|k'> = R(\omega)|k>$ because $|k'| = |k|$.

A further step is to choose a reference direction, for example the z-axis, and take $|k_0>$ parallel to this axis. A rotation about k_0 is denoted by $R(\omega_0)$. From (6.175), it follows that all $S(0,\omega_0)|k_0>$ vectors transform like $|k_0>$ under translations because $R(\omega_0)|k_0> = |k_0>$. They form an invariant subspace under rotations about the z-axis and may be labelled by m, like the irreps of R_2. Thus the vectors of this invariant subspace can be denoted by $|k_0\,m>$ and they have the property

$$S(0,\omega_0)|k_0 m> = e^{-im\omega_0}|k_0 m>. \quad (6.176)$$

From a given basis vector $|k_0\,m>$ one can generate a set $|k\,m>$ by a rotation in the xy-plane which carries k_0 into k, $R(\tilde{\omega})|k_0> = |k>$:

$$|km> = S(0,\tilde{\omega})|k_0\,m>. \quad (6.177)$$

This is an infinite set with the same m and $k = |k|$ but different directions for k. To prove that this set is invariant under E_3 one uses the fact that any rotation R can be factorized into $R = R_{xy} R_z$, where R_{xy} is a rotation about an axis lying in the xy-plane and R_z is a rotation about the z-axis. Then the irreps of E_3 can be labelled by $k = |\mathbf{k}|$ and m, so one can denote them by $D^{k,m}$. Since translations and rotations do not commute, the generators of T_3 and E_3 do not commute either (see (6.178) below) which amounts to say that the momentum and angular momentum of a system invariant under E_3 cannot be measured simultaneously. However, the operator $\mathbf{L} \cdot \mathbf{P}$, which is the component of the angular momentum along the direction of motion, commutes with $\mathbf{P}$. This operator is called the *helicity* operator. The basis $|k\,m>$ defined by (6.171) is the one where the momentum is diagonalized simultaneously with the helicity operator so that k is the magnitude of the momentum and m is the component of the angular momentum in the direction of motion. The basis $|k\,m>$ defines the so-called helicity representation.

In the particular case $k = 0$, the procedure described above does not apply, but the irreps are just those of R_3, i.e. $D^{0j} = D^j$.

The six generators of E_3 are the components of the angular momentum (6.35), and the momentum components (6.165). The resulting Lie algebra is

$$[L_i, L_j] = i\,\varepsilon_{ijk} L_k$$
$$[L_i, P_j] = i\,\varepsilon_{ijk} P_k \qquad (6.178)$$
$$[P_i, P_j] = 0$$

which shows that R_3 is a subgroup of E_3, and T_3 an invariant subgroup (equation (5.70)). These relations also suggest that the Lie algebra of E_3 can be written as the semi-direct sum of the algebra $\mathscr{L}_1$ of T_3 and $\mathscr{L}_2$ of R_3 (see Section 5.8).

7

THE POINCARÉ GROUP AND THE LORENTZ GROUP

The Poincaré group is required in relativistic situations and comprises all geometrical transformations which leave invariant the quadratic form

$$s^2(x_1, x_2) = c^2(t_1 - t_2)^2 - (\boldsymbol{x}_1 - \boldsymbol{x}_2)^2. \tag{7.1}$$

These transformations could be rotations (proper and improper), Lorentz transformations or translations in Minkowski space-time, or any combination of them. When translations are excluded the group is called the *complete homogeneous Lorentz group* L. When they are included it is called the *Poincaré* or *inhomogeneous Lorentz group*. All pure space - time translations form a group by themselves which we denote by T_4. This is an extension to four dimensions of the translation group T_3 introduced in the previous chapter. Both T_4 and L are subgroups of the Poincaré group. Moreover, T_4 is an invariant subgroup (see, for example, Cornwell 1984, Chapter 2, Section 7). T_4 and L have only the identity element in common and any element of the Poincaré group can be written as a product of an element of T_4 and an element of L. Thus the Poincaré group is isomorphic to the semi-direct product $T_4 \wedge L$ (Section 2.7). The fact that it possesses the invariant abelian subgroup T_4 implies that the Poincaré group is not semi-simple.

In the next section the Minkowski space–time notation is introduced. In Section 7.2 a general Lorentz transformation is introduced and the connection between the homogeneous Lorentz group and the groups O(3, 1) and SL(2, $\mathbb{C}$) is discussed. In Section 7.3 the generators of the homogeneous Lorentz group are obtained from the change of a scalar function under an infinitesimal transformation. These generators represent the components of a generalized angular momentum. From the invariance of the Dirac equation under a Lorentz transformation the generators acquire extra contributions. Their space part represents the components of the intrinsic angular momentum (spin) of the Dirac particle (Section 7.4).

7.1 NOTATION

The space–time coordinates of a point are represented by

$$x^\mu = (x^0, x^k) = (x^0, x^1, x^2, x^3) = (ct, x, y, z), \tag{7.2}$$

where the Greek index μ runs from 0 to 3 and the Latin index k from 1 to 3 .

A contravariant 4-vector transforms like the coordinate vector x^μ and a covariant 4-vector transforms like the gradient. The coordinates x_μ of a covariant 4-vector are obtained by changing the sign of the space components of x^μ

$$x_\mu = (x_0, x_1, x_2, x_3) = (x_0, x_k) = (ct, -x, -y, -z). \qquad (7.3)$$

The metric tensor

$$g = (g_{\mu\nu}) = \begin{pmatrix} 1 & 0 & 0 & 0 \\ 0 & -1 & 0 & 0 \\ 0 & 0 & -1 & 0 \\ 0 & 0 & 0 & -1 \end{pmatrix} \qquad (7.4)$$

is responsible for lowering and raising indices so one can write

$$x_\mu = g_{\mu\nu} x^\nu. \qquad (7.5)$$

Some properties of the metric tensor are

$$g^{\mu\nu} = g_{\mu\nu} \qquad (7.6)$$

$$g_\mu{}^\nu = g_{\mu\rho} g^{\rho\nu} = g^\mu{}_\nu = \delta_\mu{}^\nu = \delta^\mu{}_\nu = \begin{cases} 1 & \mu = \nu \\ 0 & \mu \neq \nu. \end{cases} \qquad (7.7)$$

The scalar product is defined by

$$a^\mu b_\mu = a_\mu b^\mu = a^0 b^0 - \boldsymbol{a} \cdot \boldsymbol{b} \qquad (7.8)$$

so that the norm of a vector is

$$x^2 = x_\mu x^\mu = x_0^2 - \boldsymbol{x}^2 = c^2 t^2 - \boldsymbol{x}^2. \qquad (7.9)$$

The sign of the norm divides the 4-vectors into three categories:

$$\begin{aligned} x^2 &< 0 \quad \text{space-like vector;} \\ x^2 &= 0 \quad \text{null vector;} \\ x^2 &> 0 \quad \text{time-like vector.} \end{aligned} \qquad (7.10)$$

This classification corresponds to the position of the 4-vector with respect to the light cone. For the last two categories, if $x^0 > 0$ the vector points to the future and if $x^0 < 0$ it points to the past.

7.2 LORENTZ TRANSFORMATIONS

The relation between the coordinates of an event in two frames K and K', moving uniformly with a relative velocity $\boldsymbol{v}$ is given by a *pure* Lorentz transformation or *boost* which is written

$$t' = \gamma \left(t - \frac{\boldsymbol{x} \cdot \boldsymbol{v}}{c^2} \right)$$

LORENTZ TRANSFORMATIONS

$$x'_{\parallel} = \gamma(x_{\parallel} - vt) \qquad (7.11)$$
$$x'_{\perp} = x_{\perp}$$

where

$$\gamma = (1 - \beta^2)^{-1/2}; \qquad \beta = \frac{v}{c} \qquad (7.12)$$

and $x_{\parallel}$ and $x_{\perp}$ are the components of x parallel and perpendicular to the velocity v of K' with respect to K.

In the non-relativistic limit $v \ll c$ one can take $c \to \infty$ and $\gamma \to 1$ and (7.11) reduces to a Galilean transformation. When the velocities involved approach the speed of light the transformation (7.11) is the correct one since it maintains the constancy of the velocity of light in all inertial systems, as required by the special theory of relativity (see Chapter 1). To prove this, one can start from the velocity u' in the frame K':

$$u' = \frac{dx'}{dt} \Big/ \frac{dt'}{dt}. \qquad (7.13)$$

From (7.11) its parallel and transverse components take the form

$$u'_{\parallel} = \frac{u_{\parallel} - v}{1 - \frac{(u \cdot v)}{c^2}} \qquad (7.14)$$

$$u'_{\perp} = \frac{u_{\perp}}{\gamma\left(1 - \frac{(u \cdot v)}{c^2}\right)} \qquad (7.15)$$

where u is the velocity in the frame K. Then one has

$$\begin{aligned} u'^2 &= (u'_{\parallel})^2 + (u'_{\perp})^2 \\ &= \left[u_{\parallel}^2 - 2u \cdot v + v^2 + \left(1 - \frac{v^2}{c^2}\right)(u^2 - u_{\parallel}^2)\right] \Big/ \left(1 - \frac{u \cdot v}{c^2}\right)^2 \\ &= \left[u_{\parallel}^2 \frac{v^2}{c^2} - 2u \cdot v + u^2 + v^2\left(1 - \frac{u^2}{c^2}\right)\right] \Big/ \left(1 - \frac{u \cdot v}{c^2}\right)^2 \qquad (7.16) \\ &= c^2 - c^2 \frac{\left(1 - \frac{u^2}{c^2}\right)\left(1 - \frac{v^2}{c^2}\right)}{\left(1 - \frac{u \cdot v}{c^2}\right)^2}. \end{aligned}$$

The last row has been obtained by using $u \cdot v \equiv u_{\parallel} v$. The expression (7.16) shows that if $u = c$ in the K frame one has $u' = c$ in K' too.

If we choose v in the direction of Oz, the Equations (7.11) become

$$t' = \gamma\left(t - \frac{\beta}{c}z\right)$$

$$x' = x$$
$$y' = y$$
$$z' = \gamma(z - \beta c t).$$

With the notation of Section 7.1 one gets

$$x'^0 = \gamma(x^0 - \beta x^3)$$
$$x'^1 = x^1$$
$$x'^2 = x^2$$
$$x'^3 = \gamma(x^3 - \beta x^0). \tag{7.17}$$

One can introduce a parameter ω instead of β through

$$\beta = \tanh \omega, \qquad \gamma = \cosh \omega, \qquad \beta\gamma = \sinh \omega \tag{7.18}$$

Then the last and the first equations of (7.17) become

$$x'^3 = x^3 \cosh \omega - x^0 \sinh \omega$$
$$x'^0 = -x^3 \sinh \omega + x^0 \cosh \omega. \tag{7.19}$$

Since the cosine and sine of an imaginary angle $i\omega$ are $\cosh \omega$ and $i \sinh \omega$ the transformation (7.19) can be viewed as a rotation like (6.43) but in the (x^3, ix^0) plane. For that reason ω is called the Lorentz rotation angle.

The parameter ω varies in the range

$$-\infty < \omega < \infty \tag{7.20}$$

inasmuch as $-c < v < c$ and it provides a parametrization for the Lorentz group. When $v \to \pm c$ or $\omega \to \pm\infty$ the matrix elements of the Lorentz transformation (7.19) are unbounded, which makes the Lorentz group *non-compact* (Section 5.7).

The term 'Lorentz transformation' can be used for a three-dimensional rotation as studied in Chapter 6, for a boost of the form (7.19), and for a space–time translation. In this sense a Lorentz transformation is a linear transformation which leaves invariant the quadratic form (7.1). The most general Lorentz transformation between two coordinate systems is a linear transformation of the form

$$x'^\mu = \Lambda^\mu{}_\nu x^\nu + a^\mu \tag{7.21}$$

where $\Lambda^\mu{}_\nu$ is a real matrix and the real 4-vector a^μ represents a space-time translation.

Then, according to the introduction to this chapter, one has

$$a^\mu \neq 0 \quad \text{Poincaré or inhomogeneous Lorentz group}$$
$$a^\mu = 0 \quad \text{homogeneous Lorentz group.}$$

LORENTZ TRANSFORMATIONS

In the following we shall restrict the discussion to the homogeneous Lorentz group or, briefly the Lorentz group. Then we can set $t_2 = x_2 = 0$ in (7.1) and drop the subscript 1 so we can use the quadratic form

$$s^2 = c^2 t^2 - (x^1)^2 - (x^2)^2 - (x^3)^2$$

$$= (x^0, x^1, x^2, x^3) \begin{pmatrix} 1 & 0 & 0 & 0 \\ 0 & -1 & 0 & 0 \\ 0 & 0 & -1 & 0 \\ 0 & 0 & 0 & -1 \end{pmatrix} \begin{pmatrix} x^0 \\ x^1 \\ x^2 \\ x^3 \end{pmatrix} \quad (7.22)$$

where the square matrix is the tensor (7.4). The transformation (7.21) with $a^\mu = 0$ can be written in matrix form as

$$x' = \Lambda x \quad (7.23)$$

where Λ is a 4×4 matrix and x' and x are 4×1 column matrices, as in (7.22). The transpose of (7.23) is

$$(x')^T = x^T \Lambda^T; \qquad (\Lambda^T)_\mu{}^\nu = \Lambda^\nu{}_\mu. \quad (7.24)$$

In this notation one can write (7.22) as

$$s^2 = x^T g x. \quad (7.25)$$

The invariance means

$$(x')^T g x' = x^T \Lambda^T g \Lambda x = x^T g x. \quad (7.26)$$

Hence the matrix Λ must satisfy the 'pseudo-orthogonality' constraint of type (5.90)

$$\Lambda^T g \Lambda = g \quad (7.27)$$

which, in terms of matrix elements, gives

$$\Lambda^\mu{}_\alpha \Lambda_\mu{}^\beta = \delta_\alpha{}^\beta. \quad (7.28)$$

From (7.27) one can get the inverse of Λ as

$$\Lambda^{-1} = g \Lambda^T g \quad (7.29)$$

from which the inverse transformation $x = \Lambda^{-1} x'$ reads

$$x^\beta = \Lambda_\mu{}^\beta x'^\mu. \quad (7.30)$$

The constraint (7.27) leads to

$$\det \Lambda = \pm 1. \quad (7.31)$$

The Lorentz group with $\det \Lambda = +1$ is called the *proper* Lorentz group L_+. The group with $\det \Lambda = -1$ is called the *improper* Lorentz group. It contains space inversion $x \to -x$ or time reversal $t \to -t$.

Setting $\alpha = \beta = 0$ in (7.28) and raising all indices on the left-hand side one gets

$$(\Lambda^{00})^2 = 1 + (\Lambda^{10})^2 + (\Lambda^{20})^2 + (\Lambda^{30})^2 \geq 1. \tag{7.32}$$

The inequality sign is due to the fact that all Λ^{ij} are real. One can distinguish four different types of transformations which are not continuously connected to each other, like the proper and improper rotations (Section 6.1). These transformations are:

(1) $\det \Lambda = 1,\quad \Lambda^{00} \geq 1,$ connected to the identity;
(2) $\det \Lambda = -1,\quad \Lambda^{00} \geq 1,$ contains space inversions;
(3) $\det \Lambda = -1,\quad \Lambda^{00} \leq -1,$ contains time reversal;
(4) $\det \Lambda = 1,\quad \Lambda^{00} \leq -1,$ contains space inversion and time reversal

The transformations of type (1) form a subgroup called the *proper orthochronous homogeneous Lorentz group*, $L_+^\uparrow$, because these transformations preserve the direction of time. This subgroup forms the component of L that is connected with the identity and is also known as the restricted homogeneous Lorentz group. Adding to (1) transformations of type (2) or (3) one obtains groups referred to as L_s or L_t respectively. These are also important in physics as mentioned in Chapter 1. They are related to parity and time reversal properties of the Hamiltonian. All four transformations give the *complete* homogeneous Lorentz group.

A real orthogonal matrix a that is an element of O(4) satisfies $a^T a = 1$. The condition (7.27) makes Λ a pseudo-orthogonal matrix of type O(3,1) (Table 5.2) leaving invariant the form $(x^1)^2 + (x^2)^2 + (x^3)^2 - c^2 t^2$. Thus the homogeneous Lorentz group L and the proper homogeneous Lorentz group L_+ are isomorphic to O(3,1) and SO(3,1), respectively.

The restricted homogeneous Lorentz group is also related to the group SL(2, $\mathbb{C}$). There exists a two-to-one homomorphic mapping, like that between SU(2) and SO(3) (Chapter 8), i.e. to two matrices a and $-a$ belonging to SL(2, $\mathbb{C}$) there correspond only one $\Lambda(a) = \Lambda(-a)$. To prove this one starts from the 2×2 hermitian matrix

$$V = \begin{pmatrix} x^0 + x^3 & x^1 - i x^2 \\ x^1 + i x^2 & x^0 - x^3 \end{pmatrix} \tag{7.33}$$

equivalently expressed as the sum

$$V = x^\mu \sigma_\mu \quad (\mu = 0, 1, 2, 3) \tag{7.34}$$

where x^μ is a 4-vector, σ_0 is the identity 2×2 matrix, and $\sigma_i (i = 1, 2, 3)$ are the Pauli matrices. These matrices satisfy

$$\text{tr}(\sigma_i \sigma_j) = 2\delta_{jk} \quad (j, k = 0, 1, 2, 3) \tag{7.35}$$

LORENTZ TRANSFORMATIONS

from which it follows that

$$x^\mu = \frac{1}{2}\mathrm{tr}(\sigma_\mu V). \tag{7.36}$$

Then for any $a \in \mathrm{SL}(2,\mathbb{C})$

$$V' = a V a^+ \tag{7.37}$$

is also a Hermitian matrix

$$V' = \begin{pmatrix} x'^0 + x'^3 & x'^1 - ix'^2 \\ x'^1 + ix'^2 & x'^0 - x'^3 \end{pmatrix} \tag{7.38}$$

related to another 4-vector x'^μ.

Thus one can write

$$x'^\mu = \frac{1}{2}\mathrm{tr}(\sigma_\mu V') = \frac{1}{2}\mathrm{tr}(\sigma_\mu a V a^+)$$
$$= \frac{1}{2}\mathrm{tr}(\sigma_\mu a \sigma_\nu a^+ x^\nu) = \Lambda_{\mu\nu}(a) x^\nu$$

with

$$\Lambda(a)_{\mu\nu} = \frac{1}{2}\mathrm{tr}(\sigma_\mu a \sigma_\nu a^+). \tag{7.39}$$

This shows that the transformation (7.37) induces a 4×4 matrix Λ which depends on a but not on x^μ. This transformation is actually a Lorentz transformation because $\det a = 1$ implies from (7.37) that

$$\det V' = \det V \tag{7.40}$$

which represents simply the invariance of the quadratic form (7.22). Moreover $\mathrm{SL}(2,\mathbb{C})$ is a connected group (see Section 5.7 for connectedness) which allows a to vary continuously. Then the matrix (7.39) must also belong to the connected component of L_+ which is $L_+^\uparrow$. To establish a homomorphic mapping we must show that

$$\Lambda(a_2)\Lambda(a_1) = \Lambda(a_2 a_1). \tag{7.41}$$

For this purpose we define

$$V' = a_1 V a_1^+ \rightarrow x' = \Lambda(a_1) x \tag{7.42}$$
$$V'' = a_2 V' a_2^+ \rightarrow x'' = \Lambda(a_2) x'. \tag{7.43}$$

One can also write

$$V'' = a_2 V' a_2^+ = a_2 a_1 V a_1^+ a_2^+ \rightarrow x'' = \Lambda(a_2 a_1) x. \tag{7.44}$$

From (7.42)–(7.44) one obtains (7.41).

To prove that the mapping is two-to-one we consider a transformation a belonging to the kernel (Section 2.3). For this transformation one has $aVa^+ = V$ which can be satisfied only if $a = \lambda \mathbf{1}$ where λ is a constant and $\mathbf{1}$ is the 2×2 unit matrix. For $\det a = 1$ it follows that $\lambda = \pm 1$ so that the kernel consists only of $\mathbf{1}$ and $-\mathbf{1}$. Hence the homomorphic mapping is two-to-one.

7.3 INFINITESIMAL TRANSFORMATIONS

An infinitesimal Lorentz transformation reads

$$\Lambda^\mu{}_\nu = \delta^\mu{}_\nu + \omega^\mu{}_\nu. \tag{7.45}$$

In order that equation (7.28) be satisfied, the infinitesimal parameters must have the property

$$\omega^{\mu\nu} = -\omega^{\nu\mu}. \tag{7.46}$$

Thus there are six independent parameters, ω^{01}, ω^{02}, ω^{03}, ω^{12}, ω^{23} and ω^{31}. The first three are related to pure Lorentz transformations, also called boosts, and the last three to rotations in $\mathbb{R}^3$ as defined in Section 6.1. To check the meaning of ω^{03}, for example, let us consider an infinitesimal boost in the x^3-direction. From (7.45) we have

$$\begin{aligned} x'^0 &= x^0 + \omega^0{}_3 x^3 = x^0 - \omega^{03} x^3 \\ x'^3 &= \omega^3{}_0 x^0 + x^3 = -\omega^{03} x^0 + x^3. \end{aligned} \tag{7.47}$$

On the other hand, taking β small and $\gamma \simeq 1$ in (7.18) and (7.19), we obtain the same infinitesimal transformation. By identification one obtains $\omega^{03} = \beta$ or, in general, for a boost with velocity v_i in the x^i-direction, one has

$$\omega^{0i} = \beta_i = \frac{v_i}{c} \quad (i = 1, 2, 3). \tag{7.48}$$

To show that the other parameters represent rotations let us consider a Lorentz transformation with only $\omega^1{}_2 \neq 0$, the other parameters being zero. Then (7.45) gives

$$\begin{aligned} x'^1 &= x^1 + \omega^1{}_2 x^2 = x^1 - \omega^{12} x^2 \\ x'^2 &= x^2 + \omega^2{}_1 x^1 = x^2 + \omega^{12} x^1. \end{aligned} \tag{7.49}$$

These relations represent a particular case of (6.31), namely $\omega_1 = \omega_2 = 0$ and $\omega_3 = \omega^{12}$, i.e. a rotation through ω^{12} about the Oz-axis. By analogy, one can show that ω^{23} corresponds to ω_1 and ω^{31} to ω_2 of (6.31).

According to (6.12) the infinitesimal change of a scalar function can be written as

$$S\psi_\alpha(x) = \psi_\alpha(\Lambda^{-1}x) = \left(1 - dx^\mu \frac{\partial}{\partial x^\mu}\right)\psi_\alpha \tag{7.50}$$

where the last equality is a generalization of (6.33) to four dimensions and dx^μ are given by (7.45) as

$$dx^\mu = \omega^\mu{}_\nu x^\nu = \omega^{\mu\alpha} g_{\alpha\nu} x^\nu = \omega^{\mu\alpha} x_\alpha. \tag{7.51}$$

Thus, the infinitesimal transformation S takes the form

$$S = 1 - \omega^{\mu\alpha} x_\alpha \partial_\mu \tag{7.52}$$

where

$$\partial_\mu = \frac{\partial}{\partial x^\mu}. \tag{7.53}$$

Due to the antisymmetry property (7.46) one can rewrite S as

$$S = 1 - \frac{i}{2}\omega^{\mu\alpha} L_{\mu\alpha} \tag{7.54}$$

with

$$L_{\mu\alpha} = i(x_\mu \partial_\alpha - x_\alpha \partial_\mu) \tag{7.55}$$

the generators of the Lorentz group. They are a generalization of the angular momentum operators (6.35) and have the property $L_{\mu\alpha} = -L_{\alpha\mu}$. Actually the six generators of the Lorentz group can be grouped into two categories defined as

$$L_i = \frac{1}{2}\varepsilon_{ijk} L_{jk} \tag{7.56}$$

$$K_i = L_{0i} \tag{7.57}$$

The L_i are angular momentum operators satisfying the commutation rules (6.38). They form a subalgebra associated to the subgroup SO(3) of the proper Lorentz group. The operators K_i produce boosts and are the same as the generators (6.130) of O(3,1).

The generators (7.55) satisfy the following commutation rules

$$[L_{\mu\nu}, L_{\rho\sigma}] = -i(g_{\mu\rho} L_{\nu\sigma} - g_{\mu\sigma} L_{\nu\rho} + g_{\nu\sigma} L_{\mu\rho} - g_{\nu\rho} L_{\mu\sigma}). \tag{7.58}$$

In terms of L_i and K_i given by (7.56) and (7.57), respectively, these commutation relations become

$$\begin{aligned}[L_i, L_j] &= i\,\varepsilon_{ijk} L_k \\ [L_i, K_j] &= i\,\varepsilon_{ijk} K_k \\ [K_i, K_j] &= -i\,\varepsilon_{ijk} L_k\end{aligned} \tag{7.59}$$

It is not surprising that this is the same as the algebra (6.132) of SO(3,1) because we know already that the two groups are isomorphic. The generators L_i close on themselves forming the subalgebra so(3). This is not valid for boosts, as indicated by the third commutation relation of (7.59). Acting with the generators L_i and K_i in the space of a 4-vector x^μ one obtains the generators in matrix form as

$$K_1 = -i \begin{pmatrix} 0 & -1 & 0 & 0 \\ -1 & 0 & 0 & 0 \\ 0 & 0 & 0 & 0 \\ 0 & 0 & 0 & 0 \end{pmatrix}; \quad K_2 = -i \begin{pmatrix} 0 & 0 & -1 & 0 \\ 0 & 0 & 0 & 0 \\ -1 & 0 & 0 & 0 \\ 0 & 0 & 0 & 0 \end{pmatrix};$$

$$K_3 = -i \begin{pmatrix} 0 & 0 & 0 & -1 \\ 0 & 0 & 0 & 0 \\ 0 & 0 & 0 & 0 \\ -1 & 0 & 0 & 0 \end{pmatrix}$$

$$L_1 = -\mathrm{i}\begin{pmatrix} 0 & 0 & 0 & 0 \\ 0 & 0 & 0 & 0 \\ 0 & 0 & 0 & 1 \\ 0 & 0 & -1 & 0 \end{pmatrix}; \quad L_2 = -\mathrm{i}\begin{pmatrix} 0 & 0 & 0 & 0 \\ 0 & 0 & 0 & -1 \\ 0 & 0 & 0 & 0 \\ 0 & 1 & 0 & 0 \end{pmatrix};$$

$$L_3 = -\mathrm{i}\begin{pmatrix} 0 & 0 & 0 & 0 \\ 0 & 0 & 1 & 0 \\ 0 & -1 & 0 & 0 \\ 0 & 0 & 0 & 0 \end{pmatrix}.$$
(7.60)

Removing i one obtains six real matrices forming the real Lie algebra of the Lorentz group as encountered in classical physics (see, for example, Jackson, 1975, paragraph 11.7).

The algebra (7.59) can be brought to another form by defining the linear combinations

$$X_i^\pm = \frac{1}{2}(L_i \pm \mathrm{i} K_i). \tag{7.61}$$

The commutation relations of these operators are

$$[X_i^+, X_j^+] = \mathrm{i}\,\varepsilon_{ijk} X_k^+$$
$$[X_i^-, X_j^-] = \mathrm{i}\,\varepsilon_{ijk} X_k^- \tag{7.62}$$
$$[X_i^+, X_j^-] = 0.$$

This shows that the algebra of the proper orthochronous Lorentz group breaks up into the direct sum of two su(2) algebras in the same way as the algebra of SO(3,1) breaks up into two so(3) algebras, as shown by Equations (6.148).

7.4 THE SPIN OF A DIRAC PARTICLE

The generators of the Lorentz group found in the previous section represent a generalization of the angular momentum and result from the change of a scalar function under an infinitesimal Lorentz transformation. Here, instead of a scalar, we consider a four-component spinor function. Under a Lorentz transformation, the spinor components mix among themselves. The situation is analogous to the transformation of a vector function under rotations, discussed in Section 6.1. The spinor under discussion is a solution of the Dirac equation and the generators resulting from the effect of an infinitesimal transformation form an antisymmetric tensor operator, the space part of which represents the spin of a Dirac particle.

The starting point is the invariance of the Dirac equation under a Lorentz transformation. Comprehensive presentations of the Dirac equation can be found in many books (Schweber 1961; Messiah 1964; Bjorken and Drell 1964; Sakurai 1967, etc.). Unfortunately, the conventions for the Dirac matrices differ. Here, we use the same conventions as in Bjorken and Drell (1964).

In its covariant form, the Dirac equation reads

$$[\gamma^\mu(i\partial_\mu - eA_\mu) - m]\psi = 0 \tag{7.63}$$

with $\hbar = c = 1$. The 4×4 Dirac matrices γ^μ are

$$\gamma^\mu = (\gamma^0, \gamma^1, \gamma^2, \gamma^3) = (\gamma^0, \boldsymbol{\gamma})$$

$$\gamma^0 = \beta = \begin{pmatrix} 1 & 0 \\ 0 & -1 \end{pmatrix}; \quad \gamma^i = \begin{pmatrix} 0 & \sigma_i \\ -\sigma_i & 0 \end{pmatrix} \quad (i = 1, 2, 3). \tag{7.64}$$

They satisfy the relations

$$\gamma^\mu \gamma^\nu + \gamma^\nu \gamma^\mu = 2g^{\mu\nu}\mathbf{1} \tag{7.65}$$

$$\gamma^{\mu+} = \gamma^0 \gamma^\mu \gamma^0 = \gamma_\mu \tag{7.66}$$

$$\gamma_\mu^+ = \gamma_\mu^{-1}. \tag{7.67}$$

In the following, we need a theorem known as *Pauli's fundamental theorem*: Given two different four-matrix sets γ^μ and $\hat{\gamma}^\mu$ ($\mu = 0, 1, 2, 3$) such that

$$\left\{\gamma^\mu, \gamma^\nu\right\} = 2g^{\mu\nu}\mathbf{1}$$

$$\left\{\hat{\gamma}^\mu, \hat{\gamma}^\nu\right\} = 2g^{\mu\nu}\mathbf{1} \tag{7.68}$$

there exists a non-singular 4×4 matrix with the property

$$\gamma^\mu = S\hat{\gamma}^\mu S^{-1}. \tag{7.69}$$

The matrix S is unique up to a multiplication constant. Moreover, if S is unitary and the matrices γ^μ satisfy (7.66), then the same holds for $\hat{\gamma}^\mu$.

According to the relativity principle, the Dirac equation must have the same form in two different inertial systems K and K' related by a Lorentz transformation. In the moving system, K', the Dirac equation is

$$[\gamma^\mu(i\partial'_\mu - eA'_\mu) - m]\psi'(x') = 0. \tag{7.70}$$

By analogy to the case of a vector particle (see equation (6.90)), the solution $\psi'(x')$ where $x' = \Lambda x$ can be written as

$$\psi'(x') = S\psi(\Lambda^{-1}x') \tag{7.71}$$

where S is a transformation which mixes the components of the spinor ψ among themselves. S must have an inverse S^{-1} to give the wave function $\psi(x)$ in K in terms of that of K':

$$\psi(x) = \psi(\Lambda^{-1}x') = S^{-1}\psi'(x') = S^{-1}\psi'(\Lambda x). \tag{7.72}$$

The problem is to find S. By using (7.72), we rewrite the Dirac equation (7.63) in K as

$$[\gamma^\mu(i\partial_\mu - eA_\mu) - m]S^{-1}\psi'(x') = 0. \qquad (7.73)$$

After multiplication with S, we obtain

$$[S\gamma^\mu \Lambda^\nu{}_\mu S^{-1}(i\partial'_\nu - eA'_\nu) - m]\psi'(x') = 0 \qquad (7.74)$$

where the covariant vectors ∂_μ and A_μ have undergone a transformation like (7.30) but for lower indices:

$$\partial_\mu = \frac{\partial}{\partial x^\mu} = \frac{\partial x'^\nu}{\partial x^\mu}\frac{\partial}{\partial x'^\nu} = \Lambda^\nu{}_\mu \partial'_\nu \qquad (7.75)$$

$$A_\mu = \Lambda^\nu{}_\mu A'_\nu. \qquad (7.76)$$

Now, one can define

$$\hat{\gamma}^\nu = \Lambda^\nu{}_\mu \gamma^\mu \qquad (7.77)$$

and the fundamental theorem gives

$$S\hat{\gamma}^\nu S^{-1} = \gamma^\nu \qquad (7.78)$$

which brings us back to (7.70). Thus, the Dirac equation is covariant provided S satisfies

$$\Lambda^\nu{}_\mu \gamma^\mu = S^{-1}\gamma^\nu S. \qquad (7.79)$$

Now, we consider an infinitesimal transformation similar to (7.54)

$$S = 1 - \frac{i}{4}\omega^{\mu\nu}\sigma_{\mu\nu} \qquad (7.80)$$

but, instead of $L_{\mu\nu}$, we introduce the antisymmetric operators $\frac{1}{2}\sigma_{\mu\nu}$, to be found. Replacing (7.80) in (7.79) and using the antisymmetry of $\omega^{\mu\nu}$, one obtains the relation

$$[\gamma^\nu, \sigma_{\alpha\beta}] = 2i(g^\nu{}_\alpha\gamma_\beta - g^\nu{}_\beta\gamma_\alpha). \qquad (7.81)$$

Starting from the property (7.65) of the γ matrices, one can see that this equation is satisfied by any of the six matrices

$$\sigma_{\mu\nu} = \frac{i}{2}[\gamma_\mu, \gamma_\nu] \qquad (7.82)$$

which, replaced in (7.80), gives the desired answer.

Putting together (7.54) and (7.80), one obtains the total change of a Dirac spinor under an infinitesimal transformation

$$S\psi(x) = (1 - \frac{i}{2}\omega^{\mu\nu}L_{\mu\nu})(1 - \frac{i}{4}\omega^{\mu\nu}\sigma_{\mu\nu})\psi(x), \qquad (7.83)$$

the first bracket being the change due to its argument $\Lambda^{-1}x$ and the second due to the mixing of its components. As discussed in the previous section, the parameters $\omega^{ij}(i, j = 1, 2, 3)$ are related to rotations. Then, using the definition (7.64) of the γ matrices, one obtains

THE SPIN OF A DIRAC PARTICLE

$$\sigma_{12} = i\gamma_1\gamma_2 = i\gamma^1\gamma^2 = \begin{pmatrix} \sigma_3 & 0 \\ 0 & \sigma_3 \end{pmatrix}$$

$$\sigma_{23} = i\gamma_2\gamma_3 = i\gamma^2\gamma^3 = \begin{pmatrix} \sigma_1 & 0 \\ 0 & \sigma_1 \end{pmatrix} \quad (7.84)$$

$$\sigma_{31} = i\gamma_3\gamma_1 = i\gamma^3\gamma^1 = \begin{pmatrix} \sigma_2 & 0 \\ 0 & \sigma_2 \end{pmatrix}.$$

This shows that the transformation (7.80) associated to a rotation becomes

$$S = 1 - \frac{i}{2}(\omega^{12}\sigma_{12} + \omega^{23}\sigma_{23} + \omega^{31}\sigma_{31}) = 1 - i\boldsymbol{\omega}\cdot\boldsymbol{\Sigma} \quad (7.85)$$

where

$$\boldsymbol{\Sigma} = \frac{1}{2}\begin{pmatrix} \boldsymbol{\sigma} & 0 \\ 0 & \boldsymbol{\sigma} \end{pmatrix} \quad (7.86)$$

is the spin *vector operator* of the Dirac particle. Thus, from (7.83) one finds that the generators associated to a rotation are the components of the total angular momentum of a Dirac particle

$$\boldsymbol{J} = \boldsymbol{L} + \boldsymbol{\Sigma}. \quad (7.87)$$

This equation represents a generalization of (6.95) from a three-component spinor to a four-component spinor.

Similarly, the boost operators $\boldsymbol{K}$ acquire new contributions too, due to the transformation of 4-vector spinor components among themselves under a pure Lorentz transformation. From (7.80), in the case of a pure boost, one has

$$S = 1 - \frac{i}{2}\omega^{0i}\sigma_{0i} \quad (7.88)$$

with

$$\sigma_{0i} = \frac{1}{i}\begin{pmatrix} 0 & \sigma_i \\ \sigma_i & 0 \end{pmatrix} \quad (i = 1, 2, 3)$$

so that the new generators resulting from (7.83) are

$$N_i = K_i + \frac{i}{2}\sigma_{0i} \quad (7.89)$$

where K_i are defined by (7.57). The generators (7.87) and (7.89) can be put together in the general form

$$J_{\alpha\beta} = L_{\alpha\beta} + \frac{1}{2}\sigma_{\alpha\beta} = i(x_\alpha\partial_\beta - x_\beta\partial_\alpha) + \frac{1}{2}\sigma_{\alpha\beta}. \quad (7.90)$$

The finite form of a Lorentz transformation becomes

$$S(\omega) = \exp(-\frac{i}{2}\omega^{\mu\nu}J_{\mu\nu}) \quad (7.91)$$

which is consistent with (7.83).

7.5 IRREDUCIBLE REPRESENTATIONS

The relations (7.62) are equally valid for $X_i^\pm$ of (7.61) redefined as

$$X_i^\pm = \frac{1}{2}(J_i \pm iN_i) \tag{7.92}$$

This shows that the algebra of the restricted Lorentz group reduces to the direct sum of two su(2) algebras. It follows that a finite dimensional irreducible representation can be obtained by direct products of irreps of SU(2) and can be labelled with a pair of indices j and j', each associated to an su(2) algebra. Accordingly, j and j' take values 0, $\frac{1}{2}$, 1, $\frac{3}{2}$, etc., as for SU(2). These representations, for which we use the notation $D^{(j,j')}$, have dimension $(2j+1)(2j'+1)$. One can show that the representation $D^{(j,j')}$ is single valued if $j + j'$ is integer, and double valued otherwise.

On the other hand, as specified in Section 5.10, for a semi-simple group of rank ℓ, one can construct ℓ independent invariants, the eigenvalues of which completely specify the irreducible representations of that group. The restricted homogeneous Lorentz group has rank $\ell = 2$ and one can construct the invariants

$$\begin{aligned} \frac{1}{2} J^{\mu\nu} J_{\mu\nu} &= \boldsymbol{J}^2 - \boldsymbol{N}^2 \\ \frac{1}{2} \varepsilon^{\mu\nu\rho\sigma} J_{\mu\nu} J_{\rho\sigma} &= \boldsymbol{J} \cdot \boldsymbol{N} \end{aligned} \tag{7.93}$$

with the metric tensor $\varepsilon^{\mu\nu\rho\sigma}$ defined as in Equation (5.185). Using (7.92), one has

$$\boldsymbol{J}^2 - \boldsymbol{N}^2 = 2[(X^+)^2 + (X^-)^2] \tag{7.94}$$
$$\boldsymbol{J} \cdot \boldsymbol{N} = -i[(X^+)^2 - (X^-)^2]. \tag{7.95}$$

The basis of the subalgebras X_i^+ and X_i^- are $|jm>$ and $|j'm'>$, respectively, and

$$(X^+)^2 |jm> = j(j+1)|jm> \tag{7.96}$$
$$(X^-)^2 |j'm'> = j'(j'+1)|jm>. \tag{7.97}$$

Thus, in the direct product space $|jm> |j'm'>$, the eigenvalue equations of (7.94) and (7.95) are

$$\begin{aligned} (\boldsymbol{J}^2 - \boldsymbol{N}^2)|jm> |j'm'> &= 2[j(j+1) + j'(j'+1)]|jm> |j'm'> \\ \boldsymbol{J} \cdot \boldsymbol{N}|jm> |j'm'> &= -i[j(j+1) - j'(j'+1)]|jm> |j'm'>. \end{aligned} \tag{7.98}$$

The two indices j and j' of $D^{(j,j')}$ are therefore related to the eigenvalues of the two invariants (7.93).

The representation $D^{(0,0)}$ can be obtained in an invariant subspace called a scalar subspace, for example s^2 of (7.22) or any power of it. The representations $D^{(1/2,0)}$ and $D^{(0,1/2)}$ are called Weyl representations and are associated to a two-component spinor and a conjugate spinor, respectively. In quantum field theories, they are used to describe neutrinos. In these cases, an explicit matrix representation of the

IRREDUCIBLE REPRESENTATIONS

infinitesimal generators can be given in terms of the Pauli matrices (Cornwell 1984, Chapter 17).

$$J_k^{(1/2,0)} = \frac{1}{2}i\sigma_k \qquad J_k^{(0,1/2)} = \frac{1}{2}i\sigma_k$$
$$N_k^{(1/2,0)} = \frac{1}{2}\sigma_k \qquad N_k^{(0,1/2)} = -\frac{1}{2}\sigma_k. \tag{7.99}$$

A 4-vector transforms under the single-valued representation $D^{(1/2,\,1/2)}$. To check this, we can set $J = L$ and $N = K$ in (7.92) and, using the matrices (7.60), we get

$$L^2 - K^2 = 3 \qquad L \cdot K = 0$$

consistent with (7.98). The representation, used to describe the electron, is the reducible representation $D^{(1/2,0)} + D^{(0,1/2)}$ acting on a four-component spinor. The reason is that one must include the parity operator (see, for example, Bjorken and Drell 1964). Under parity $L \to L, K \to -K$, so that X^+ and X^- are interchanged and the representation turns out to be symmetric in j and j'. Then one can also argue that the representation associated with the electromagnetic tensor $F_{\mu\nu}$ is $D^{(1,0)} + D^{(0,\,1)}$.

The finite dimensional representations discussed above, except for $D^{(0,0)}$, are *non-unitary*. From (7.92), one can redefine J and N as

$$J = X^+ + X^-$$
$$N = -i(X^+ - X^-). \tag{7.100}$$

The generators $X_i^\pm$ are Hermitian, from which it follows that J is Hermitian and N anti-Hermitian. Hence, when exponentiated, they produce *anti-unitary* matrices. The non-unitarity is a consequence of the fact that the Lorentz group is non-compact. The group also has unitary irreducible representation, but these are infinite dimensional. For an account of infinite dimensional representations, see, for example, Gel'fand, Minlos, and Shapiro (1963).

Exercise 7.1 Show that the transformation

$$S = e^{i\omega n \cdot K}$$

boosts the 4-vector $(1, \mathbf{0})$ into $(\cosh \omega, -\mathbf{n} \sinh \omega)$.

Solution Let us consider the particular case of a boost along the x^1-axis. Then

$$S = e^{i\omega K_1} = e^{\omega I_1} = \cosh \omega I_1 + \sinh \omega I_1 \tag{1}$$

where

$$I_1 = \begin{pmatrix} 0 & -1 & 0 & 0 \\ -1 & 0 & 0 & 0 \\ 0 & 0 & 0 & 0 \\ 0 & 0 & 0 & 0 \end{pmatrix} \tag{2}$$

results from K_1 of (7.60) by removing $-i$. The matrix I_1 satisfies

$$I_1^3 = I_1. \tag{3}$$

One can consider the expansion

$$\cosh \omega I_1 = \frac{1}{2} \sum_n \left(\frac{\omega^n I_1^n}{n!} + \frac{(-)^n \omega^n I_1^n}{n!} \right) \tag{4}$$

and, by using (3), one gets

$$\cosh \omega I_1 = 1 - I_1^2 + I_1^2 \left(1 + \frac{\omega^2}{2!} + \frac{\omega^4}{4!} + \cdots \right) = 1 - I_1^2 + I_1^2 \text{ch } \omega. \tag{5}$$

Similarly, one can prove that

$$\sinh \omega I_1 = I_1 \sinh \omega \tag{6}$$

so that

$$S = 1 - I_1^2 + I_1^2 \cosh \omega + I_1 \sinh \omega \tag{7}$$

where

$$I_1^2 = \begin{pmatrix} 1 & 0 & 0 & 0 \\ 0 & 1 & 0 & 0 \\ 0 & 0 & 0 & 0 \\ 0 & 0 & 0 & 0 \end{pmatrix}. \tag{8}$$

Acting with S on the 4-vector $(1, 0, 0, 0)$ gives

$$\begin{pmatrix} x'^0 \\ x'^1 \\ x'^2 \\ x'^3 \end{pmatrix} = \begin{pmatrix} \cosh \omega & -\sinh \omega & 0 & 0 \\ -\sinh \omega & \cosh \omega & 0 & 0 \\ 0 & 0 & 1 & 0 \\ 0 & 0 & 0 & 1 \end{pmatrix} \begin{pmatrix} 1 \\ 0 \\ 0 \\ 0 \end{pmatrix}$$

i.e. one obtains the 4-vector $x'^0 = \cosh \omega$, $x'^1 = -\sinh \omega$, $x'^2 = 0$, $x'^3 = 0$. The proof can be generalized to boosts along the x^2 and x^3 axes to obtain the desired answer.

Note that $S = e^{i\omega \mathbf{n} \cdot \mathbf{K}}$ is the finite form of (6.131) with $\boldsymbol{\beta} \to \omega \mathbf{n}$ containing only boosts (no rotations).

7.6 THE POINCARÉ GROUP

The Poincaré group includes both Lorentz transformations and space–time translations. In the previous sections, we dealt with (homogeneous) Lorentz transformations only, for which $\alpha_\mu = 0$. Here, we consider $\alpha_\mu \neq 0$ in order to allow for space–time translations.

THE POINCARÉ GROUP

For the moment, let us discuss pure translations. Under an infinitesimal translation with a^μ small, a 4-vector acquires the components

$$Tx^\mu = x'^\mu = x^\mu + a^\mu \tag{7.101}$$

which is a generalization of space translations (6.161) to Minkowski space–time. As for rotations, we take the active point of view (Section 6.1) expressed by Equation (6.12), which contains the inverse transformation. Here, the inverse T^{-1} of the transformation (7.101) is $x'^\mu = x^\mu - a^\mu$. Thus, in analogy to (6.12), the infinitesimal change of a scalar function under (7.101) reads

$$S_T \psi(x) = \psi(T^{-1}x) = \psi(x^\mu - a^\mu) = \left(1 - a^\mu \frac{\partial}{\partial x^\mu}\right)\psi(x)$$
$$= (1 + ia^\mu P_\mu)\psi(x) \tag{7.102}$$

where

$$P_\mu = i\partial_\mu \tag{7.103}$$

are the generators of the translation group T_4 in Minkowski space–time. They represent the covariant components of the energy–momentum 4-vector. Their algebra is obviously

$$[P_\mu, P_\nu] = 0. \tag{7.104}$$

The translation group T_4 is an invariant abelian subgroup of the Poincaré group, or else the Poincaré group is the semi-direct product $T_4 \wedge L$ of T_4 and the Lorentz group L. Thus, the Poincaré group is not semi-simple (Section 5.9). Its algebra contains the commutation relations (7.58), (7.104) and

$$[P_\mu, L_{\alpha\beta}] = i(g_{\alpha\mu}P_\beta - g_{\beta\mu}P_\alpha) \tag{7.105}$$

where $L_{\alpha\beta}$ is defined by (7.55) (see Exercise 7.3).

The finite form of the transformation S_T of (7.102) is

$$S_T(a) = e^{ia^\mu P_\mu} \tag{7.106}$$

with four parameters a^μ.

The commutation relations (7.105) indicate that Lorentz transformations do not commute with translations. However, these two transformations have a simple multiplication law. If $T(a)$ denotes a translation by the 4-vector a, Λ a Lorentz transformation, and x an arbitrary 4-vector, one can write

$$T(a)\Lambda x = \Lambda x + a \tag{7.107}$$

and

$$\Lambda T(a)x = \Lambda(x + a) = \Lambda x + \Lambda a = T(\Lambda a)\Lambda x. \tag{7.108}$$

This shows that the most general Poincaré transformation can be written as a product of a translation and a Lorentz transformation with the latter acting first.

Based on (7.108), one can write the product of two inhomogeneous Lorentz transformations $P(a_1, \Lambda_1)$ and $P(a_2, \Lambda_2)$ as

$$P(a_1, \Lambda_1)P(a_2, \Lambda_2) = T(a_1)\Lambda_1 T(a_2)\Lambda_2$$
$$= T(a_1)T(\Lambda_1 a_2)\Lambda_1\Lambda_2 = T(a_1 + \Lambda_1 a_2)\Lambda_1\Lambda_2.$$

Thus,

$$P(a_1, \Lambda_1)P(a_2, \Lambda_2) = P(a_1 + \Lambda_1 a_2, \Lambda_1\Lambda_2). \qquad (7.109)$$

The Poincaré group is very similar to the Euclidean group (Section 6.3) where the rotations have been replaced by Lorentz transformations and the space translations by space - time translations. So, instead of the six parameters of the Euclidean group, there are now *ten* parameters, six for L and four for T.

The Poincaré group has two quadratic invariant operators which commute with all its generators. These are

$$P^2 = P_\mu P^\mu \qquad (7.110)$$

and

$$W^2 = W_\mu W^\mu \qquad (7.111)$$

where

$$W^\mu = \frac{1}{2}\varepsilon^{\mu\nu\rho\sigma} J_{\nu\rho} P_\sigma \qquad (7.112)$$

where $\varepsilon^{\mu\nu\rho\sigma}$ is a metric tensor in four upper indices, defined as in Equation (5.185), and $J_{\nu\rho}$ the generators (7.90) of the Lorentz group.

The first of these invariants is the square of the 4-momentum and its eigenvalue is related to the particle mass m, also an invariant, through

$$m^2 c^2 = p^2 = \left(\frac{E}{c}\right)^2 - \mathbf{p}^2. \qquad (7.113)$$

One can easily check that P^2 is translationally and Lorentz invariant or, in other words, it commutes with all the generators

$$[P^2, P_\mu] = 0 \qquad (7.114)$$
$$[P^2, J_{\rho\sigma}] = 0. \qquad (7.115)$$

The other invariant is defined in terms of W_μ, a covariant spin operator, called the Pauli–Lubanski vector. W^2 also commutes with all the generators of the Poincaré group

$$[W^2, P_\mu] = 0 \qquad (7.116)$$
$$[W^2, J_{\rho\sigma}] = 0. \qquad (7.117)$$

THE POINCARÉ GROUP

Let us consider the case of massive particles. From the definition of W_μ, one can see that, for a particle of mass m, in a rest-frame eigenstate of P_μ, the eigenvalue of P_μ is $p_\mu = (m, \mathbf{0})$ and the action of the operator W_μ becomes equivalent to taking

$$W_0 = 0 \qquad W_i = \frac{1}{2} m \varepsilon_{ijk} J_{jk} = m J_i. \qquad (7.118)$$

Hence[†]

$$W^2 = -m^2 J^2. \qquad (7.119)$$

It follows that the eigenvalues of P^2 and W^2 are m^2 and $-m^2 s(s+1)$ where s is the spin of the particle. Thus, m and s specify the irreducible representations of the Poincaré group. To distinguish between vectors of the same invariant subspace, one must find additional indices. They are eigenvalues of operators which commute with P^2 and W^2 and among themselves.

From the commutation relations (7.105) extended to $J_{\alpha\beta}$, it follows that

$$[W^\alpha, P_\beta] = 0 \qquad (7.120)$$

but

$$[W^\alpha, W^\beta] \neq 0 \quad \text{for } \alpha \neq \beta. \qquad (7.121)$$

Consequently, there is a set of five mutually commuting operators which commute with P^2 and W^2. These may be taken to be, for example P_1, P_2, P_3, P_4, and W_3. Their corresponding eigenvalues $p_\mu (\mu = 0, 1, 2, 3)$ and $s = s_3$ can be used to label the basis vectors of an irrep by (m,s). Since the mass m is already fixed, it is enough to specify $\mathbf{p} = (p_1, p_2, p_3), p_0 = E/c$ being fixed by (7.113). The role of W_3 is equivalent to that of the helicity operator $\mathbf{J} \cdot \hat{\mathbf{p}}$ with $\hat{\mathbf{p}} = \mathbf{p}/|\mathbf{p}|$ which also commutes with $\mathbf{p}$ (see Exercise 7.2). The helicity, denoted here by λ, is the component of the spin along the direction of motion and it was introduced first in relation to the Euclidean group E_3. Then the complete labelling of the momentum eigenstate of a particle of mass m and spin s is $|m, s, \mathbf{p}, \lambda>$ with

$$P_\mu |m, s, \mathbf{p}, \lambda> = p_\mu |m, s, \mathbf{p}, \lambda> \qquad (7.122)$$
$$\mathbf{J} \cdot \mathbf{P} |m, s, \mathbf{p}, \lambda> = \lambda |\mathbf{p}| |m, s, \mathbf{p}, \lambda>. \qquad (7.123)$$

These momentum eigenstates form the infinite dimensional space of the Poincaré group. The translation operator (7.106) produces a phase change

$$S_T |m, s, \mathbf{p}, \lambda> = e^{i p \cdot a} |m, s, \mathbf{p}, \lambda>. \qquad (7.124)$$

The action of a homogeneous Lorentz transformation is a bit more complicated. To put it in a simple form, we first note that a state $|m, s, \mathbf{p}, \lambda>$ can be obtained from a rest-frame state $|m, s, 0, \lambda>$ by a Lorentz transformation denoted by Λ_{p0}. In fact, in the rest-frame, the helicity operator is indeterminate, so one can arbitrarily choose $\lambda = s_3$, where s_3 is the spin component along the z-direction. Then the role of Λ_{p0} is

---

[†] Equation (7.119) is in fact valid in any frame, see, for example, Itzykson and Zuber 1980, p. 53.

first to rotate the z-axis into the direction of $\boldsymbol{p}$ and then to perform a boost $S = e^{i\chi \cdot K}$ in the $\boldsymbol{p}$ direction with sinh $\chi = |\boldsymbol{p}|/m$ in units $c = 1$:

$$|m, s, \boldsymbol{p}, \lambda> = \Lambda_{p0} |m, s, \boldsymbol{0}, s_3>. \tag{7.125}$$

We next consider the action of the transformation $\Lambda_{p'p}$ which takes the momentum p into p'

$$\begin{aligned}\Lambda_{p'p}|m, s, \boldsymbol{p}, \lambda> &= \Lambda_{p'p}\Lambda_{p0} |m, s, \boldsymbol{0}, \lambda> \\ &= \Lambda_{p'0}(\Lambda_{p'0}^{-1}\Lambda_{p'p}\Lambda_{p0}) |m, s, \boldsymbol{0}, \lambda>.\end{aligned} \tag{7.126}$$

In the bracket there is a succession of operations which take the rest-frame momentum back into itself, which means that $p_\mu = (m, \boldsymbol{0})$ remains invariant. This operation is known as *Wigner rotation*. Its representations are $D^s_{\lambda'\lambda}$. One can write

$$\begin{aligned}\Lambda_{p'p} |m, s, \boldsymbol{p}, \lambda> &= \Lambda_{p'0} \sum_{\lambda'} D^s_{\lambda'\lambda} |m, s, \boldsymbol{0}, \lambda'> \\ &= \sum_{\lambda'} D^s_{\lambda'\lambda} |m, s, \boldsymbol{p}', \lambda'>.\end{aligned} \tag{7.127}$$

The technique described above corresponds to defining the so-called *little group* of the momentum $p_\mu = (m, \boldsymbol{0})$. (For little groups, see, for example, Hamermesh 1962, Chapter 12.) In the above case, the little group is isomorphic to SU(2). Defining little groups is a step required by the method of induced representations used for constructing representations of a group from the representations of its subgroups.

For massless particles, the situation is different. One can start from the 4-vector $p_\mu = (p_3, 0, 0, p_3)$. It turns out that the little group is isomorphic to the Euclidean group E_2, which itself has a semi-direct structure, and its unitary irreducible representations may also be determined by the method of induced representations (see, for example, Cornwell 1984, Chapter 17, Section 8).

Under a space rotation, helicity is invariant because it is a scalar product. For massless particles, it is not only rotationally invariant, but also Lorentz invariant (see Lee 1981, Chapter 3). But under parity, it changes sign because $\boldsymbol{J} \cdot \boldsymbol{P}$ is a pseudoscalar. Thus, in a process which conserves parity, a massless particle must possess both helicities $\pm \lambda$ and its eigenstates produce a reducible representation of the Poincaré group.

Exercise 7.2 Prove that the helicity and the momentum operators commute with each other, i.e. that

$$[\boldsymbol{J} \cdot \boldsymbol{P}, P_j] = 0 \quad \text{for } j = 1, 2, 3. \tag{1}$$

Solution Using the identity

$$[AB, C] = A[B, C] + [A, C]B \tag{2}$$

and

$$[P_i, P_j] = 0 \tag{3}$$

the commutation relation (1) can be rewritten as

$$[\boldsymbol{J}\cdot\boldsymbol{P}, P_j] = \sum_i [J_i, P_j]P_i. \tag{4}$$

Using a definition of type (7.56) for J_i and the commutation relations (7.105) extended to $J_{\alpha\beta}$, one gets

$$\begin{aligned}
[\boldsymbol{J}\cdot\boldsymbol{P}, P_j] &= \frac{1}{2}\sum_{imn} \varepsilon_{imn}[J_{mn}, P_j]P_i \\
&= -\frac{i}{2}\sum_{imn} \varepsilon_{imn}(g_{mj}P_n - g_{nj}P_m)P_i \\
&= -\frac{i}{2}g_{jj}\left(\sum_{in} \varepsilon_{ijn}P_nP_i - \sum_{im} \varepsilon_{imj}P_mP_i\right) \\
&= -ig_{jj}\sum_{in} \varepsilon_{ijn}P_nP_i.
\end{aligned} \tag{5}$$

For any given j, one can show that the expression (5) is zero, due to (3).

SUPPLEMENTARY EXERCISE

7.3 The contravariant energy–momentum 4-vector P^μ is defined as

$$P^\mu = (P^0, \boldsymbol{P}) = (H, \boldsymbol{P}) \tag{1}$$

where H is the Hamiltonian operator and $\boldsymbol{P}$ is the momentum operator. In this notation, show that the commutation relations (7.105) take the form

$$[H, L^i] = 0 \tag{2}$$
$$[P^i, L^j] = i\varepsilon_{ijk}P^k \tag{3}$$
$$[H, K^i] = iP^i \tag{4}$$
$$[P^i, K^j] = i\delta_{ij}H. \tag{5}$$

Note that, together with the relation (7.59), they give the algebra of the Poincaré group.

8

UNITARY GROUPS

As mentioned in Section 5.6, the linear transformations which leave invariant the quadratic Hermitian form

$$\sum_{i=1}^{n} x_i x^*_i$$

form a unitary group U(n). Its elements are unitary matrices.

Unitary groups have extensive applications in physics. They describe intrinsic properties of quantum particles or collective degrees of freedom in composite systems. For example, the group SU(3) was used by Elliott to explain rotational levels of deformed nuclei (Elliott 1958, Elliott and Harvey 1963; for a review, see Hecht 1964).

The group SU(2) is related to spin or isospin degrees of freedom. The group SU(3) is related to the combination of isospin and hypercharge. It was used by Gell-Mann and Ne'eman to classify elementary particles (see Gell-Mann 1961, 1962; Ne'eman 1961; Gell-Mann and Ne'eman 1964). Within SU(3), one can introduce three flavours: up, down, and strange. Adding a new flavour, called charm, one can use SU(4) to extend the classification of elementary particles by incorporating charmed particles as well. Flavour symmetry is supported by the idea that hadrons are composite quark systems and that quark–quark interaction is flavour independent. The SU(3) classification is consistent with the existence of up, down, and strange quarks. In the SU(4) classification, one needs a fourth quark, the c quark, supported by experimental evidence.

The evidence for a fifth quark, the b quark, and some support for the existence of a sixth quark, the t quark, could enlarge the flavour symmetry group to SU(5) and SU(6). However, the amount of symmetry breaking increases considerably with the number of flavours, starting with the charmed quark.

The group SU(4) was introduced a long time ago, as a symmetry group for nuclei (see Wigner and Feenberg 1941). It is based on the fact that each nucleon has four independent intrinsic states corresponding to two possible spin states $\pm\frac{1}{2}$ and two possible isospin states $\pm\frac{1}{2}$. The adequate group is $SU_S(2) \times SU_I(2)$ where S and I stand for spin and isospin. This is a subgroup of SU(4) and the required irreps of SU(4) are generally composed of direct products of irreps of $SU_S(2)$ and $SU_I(2)$. They describe the so-called supermultiplets in nuclei.

In an analogous way, one can combine spin and SU(3)-flavour in hadrons, seen as composite quark systems. Then, one gets the direct product group $SU_S(2) \times SU_F(3)$

which is a subgroup of SU(6). The irreps of SU(6) give rise to supermultiplets of baryons (see, for example, Close 1979 or Nachtmann 1990).

The group $SU(2) \times U(1)$ provides the framework of the electroweak interactions and the SU(3)-colour is the gauge group for the strong interactions (see Chapter 9).

Unitary groups are also used in the classification of many-particle wave functions in atomic or nuclear physics (see, for example, Harmermesh 1962, Wybourne 1970, Talmi 1993).

In the following, we shall describe some general mathematical properties of $U(N)$ and $SU(N)$ and present in some detail the groups SU(2), SU(3), and SU(4). As an application, the classification of elementary particles is considered.

8.1 GENERAL PROPERTIES

An $n \times n$ unitary matrix u can generally be written as

$$u = e^{ih} \tag{8.1}$$

where the matrix h is Hermitian:

$$h = h^+ \tag{8.2}$$

which follows from the unitarity condition $uu^+ = 1$. The hermiticity of h implies that its n diagonal elements are real and the off-diagonal elements contain $\frac{1}{2}n(n-1)$ complex independent parameters. In all, there are n^2 real independent parameters, which is the order of $U(n)$. The unitarity condition also implies

$$|\det u|^2 = 1. \tag{8.3}$$

The matrix u can be brought to its diagonal form u' by a unitary matrix v:

$$u' = vuv^{-1}.$$

Then, one can write

$$\det u = \det u' = \begin{vmatrix} e^{ih'_{11}} & & & \\ & e^{ih'_{22}} & & \\ & & \ddots & \\ & & & e^{ih'_{nn}} \end{vmatrix} \tag{8.4}$$

$$= e^{i \operatorname{tr} h'} = e^{i \operatorname{tr}(vhv^{-1})} = e^{i \operatorname{tr} h}.$$

As h is a Hermitian matrix, it follows that

$$\operatorname{tr} h = \alpha = \text{real} \tag{8.5}$$

so, together with (8.4), this gives

$$\det u = e^{i\alpha}. \tag{8.6}$$

For $\alpha = 0 \pmod{2\pi}$, one has det $u = 1$, i.e. a *special* unitary transformation, which is an element of SU(n). The group SU(n) has $n^2 - 1$ parameters because α has been specified. Thus, an element of SU(n) is defined by

$$u_0 = e^{ih_0}, \qquad \text{tr } h_0 = 0, \qquad \det u_0 = 1. \tag{8.7}$$

Then, an element of U(n) can be written as

$$u = e^{ih} = e^{i(h_0 + \alpha/n \, 1_n)} = e^{i(\alpha/n)1_n} u_0 \tag{8.8}$$

where 1_n is the $n \times n$ unit matrix. The matrices $e^{i(\alpha/n)1_n}$ are unitary and form a one-parameter Lie group, the group U(1). Although an element of U(n) is given by the product of an element of U(1) and an element of SU(n), the group U(n) is not the direct product of U(1) and SU(n) because these groups have two elements in common, 1_n and -1_n, while a direct product group requires only the identity element to be in common. However, as mentioned in Section 5.8, the algebra of U(n) is isomorphic to the algebra of U(1) $\times$ SU(n).

The group U(1) is important in physics. Its generator $F_0 \sim 1_n$ represents the particle number operator. One can easily understand this by writing the generators in a second quantization form (see Section 8.6). If N is the eigenvalue of F_0, the action of U(1) on a wave function ψ introduces a phase

$$\text{U}(1)\,\psi \to e^{iNa^0}\psi \tag{8.9}$$

where a^0 is related to the parameter α (Equation (8.13) below). So, the invariance with respect to U(1) leads to the conservation of the particle number, for example, the baryon number, the lepton number, or the charge.

In terms of the group parameters a^ρ the Hermitian matrix h of (8.1) can be written as

$$h = \sum_{\rho=0}^{n^2-1} a^\rho \frac{\lambda_\rho}{2}. \tag{8.10}$$

The matrices λ_ρ form a complete set. For example, for U(2), they are: $1_2, \sigma_x, \sigma_y, \sigma_z$ (see (8.20)) and for U(3) they are: 1_3 and the eight Gell-Mann matrices to be introduced below (see (8.91)). These matrices are normalized such that

$$\text{tr}(\lambda_i \lambda_j) = 2\delta_{ij}. \tag{8.11}$$

In particular, this gives

$$\lambda_0 = \sqrt{\frac{2}{n}} 1_n. \tag{8.12}$$

which amounts to

$$a^0 = \text{tr}(h\lambda_0) = \alpha\sqrt{\frac{2}{n}}. \tag{8.13}$$

For SU(n), the Hermitian matrix h_0 contains all terms of (8.10) except λ_0:

$$h_0 = \sum_{\rho=1}^{n^2-1} a^\rho \frac{\lambda_\rho}{2}. \qquad (8.14)$$

The property tr $h_0 = 0$ (Equation (8.7)) implies

$$\mathrm{tr}\,\lambda_\rho = 0 \qquad (8.15)$$

because a^ρ are arbitrary, independent parameters.

8.2 THE GROUP SU(2)

In this section, we shall discuss the parametrization of the group SU(2) and will prove explicitly the homomorphism of SU(2) with R_3.

An element of SU(2) is a 2×2 unitary matrix

$$u = \begin{pmatrix} a & b \\ c & d \end{pmatrix} \qquad (8.16)$$

where a, b, c and d are complex, i.e. there are eight real parameters. One also has

$$\det u = 1.$$

The inverse of (8.16) is

$$u^{-1} = \begin{pmatrix} d & -b \\ -c & a \end{pmatrix}.$$

The unitarity condition $u^+ = u^{-1}$ requires $d = a^*$ and $c = -b^*$ so that u can be put in the form

$$u = \begin{pmatrix} a & b \\ -b^* & a^* \end{pmatrix} \qquad (8.17)$$

which contains only four parameters. The condition $\det u = 1$ brings the constraint

$$a\,a^* + b\,b^* = 1 \qquad (8.18)$$

so that one is left with three independent parameters, for example $|a|$, arg a and arg b.

For an infinitesimal transformation, it is useful to introduce the following parametrization

$$a = 1 - \frac{i}{2}a_{11}, \qquad b = -\frac{1}{2}b_{12} - \frac{i}{2}a_{12} \qquad (8.19)$$

which is consistent with $\det u = 1$.

Then, one can write

$$\begin{aligned} u &= \begin{pmatrix} 1 & 0 \\ 0 & 1 \end{pmatrix} - \frac{i}{2}a_{11}\begin{pmatrix} 1 & 0 \\ 0 & -1 \end{pmatrix} - \frac{i}{2}a_{12}\begin{pmatrix} 0 & 1 \\ 1 & 0 \end{pmatrix} - \frac{b_{12}}{2}\begin{pmatrix} 0 & 1 \\ -1 & 0 \end{pmatrix} \\ &= 1_2 - \frac{i}{2}a_{11}\sigma_z - \frac{i}{2}a_{12}\sigma_x - \frac{i}{2}b_{12}\sigma_y \end{aligned} \qquad (8.20)$$

where $\mathbf{1}_2$ is the 2×2 unit matrix and σ_x, σ_y, and σ_z are the Pauli matrices

$$\sigma_x = \begin{pmatrix} 0 & 1 \\ 1 & 0 \end{pmatrix}; \qquad \sigma_y = \begin{pmatrix} 0 & -i \\ i & 0 \end{pmatrix}; \qquad \sigma_z = \begin{pmatrix} 1 & 0 \\ 0 & -1 \end{pmatrix}. \tag{8.21}$$

Using (5.47), one can identify the three generators of SU(2) with the three components of the spin

$$J_i = S_i = \frac{1}{2}\sigma_i \quad (i = x, y, z). \tag{8.22}$$

A finite transformation reads

$$u = e^{-i\boldsymbol{\omega} \cdot \boldsymbol{J}} \tag{8.23}$$

from which one can obtain (8.20) once more by taking $\boldsymbol{\omega} = (a_{12}, b_{12}, a_{11})$. This has the same form as the rotation operator (6.37) but with $\boldsymbol{J} = \boldsymbol{S}$, instead of $\boldsymbol{J} = \boldsymbol{L}$. From the Pauli matrices, it follows that the algebra of SU(2) is the same as that of the rotation group, as given by (6.38). However, the two groups are not necessarily isomorphic. In the next section, we shall show that there is a homomorphic relation between them.

8.3 THE HOMOMORPHISM OF SU(2) WITH R_3

In this section, we shall prove that SU(2) is homomorphic to R_3

$$\text{SU}(2) \to R_3.$$

To prove homomorphism, one has to deal with a finite transformation like (8.17) and relate it to a rotation. For this purpose, let us introduce the complex vector

$$\boldsymbol{w} = \boldsymbol{u} + i\boldsymbol{v} \tag{8.24}$$

having the property

$$\boldsymbol{w} \cdot \boldsymbol{w} = w_1^2 + w_2^2 + w_3^2 = 0. \tag{8.25}$$

Consider a real rotation R of angles α, β, γ as defined in Section 6.1. Under this transformation, the vector $\boldsymbol{w}$ becomes

$$\boldsymbol{w}' = R\boldsymbol{w} \tag{8.26}$$

with

$$\boldsymbol{u}' = R\boldsymbol{u}, \qquad \boldsymbol{v}' = R\boldsymbol{v}'. \tag{8.27}$$

By definition, a rotation leaves invariant the length of any vector, so that

$$\boldsymbol{w}' \cdot \boldsymbol{w}' = (w_1')^2 + (w_2')^2 + (w_3')^2 = 0. \tag{8.28}$$

Now, we wish to parametrize the components w_i ($i = 1, 2, 3$) in the following way. Note that (8.25) also means that

$$\boldsymbol{u} \cdot \boldsymbol{u} = \boldsymbol{v} \cdot \boldsymbol{v}, \qquad \boldsymbol{u} \cdot \boldsymbol{v} = 0. \tag{8.29}$$

These relations indicate that the vector $\boldsymbol{w}$ depends in fact on four quantities only. These could be the three components of $\boldsymbol{u}$ and an angle φ in a plane perpendicular on $\boldsymbol{u}$ which indicates the orientation of $\boldsymbol{v}$ in this plane. For our purpose, it is more convenient to introduce two complex quantities ξ and η—four real variables—and relate them to w_i in the following way

$$w_1 - iw_2 = \eta^2 \tag{8.30}$$
$$w_1 + iw_2 = -\xi^2 \tag{8.31}$$
$$w_3 = \xi\eta. \tag{8.32}$$

These relations are consistent with (8.25). The signs of ξ and η are arbitrary, but their relative sign is determined by (8.32).

A general rotation about a given axis is defined in (6.41) as a succession of three distinct rotations: a rotation of angle γ about the z-axis, followed by a rotation of β around the original y-axis, and a rotation of α, again about the z-axis. Let us start with the last rotation. Following (6.43), one has

$$w'_1 = w_1 \cos\alpha + w_2 \sin\alpha \tag{8.33}$$
$$w'_2 = -w_1 \sin\alpha + w_2 \cos\alpha \tag{8.34}$$
$$w'_3 = w_3 \tag{8.35}$$

which leads to

$$\eta'^2 = w'_1 - iw'_2 = e^{i\alpha}(w_1 - iw_2) = e^{i\alpha}\eta^2 \tag{8.36}$$
$$-\xi'^2 = w'_1 + iw'_2 = -e^{-i\alpha}\xi^2. \tag{8.37}$$

By taking the square root of (8.36) and (8.37), one obtains

$$\begin{pmatrix} \xi' \\ \eta' \end{pmatrix} = \pm \begin{pmatrix} e^{-i/2\alpha} & 0 \\ 0 & e^{i/2\alpha} \end{pmatrix} \begin{pmatrix} \xi \\ \eta \end{pmatrix} \tag{8.38}$$

where the signs are consistent with (8.35). Note that the resulting matrix

$$u_\alpha = \begin{pmatrix} e^{-i/2\alpha} & 0 \\ 0 & e^{i/2\alpha} \end{pmatrix} \tag{8.39}$$

is unitary and two distinct matrices $\pm u_\alpha$ correspond to the same rotation of angle α.

Under a rotation of angle β about the y-axis, the transformation is similar to (6.54):

$$w'_1 = w_1 \cos\beta - w_3 \sin\beta \tag{8.40}$$
$$w'_2 = w_2 \tag{8.41}$$
$$w'_3 = w_1 \sin\beta + w_3 \cos\beta \tag{8.42}$$

from which one gets

$$\xi'^2 = \left(\xi\cos\tfrac{\beta}{2} + \eta\sin\tfrac{\beta}{2}\right)^2 \tag{8.43}$$

$$\eta'^2 = \left(\xi\sin\tfrac{\beta}{2} - \eta\cos\tfrac{\beta}{2}\right)^2 \tag{8.44}$$

$$\xi'\eta' = \left(\xi\cos\tfrac{\beta}{2} + \eta\sin\tfrac{\beta}{2}\right)\left(-\xi\sin\tfrac{\beta}{2} + \eta\cos\tfrac{\beta}{2}\right). \tag{8.45}$$

It turns out that

$$\begin{pmatrix}\xi'\\\eta'\end{pmatrix} = \pm\begin{pmatrix}\cos\beta/2 & \sin\beta/2\\ -\sin\beta/2 & \cos\beta/2\end{pmatrix}\begin{pmatrix}\xi\\\eta\end{pmatrix}. \tag{8.46}$$

One obtains a unitary (real) 2×2 matrix

$$u_\beta = \begin{pmatrix}\cos\beta/2 & \sin\beta/2\\ -\sin\beta/2 & \cos\beta/2\end{pmatrix} \tag{8.47}$$

and again both $+u_\beta$ and $-u_\beta$ are related to the same rotation of angle β. The matrix associated with a rotation of angle γ about the z-axis is like (8.39) but depending on γ instead of α

$$u_\gamma = \begin{pmatrix}e^{-i/2\gamma} & 0\\ 0 & e^{i/2\gamma}\end{pmatrix}. \tag{8.48}$$

Therefore, the homomorphism between SU(2) and R_3 means that two distinct 2×2 unitary matrices of opposite sign correspond to the same rotation R:

$$\pm u \to R \tag{8.49}$$

where

$$u = u_\alpha\, u_\beta\, u_\gamma = \begin{pmatrix}e^{-i/2(\alpha+\gamma)}\cos\tfrac{\beta}{2} & e^{-i/2(\alpha-\gamma)}\sin\tfrac{\beta}{2}\\ -e^{i/2(\alpha-\gamma)}\sin\tfrac{\beta}{2} & e^{-i/2(\alpha+\gamma)}\cos\tfrac{\beta}{2}\end{pmatrix}. \tag{8.50}$$

This can be identified with the special unitary matrix (8.17) as follows

$$\begin{aligned}|a| &= \cos\tfrac{\beta}{2}, & \arg a &= -\tfrac{1}{2}(\alpha+\gamma).\\ |b| &= \sin\tfrac{\beta}{2}, & \arg b &= -\tfrac{1}{2}(\alpha-\gamma).\end{aligned} \tag{8.51}$$

The matrix (8.50) is the transpose of the fundamental representation $D^{1/2}$ of SU(2). Indeed, following the definition (6.42) and putting

$$\xi = \chi_{\tfrac{1}{2}\tfrac{1}{2}}, \qquad \eta = \chi_{\tfrac{1}{2}-\tfrac{1}{2}} \tag{8.52}$$

one can make the identification

$$D^{1/2}(\alpha,\beta,\gamma) = u^{\mathrm{T}} \tag{8.53}$$

MULTIPLETS

where u^T is the transpose of the matrix (8.50). The transposition appears because, in (6.42), the summation is on the first index and u has been introduced with a summation over the second index. Then, the matrix u_β of (8.47) is the transpose of the reduced rotation matrix d^j of (6.59):

$$d^{1/2}(\beta) = u_\beta^T \qquad (8.54)$$

which is consistent with the expression (6.60) for $j = \frac{1}{2}$. Due to (8.23), one can incorporate the spin (8.22) into (6.37) and take the three generators of the rotation group as the three components of the total angular momentum

$$J = L + S. \qquad (8.55)$$

So, when a particle has a spin, the rotation operator (6.37) contains the total angular momentum. On that basis, the representation theory of SU(2) and R_3 is the same and the results of Chapter 6 hold here too.

Exercise 8.1 Derive the reduced rotation matrix $d^{1/2}(\beta)$ directly from the transformation (8.23).

Solution One considers a unitary transformation with $\omega_1 = \omega_3 = 0$ and $\omega_2 = \beta$. Then, (8.23) reads

$$u = e^{-i/2\beta\sigma_y} = \cos\frac{\beta}{2} - i\sigma_y \sin\frac{\beta}{2}. \qquad (1)$$

The invariant subspace of $d^{1/2}$ has the components (8.52)

$$\xi = \chi_{\frac{1}{2}\frac{1}{2}} = \uparrow, \qquad \eta = \chi_{\frac{1}{2}-\frac{1}{2}} = \downarrow. \qquad (2)$$

One gets

$$\sigma_y \uparrow = i \downarrow \qquad (3)$$
$$\sigma_y \downarrow = -i \uparrow \qquad (4)$$

from which it follows that

$$<\uparrow |u| \uparrow> = \cos\frac{\beta}{2}, \qquad <\uparrow |u| \downarrow> = -\sin\frac{\beta}{2}$$
$$<\downarrow |u| \uparrow> = \sin\frac{\beta}{2}, \qquad <\downarrow |u| \downarrow> = \cos\frac{\beta}{2} \qquad (5)$$

i.e. the matrix (8.54)

8.4 MULTIPLETS

A multiplet is a set of quantum states which form an invariant subspace for an irreducible representation of a group. The states in a multiplet are related to each other through the group transformations but are degenerate. This follows immediately from the invariance of the Hamiltonian under a particular group of transformations as it was illustrated, for example, for the rotation group (Section 6.1). In atomic and nuclear physics, there are angular momentum multiplets which follow from the SU(2) invariance. As mentioned in the introduction of this chapter, the SU(2) group is related either to spin S or to isospin I. The mathematical description of isospin is completely analogous to that of spin, as given in the previous section. To distinguish between them, we shall use the notation $SU_S(2)$ for spin and $SU_I(2)$ for isospin.

The two isospin states $I = \frac{1}{2}, I_z = \pm\frac{1}{2}$ have the same properties as the spin states $S = \frac{1}{2}, S_z = \pm\frac{1}{2}$. As for spin, one can define two isospin space basis vectors

$$u = \begin{pmatrix} 1 \\ 0 \end{pmatrix} \qquad d = \begin{pmatrix} 0 \\ 1 \end{pmatrix}. \tag{8.56}$$

By analogy with (8.22), the generators of the $SU_I(2)$ transformations are

$$I_i = \frac{1}{2} \tau_i \quad (i = x, y, z) \tag{8.57}$$

where

$$\tau_x = \begin{pmatrix} 0 & 1 \\ 1 & 0 \end{pmatrix}, \qquad \tau_y = \begin{pmatrix} 0 & -i \\ i & 0 \end{pmatrix} \qquad \tau_z = \begin{pmatrix} 1 & 0 \\ 0 & -1 \end{pmatrix}. \tag{8.58}$$

They have the same algebra as R_3, given by Equation (6.38), and a finite transformation reads

$$u_I = e^{-i\,\boldsymbol{\omega}\cdot\boldsymbol{I}} \tag{8.59}$$

where ω is a rotation angle in the isospin space.

The group $SU_I(2)$ is a symmetry group of the strong interactions, which means that

$$[H_{\text{strong}}, \boldsymbol{I}] = 0. \tag{8.60}$$

Then, an isomultiplet contains $2I + 1$ particles. They are regarded as different states of the same particle and should have equal masses. Small differences are, however, observed. They are caused by electromagnetic or other interactions. For $I = \frac{1}{2}$, examples of isomultiplets are the baryons (p, n) or the mesons (K^+, K^0). The pions (π^+, π^0, π^-) form an isomultiplet with $I = 1$. For other isomultiplets which occur in nature, see Section 8.6.

The extension to many particles is similar to the treatment of the total angular momentum of a many-particle system. Isospins of n particles couple together to give

$$\boldsymbol{I} = \sum_{i=1}^{n} \boldsymbol{I}(i). \tag{8.61}$$

An interesting example of a many-particle system from the isospin point of view is the nucleus. For a discussion, see for example Elliott and Dawber (1979), Section 10.1.

8.5 G-PARITY

The G-parity was introduced by Lee and Yang (1956). It represents the interplay between the isospin symmetry and the particle–antiparticle invariance of the strong interaction. The G-parity is conserved by strong interactions and its operator is defined by

$$G = Ce^{-i\pi I_2} \tag{8.62}$$

where the second factor is a rotation of π about the y-axis in the isospin space (see (8.59)) and C is the particle–antiparticle conjugation operator in relativistic quantum mechanics (Bjorken and Drell 1964). In non-relativistic quantum mechanics, to which we refer below, C would amount to the complex conjugate operation. For the present discussion, complex conjugation is the relevant operation. To understand how (8.62) arises, we have to introduce the *conjugate representations* of SU(2).

Conjugate representations

General properties of conjugate representations were introduced in Section 5.11. Here, we consider the conjugate of the fundamental representation. We start from the matrix (8.17) and take its complex conjugate

$$u^* = \begin{pmatrix} a^* & b^* \\ -b & a \end{pmatrix}. \tag{8.63}$$

The question which arises is whether or not u^* is equivalent to u, and the answer is yes. One can find indeed a similarity transformation between u^* and u

$$u^* = T\,u\,T^{-1} \tag{8.64}$$

where

$$T = \begin{pmatrix} 0 & -1 \\ 1 & 0 \end{pmatrix}. \tag{8.65}$$

Let us now take the spinor

$$\xi = \begin{pmatrix} \xi_1 \\ \xi_2 \end{pmatrix} \tag{8.66}$$

which transforms under (8.17) into $\xi' = \begin{pmatrix} \xi'_1 \\ \xi'_2 \end{pmatrix}$, where

$$\begin{aligned} \xi'_1 &= a\xi_1 + b\xi_2 \\ \xi'_2 &= -b^*\xi_1 + a^*\xi_2. \end{aligned} \tag{8.67}$$

Taking the complex conjugate of (8.67), one finds that the result can be put into the form

$$\begin{pmatrix} \xi_2'^* \\ -\xi_1'^* \end{pmatrix} = \begin{pmatrix} a & b \\ -b^* & a^* \end{pmatrix} \begin{pmatrix} \xi_2^* \\ -\xi_1^* \end{pmatrix} \tag{8.68}$$

which shows that the spinors ξ and

$$\bar{\xi} = -\begin{pmatrix} \xi_2^* \\ -\xi_1^* \end{pmatrix} \tag{8.69}$$

transform in the same way. They are related by

$$\bar{\xi} = CT\xi \tag{8.70}$$

where C is the complex conjugate operation and T is defined by (8.65). One can write (see Exercise 8.1)

$$e^{-i\pi I_2} = \cos\frac{\pi}{2} - i\tau_2 \sin\frac{\pi}{2} = -i\tau_2 = \begin{pmatrix} 0 & -1 \\ 1 & 0 \end{pmatrix} = T. \tag{8.71}$$

Thus, one can identify CT with the G-parity operator (8.62) if one admits that C transforms particles into antiparticles $Cp = \bar{p}$, $Cn = \bar{n}$, etc. The proton p and the neutron n form an isospin doublet, where p corresponds to $\phi_{\frac{1}{2}\frac{1}{2}}$ and n to $\phi_{\frac{1}{2}-\frac{1}{2}}$ in the isospin space. The quarks u and d form an isodoublet in the same way. Then, one obtains

and

$$G\begin{pmatrix} p \\ n \end{pmatrix} = -\begin{pmatrix} \bar{n} \\ -\bar{p} \end{pmatrix} \qquad G\begin{pmatrix} \bar{n} \\ -\bar{p} \end{pmatrix} = \begin{pmatrix} p \\ n \end{pmatrix} \tag{8.72}$$

$$G\begin{pmatrix} u \\ d \end{pmatrix} = -\begin{pmatrix} \bar{d} \\ -\bar{u} \end{pmatrix} \qquad G\begin{pmatrix} \bar{d} \\ -\bar{u} \end{pmatrix} = \begin{pmatrix} u \\ d \end{pmatrix} \tag{8.73}$$

i.e. $\bar{n}$ or $\bar{d}$ correspond to $\phi_{\frac{1}{2}\frac{1}{2}}$ and $-\bar{p}$ or $-\bar{u}$ correspond to $\phi_{\frac{1}{2}-\frac{1}{2}}$ in the isospin space.

To each irrep of SU(2) one can associate a Young diagram. The fundamental representation u and its complex conjugate are both 2×2 matrices and are equivalent. This means they are described by the same Young diagram

$$u \to \square \qquad \text{or} \quad (2)$$
$$u^* \to \square \qquad \text{or} \quad (\bar{2}). \tag{8.74}$$

The figure in the bracket represents the dimension of the corresponding irrep. From a physical point of view, it is sometimes useful to distinguish between u and u^* (particles and antiparticles). For this reason, u^* is designated by $(\bar{2})$ or (2^*) instead of (2).

The direct product of (2) and $(\bar{2})$ gives

$$(2) \times (\bar{2}) = (3) + (1) \qquad (8.75)$$

$$I = \tfrac{1}{2} \qquad I = \tfrac{1}{2} \qquad I = 1 \qquad I = 0$$

It corresponds to the coupling of a particle and an antiparticle, both of isospin $I = \tfrac{1}{2}$ to either $I = 1$ or $I = 0$.

Exercise 8.2 Consider a system made out of a quark u or d and an antiquark $\bar{u}$ or $\bar{d}$ (the non-strange mesons). Write the explicit form of a state of isospin $I = 1, I_3 = 0$.

Solution As indicated above, the identification of the u and d quarks and their antiparticles in the isospin space is

$$u = \varphi_{\frac{1}{2}\frac{1}{2}} \qquad \bar{d} = \bar{\varphi}_{\frac{1}{2}\frac{1}{2}}$$
$$d = \varphi_{\frac{1}{2}-\frac{1}{2}} \qquad -\bar{u} = \bar{\varphi}_{\frac{1}{2}-\frac{1}{2}}. \qquad (1)$$

A state of $I = 1, I_3 = 0$ is obtained by the following coupling

$$|I = 1, I_3 = 0 \rangle = \sum C^{\frac{1}{2}\frac{1}{2}1}_{t_1 t_2 0} \varphi_{\frac{1}{2}t_1} \bar{\varphi}_{\frac{1}{2}t_2}$$
$$= C^{\frac{1}{2}\frac{1}{2}1}_{\frac{1}{2}-\frac{1}{2}0} (\varphi_{\frac{1}{2}\frac{1}{2}} \bar{\varphi}_{\frac{1}{2}-\frac{1}{2}} + \varphi_{\frac{1}{2}-\frac{1}{2}} \bar{\varphi}_{\frac{1}{2}\frac{1}{2}}) \qquad (2)$$

where the symmetry property (6.77) of the CG coefficient has been used. From Table 6.1 and (1), one gets

$$|I = 1, I_3 = 0 \rangle = \frac{1}{\sqrt{2}} (d\bar{d} - u\bar{u}). \qquad (3)$$

This expression is part of the flavour state of the neutral pion π^0, see Table 8.1.

The quantum number G

The G-parity operator commutes with $\boldsymbol{I}$ (Lee 1981, Chapter 11)

$$[G, \boldsymbol{I}] = 0. \qquad (8.76)$$

The strong interaction Hamiltonian H_{strong} is both isospin- and C-invariant, thus G-invariant. Therefore, one can construct eigenstates of H_{strong} with definite I, I_3 and G.
From (8.72) or (8.73), one obtains

$$G^2 \begin{pmatrix} p \\ n \end{pmatrix} = -\begin{pmatrix} p \\ n \end{pmatrix} \qquad G^2 \begin{pmatrix} u \\ d \end{pmatrix} = -\begin{pmatrix} u \\ d \end{pmatrix} \qquad (8.77)$$

which shows that for $I = \frac{1}{2}$ one has

$$G^2 = -1. \tag{8.78}$$

The generalization of this property to an isospinor associated with a given I is

$$G^2 = (-)^{2I}. \tag{8.79}$$

It turns out that for pions

$$G |\pi> = -|\pi> \tag{8.80}$$

irrespective of their charge. This means that in a strong process the number of pions must be either even or odd, both in the initial and final states. For the ρ meson, one has

$$G |\rho> = |\rho> \tag{8.81}$$

for 0 or ± 1 charge. G-conservation requires that ρ cannot decay into an odd number of pions. However, isospin symmetry and G-symmetry are violated by the electromagnetic interaction.

The pions $\pi^\pm$, π^0 form an isomultiplet with $I = 1$ (Section 8.4). Let us consider π^0 which has $I_3 = 0$. Its composition in terms of quarks is given in Exercise 8.2. Now, if one wants to write a state of explicit G-parity, one must take the combination

$$\begin{aligned}\pi^0 &= \frac{1}{2}[(d\bar{d} - u\bar{u}) - G(d\bar{d} - u\bar{u})] \\ &= \frac{1}{2}(d\bar{d} - u\bar{u} + \bar{d}d - \bar{u}u)\end{aligned} \tag{8.82}$$

which obeys (8.80). Flavour states of explicit G-parity are given in Table 8.1 for the pseudoscalar and vector mesons containing u, d, and s quarks.

8.6 THE GROUP SU(3)

First, we shall derive the generators of the SU(3) group following the general procedure (Section 5.1) based on an infinitesimal transformation. The 3×3 matrices obtained from acting with these generators in a three-dimensional space give rise to the Gell-Mann matrices. The connection between the isotropic harmonic oscillator and SU(3) is also discussed.

Infinitesimal transformations

We start from the fundamental representation which is a unitary 3×3 matrix. In the neighbourhood of unity, it can be written as

$$u = \mathbf{1}_3 + i\rho \tag{8.83}$$

where $\mathbf{1}_3$ is the 3×3 unit matrix. Here, ρ is a 3×3 matrix with complex entries satisfying the property

$$\rho = \rho^+ \tag{8.84}$$

THE GROUP SU(3)

Table 8.1 Flavour states of given G-parity for pseudoscalar and vector mesons containing u, d, and s quarks.

Flavour wave function	Meson
$\frac{1}{\sqrt{2}}(u\bar{s} \pm \bar{s}u)$	K^+ (K^{*+})
$\frac{1}{\sqrt{2}}(d\bar{s} \pm \bar{s}d)$	K^0 (K^{0*})
$\frac{1}{\sqrt{2}}(s\bar{u} \pm \bar{u}s)$	K^- (K^{*-})
$\frac{1}{\sqrt{2}}(s\bar{d} \pm \bar{d}s)$	$\bar{K}^0$ ($\bar{K}^{0*}$)
$\frac{1}{\sqrt{2}}(u\bar{d} \pm \bar{d}u)$	π^+ (ρ^+)
$\frac{1}{\sqrt{2}}(d\bar{u} \pm \bar{u}d)$	π^- (ρ^-)
$\frac{1}{2}[(d\bar{d} - u\bar{u}) \pm (\bar{d}d - \bar{u}u)]$	π^0 (ρ^0)
$\frac{1}{2\sqrt{3}}[(u\bar{u} + d\bar{d} - 2s\bar{s}) \pm (\bar{u}u + \bar{d}d - 2\bar{s}s)]$	η_8^0 (ω_8^0)
$\frac{1}{\sqrt{6}}[(u\bar{u} + d\bar{d} + s\bar{s}) \pm (\bar{u}u + \bar{d}d + \bar{s}s)]$	η_1^0 (ω_1^0)

required by the unitarity condition. Then, if $\rho_{ij} = a_{ij} + i\, b_{ij}$, one must have

$$a_{ij} = a_{ji}, \qquad b_{ij} = -b_{ji} \tag{8.85}$$

which means that $b_{ii} = 0$.

It is convenient to work directly in a spherical basis where the coordinates of a point are x_1, x_0, x_{-1} given by (6.45)–(6.47). The infinitesimal unitary transformation introduced above changes x_i into x_i' ($i = 1, 0, -1$) by

$$\begin{pmatrix} x_1' \\ x_0' \\ x_{-1}' \end{pmatrix} = \begin{pmatrix} 1 + ia_{11} & ia_{10} - b_{10} & ia_{1-1} - b_{1-1} \\ ia_{10} + b_{10} & 1 + ia_{00} & ia_{0-1} - b_{0-1} \\ ia_{1-1} + b_{1-1} & ia_{0-1} + b_{0-1} & 1 + ia_{-1-1} \end{pmatrix} \begin{pmatrix} x_1 \\ x_0 \\ x_{-1} \end{pmatrix}. \tag{8.86}$$

276   UNITARY GROUPS

Here, we deal with special transformation, i.e. det $u = 1$. The parameters of a special infinitesimal transformation must therefore fulfil the constraint (second-order terms in ρ_{ij} are neglected)
$$a_{11} + a_{00} + a_{-1\,-1} = 0. \tag{8.87}$$
Thus, eliminating a_{-1-1}, the matrix in (8.86) depends on eight independent parameters $a_{11}, a_{00}, a_{10}, b_{10}, a_{1-1}, b_{1-1}, a_{0-1}$ and b_{0-1}, and the order of SU(3) is eight.

Under an infinitesimal transformation, a scalar function F changes to

$$F(x'_1, x'_0, x'_{-1}) = SF(x_1, x_0, x_{-1}) = F(x_1, x_0, x_{-1}) + \sum_i (x'_i - x_i) \frac{\partial F}{\partial x_i}$$

$$= \left[1 + ia_{11}\left(x_1 \frac{\partial}{\partial x_1} - x_{-1}\frac{\partial}{\partial x_{-1}}\right) + ia_{00}\left(x_0 \frac{\partial}{\partial x_0} - x_{-1}\frac{\partial}{\partial x_{-1}}\right) \right.$$
$$+ ia_{10}\left(x_1 \frac{\partial}{\partial x_0} + x_0 \frac{\partial}{\partial x_1}\right) + ia_{1-1}\left(x_1 \frac{\partial}{\partial x_{-1}} + x_{-1}\frac{\partial}{\partial x_1}\right) + ia_{0-1}\left(x_0 \frac{\partial}{\partial x_{-1}} + x_{-1}\frac{\partial}{\partial x_0}\right)$$
$$+ b_{10}\left(x_1 \frac{\partial}{\partial x_0} - x_0 \frac{\partial}{\partial x_1}\right) + b_{1-1}\left(x_1 \frac{\partial}{\partial x_{-1}} - x_{-1}\frac{\partial}{\partial x_1}\right)$$
$$\left. + b_{0-1}\left(x_0 \frac{\partial}{\partial x_{-1}} - x_{-1}\frac{\partial}{\partial x_0}\right) \right] F. \tag{8.88}$$

From here, according to (5.39), we identify the following infinitesimal operators

$$\widehat{X}_{11} = x_1 \frac{\partial}{\partial x_1} - x_{-1}\frac{\partial}{\partial x_{-1}}, \qquad \widehat{X}_{00} = x_0 \frac{\partial}{\partial x_0} - x_{-1}\frac{\partial}{\partial x_{-1}}$$

$$\widehat{X}_{10} = x_1 \frac{\partial}{\partial x_0} + x_0 \frac{\partial}{\partial x_1}, \qquad \widehat{Y}_{10} = -i\left(x_1 \frac{\partial}{\partial x_0} - x_0 \frac{\partial}{\partial x_1}\right)$$

$$\widehat{X}_{1-1} = x_1 \frac{\partial}{\partial x_{-1}} + x_{-1}\frac{\partial}{\partial x_1}, \qquad \widehat{Y}_{1-1} = -i\left(x_1 \frac{\partial}{\partial x_{-1}} - x_{-1}\frac{\partial}{\partial x_1}\right) \tag{8.89}$$

$$\widehat{X}_{0-1} = x_0 \frac{\partial}{\partial x_{-1}} + x_{-1}\frac{\partial}{\partial x_0}, \qquad \widehat{Y}_{0-1} = -i\left(x_0 \frac{\partial}{\partial x_{-1}} - x_{-1}\frac{\partial}{\partial x_0}\right).$$

To each of these operators, one can associate a matrix. For example,

$$\widehat{X}_{11} \begin{pmatrix} x_1 \\ x_0 \\ x_{-1} \end{pmatrix} = \begin{pmatrix} 1 & 0 & 0 \\ 0 & 0 & 0 \\ 0 & 0 & -1 \end{pmatrix} \begin{pmatrix} x_1 \\ x_0 \\ x_{-1} \end{pmatrix}, \quad \text{etc.} \tag{8.90}$$

The resulting matrices are related to the eight Gell-Mann matrices λ_i ($i = 1, \ldots, 8$) as follows

$$X_{11} + X_{00} = \begin{pmatrix} 1 & 0 & 0 \\ 0 & 1 & 0 \\ 0 & 0 & -2 \end{pmatrix} = \sqrt{3}\,\lambda_8, \quad X_{11} - X_{00} = \begin{pmatrix} 1 & 0 & 0 \\ 0 & -1 & 0 \\ 0 & 0 & 0 \end{pmatrix} = \lambda_3$$
$$\tag{8.91}$$

$$X_{10} = \begin{pmatrix} 0 & 1 & 0 \\ 1 & 0 & 0 \\ 0 & 0 & 0 \end{pmatrix} = \lambda_1, \qquad Y_{10} = \begin{pmatrix} 0 & -i & 0 \\ i & 0 & 0 \\ 0 & 0 & 0 \end{pmatrix} = \lambda_2$$

$$X_{1-1} = \begin{pmatrix} 0 & 0 & 1 \\ 0 & 0 & 0 \\ 1 & 0 & 0 \end{pmatrix} = \lambda_4, \qquad Y_{1-1} = \begin{pmatrix} 0 & 0 & -i \\ 0 & 0 & 0 \\ i & 0 & 0 \end{pmatrix} = \lambda_5$$

$$X_{0-1} = \begin{pmatrix} 0 & 0 & 0 \\ 0 & 0 & 1 \\ 0 & 1 & 0 \end{pmatrix} = \lambda_6, \qquad Y_{0-1} = \begin{pmatrix} 0 & 0 & 0 \\ 0 & 0 & -i \\ 0 & i & 0 \end{pmatrix} = \lambda_7$$

Notice that Gell-Mann matrices are traceless and one can check that they are normalized such as to satisfy (8.11). To these matrices one can add λ_0 defined according to (8.12):

$$\lambda_0 = \sqrt{\frac{2}{3}} \begin{pmatrix} 1 & 0 & 0 \\ 0 & 1 & 0 \\ 0 & 0 & 1 \end{pmatrix}. \tag{8.92}$$

This matrix corresponds to the infinitesimal generator of U(1)

$$\widehat{N} = x_1 \frac{\partial}{\partial x_1} + x_0 \frac{\partial}{\partial x_0} + x_{-1} \frac{\partial}{\partial x_{-1}} \tag{8.93}$$

which can be obtained as above, but neglecting the constraint (8.87).

By direct calculations, one can find that the su(3) Lie algebra is

$$[\lambda_k, \lambda_\ell] = 2i f_{k\ell m} \lambda_m \tag{8.94}$$

where $f_{k\ell m}$ are antisymmetric under exchange of any two indices. The only non-vanishing values are permutations of

$$f_{123} = 1, \qquad f_{147} = f_{165} = f_{246} = f_{257} = f_{345} = f_{376} = \frac{1}{2},$$
$$f_{458} = f_{678} = \frac{\sqrt{3}}{2}. \tag{8.95}$$

If λ_0 is excluded, they also satisfy

$$\{\lambda_k, \lambda_\ell\} = 2 d_{k\ell m} \lambda_m + \frac{4}{3} \delta_{k\ell} \mathbf{1}_3 \quad \text{(without } \lambda_0\text{)} \tag{8.96}$$

and if λ_0 is included, they satisfy

$$\{\lambda_k, \lambda_\ell\} = 2 d_{k\ell m} \lambda_m \quad \text{(with } \lambda_0\text{)} \tag{8.97}$$

where the constants $d_{k\ell m}$ are symmetric under permutations of any two indices. Their non-vanishing values are given in Table 8.2.

Similarly to SU(2), one can introduce the generators

$$\widehat{F}_i = \frac{1}{2} \lambda_i. \tag{8.98}$$

Table 8.2 Non-vanishing values of the constants $d_{k\ell m}$ as defined by (8.96) or (8.97).

$k\ell m$	$d_{k\ell m}$	$k\ell m$	$d_{k\ell m}$	$k\ell m$	$d_{k\ell m}$
000	$\sqrt{\frac{2}{3}}$	118	$\frac{\sqrt{3}}{3}$	366	$-\frac{1}{2}$
011	$\sqrt{\frac{2}{3}}$	146	$\frac{1}{2}$	377	$-\frac{1}{2}$
022	$\sqrt{\frac{2}{3}}$	157	$\frac{1}{2}$	448	$-\frac{\sqrt{3}}{6}$
033	$\sqrt{\frac{2}{3}}$	228	$\frac{\sqrt{3}}{3}$	558	$-\frac{\sqrt{3}}{6}$
044	$\sqrt{\frac{2}{3}}$	247	$-\frac{1}{2}$	668	$-\frac{\sqrt{3}}{6}$
055	$\sqrt{\frac{2}{3}}$	256	$\frac{1}{2}$	778	$-\frac{\sqrt{3}}{6}$
066	$\sqrt{\frac{2}{3}}$	338	$\frac{\sqrt{3}}{3}$	888	$-\frac{\sqrt{3}}{6}$
077	$\sqrt{\frac{2}{3}}$	344	$\frac{1}{2}$		
088	$\sqrt{\frac{2}{3}}$	355	$\frac{1}{2}$		

The commutation relations of $\widehat{F}_i$ can be obtained straightforwardly from (8.94):

$$[\widehat{F}_k, \widehat{F}_\ell] = \mathrm{i}\, f_{k\ell m}\, \widehat{F}_m \tag{8.99}$$

which represents another form of the su(3) Lie algebra.

Using the operators $\widehat{F}_i$, one can write the transformation S defined by (8.88) as

$$S = 1 + \mathrm{i}\,\delta\,\theta_i \widehat{F}_i \tag{8.100}$$

Its finite form

$$S = \mathrm{e}^{\mathrm{i}\theta_i \widehat{F}_i} \tag{8.101}$$

is a unitary transformation because θ_i are real and $\widehat{F}_i$ are Hermitian.

The root diagram of SU(3)

In Section 5.9, the standard form of any semi-simple Lie algebra was introduced. The standard form contains generators of type H_i which commute among themselves and are associated to zero-roots and generators of type E_α associated to non-vanishing roots α. The correspondence between the generators (8.98) and the standard form is given in Table 8.3 in columns 1 and 2. Column 3 reproduces the Gell-Mann notation and column 4 gives the notation used in particle physics, where the ladder operators $\widehat{I}_\pm, \widehat{V}_\pm, \widehat{U}_\pm$ are currently used. The last column gives the SU(3) generators in terms of operators $A_{ij} = a_i^+ a_j$, obtained below from the isotropic harmonic oscillator in the second quantization.

The six roots $\pm\alpha, \pm\beta, \pm(\alpha+\beta)$ can be identified in Fig. 8.1, which reproduces the diagram A_2 of Fig. 5.3. According to the general theory developed in Section 5.9, this root diagram indicates that SU(3) is a group of rank $\ell = 2$ and two consecutive roots have equal length and make an angle of 60°. For convenience, the vertical axis corresponds to H_1 and the horizontal axis to H_2. Then, the root vectors, of unit length, have the following components

$$\begin{aligned}
\alpha_1 &= 0 & \beta_1 &= \frac{\sqrt{3}}{2} & (\alpha+\beta)_1 &= \frac{\sqrt{3}}{2} \\
\alpha_2 &= 1 & \beta_2 &= -\frac{1}{2} & (\alpha+\beta)_2 &= \frac{1}{2}.
\end{aligned} \qquad (8.102)$$

The value of the structure constants $N_{\alpha\beta}$ of (5.126c) are determined as explained in Section 5.9 with the phase convention that $N_{\alpha\beta}$ is positive:

$$\begin{aligned}
N_{\alpha\beta} &= N_{-\alpha-\beta,\alpha} = N_{\alpha+\beta,-\beta} = \frac{1}{\sqrt{2}} \\
N_{-a,-\beta} &= N_{\alpha+\beta,-\alpha} = N_{-\alpha-\beta,\beta} = -\frac{1}{\sqrt{2}}.
\end{aligned} \qquad (8.103)$$

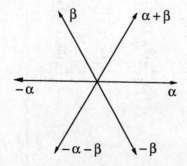

Figure 8.1 Root diagram of SU(3).

Table 8.3 SU(3) generators in the standard form of Equations (5.126) and connection to other bases. For root notation, see Fig. 8.1.

Standard form	Equations (8.89)	Equation (8.98)	Elementary particles	Harmonic oscillator
H_1	$\frac{1}{2\sqrt{3}}\left(\widehat{X}_{11} + \widehat{X}_{00}\right)$	$\widehat{F}_8$	$\frac{\sqrt{3}}{2}\widehat{Y}$	$\frac{1}{2\sqrt{3}}(A_{xx} + A_{yy} - 2A_{zz})$
H_2	$\frac{1}{2}\left(\widehat{X}_{11} - \widehat{X}_{00}\right)$	$\widehat{F}_3$	$\widehat{I}_3$	$\frac{1}{2}(A_{xx} - A_{yy})$
$E_{\pm\alpha}$	$\frac{1}{2\sqrt{2}}\left(\widehat{X}_{10} \pm i\widehat{Y}_{10}\right)$	$\frac{1}{\sqrt{2}}\left(\widehat{F}_1 \pm i\widehat{F}_2\right)$	$\frac{1}{\sqrt{2}}\widehat{I}_\pm$	$\widehat{I}_+ = A_{xy}, \quad \widehat{I}_- = A_{yx}$
$E_{\pm(\alpha+\beta)}$	$\frac{1}{2\sqrt{2}}\left(\widehat{X}_{1,-1} \pm i\widehat{Y}_{1,-1}\right)$	$\frac{1}{\sqrt{2}}\left(\widehat{F}_4 \pm i\widehat{F}_5\right)$	$\frac{1}{\sqrt{2}}\widehat{V}_\pm$	$\widehat{V}_+ = A_{xz}, \quad \widehat{V}_- = A_{zx}$
$E_{\pm\beta}$	$\frac{1}{2\sqrt{2}}\left(\widehat{X}_{0,-1} \pm i\widehat{Y}_{0,-1}\right)$	$\frac{1}{\sqrt{2}}\left(\widehat{F}_6 \pm i\widehat{F}_7\right)$	$\frac{1}{\sqrt{2}}\widehat{U}_\pm$	$\widehat{U}_+ = A_{yz}, \quad \widehat{U}_- = A_{zy}$

THE GROUP SU(3)

With these conventions, and taking $g_{ik} = \delta_{ik}$, the su(3) algebra (8.99) can be rewritten in the standard form (5.126)

$$[H_1, H_2] = 0,$$

$$[H_1, E_{\pm\alpha}] = 0, \quad [H_1, E_{\pm\beta}] = \pm\frac{\sqrt{3}}{2}E_{\pm\beta}, \quad [H_1, E_{\pm(\alpha+\beta)}] = \pm\frac{\sqrt{3}}{2}E_{\pm(\alpha+\beta)}$$

$$[H_2, E_{\pm\alpha}] = \pm E_\alpha, \quad [H_2, E_{\pm\beta}] = \mp\frac{1}{2}E_{\pm\beta}, \quad [H_2, E_{\pm(\alpha+\beta)}] = \pm\frac{1}{2}E_{\pm(\alpha+\beta)}$$

$$[E_\alpha, E_{-\alpha}] = H_2, \quad [E_\beta, E_{-\beta}] = \frac{\sqrt{3}}{2}H_1 - \frac{1}{2}H_2, \quad [E_{\alpha+\beta}, E_{-\alpha-\beta}] = \frac{\sqrt{3}}{2}H_1 + \frac{1}{2}H_2$$

$$[E_\alpha, E_\beta] = \frac{1}{\sqrt{2}}E_{\alpha+\beta}, \quad [E_{-\alpha}, E_{-\beta}] = -\frac{1}{\sqrt{2}}E_{-\alpha-\beta}; \quad \text{etc.} \qquad (8.104)$$

For the purpose of application to elementary particles, it is useful to rewrite the above relations in terms of the operators given in column 4 of Table 8.3. Here, we give the following list

$$[\hat{Y}, \hat{I}_3] = 0 \qquad (8.105a)$$

$$[\hat{Y}, \hat{I}_\pm] = 0, \quad [\hat{Y}, \hat{U}_\pm] = \pm\hat{U}_\pm, \quad [\hat{Y}, \hat{V}_\pm] = \pm\hat{V}_\pm \qquad (8.105b)$$

$$[\hat{I}_3, \hat{I}_\pm] = \pm\hat{I}_\pm, \quad [\hat{I}_3, \hat{U}_\pm] = \mp\frac{1}{2}\hat{U}_\pm, \quad [\hat{I}_3, \hat{V}_\pm] = \pm\frac{1}{2}\hat{V}_\pm \qquad (8.105c)$$

$$[\hat{I}_+, \hat{I}_-] = 2\hat{I}_3, \quad [\hat{U}_+, \hat{U}_-] = \frac{3}{2}\hat{Y} - \hat{I}_3, \quad [\hat{V}_+, \hat{V}_-] = \frac{3}{2}\hat{Y} + \hat{I}_3 \qquad (8.105d)$$

$$[\hat{U}_+, \hat{V}_-] = \hat{I}_-, \quad [\hat{I}_-, \hat{V}_+] = \hat{U}_+, \quad [\hat{I}_+, \hat{U}_+] = \hat{V}_+ \qquad (8.105e)$$

$$[\hat{U}_+, \hat{V}_+] = [\hat{U}_+, \hat{I}_-] = [\hat{V}_+, \hat{I}_+] = 0. \qquad (8.105f)$$

The other missing commutators can be obtained from (8.105) by using

$$(\hat{I}_+)^\dagger = \hat{I}_-, \quad (\hat{U}_+)^\dagger = \hat{U}_-, \quad (\hat{V}_+)^\dagger = \hat{V}_- \qquad (8.106)$$

consistent with Section 5.9.

The three-dimensional harmonic oscillator

The isotropic harmonic oscillator is a well-known example of a Hamiltonian with an accidental degeneracy, like the hydrogen atom discussed in Section 6.2. The group U(3) is its largest symmetry group due to the dependence of type r^2 of V. This is another example of the so-called dynamical symmetries introduced in Section 6.2. The harmonic oscillator is used in describing small departures from an equilibrium position. In nuclear physics, it is used as a first approximation to the motion of individual nucleons in the nucleus (shell model) and in subnuclear physics as an approximation of the linear confinement potential between a quark and an antiquark. The results derived below can be straightforwardly generalized to an n-dimensional harmonic oscillator, the symmetry group of which is U(n). The dynamical symmetry of the isotropic harmonic oscillator is discussed both classically and quantum mechanically in Schiff (1968), Section 30. For the SU(3) classification of shell model states (the Elliott model) the reader is referred to the review paper of Hecht (1964). Applications to quark systems will be discussed in Chapter 10.

If one measures distances in units of the oscillator length parameter $b = \left(\dfrac{\hbar}{m\omega}\right)^{\frac{1}{2}}$ and energies in units of $\hbar\omega$, the harmonic oscillator Hamiltonian of a particle of mass m reads

$$H = \frac{1}{2}\left(-\nabla^2 + r^2\right). \tag{8.107}$$

It is convenient to work in Cartesian coordinates and introduce the creation operators

$$a_x^+ = \frac{1}{\sqrt{2}}\left(x - \frac{\partial}{\partial x}\right), \quad a_y^+ = \frac{1}{\sqrt{2}}\left(y - \frac{\partial}{\partial y}\right), \quad a_z^+ = \frac{1}{\sqrt{2}}\left(z - \frac{\partial}{\partial z}\right) \tag{8.108}$$

and their corresponding adjoints, the annihilation operators

$$a_x = \frac{1}{\sqrt{2}}\left(x + \frac{\partial}{\partial x}\right), \quad a_y = \frac{1}{\sqrt{2}}\left(y + \frac{\partial}{\partial y}\right), \quad a_z = \frac{1}{\sqrt{2}}\left(z + \frac{\partial}{\partial z}\right) \tag{8.109}$$

in terms of which the Hamiltonian (8.107) becomes

$$H = \Sigma\, a_i^+ a_i + \frac{3}{2} \tag{8.110}$$

which is an example of the quadratic Hermitian form (5.79).

The above operators satisfy the commutation relations

$$[a_i, a_j] = [a_i^+, a_j^+] = 0, \quad [a_i, a_j^+] = \delta_{ij} \tag{8.111}$$

$$[a_i, H] = a_i, \quad [a_i^+, H] = -a_i^+. \tag{8.112}$$

The ground state of H

$$|0> = \frac{1}{\pi^{3/4}} \exp\left[-\frac{1}{2}(x^2 + y^2 + z^2)\right] \tag{8.113}$$

satisfies the condition

$$a_i |0> = 0.$$

An excited state corresponding to $N = N_x + N_y + N_z$ quanta of excitation is

$$\psi_N = (a_x^+)^{N_x} (a_y^+)^{N_y} (a_z^+)^{N_z} |0>. \tag{8.114}$$

The degeneracy of a level with N quanta is $\frac{1}{2}(N+1)(N+2)$, which represents the sum over $(2\ell+1)$-fold degeneracy, with $\ell = N, N-2, \ldots, 1$ or 0 depending whether N is odd or even. Thus the harmonic oscillator has a greater degeneracy than expected from spherical symmetry and this comes from the invariance of H under

U(3), and R_3 is only a subgroup of U(3). The invariance is expressed by the commutation relations

$$[a_i^+ a_j, H] = 0 \qquad (8.115)$$

which can be straightforwardly derived from (8.111) and (8.112). Thus, there are nine operators

$$A_{ij} = a_i^+ a_j \qquad (8.116)$$

which commute with H. They represent the nine generators of U(3) and their relation to the eight generators of SU(3) defined in the previous section is shown in Table 8.3.

The ninth operator

$$H_0 = a_x^+ a_x + a_y^+ a_y + a_z^+ a_z \qquad (8.117)$$

corresponds to (8.93) and gives U(1) transformations which obviously conserve the number of quanta. It commutes with all the other operators because $H_0 = H - \frac{3}{2}$. In terms of A_{ij}, a unitary operator takes the form

$$U = \exp\left(\sum_{i,j} C_{ij} A_{ij}\right) \qquad (8.118)$$

where the parameters C_{ij} must satisfy the condition

$$C_{ij} = -C_{ji}^* \qquad (8.119)$$

required by $U^+ = U^{-1}$. Thus, taking $C_{ij} = a_{ij} + i\, b_{ij}$, one obtains $a_{ij} = -a_{ji}, b_{ij} = b_{ji}$, i.e. nine independent parameters as required by U(3) (see equation (5.82)). Under U, a creation operator transforms as

$$(a_i^+) = U a_i^+ U^{-1} = \Sigma(\exp C)_{ij}\, a_j^+ = \Sigma\, u_{ij}\, a_j^+ \qquad (8.120)$$

where C is the matrix introduced above and u is a unitary matrix which is an element of U(3). In applications to nuclear physics (the Elliott model), one takes linear combinations of A_{ij} which are irreducible tensor operators with respect to rotations. The reason is that in this form they have a physical significance for describing deformed nuclei. The $L=0$ tensor operator is given by (8.117) and represents the particle number operator, the $L=1$ tensor operator corresponds to the angular momentum, and the $L=2$ operator corresponds to the quadrupole operator $Q_\mu (\mu = 0, \pm 1, \pm 2)$. The quadrupole operator generates the quadrupole–quadrupole interaction between nucleons, which is diagonal in an SU(3) basis (see Elliott and Dawber 1979, Chapter 19).

From (8.111), one can show that the operators A_{ij} obey the following commutation relations

$$[A_{ij}, A_{k\ell}] = A_{i\ell}\delta_{jk} - A_{kj}\delta_{i\ell}. \qquad (8.121)$$

The validity of this relation is not only restricted to U(3). It also represents the algebra of U(n) provided the definition of A_{ij} is extended to an n-dimensional isotropic harmonic oscillator ($i, j = 1, 2, \ldots, n$). This means that from algebraic point of view the extension from three to n dimensions is straightforward because the symmetry group of the n-dimensional harmonic oscillator is U(n).

The restriction from U(n) to SU(n) can be made by defining the operators

$$A'_{ij} = A_{ij} - \frac{1}{n}\delta_{ij}H_0 \tag{8.122}$$

where H_0 is a generalization of (8.117) to n dimensions. The definition (8.122) implies that

$$\sum_{i=1}^{n} A'_{ii} = 0 \tag{8.123}$$

i.e. only $n - 1$ of these operators are linearly independent, giving a total number of $n^2 - 1$ generators for SU(n).

The weight diagram of the fundamental representation

In Section 5.10, we introduced the concept of weights and invariant operators. They can be used to obtain a complete labelling of the invariant subspace of a representation. One can label an irrep either by its highest weight or by the eigenvalues of the invariant operators. In any case, one needs the eigenvalues of H_i. Before defining the invariant operators of SU(3) and discussing the problem of labelling, let us practise first by drawing the weight diagram of the fundamental representation.

We recall that the weight of a representation is the set of eigenvalues $m_1, m_2, \ldots, m_\ell$ of H_i, as defined by Equations (5.168). For SU(3), $\ell = 2$ so that, as for roots, one can draw a plane diagram containing points representing values of the set (m_1, m_2).

Instead of H_1 and H_2 we use

$$\widehat{Y} = \frac{2}{\sqrt{3}}H_1, \quad \widehat{I}_3 = H_2 \tag{8.124}$$

defined in Table 8.3. These are operators relevant for applications to elementary particle physics. In the three-dimensional space of the fundamental representation, we introduce the linearly independent basis

$$u = \begin{pmatrix} 1 \\ 0 \\ 0 \end{pmatrix}, \quad d = \begin{pmatrix} 0 \\ 1 \\ 0 \end{pmatrix}, \quad s = \begin{pmatrix} 0 \\ 0 \\ 1 \end{pmatrix}. \tag{8.125}$$

The terminology is related to the quarks u, d, and s because, as we shall see below, their eigenvalues correspond to the quark quantum numbers. Using matrices (8.91),

THE GROUP SU(3)

one immediately finds

$$\widehat{Y}u = \frac{1}{3}u \quad \widehat{Y}d = \frac{1}{3}d \quad \widehat{Y}s = -\frac{2}{3}s$$
$$\widehat{I}_3 u = \frac{1}{2}u \quad \widehat{I}_3 d = -\frac{1}{2}d \quad \widehat{I}_3 s = 0$$
(8.126)

In the weight diagram drawn in Fig. 8.2, each point represents a set (m_1, m_2) of eigenvalues associated to a particular basis vector defined by (8.125).

By further use of matrices (8.91) and Table 8.3, one finds

$$d = \widehat{I}_- u \quad u = \widehat{I}_+ d$$
$$s = \widehat{V}_- u \quad u = \widehat{V}_+ s \qquad (8.127)$$
$$s = \widehat{U}_- d \quad d = \widehat{U}_+ s$$

and

$$\widehat{U}_- u = \widehat{I}_+ u = \widehat{V}_+ u = 0. \qquad (8.128)$$

Relations (8.127) show that the ladder operators allow us to shift from one point to another so as to cover all points of the diagram. This remains valid for any diagram. The action of the ladder operators is schematically drawn in Fig. 8.3.

The relations (8.128) indicate that u corresponds to the highest weight, beyond which there is no other point. They are equivalent to the highest weight definition introduced in Section 5.10. Indeed, the differences $\left(\frac{1}{2}, \frac{1}{3}\right) - \left(-\frac{1}{2}, \frac{1}{3}\right) = (1, 0)$ and $\left(\frac{1}{2}, \frac{1}{3}\right) - \left(0, -\frac{2}{3}\right) = (\frac{1}{2}, 1)$ are positive, meaning that the first non-vanishing component is positive, so that u has a higher weight than d or s.

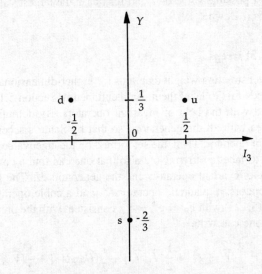

Figure 8.2 Weight diagram of the fundamental representation of SU(3).

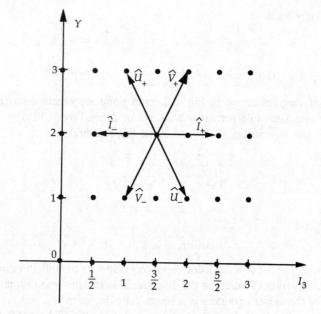

Figure 8.3 Action of the ladder operators in the I_3-Y plane.

The definition of the highest weight is a matter of convention, but once fixed, it must be followed for all diagrams. One could define alternatively the highest weight to have the largest possible value of Y and for this Y the largest value of I_3 (see, for example, Elliott and Dawber 1979).

Labelling of SU(3) irreps

To go further in constructing weight diagrams for higher-dimensional representations of SU(3), one needs first to label them. As mentioned in Section 5.10, this operation is usually realized with the help of invariant operators. By definition, an invariant operator commutes with all the generators, so that by Schur's second lemma it is a multiple of the unit operator. Then, the set formed by the eigenvalues of the invariant operators is used to label an irrep. Note also that one can find a common system of eigenvectors for the invariant operators and the generators H_i. The group SU(3) has two invariant operators, a quadratic operator F^2, and a cubic operator G^3. F^2 is the Casimir operator (5.173) with $g_{ik} = g^{ik} = \delta_{ik}$, consistent with the previous section, so using Table 8.3, one can write

$$F^2 = \Sigma \widehat{F}_i \widehat{F}_i = \frac{3}{4} \widehat{Y}^2 + \widehat{I}_3^2 + \frac{1}{2}\{\widehat{I}_+, \widehat{I}_-\} + \frac{1}{2}\{\widehat{U}_+, \widehat{U}_-\} + \frac{1}{2}\{\widehat{V}_+, \widehat{V}_-\}. \quad (8.129)$$

The operator G^3 is given by (5.177) with $n = 3$. One can show that it can be rewritten in the form

$$G^3 = 8\, \Sigma d_{ijk}\, \widehat{F}_i\, \widehat{F}_j\, \widehat{F}_k. \tag{8.130}$$

The expression of F^2 and G^3 in terms of the generators A_{ij} of (8.116) can be found in Elliott and Dawber (1979), Section 11.10. The eigenvalues F^2 and G^3 in the invariant subspace of an irrep provide a set of two indices to label that particular irrep. However, in practice one uses an alternative label obtained by associating a Young diagram to an irrep. This subject has been generally discussed in Section 5.11. For SU(3), any Young diagram contains a maximum of three rows and the representations with the partition $(f_1 + s, f_2 + s, f_3 + s)$ with $s = 0, 1, 2, \ldots, n$ are equivalent among themselves. This means that by adding a finite number of columns of three rows each to any diagram, one obtains equivalent representations (see the example (5.187)). In other words, it is enough to specify the length of the first and second row as below

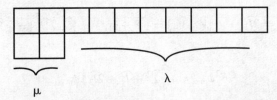

which means that only two labels $\lambda = f_1 - f_2$ and $\mu = f_2 - f_3$ are needed. Then, the notation for an irrep is (λ, μ). For example, for the fundamental representation, one uses

$\square$ or $(\lambda\mu) = (10)$.

The eigenvalue of F^2 or G^3 is the same for any of the vectors of the invariant subspace of an irrep. It is convenient to calculate it starting from the highest weight vector $\phi_{\max}$. We choose the following definition

$$\widehat{I}_+\phi_{\max} = \widehat{V}_+\phi_{\max} = \widehat{U}_-\phi_{\max} = 0 \tag{8.131}$$

$$\widehat{Y}\phi_{\max} = \frac{1}{3}(\lambda - \mu)\phi_{\max} \tag{8.132}$$

$$\widehat{I}_3\phi_{\max} = \frac{1}{2}(\lambda + \mu)\phi_{\max}. \tag{8.133}$$

Note that (8.126) and (8.128) are consistent with this definition. It is easy to show that, for an irrep $(\lambda\mu)$, the eigenvalue equation for F^2 is (see Exercise 8.3)

$$F^2\phi_{\max} = \frac{1}{3}(\lambda^2 + \mu^2 + \lambda\mu + 3\lambda + 3\mu)\phi_{\max} \tag{8.134}$$

and it is rather lengthy to find that (see, for example, Elliott and Dawber[†] (1979), Section 11.10)

$$G^3\phi_{\max} = \frac{1}{9}(\lambda - \mu)(2\lambda + \mu + 3)(\lambda + 2\mu + 3)\phi_{\max}. \qquad (8.135)$$

Exercise 8.3 Starting from the definition (8.131)–(8.133), derive Equation (8.134).

Solution Using the definition (8.129) of F^2 and the property (8.131) of the highest weight vector $\phi_{\max}$, one can write

$$F^2\phi_{\max} = \left(\frac{3}{4}\widehat{Y}^2 + \widehat{I}_3^2 + \frac{1}{2}\widehat{I}_+\widehat{I}_- + \frac{1}{2}\widehat{V}_+\widehat{V}_- + \frac{1}{2}\widehat{U}_-\widehat{U}_+\right)\phi_{\max} \qquad (1)$$

or, alternatively,

$$F^2\phi_{\max} = \left\{\frac{3}{4}\widehat{Y}^2 + \widehat{I}_3^2 + \frac{1}{2}[\widehat{I}_+, \widehat{I}_-] + \frac{1}{2}[\widehat{V}_+, \widehat{V}_-] - \frac{1}{2}[\widehat{U}_+, \widehat{U}_-]\right\}\phi_{\max}. \qquad (2)$$

The commutators appearing in (2) are listed in (8.105)

$$[\widehat{I}_+, \widehat{I}_-] = 2\widehat{I}_3, \qquad [\widehat{V}_+, \widehat{V}_-] = \frac{3}{2}\widehat{Y} + \widehat{I}_3, \qquad [\widehat{U}_+, \widehat{U}_-] = \frac{3}{2}\widehat{Y} - \widehat{I}_3 \qquad (3)$$

so one gets

$$F^2\phi_{\max} = \left(\frac{3}{4}\widehat{Y}^2 + \widehat{I}_3^2 + 2\widehat{I}_3\right)\phi_{\max} \qquad (4)$$

which, together with (8.132) and (8.133), leads to (8.134):

$$F^2\phi_{\max} = \frac{1}{3}(\lambda^2 + \mu^2 + \lambda\mu + 3\lambda + 3\mu)\phi_{\max}. \qquad (5)$$

In general, the complete labelling of a basis vector of an irrep is performed by considering a chain of subgroups (see, for example, the permutation group, Chapter 4). Here, we consider the chain

$$SU(3) \supset SU_I(2) \times U_Y(1) \qquad (8.136)$$

called a canonical chain. Let us explain the indices I and Y. The group SU(3) has three SU(2) subgroups, corresponding to the I-spin, U-spin and V-spin, respectively. As can be inferred from (8.105), all three satisfy the SU(2) commutation relations (6.38), provided one defines

$$\widehat{V}_3 = \frac{3}{4}\widehat{Y} + \frac{1}{2}\widehat{I}_3, \qquad \widehat{U}_3 = \frac{3}{4}\widehat{Y} - \frac{1}{2}\widehat{I}_3. \qquad (8.137)$$

Here, we choose the $SU_I(2)$ based on the I-spin (or isospin). Its complete set of commuting operators is formed of $\widehat{I}^2$ and $\widehat{I}_3$. The generator of $U_Y(1)$ is the

---

[†]Note that our $\widehat{V}_\pm$ is $\widehat{V}_\mp$ in Elliot and Dawber's book

hypercharge operator $\widehat{Y}$ proportional to H_1 of SU(3). The quantum numbers I and Y will be discussed in the next subsection. One has

$$[\widehat{I}^2, \widehat{I}_3] = [\widehat{I}^2, \widehat{Y}] = [\widehat{Y}, \widehat{I}_3] = 0. \tag{8.138}$$

The invariants F^2 and G^3 commute with all the generators, hence also with $\widehat{I}^2$, $\widehat{Y}$, and $\widehat{I}_3$. Then, their simultaneous eigenfunctions form basis vectors for an irrep $(\lambda\,\mu)$ and these can be labelled by the set $(\lambda\,\mu)\, I\, I_3\, Y$. By explicit construction, one can show that this is a complete set. One starts from the highest weight vector and by applying the ladder operators $\widehat{I}_\pm$, $\widehat{U}_\pm$ and $\widehat{V}_\pm$ and their commutation relations, one can cover the whole invariant subspace of $(\lambda\,\mu)$. The resulting number of independent functions of $(\lambda\,\mu)$ symmetry gives the dimension of the $(\lambda\,\mu)$ irrep which can be obtained from Appendix D:

$$d^{SU(3)}_{(\lambda\,\mu)} = \frac{1}{2}(\lambda+1)(\mu+1)(\lambda+\mu+2). \tag{8.139}$$

For fixed λ and μ such as $\lambda > \mu$, possible values of Y and I are shown in Table 8.4.

Let us discuss the explicit action of the ladder operators $\widehat{I}_\pm$, $\widehat{U}_\pm$ and $\widehat{V}_\pm$ on a basis vector $\phi(I, I_3, Y)$ belonging to an irrep (λ, μ). For this purpose, we need the commutation relations (8.105b, c). For example, acting with the commutators containing $\widehat{V}_\pm$ on $\phi(I, I_3, Y)$, one obtains

$$\widehat{Y}\,\widehat{V}_\pm\,\phi = (Y\pm 1)\,\widehat{V}_\pm\,\phi$$

$$\widehat{I}_3\,\widehat{V}_\pm\,\phi = (I_3\pm\frac{1}{2})\,\widehat{V}_\pm\,\phi$$

which shows that $\widehat{V}_+(\widehat{V}_-)$ raises (lowers) Y by 1 and I_3 by $\frac{1}{2}$. In summary, one gets

$$\begin{aligned}\widehat{I}_\pm \;&:\; \Delta I_3 = \pm 1,\quad \Delta Y = 0,\quad \Delta I = 0 \\ \widehat{V}_\pm \;&:\; \Delta I_3 = \pm\frac{1}{2},\quad \Delta Y = \pm 1,\quad |\Delta I| = \frac{1}{2} \\ \widehat{U}_\pm \;&:\; \Delta I_3 = \mp\frac{1}{2},\quad \Delta Y = \pm 1,\quad |\Delta I| = \frac{1}{2}.\end{aligned} \tag{8.140}$$

For $\widehat{I}_\pm$, the change $\Delta I = 0$ means that $\widehat{I}_\pm$ has non-vanishing matrix elements only within the same isomultiplet. For $\widehat{V}_\pm$, $\widehat{U}_\pm$ the change in I is expected from the change in ΔI_3. These operators act between members of distinct isomultiplets.

The rules (8.140) are illustrated in Fig. 8.3. The general formulae for the action of the ladder operators on a basis vector $\phi(I, I_3, Y)$ of an irrep $(\lambda\,\mu)$ are (de Swart 1963, 1966)

$$\widehat{I}_\pm\phi(I, I_3, Y) = \sqrt{I(I+1) - I_3(I_3\pm 1)}\,\phi(I, I_3\pm 1, Y) \tag{8.141a}$$

$$\widehat{V}_\pm\phi(I, I_3, Y) = a_\pm\phi(I+\frac{1}{2}, I_3\pm\frac{1}{2}, Y\pm 1) + b_\pm\phi(I-\frac{1}{2}, I_3\pm\frac{1}{2}, Y\pm 1)$$

$$\tag{8.141b}$$

Table 8.4 Possible values of Y and I illustrated for $\lambda > \mu$ (after Hecht 1964).

Y	I				
$-\frac{1}{3}(2\lambda+\mu)$	$\frac{\mu}{2}$				
$-\frac{1}{3}(2\lambda+\mu)+1$	$\frac{\mu}{2}-\frac{1}{2}$	$\frac{\mu}{2}+\frac{1}{2}$			
$-\frac{1}{3}(2\lambda+\mu)+2$	$\frac{\mu}{2}-1$	$\frac{\mu}{2}$	$\frac{\mu}{2}+1$		
.	.	.			
.	.	.			
$-\frac{1}{3}(2\lambda+\mu)+\mu$	0	1	2	$\cdots$	μ
$-\frac{2}{3}(\lambda-\mu)+1$	$\frac{1}{2}$	$\frac{3}{2}$	$\frac{5}{2}$		$\mu+\frac{1}{2}$
$-\frac{2}{3}(\lambda-\mu)+2$	1	2			$\mu+1$
.	.				.
.	.				.
$\frac{1}{3}(\lambda-\mu)$	$\frac{\lambda}{2}-\frac{\mu}{2}$		$\cdots$		$\frac{\lambda}{2}+\frac{\mu}{2}$
$\frac{1}{3}(\lambda-\mu)+1$	$\frac{\lambda}{2}-\frac{\mu}{2}+\frac{1}{2}$				$\frac{\lambda}{2}+\frac{\mu}{2}-\frac{1}{2}$
.	.				.
.	.				.
$\frac{1}{3}(\lambda+2\mu)-2$	$\frac{\lambda}{2}-1$	$\frac{\lambda}{2}$	$\frac{\lambda}{2}+1$		
$\frac{1}{3}(\lambda+2\mu)-1$	$\frac{\lambda}{2}-\frac{1}{2}$	$\frac{\lambda}{2}+\frac{1}{2}$			
$\frac{1}{3}(\lambda+2\mu)$	$\frac{\lambda}{2}$				

$$\widehat{U}_{\pm}\phi(I, I_3, Y) = c_{\pm}\phi(I + \frac{1}{2}, I_3 \mp \frac{1}{2}, Y \pm 1) + d_{\pm}\phi(I - \frac{1}{2}, I_3 \mp \frac{1}{2}, Y \pm 1) \quad (8.141c)$$

with

$a_+ =$

$$\left\{ \frac{(I+I_3+1)[\frac{1}{3}(\lambda-\mu)+I+\frac{1}{2}Y+1][\frac{1}{3}(\lambda+2\mu)+I+\frac{1}{2}Y+2][\frac{1}{3}(2\lambda+\mu)-I-\frac{1}{2}Y]}{2(I+1)(2I+1)} \right\}^{1/2}$$

(8.142)

and

$b_+ =$

$$\left\{ \frac{(I-I_3)[\frac{1}{3}(\mu-\lambda)+I-\frac{1}{2}Y][\frac{1}{3}(\lambda+2\mu)-I+\frac{1}{2}Y+1][\frac{1}{3}(2\lambda+\mu)+I-\frac{1}{2}Y+1]}{2I(2I+1)} \right\}^{1/2}$$

(8.143)

The other coefficients are related to a_+ and b_+ as shown below. The coefficients $a_{\pm}$ and $b_{\pm}$ are in fact the matrix elements of $\widehat{V}_{\pm}$. They depend on I, I_3 and Y so one can write

$$a_{\pm}(I, I_3, Y) = \langle \phi(I + \frac{1}{2}, I_3 \pm \frac{1}{2}, Y \pm 1)|\widehat{V}_{\pm}|\phi(I, I_3, Y)\rangle \quad (8.144)$$

$$b_{\pm}(I, I_3, Y) = \langle \phi(I - \frac{1}{2}, I_3 \pm \frac{1}{2}, Y \pm 1)|\widehat{V}_{\pm}|\phi(I, I_3, Y)\rangle \quad (8.145)$$

and $c_{\pm}$, $d_{\pm}$ are

$$c_{\pm} = \langle \phi(I + \frac{1}{2}, I_3 \mp \frac{1}{2}, Y \pm 1)|\widehat{U}_{\pm}|\phi(I, I_3, Y)\rangle \quad (8.146)$$

$$d_{\pm} = \langle \phi(I - \frac{1}{2}, I_3 \mp \frac{1}{2}, Y \pm 1)|\widehat{U}_{\pm}|\phi(I, I_3, Y)\rangle. \quad (8.147)$$

Using the property (8.106), one can show that

$$a_-(I, I_3, Y) = b_+(I + \frac{1}{2}, I_3 - \frac{1}{2}, Y - 1) \quad (8.148)$$

$$b_-(I, I_3, Y) = a_+(I - \frac{1}{2}, I_3 - \frac{1}{2}, Y - 1) \quad (8.149)$$

$$c_+(I, I_3, Y) = [(I + \frac{1}{2})(I + \frac{3}{2}) - (I_3 + \frac{1}{2})(I_3 - \frac{1}{2})]^{1/2} a_+(I, I_3, Y)$$
$$\quad - [I(I+1) - I_3(I_3 - 1)]^{1/2} a_+(I, I_3 - 1, Y) \quad (8.150)$$

$$d_+(I, I_3, Y) = [(I + \frac{1}{2})(I - \frac{1}{2}) - (I_3 + \frac{1}{2})(I_3 - \frac{1}{2})]^{1/2} b_+(I, I_3, Y) \qquad (8.151)$$
$$- [I(I+1) - I_3(I_3 - 1)]^{1/2} b_+(I, I_3 - 1, Y)$$
$$c_-(I, I_3, Y) = d_+(I + \frac{1}{2}, I_3 + \frac{1}{2}, Y - 1) \qquad (8.152)$$
$$d_-(I, I_3, Y) = c_+(I - \frac{1}{2}, I_3 + \frac{1}{2}, Y - 1). \qquad (8.153)$$

To define uniquely these matrix elements, some relative phase convention must be made. For the states in the same multiplet, the standard Condon and Shortley phase convention has been chosen. Accordingly, the non-zero matrix elements of $\hat{I}_\pm$ are positive. The relative phases between different isomultiplets are defined by the requirement that the non-zero matrix elements $a_\pm$ and $b_\pm$ of the operator $\widehat{V}_\pm$ are real and positive. Table 4.1 is based on this phase convention. Other phase conventions do exist, for example the Gel'fand–Biedenharn phase convention of the unitary group according to which the matrix elements of the generators $A_{i,\,i-1}$ defined by (8.116) are positive (see Chen 1989, Section 7.5.1). For SU(3), this means that the matrix elements of $\hat{I}_-$ and $\widehat{U}_-$ are positive.

At this stage, we have all the elements to present some general characteristics of weight diagrams of SU(3). Recall that each irrep has its own diagram. Due to the analogous role played by the three equivalent su(2) subalgebras of su(3), the weight diagrams can have either a triangular (Fig. 8.2) or a hexagonal shape (Figs 8.4 or 8.5), obtained from the intersection of I-, V- and U-multiplet lines.

Each basis vector of an invariant subspace is associated to a point (I_3, Y) which lies on or within the hexagon. Since $\hat{I}_3$, $\widehat{V}_3$ and $\widehat{U}_3$ are spin projection operators, their eigenvalues must be $0, \pm\frac{1}{2}, 1, \pm\frac{3}{2}$, etc and from (8.137), it follows that

$$\widehat{Y} = \frac{2}{3}(\widehat{V}_3 - \widehat{U}_3) \qquad (8.154)$$

hence the eigenvalues of $\widehat{Y}$ must be $0, \pm\frac{1}{3}, \pm\frac{2}{3}, 1$, etc., consistent with Table 8.4. Each diagram has three symmetry axes, one is perpendicular on I_3, the other on V_3 and the third on U_3. The symmetry axes intersect at 120 if the scale of Y is reduced by $\sqrt{3/2}$ (see definition (8.124) of $\widehat{Y}$).

One can now draw the contour of the weight diagram associated to an irrep $(\lambda\ \mu)$. Figure 8.5 shows the case $\lambda > \mu$.

First, one draws the point $A = (\frac{1}{2}(\lambda + \mu), \frac{1}{3}(\lambda - \mu))$ corresponding to $\phi_{\max}$ of (8.131)–(8.133). According to Fig. 8.3, the point B can be reached by acting on $\phi_{\max}$ with $\widehat{U}_+$ a number of times n to give $\phi_B = (\widehat{U}_+)^n \phi_{\max}$. The line AB is a U-spin multiplet with $U = $ const and $-U \le U_3 \le U$. Its ends, A and B, correspond to $-U = U_{\max}$ and U, respectively. Putting $Y_{\max} = \frac{1}{3}(\lambda - \mu)$ and $(I_3)_{\max} = \frac{1}{2}(\lambda + \mu)$ in (8.137), one gets

THE GROUP SU(3)

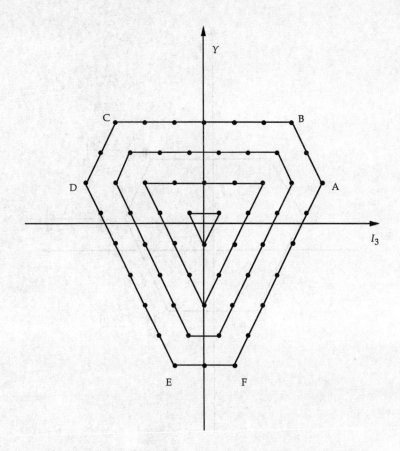

Figure 8.4 Typical weight diagram of a representation of SU(3). Here, $\lambda = 6$, $\mu = 2$.

$$U_{\max} = \frac{3}{4}Y_{\max} - \frac{1}{2}(I_3)_{\max} = -\frac{1}{2}\mu, \qquad V_{\max} = \frac{\lambda}{2} \qquad (8.155)$$

i.e. point A has $U = \frac{\mu}{2}$ and on the line AB there are $2U + 1 = \mu + 1$ points. Therefore, to reach B from A, one needs to apply $\widehat{U}_+$ on $\phi_{\max}$ $n = \mu$ times. Using the commutation relations, $[\hat{I}_3, \widehat{U}_+] = -\frac{1}{2}\widehat{U}_+$ and $[\widehat{Y}, \widehat{U}_+] = \widehat{U}_+$, μ times one finds

$$\widehat{I}_3 \phi_B = \left[(I_3)_{\max} - \frac{\mu}{2}\right]\phi_B = \frac{\lambda}{2}\phi_B$$

$$\widehat{Y}\phi_B = [Y_{\max} + \mu]\phi_B = \frac{1}{3}(\lambda + 2\mu)\phi_B \qquad (8.156)$$

where $\phi_B \sim (\widehat{U}_+)^\mu \phi_{\max}$. In a similar way, one can discuss the AF line which is a V-multiplet. From (8.155), one can see that $\phi_F \sim (\widehat{V}_-)^\lambda \phi_{\max}$ and from the

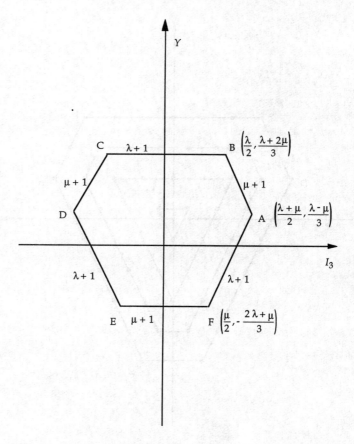

Figure 8.5 The contour of a $(\lambda\mu)$ weight diagram with $\lambda > \mu$. The number of points (weights) to be drawn on each side—ends included—are indicated.

corresponding commutations relations (8.105), one finds

$$\widehat{I}_3\phi_F = [(I_3)_{max} - \frac{\lambda}{2}]\phi_F = \frac{\mu}{2}\phi_F$$
$$\widehat{Y}\phi_F = [Y_{max} - \lambda]\phi_F = -\frac{1}{3}(2\lambda + \mu)\phi_F. \tag{8.157}$$

The point C is reached from B by applying $\hat{I}_-$ on ϕ_B λ times and E is reached from F by applying $\hat{I}_-$ on ϕ_F μ times. The point D can be obtained from A by reflection symmetry and it can be reached from A by applying $\hat{I}_-$ on ϕ_{max} $\lambda + \mu$ times. Also for every point on AB or AF, there is an I-spin multiplet ending on CD and DE, respectively. From every point on ABC one can build a V-spin multiplet ending on

DEF and to every point on AFE, a U-spin multiplet ending on BCD. In this way, starting from the point A, one can generate a lattice like that in Fig. 8.4 where each basis vector of the invariant subspace of an irrep $(\lambda\ \mu)$ is represented by a point. Conversely, points inside the hexagon do not always correspond to a single basis vector. Thus, inner points have a multiplicity different from 1. The points A, B, and F correspond to the right, lower, and upper corners, respectively, of the parallelogram appearing in Table 8.4. We have already mentioned that there is some arbitrariness in defining the highest weight. While, here, A corresponds to the highest weight, our point B gives the highest weight of Elliott and Dawber (1979).

Let us illustrate the multiplicity problem on the diagram drawn in Fig. 8.4. On its contour (the largest hexagon), there is only one basis vector corresponding to a given point. Inside its contour, on the first hexagon each point corresponds to two vectors. If it were a second, it would have corresponded to three points, and so on. After that, the hexagon becomes a triangle and from that state on the number of independent basis vectors at each point remains the same. In general, for the case where $\lambda > \mu$ there are μ hexagons so that the multiplicity of points on any triangle is $\mu + 1$. An interesting case is the representation with $\lambda = 1$, $\mu = 1$ because this representation is used in the classification of hadrons. Its weight diagram is drawn in Fig. 8.6.

Complex conjugate representations

In Section 5.11, we introduced the notions of contragradient and complex conjugate representations. For unitary representations, the two are identical. As SU(n) is a

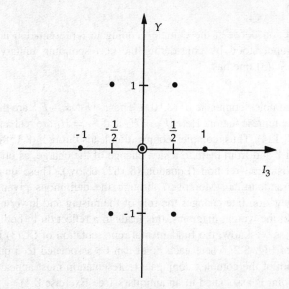

Figure 8.6 The weight diagram of the $(\lambda\mu) = (11)$ representation of SU(3). The weight 0 has multiplicity 2, indicated as a point surrounded by a circle.

compact group, all its representations are unitary or equivalent to unitary representations, so one can refer to complex conjugate representations D^* only. Although D and D^* have the same dimension, they are not equivalent in general. If their Young diagrams are different, this means that they are not equivalent because there is a one-to-one correspondence between a representation and a Young diagram, as we already know from Section 5.11, where it was stated that if a representation is described by the partition $[f_1, f_2, ..., f_n]$, its contragradient representation is associated to $[f_1 - f_n, f_1 - f_{n-1}, ..., f_1 - f_2]$, i.e. they are such that, turning the Young diagram of the contragradient representation upside down next to the original representations, one obtains a rectangular diagram with n rows and f_1 columns. An example was shown in (5.188).

An important case, with physical implications, is the complex conjugate representation of the fundamental representation. As for SU(2), discussed previously, the complex conjugate representation is associated with the antiparticle of the particle described by the fundamental representation. In atomic or nuclear shell models, it describes a hole in a closed shell. The fundamental representation has the partition $[f_1] = [1]$. Then, according to the above rule, the complex conjugate representation has the partition $[1^{n-1}]$. This is consistent with our discussion for SU(2) in Section 8.5. Next, the basis states (8.125) representing the u, d, and s quarks span the irrep $\square$ of SU(3). Then, the antiquark states $\bar{u}$, $\bar{d}$ and $\bar{s}$ span the irrep $\dbox$. In a basis formed of u, d, s and c quark states, one can introduce the fundamental representation of SU(4). Then, the antiquark states $\bar{u}$, $\bar{d}$, $\bar{s}$ and $\bar{c}$ span the representation $\tbox$ of SU(4), and so on.

The connection between the complex conjugate representation and antiparticles can also be understood by considering the corresponding unitary operator. For example, for SU(3) one has

$$S^* = e^{-i\theta_i \widehat{F_i}^*} \tag{8.158}$$

which is the complex conjugate of (8.101). Then, $-\widehat{F}_i^* = -\widehat{F}_i^T$ are the generators of the antiparticle representation. Here, $\widehat{F}_3 = H_2$ and $\widehat{F}_8 = H_1$ are real and are related to $\hat{I}_3$ and $\hat{Y}$ by (8.124). Thus, complex conjugation leads through (8.158) to a change in sign of I_3 and Y and from there to a sign change of the charge, as obtained from the Gell-Mann–Nishijima relation (Equation (8.162) below). These quantum numbers represent the antiparticle. Moreover, through the definitions given in Table 8.3, complex conjugation interchanges the role of the raising and lowering operators $\hat{I}_\pm$, $\widehat{U}_\pm$ and $\widehat{V}_\pm$. In the weight diagrams, this leads to a reflection in both axes I_3 and Y. For example, as we know, the fundamental representation of SU(3) is described by the diagram of Fig. 8.2, where each point can be associated to a quark. Then, the weight diagram of the complex conjugate representation must appear as in Fig. 8.7, where each point is associated to an antiquark (see Exercise 8.4).

Thus, reflection in both axes of the weight diagram leads to the weight diagram of the complex conjugate representation. On the other hand, for a representation $(\lambda \, \mu)$ of

SU(3), one has $\lambda = f_1 - f_2$, $\mu = f_2 - f_3$, so that the complex conjugate representation by definition must have $f_1' = f_1 - f_3 = \lambda + \mu$ and $f_2' = f_1 - f_2 = \lambda$. In terms of Young diagrams, this means

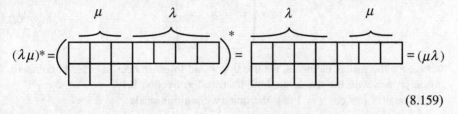

(8.159)

The formula (8.139) for the dimension of an SU(3) irrep is symmetric in λ and μ and gives the same value for $(\lambda\mu)^*$ and $(\lambda\mu)$, as expected.

Then if, instead of $(\lambda\mu)$, one uses the dimension (d) as a label, the complex conjugate representation is labelled by $\bar{d}$ (or d^*). For example, the fundamental representation is sometimes labelled by 3 and its complex conjugate by $\bar{3}$. With this notation, one can write the direct product

$$3 \times \bar{3} = \bar{3} \times 3 = 8 + 1. \tag{8.160}$$

This will be used in the classification of mesons later in this section.

Exercise 8.4 Derive explicitly the complex conjugate matrix representation of the fundamental representation of SU(3) and find its weight diagram.

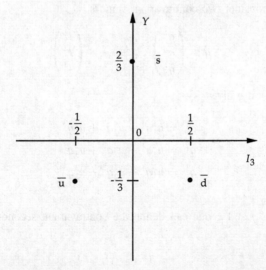

Figure 8.7 The weight diagram of the complex conjugate representation $\bar{3}$ of SU(3).

Solution We start from the fundamental representation

$$u = \begin{pmatrix} u_1^1 & u_2^1 & u_3^1 \\ u_1^2 & u_2^2 & u_3^2 \\ u_1^3 & u_2^3 & u_3^3 \end{pmatrix} \qquad (1)$$

where, for the matrix elements, we use upper and lower indices in order to compare this approach with the tensor method discussed in Section 8.10.

This matrix has det $u = 1$ and the unitary condition reads

$$(u^{-1})_i^j = (u_j^i)^* \qquad (2)$$

or, equivalently,

$$(u^{-1})_i^j u_j^k = \delta_i^k. \qquad (3)$$

We wish to find the matrix elements $(u_j^i)^*$ in terms of u_j^i. Putting $i = 1$ and $k = 1, 2,$ or 3 in (3), we obtain a system of three equations with the unknowns $(u^{-1})_1^1$, $(u^{-1})_1^2$ and $(u^{-1})_1^3$. The solution is immediate:

$$(u^{-1})_1^1 = (u_1^1)^* = u_2^2 u_3^3 - u_3^2 u_2^3 \qquad (4)$$
$$(u^{-1})_1^2 = (u_2^1)^* = u_1^3 u_3^2 - u_1^2 u_3^3 \qquad (5)$$
$$(u^{-1})_1^3 = (u_3^1)^* = u_1^2 u_2^3 - u_1^3 u_2^2 \qquad (6)$$

where det $u = 1$ has been used. The other matrix elements of the complex conjugate representation can be found in a similar way.

Now, let us consider two contravariant spinors

$$q = \begin{pmatrix} q^1 \\ q^2 \\ q^3 \end{pmatrix} \quad \text{and} \quad t = \begin{pmatrix} t^1 \\ t^2 \\ t^3 \end{pmatrix}. \qquad (7)$$

The action of u on q gives

$$\begin{aligned} q^{1'} &= u_1^1 q^1 + u_2^1 q^2 + u_3^1 q^3 \\ q^{2'} &= u_1^2 q^1 + u_2^2 q^2 + u_3^2 q^3 \\ q^{3'} &= u_1^3 q^1 + u_2^3 q^2 + u_3^3 q^3 \end{aligned} \qquad (8)$$

and similarly for t.

Starting from q and t, one can define the contravariant second-rank tensor (see Section 8.10)

$$T^{ij} = q^i t^j \qquad (9)$$

The linear combinations

$$\eta_1 = q^2 t^3 - q^3 t^2, \qquad \eta_2 = q^3 t^1 - q^1 t^3, \qquad \eta_3 = q^1 t^2 - q^2 t^1 \tag{10}$$

form a vector, or a covariant tensor of rank one. This is a particular case of the contraction procedure described in Section 8.10. Accordingly, one can write

$$\eta_i = \varepsilon_{ijk} q^j t^k. \tag{11}$$

Under the unitary transformation (1), η_i transforms as

$$\eta'_1 = q^{2'} t^{3'} - q^{3'} t^{2'}, \quad \text{etc.} \tag{12}$$

Using (8) and its analogue for t, one can show that

$$\eta'_1 = (u_1^1)^* \eta_1 + (u_2^1)^* \eta_2 + (u_3^1)^* \eta_3, \quad \text{etc.} \tag{13}$$

which shows that the components of η_i form the invariant subspace of the complex conjugate representation.

Now that we have obtained the basis vectors of the conjugate representation, we can find its weight diagram. For simplicity, one can take

$$q \equiv t = \begin{pmatrix} u \\ d \\ s \end{pmatrix}.$$

The action of $\widehat{Y}$ and $\hat{I}_3$ on u, d, and s being defined in (8.126), one can easily get the eigenvalues (weights) of η_i, recalling that Y and I_3 are additive quantum numbers. The result is presented in the table below:

η_i	Y	I_3	$\bar{q}$
$\frac{1}{\sqrt{2}}(ud - du)$	$\frac{2}{3}$	0	$\bar{s}$
$\frac{1}{\sqrt{2}}(us - su)$	$-\frac{1}{3}$	$\frac{1}{2}$	$\bar{d}$
$\frac{1}{\sqrt{2}}(ds - sd)$	$-\frac{1}{3}$	$-\frac{1}{2}$	$\bar{u}$

where η_i have been normalized. The last column indicates the antiquarks with the same quantum numbers (see Fig. 8.7). This proves that the complex conjugate representation can be defined in the space of antiparticles because there is a one-to-one correspondence between a representation and a weight diagram.

Note that, by construction, η_i are antisymmetric functions, so that the corresponding Young diagrams is ⊟. Therefore, we have given a practical proof that

$$u^* \to \boxed{}$$

in agreement with the discussion at the beginning of this subsection.

Classification of hadrons

Hadrons are strongly interacting particles. They are divided into two categories: baryons, which have half-integer spin, and mesons, with integer spin. The ground state baryons are heavier than the ground state mesons (Tables 8.5 and 8.6). One can distinguish between the two categories through the baryon number B, which is 1 for baryons, -1 for antibaryons, and zero for mesons. There is a wide variety in each category and a classification was necessary. First, they were classified into isomultiplets, and several isomultiplets were put together to form SU(3) multiplets.

Although the first classification, due to Ohnuki (1960), was for mesons, we discuss baryons first and present the meson classification in the next subsection.

In Table 8.5, we list the low-mass baryons under consideration, together with some of their most important properties. In 1961, Gell-Mann and Ne'eman identified the baryons p, n, Λ, Σ, and Ξ as members of an SU(3) octet. This brought the theory of 'eight-fold way' which was the decisive step in the classification of elementary particles (see Gell-Mann and Ne'eman 1964). At that time, a triplet of 'baryon' fields was introduced as a mathematical device for the construction of the octet and decuplet representations of SU(3) which allowed the classification of all the then-known hadrons.

In 1964, Gell-Mann and Zweig introduced, independently, hypothetical particles of three varieties, the u, d, and s quarks, which transform according to the fundamental representation of SU(3) and have spin $\frac{1}{2}$. The baryons were described as three-quark states and mesons as quark–antiquark states. At present, there is convincing evidence that baryons are composite systems, so that one can assume they are formed of three quarks. One can then relate their classification to this type of structure. Let us consider the low-lying quarks, u, d, and s, to be referred to as distinct quantum states called flavours. They form a basis for the fundamental representation $\lambda\mu = (10)$ of $SU_F(3)$ where F stands for flavour. Three-quark states are obtained from the direct product

$$\underset{(10)}{\overset{3}{\square}} \times \underset{(10)}{\overset{3}{\square}} \times \underset{(10)}{\overset{3}{\square}} = \left(\underset{(20)}{\overset{6}{\square\square}} + \underset{(01)}{\overset{3}{\square\atop\square}} \right) \times \underset{(10)}{\overset{3}{\square}}$$

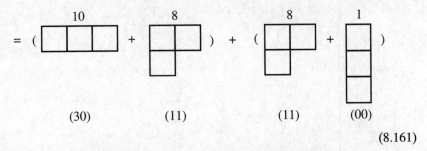

$$\tag{8.161}$$

which generates three non-equivalent SU(3) representations. They can be identified either by their $(\lambda\mu)$ label equal to (30), (11), or (00), or by their dimension calculated from the formula (8.139) and indicated above each diagram.

The irrep (11) appears with multiplicity 2. In fact, the first eight particles of Table 8.5 with $J^\pi = \frac{1}{2}^+$ are associated with the two octet representations. Through the permutation group S_3, we can understand that each of these particles has to be described by two types of states, ψ^ρ and ψ^λ (see Table 4.1 and Section 4.11) and that is why two octet representations appear in (8.161). Each of the eight particles correspond to a basis vector of the $(\lambda\mu) = (11)$ representation of SU(3), so that to each elementary particle of $J^\pi = \frac{1}{2}^+$, one can associate a point of the weight diagram represented in Fig. 8.6. This is plotted in Fig. 8.8(a). In Fig. 8.8(b), we show the weight diagram of the representation $(\lambda\mu) = (30)$ of dimension 10, where each point represents one of the ten particles of $J^\pi = \frac{3}{2}^+$ of Table 8.5. At the time this classification was proposed, all these particles had already been observed, except for the Ω^-. The discovery of this particle, in 1964, confirmed the predictions of Gell-Mann and Ne'eman, both for the quantum numbers and the mass of Ω^- derived from Gell-Mann–Okubo mass formula introduced below.

In addition to the baryons of Fig. 8.8, there also exists a flavour singlet baryon, associated to the representation (00). This is the $\Lambda(1405)$ particle of $J^\pi = \frac{1}{2}^-$ represented by the last row of Table 8.5. With respect to the others, this particle has negative parity, which can be understood in a quark model as an excited $L = 1$ state, while the octet and decuplet baryons of Fig. 8.8 are in their ground state of $L = 0$. Being an excited state, the lifetime of the $\Lambda(1405)$ baryon is of the order of 10^{-23} s, as for decuplet resonances.[2]

In non-relativistic quark models, the $J^\pi = \frac{3}{2}^+$ baryons, except the Ω^-, are also seen as excited states of the ground state baryons. The mass difference is explained by the colour hyperfine interaction acting between the quarks (see the next chapter). The quarks are fermions of spin $\frac{1}{2}$ and three quarks can couple to either $S = \frac{1}{2}$ or $\frac{3}{2}$. The $\Lambda(1405)$ particle, viewed as an $L = 1$ orbital excitation, must have $S = \frac{1}{2}$ to be consistent with Fermi–Dirac statistics.

Each horizontal line in each of the diagrams of Fig. 8.8 represents an isomultiplet and each isomultiplet has a definite hypercharge Y. The idea of hypercharge was introduced independently by Gell-Mann and Nakano and Nishijima in 1953 to

---

[2] The lifetime for a strong decay can be estimated by $\tau_{\text{strong}} = $ (hadron radius) / (speed of light) $\sim 10^{-13}$ cm $/3.10^{10}$ cm s^{-1} $\sim 10^{-23}$ s.

Table 8.5 Properties of low-mass baryons (Particle Data Group 1992).

Particle	Mass (MeV)	J^P	I	I_3	Y	Mean life τ (s)	Main decay modes
p	938.27	$\frac{1}{2}^+$	$\frac{1}{2}$	$\frac{1}{2}$	1	$> 1.6 \times 10^{25}$ years	
n	939.57	$\frac{1}{2}^+$	$\frac{1}{2}$	$-\frac{1}{2}$	1	889	p $e^- \bar{\nu}_e$(100%)
Λ	1115.6	$\frac{1}{2}^+$	0	0	0	2.6×10^{-10}	p π^-(64.1%), n π^0(35.7%)
Σ^+	1184.4	$\frac{1}{2}^+$	1	1	0	0.8×10^{-10}	p π^0(51.6%), n π^+(48.3%)
Σ^0	1192.5	$\frac{1}{2}^+$	1	0	0	7.4×10^{-20}	$\Lambda \gamma$(100%)
Σ^-	1197.4	$\frac{1}{2}^+$	1	-1	0	1.5×10^{-10}	nπ^-(99.85%)
Ξ^0	1314.9	$\frac{1}{2}^+$	$\frac{1}{2}$	$\frac{1}{2}$	-1	2.9×10^{-10}	$\Lambda \pi^0$(100%)
Ξ^-	1321.3	$\frac{1}{2}^+$	$\frac{1}{2}$	$-\frac{1}{2}$	-1	1.6×10^{-10}	$\Lambda \pi^-$(100%)

Particle	Mass	J^P	I	I_3	S	Lifetime (s)	Decay modes
Δ^{++}	1232	$\frac{3}{2}^+$	$\frac{3}{2}$	$\frac{3}{2}$	1	5.5×10^{-24}	$p\pi^+$ ⎫
Δ^{+}				$\frac{1}{2}$			$p\pi^0$, $n\pi^+$ ⎬ 99.4%
Δ^{0}				$-\frac{1}{2}$			$p\pi^-$, $n\pi^0$ ⎬
Δ^{-}				$-\frac{3}{2}$			$n\pi^-$ ⎭
Σ^{+*}	1382.8	$\frac{3}{2}^+$	1	1	0	1.8×10^{-23}	$\Lambda\pi(88\%)$, $\Sigma\pi(12\%)$
Σ^{0*}	1383.7	$\frac{3}{2}^+$	1	0	0		
Σ^{-*}	1387.2	$\frac{3}{2}^+$	1	-1	0		
Ξ^{0*}	1531.8	$\frac{3}{2}^+$	$\frac{1}{2}$	$\frac{1}{2}$	-1	6.9×10^{-23}	$\Xi\pi(100\%)$
Ξ^{-*}	1535.0	$\frac{3}{2}^+$	$\frac{1}{2}$	$-\frac{1}{2}$	-1		
Ω^{-}	1672.4	$\frac{3}{2}^+$	0	0	-2	0.8×10^{-10}	$\Lambda K^-(67.8\%)$, $\Xi\pi(32.2\%)$
$\Lambda(1405)$	1407	$\frac{1}{2}^-$	0	0	0	1.3×10^{-23}	$\Sigma\pi(100\%)$

describe unexplained phenomena like K-meson pair production and was later accepted as a concept. The hypercharge is related to the electric charge Q and the isotopic spin projection I_3 through the Gell-Mann and Nishijima relation

$$Q = I_3 + \frac{Y}{2}. \tag{8.162}$$

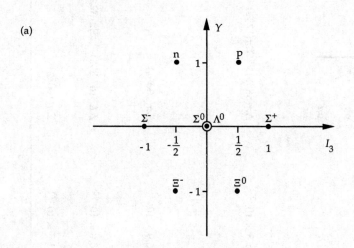

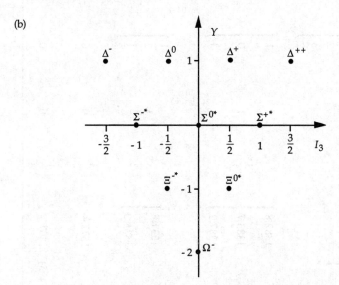

Figure 8.8 (a) The baryon octet ($J^\pi = \frac{1}{2}^+$). (b) The baryon decuplet ($J^\pi = \frac{3}{2}^+$).

An equivalent description for hypercharge is the strangeness S related to Y by

$$S = Y - B \qquad (8.163)$$

where B is the baryon number. Note that the proton, the neutron, and the Δ particle are non-strange ($S = 0$), while Σ, Λ, Ξ and Ω are strange ($S \neq 0$).

The experiments show that the baryon number is conserved in all known reactions. This leads to the postulate of baryon number conservation. At present, in the grand unification theories, it is expected that the baryon number is weakly violated, so that the proton becomes unstable. Until now, however, proton decay has not been seen experimentally. The charge is also conserved in any process. From (8.162) and (8.163), it follows that any interaction that conserves I_3 also conserves Y and S because Q and B are always conserved. The strong and electromagnetic interactions conserve S (or Y) but the weak interaction does not. An immediate consequence is that a strange particle can decay into non-strange particles only through the weak interaction. Then, the lifetime of these particles is much longer than that of those decaying strongly (see Table 8.5). Conversely, in the collision of nucleons with pions, which are also non-strange (see the next subsection), pairs of strange particles with $+S$ and $-S$ can be produced. For example,

$$\pi^- + p \to \Lambda^0 + K^0$$

where Λ^0 and K^0 are strange and decay weakly. Actually, the property of strangeness, introduced by Pais in 1952, was due to its conservation in strong interactions and its violation in weak processes.

It is interesting to note that in a U-spin multiplet the charge is constant. This can be understood through the formula (8.141c) which shows that $\widehat{U}_+(\widehat{U}_-)$ lowers (raises) I_3 by $\frac{1}{2}$ and raises (lowers) Y by 1. These two changes cancel out in the Gell-Mann–Nishijima relation (8.162).

Recall that the quark content of the octet and decuplet baryons can be found in Tables 4.1 and 4.2, respectively. One can check that their quantum numbers Y and I_3 are consistent with those of quarks, as given by (8.126). The flavour part of the $\Lambda(1405)$ particle is a Slater determinant made of u, d, and s states.

To each SU(3) baryon multiplet there corresponds a distinct antibaryon mutliplet. The weight diagram of an antibaryon multiplet is obtained by the change in sign $Y \to -Y$ and $I_3 \to -I_3$ and by reflection in both Y- and I_3-axes, as explained above.

If the $SU_F(3)$ symmetry was exact, i.e. $[H_{\text{strong}}, \widehat{F}_i] = 0$, all particles in the same $SU_F(3)$ multiplet would have to have the same mass by virtue of Schur's lemmas. Tables 8.5 and 8.6 show this is not the case. Thus, $SU_F(3)$ is violated by the strong interaction, as we shall see below.

Classification of mesons

As already mentioned, the mesons were also classified into SU(3) multiplets and, to understand this classification, let us consider, as for baryons, their quark structure. In the quark model, a meson is described as a state which transforms as a quark–antiquark pair under an SU(3) transformation. The direct product of a quark q and an antiquark $\bar{q}$ representations gives

$$(8.164)$$

For SU(3), this is equivalent to (8.160). Thus, one can expect the mesons to be either members of an SU(3) octet or simply singlet states. As such, the lighter known mesons would be described by the wave functions exhibited in Table 8.1 and the corresponding weight diagrams are those presented in Fig. 8.9 (a) and (b).

However, the experimental situation is a bit more complicated. Instead of η_1^0 and η_8^0, in reality one finds a mixture of them, called η and η' and ω_1^0 and ω_8^0 are also found in mixtures called ω and ϕ (see below). For this reason, one can group them in nonets $(8+1)$.

From the weight diagram of pseudoscalar mesons, Fig. 8.9(a), the quark weight diagram, Fig. 8.2, and the antiquark weight diagram, Fig. 8.7, one can read off the quark content of the mesons because Y and I_3 are additive quantum numbers. One finds immediately

$$\pi^+ \sim \theta\bar{\delta}, \qquad \pi^- \sim \delta\bar{\theta} \qquad (I = +) \tag{8.165}$$

$$K^+ \sim u\bar{s}, \qquad K^- \sim s\bar{u} \qquad (I = \frac{1}{2}) \tag{8.166}$$

$$K^0 \sim d\bar{s}, \qquad \bar{K}^0 \sim s\bar{d} \qquad (I = \frac{1}{2}). \tag{8.167}$$

For π^0, we need a bit of help to be found in Exercise 8.2, which provides a state of $I = 1, I_3 = 0$:

$$\pi^0 \sim \frac{1}{\sqrt{2}}(d\bar{d} - u\bar{u}). \tag{8.168}$$

The meson η_1^0 is a flavour singlet, so it must correspond to the $SU_F(3)$ scalar

$$\eta_1^0 \sim |(00)I = 0, I_3 = 0, Y = 0> = \frac{1}{\sqrt{3}}\sum_{i=1}^{3}|q_i>|\bar{q}_i> = \frac{1}{\sqrt{3}}(u\bar{u} + d\bar{d} + s\bar{s}). \tag{8.169}$$

By the orthogonality relations

$$<(11), I = 0, I_3 = 0, Y = 0 |(11), I = 1, I_3 = 0, Y = 0> = 0$$

$$<(11), I = 0, I_3 = 0, Y = 0 |(00), I = 0, I_3 = 0, Y = 0> = 0$$

one finds

$$\eta_8^0 \sim |(11)I=0, I_3=0, Y=0> = \frac{1}{\sqrt{6}}(u\bar{u}+d\bar{d}-2s\bar{s}).$$

(8.170)

Table 8.1 contains all the above functions, modified according to G-parity. In many applications, G-parity can be ignored and one can use the above function for simplicity.

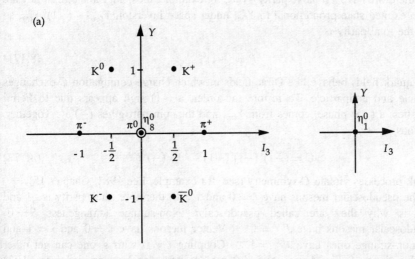

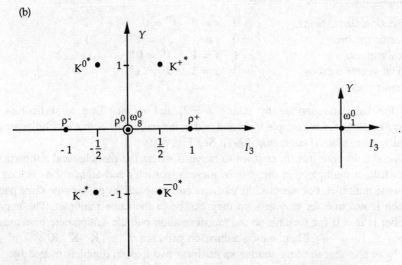

Figure 8.9 (a) Pseudoscalar mesons $J^P = 0^-$. (b) Vector mesons $J^P = 1^-$.

The light pseudoscalar and vector mesons found in nature are listed in Table 8.6, together with some of their properties. The mesons are characterized by several quantum numbers, some of them being common to baryons as the isospin I, the angular momentum J, and the parity P. Others are typical for mesons, like the G-parity discussed in Section 8.5 and charge conjugation C discussed below.

For baryons in the ground state, namely the octet and decuplet members, the relative parity is $P = +1$. The mesons are composed of a particle and an antiparticle, and therefore have a relative parity -1 as, for example, those in Table 8.6. If they are in an excited state proportional to $Y_{\ell m}$, under space inversion $Y_{\ell m} \to (-1)^\ell Y_{\ell m}$, so that the total parity is

$$P = (-1)^{\ell+1}. \tag{8.171}$$

The quark fields behave like Dirac fields in which charge conjugation C exchanges particle and antiparticle. To restore the order, a (-1) sign appears due to Fermi statistics, a $(-)^\ell$ phase, comes from $Y_{\ell m}$, and the spin part gives $(-)^{s+1}$. Together, one has

$$C = (-1)^1 (-1)^\ell (-1)^{s+1} = (-)^{\ell+s}. \tag{8.172}$$

Weak processes violate C-symmetry (see, for example, Lee 1981, Chapter 15).

The pseudoscalar mesons have $\ell = 0$ and $s = 0$; therefore, their parity is -1 and that is why they are called pseudoscalar. Non-strange (strangeness $S = 0$) pseudoscalar mesons have $J^{PC} = 0^{-+}$. Vector mesons have $\ell = 0$ and $s = 1$ and the non-strange ones have $J^{PC} = 1^{--}$. Coupling $\ell \neq 0$ with s, one can get other combinations of J^P. Moreover, $\ell \neq 0$ mesons can also be classified into SU(3) multiplets. So, in general, one expects several types of SU(3) multiplets. For example, for $\ell = 0$ or 1, one has:

pseudoscalar mesons	$\ell = 0$	$s = 0$	$J^P = 0^-$
vector mesons	$\ell = 0$	$s = 1$	$J^P = 1^-$
scalar mesons	$\ell = 1$	$s = 1$	$J^P = 0^+$
axial vector mesons	$\ell = 1$	$s = 1$	$J^P = 1^+$
tensor mesons	$\ell = 1$	$s = 1$	$J^P = 2^+$

and the list can continue by taking $\ell = 2$, and so on. Due to difficulties in measurements of, for example mass and width, the multiplets with $\ell = 1$ or larger are usually incomplete (Pennington 1991). See Fig. 8.10.

One should note that, in contrast to baryons where the particles and antiparticles have distinct multiplets, in the case of mesons, particles and antiparticles belong to the same multiplet. For mesons, the particles and antiparticles possess the same parity (which is not true for baryons), so they can be in the same multiplet. The baryon number is $B = 0$ for mesons, so the transformation particle–antiparticle here means $S \to -S$, $I_3 \to -I_3$. Then, meson-antimeson pairs are $\pi^+ \pi^-$, $K^+ K^-$, $K^0 \overline{K}^0$, $\rho^+ \rho^-$, etc. Note also that the four strange kaons form two isospin doublets related by

Table 8.6 Properties of (a) pseudoscalar mesons $J^P = 0^-$

Particle	Mass (MeV)	I	I_3	Y	Mean life τ (s)	Main decay modes	
$\pi^\pm$	139.57	1	± 1	0	2.6×10^{-8}	$\pi^+ \to \mu^+ \nu_\mu\,(99.988\%)$	$\pi^- \to \mu^- \bar\nu_\mu\,(99.988\%)$
π^0	134.97	1	0	0	8.4×10^{-17}	$\pi^0 \to 2\gamma\,(98.8\%)$	$\pi^0 \to e^+ e^- \gamma\,(1.2\%)$
$K^\pm$	493.65	$\tfrac{1}{2}$	$\pm\tfrac{1}{2}$	± 1	1.2×10^{-8}	$K^+ \to \mu^+ \nu_\mu\,(63.5\%)$	$K^- \to \mu^- \bar\nu_\mu\,(63.5\%)$
						$\quad\to \pi^+ \pi^0\,(21.2\%)$	$\quad\to \pi^- \pi^0\,(21.2\%)$
						$\quad\to \pi^+ \pi^+ \pi^-\,(5.6\%)$	$\quad\to \pi^- \pi^- \pi^+\,(5.6\%)$
						$\quad\to \pi^+ \pi^0 \pi^0\,(1.7\%)$	$\quad\to \pi^- \pi^0 \pi^0\,(1.7\%)$
K^0	497.67	$\tfrac{1}{2}$	$-\tfrac{1}{2}$	1	50% $K_S^0\,(8.9\times 10^{-11})$	$K_S^0 \to \pi^+ \pi^-\,(68.6\%)$	$K_L^0 \to 3\pi^0\,(21.6\%)$
$\bar K^0$	497.67	$\tfrac{1}{2}$	$\tfrac{1}{2}$	-1	$+50\%$ $K_L^0\,(5.2\times 10^{-8})$	$\quad\to \pi^0 \pi^0\,(31.4\%)$	$\quad\to \pi^+ \pi^- \pi^0\,(12.4\%)$
							$\quad\to \pi^\pm \mu^\mp \nu_\mu\,(27\%)$
							$\quad\to \pi^\pm e^\mp \nu_\mu\,(38.7\%)$
η	547.45	0	0	0	$\sim 5 \times 10^{-19}$	$\eta \to 2\gamma\,(38.9\%)$	
						$\quad\to 3\pi^0\,(31.9\%)$	
						$\quad\to \pi^+ \pi^- \pi^0\,(23.6\%)$	
						$\quad\to \pi^+ \pi^- \gamma\,(4.9\%)$	
η'	957.75	0	0	0	$\sim 3 \times 10^{-21}$	$\eta' \to \pi^+ \pi^- \eta\,(44.1\%)$	
						$\quad\to \rho^0 \gamma\,(30.0\%)$	
						$\quad\to 2\pi^0 \eta\,(20.6\%)$	
						$\quad\to \omega\gamma\,(3.0\%)$	
						$\quad\to 2\gamma\,(2.1\%)$	

Table 8.6 Properties of (b) vector mesons $J^P = 1^-$ (Particle Data Group 1992).

Particle	Mass (MeV)	I	I_3	Y	Width (MeV)	Main decay modes
$\rho^\pm$	770	1	± 1	0	151.5	$\sim \pi\pi$ (100%)
ρ^0		1	0	0		
ω	782	0	0	0	8.43	$\pi^+\pi^-\pi^0$ (88.8%)
						$\pi^0\gamma$ (8.5%)
						$\pi^+\pi^-$ (2.2%)
$K^{\pm *}$	891.6	$\tfrac{1}{2}$	$\pm\tfrac{1}{2}$	± 1	49.8	$\sim K\pi$ (100%)
K^{0*}	896.1	$\tfrac{1}{2}$	$-\tfrac{1}{2}$	1	50.5	
$\overline{K}^{0*}$		$\tfrac{1}{2}$	$+\tfrac{1}{2}$	-1		
ϕ	1019.4	0	0	0	4.43	K^+K^- (49.1%)
						$K^0_L K^0_S$ (34.4%)
						$\rho\pi$ (12.9%)
						$\pi^+\pi^-\pi^0$ (1.9%)
						$\eta\gamma$ (1.3%)

THE GROUP SU(3)

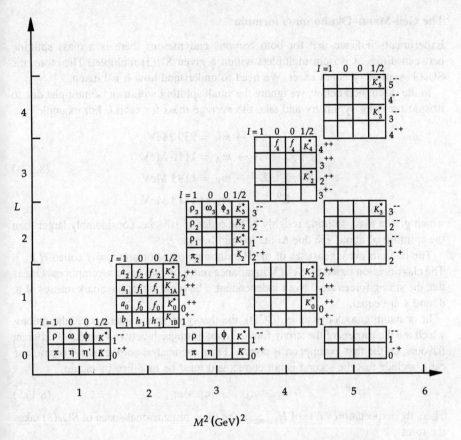

Figure 8.10 Chew–Frautschi plot of expected $q\bar{q}$ states showing orbital angular momentum, L, against mass squared. The named boxes indicate known mesons (after Pennington 1991, updated to Particle Data Group 1992 notations). The K_{1A} and K_{1B} are nearly 45° mixed states of the $K_1(1270)$ and $K_1(1400)$ (for history, see Chew and Frautschi 1961).

$$G\begin{pmatrix} K^+ \\ K^0 \end{pmatrix} = -\begin{pmatrix} \bar{K}^0 \\ -K^- \end{pmatrix} \quad \text{or} \quad G\begin{pmatrix} \bar{K}^0 \\ -K^- \end{pmatrix} = \begin{pmatrix} K^+ \\ K^0 \end{pmatrix} \tag{8.173}$$

just as $\begin{pmatrix} p \\ n \end{pmatrix}$ and $\begin{pmatrix} \bar{n} \\ -\bar{p} \end{pmatrix}$ are related through (8.72).

Using (8.173), one can check that the flavour function of the K mesons from Table 8.1 are consistent among themselves.

Looking at the meson masses in Table 8.6, one can see that neither the 0^- nor the 1^- mesons are degenerate, as expected from exact $SU_F(3)$ symmetry. The mass splitting will be discussed in the next subsection.

The Gell-Mann–Okubo mass formula

Experiments indicate that for both baryons and mesons there is a mass splitting between different isospin multiplets within a given SU(3) multiplet. Therefore, the $SU_F(3)$ symmetry is not exact. We need to understand how it is broken.

In the discussion below, we ignore the small splitting within an isomultiplet due to isospin breaking symmetry and take the average mass for each I. For example,

$$\begin{aligned} \text{p, n} &\to m_N = 939 \text{ MeV} \\ \Lambda &\to m_\Lambda = 1116 \text{ MeV} \\ \Sigma^+, \Sigma^0, \Sigma^- &\to m_\Sigma = 1193 \text{ MeV} \\ \Xi^0, \Xi^- &\to m_\Xi = 1318 \text{ MeV} \end{aligned} \tag{8.174}$$

which give a mass splitting roughly of the order of 10% i.e. considerably larger than the splitting of about 1% due to the isospin breaking.

The hadrons are eigenstates of the Hamiltonian operator previously called H_{strong}. The classification based on $SU_F(3)$ invariance relies on two basic assumptions. One is that the strong forces are flavour independent. The other is that the quark masses of u, d, and s are equal.

In quantum chromodynamics (QCD), the theory of strong interactions, the gluons, which are the carriers of the strong force, do not distinguish between quarks of different flavours, so the first assumption is correct. From the analysis of the mass spectra, one can conclude that the second is not correct and must be modified by taking

$$m_u \cong m_d, \qquad m_s > m_u. \tag{8.175}$$

Then, the expectation value of H_{strong} in the three-dimensional space of $SU_F(3)$ takes the form

$$\begin{aligned} <H_{\text{strong}}> &= \begin{pmatrix} m_u & 0 & 0 \\ 0 & m_d & 0 \\ 0 & 0 & m_s \end{pmatrix} \\ &= \frac{2m_u + m_s}{3} \begin{pmatrix} 1 & 0 & 0 \\ 0 & 1 & 0 \\ 0 & 0 & 1 \end{pmatrix} + \frac{m_u - m_s}{3} \begin{pmatrix} 1 & 0 & 0 \\ 0 & 1 & 0 \\ 0 & 0 & -2 \end{pmatrix} \\ &= \frac{2m_u + m_s}{3} \mathbf{1} + \frac{m_u - m_s}{\sqrt{3}} \lambda_8. \end{aligned} \tag{8.176}$$

This suggests that $\langle H_{\text{strong}} \rangle$ can be split into two terms

$$\langle H_{\text{strong}} \rangle = H_0 + H_8 \tag{8.177}$$

where $H_0 \sim \mathbf{1}$ is $SU_F(3)$ invariant and H_8 is a symmetry breaking term with specific transformation properties under $SU_F(3)$ carried by λ_8, which is a member of an octet (see the adjoint representation, Section 8.9). For completeness, the λ_8 term must be supplemented by another one with analogous properties. The most general form is (Nachtmann 1990)

$$H_8 = x\widehat{F}_8 + y d_{8ab}\,\widehat{F}_a\,\widehat{F}_b \tag{8.178}$$

where x and y are parameters to be fixed, $\widehat{F}_i$ are the SU(3) generators, Equation (8.98), and d_{8ab} are the constants of Table 8.2. The second term of (8.178) can be rewritten as

$$\begin{aligned} d_{8ab}\,\widehat{F}_a\widehat{F}_b &= \frac{1}{\sqrt{3}}\left(\widehat{F}_1^2 + \widehat{F}_2^2 + \widehat{F}_3^2\right) - \frac{\sqrt{3}}{6}\left(\widehat{F}_4^2 + \widehat{F}_5^2 + \widehat{F}_6^2 + \widehat{F}_7^2\right) - \frac{1}{\sqrt{3}}\,\widehat{F}_8^2 \\ &= \frac{\sqrt{3}}{2}\left(\widehat{F}_1^2 + \widehat{F}_2^2 + \widehat{F}_3^2\right) - \frac{\sqrt{3}}{6}\,\widehat{F}_8^2 - \frac{\sqrt{3}}{6}\widehat{F}^2 \\ &= \frac{\sqrt{3}}{2}\left(\widehat{I}^2 - \frac{1}{4}\widehat{Y}^2 - \frac{1}{3}F^2\right) \end{aligned} \tag{8.179}$$

which is indeed isospin and hypercharge invariant as it had to be. The last term contains the Casimir operator (8.129) which is an SU(3) invariant. The rest can be treated perturbatively. For a baryon B of given Y and I belonging to a $(\lambda\mu)$ multiplet, first-order perturbation theory leads to

$$m_B = <B|H_{\text{strong}}|B> = m_0 + \delta m_1\,Y + \delta m_2\left[I(I+1) - \frac{1}{4}\,Y^2\right] \tag{8.180}$$

which is the well-known Gell-Mann–Okubo mass formula. The parameter m_0 incorporates the expectation value of H_0 and F^2 for a given $(\lambda\mu)$, $\delta m_1 = x$ and $\delta m_2 = \frac{\sqrt{3}}{2}y$.

For the baryon octet, where the four entries of (8.174) have to be fitted, one can, instead, eliminate the three parameters and obtain a linear relation between masses:

$$\frac{1}{2}(m_N + m_\Xi) = \frac{1}{4}(3\,m_\Lambda + m_\Sigma). \tag{8.181}$$

This relation is satisfied within a few MeV.

For the decuplet, the Gell-Mann–Okubo formula can be brought to a simpler form than (8.180):

$$m_B = m_0 + \delta m_1\,Y \tag{8.182}$$

which reflects the nearly equal spacing between the isomultiplet masses

$$m_\Omega - m_{\Xi^*} \cong 139 \text{ MeV}$$
$$m_{\Xi^*} - m_{\Sigma^*} \cong 149 \text{ MeV}$$
$$m_{\Sigma^*} - m_\Delta \cong 152 \text{ MeV}$$

At the time when Gell-Mann and Ne'eman proposed the SU(3) classification, the Ω^- particle had not yet been discovered. Its existence was postulated because in the decuplet a place was empty and its mass was predicted within few MeV from (8.182).

For the mass splitting of mesons, one uses the same Hamiltonian, H_{strong}, but this is no longer diagonal in the space of pseudoscalar or vector meson states because H_8, being an octet, allows a coupling between octet and singlet states, as for example between (8.169) and (8.170). Also, in (8.178) one must set $x = 0$, the reason being related to the fact that mesons and anti-mesons belong to the same multiplet. Then, by the *CPT* theorem, particle and antiparticle have the same mass, but they have opposite hypercharge, so a linear term in $\widehat{Y}$ cannot appear in H_{strong}. Let us look at the pseudoscalar sector, for example. The diagonal matrix elements of H_{strong} are

$$< \eta_1 |H_{\text{strong}}|\eta_1 > = < \eta_1|H_0|\eta_1 > = m_1 \tag{8.183}$$

$$< M_8 |H_{\text{strong}}|M_8 > = < M_8 |H_0 + H_8 |M_8 > = m_8 + \delta m\left[I(I+1) - \frac{Y^2}{4}\right] \tag{8.184}$$

where $M_8 = \pi$, K or η_8. Hence

$$m_8 + 2\,\delta m = m_\pi$$

$$m_8 + \frac{1}{2}\delta m = m_K$$

which give

$$m_8 = \frac{1}{3}(4m_K - m_\pi) \qquad \delta m = \frac{2}{3}(m_\pi - m_K). \tag{8.185}$$

With $m_\pi = 138$ MeV and $m_K = 496$ MeV, one obtains

$$m_8 = 615.3 \text{ MeV}, \qquad \delta m = -238.7 \text{ MeV}. \tag{8.186}$$

The term H_8 allows coupling between states with the same I and Y but not between different isomultiplets. So, in the nonet meson space the matrix of H_{strong} is diagonal except for the subspace formed by η_1 and η_8, where one has to solve the eigenvalue problem

$$\begin{pmatrix} m_8 & < \eta_8|H_8|\eta_1 > \\ < \eta_1|H_8|\eta_8 > & m_1 \end{pmatrix}\begin{pmatrix} c_1 \\ c_2 \end{pmatrix} = \lambda \begin{pmatrix} c_1 \\ c_2 \end{pmatrix} \tag{8.187}$$

with $c_1^2 + c_1^2 = 1$. Then, the eigenstates can be written as

$$|\eta> = \cos\theta_p\, |\eta_8> + \sin\theta_p\, |\eta_1>$$
$$|\eta'> = -\sin\theta_p\, |\eta_8> + \cos\theta_p\, |\eta_1> \tag{8.188}$$

with θ_p to be found from experiment if the eigenstates λ are identified with the physical particles η and η'. One finds $\theta_p \cong -24°$ (see Exercise 8.5).

The mesons are bosons, so they obey the Klein–Gordon equation which contains the energy (mass) squared. If the above treatment is applied to the square of the mass instead of the mass itself, the resulting mixing angle is $\theta \cong -11°$.

By analogy to (8.188), one can write for the vector mesons

$$|\omega> = \cos\theta_v\, |\omega_8> + \sin\theta_v\, |\omega_1>$$
$$|\phi> = -\sin\theta_v\, |\omega_8> + \cos\theta_v\, |\omega_1>. \tag{8.189}$$

It turns out that the angle required by experiment is $\theta_v \cong \arctan \frac{1}{\sqrt{2}}$ which leads to an 'ideal' mixing where only ϕ contains strange quarks

$$\omega \sim \frac{1}{\sqrt{2}}(u\bar{u} + d\bar{d})$$
$$\phi \sim s\bar{s}.$$
(8.190)

Exercise 8.5 Calculate the mixing angle θ_p of the $\eta - \eta'$ system defined in Equations (8.188)

Solution Setting successively $\lambda = m_\eta$ and $\lambda = m_{\eta'}$ in (8.187), one obtains

$$m_8 \cos \theta_p + <\eta_1|H_8|\eta_8> \sin \theta_p = m_\eta \cos \theta_p$$
$$-m_8 \sin \theta_p + <\eta_1|H_8|\eta_8> \cos \theta_p = -m_{\eta'} \sin \theta_p. \quad (1)$$

By eliminating the coupling $<\eta_8|H_8|\eta_1>$, one obtains

$$m_8 = m_\eta \cos^2 \theta_p + m_{\eta'} \sin^2 \theta_p. \quad (2)$$

One can substitute this expression in the left-hand-side of (8.185) to get

$$4m_K - m_\pi = 3(m_\eta \cos^2 \theta_p + m_{\eta'} \sin^2 \theta_p). \quad (3)$$

This gives

$$\tan^2 \theta_p = \frac{3m_\eta - 4m_K + m_\pi}{4m_K - m_\pi - 3m_{\eta'}} \cong 0.194 \quad (4)$$

for $m_\pi = 138$ MeV, $m_K = 496$ MeV, $m_\eta = 549$ MeV, and $m_{\eta'} = 958$ MeV. Hence, with some phase convention, one obtains

$$\theta_p \cong -23.7°. \quad (5)$$

From the trace invariance

$$m_1 + m_8 = m_\eta + m_{\eta'}$$

and (8.186), one finds $m_1 = 892$ MeV, and from (1)

$$<\eta_1|H_8|\eta_8> \cong 150 \text{ MeV}.$$

8.7 BEYOND SU(3)

As we shall see below and in subsequent sections, the unitary groups beyond SU(3) have extensive applications to particle physics. They are also used in nuclear physics models (Arima and Iachello 1975).

In these discussions, we quite often use arguments which rely on the dimension of an irrep of SU(n)

$$d^{SU(n)}_{[f]} = \prod_{i<j}^{n} \frac{f_i - f_j + j - i}{j - i} \quad (8.191)$$

derived in the Appendix D. Here, the representation is labelled by the partition $[f] = [f_1, f_2, ..., f_n]$.

For practical purposes in Table 8.7 we list values of $d_{[f]}^{SU(n)}$, $n = 3, 4, ..., 12$, for all allowed Young diagrams containing up to six boxes.

Quarks with flavour and spin

In the previous sections, we considered the flavour degree of freedom with three distinct flavours u, d, and s, and SU(3) transformations in this space. The quarks are

Table 8.7 Dimension of irreps of SU(N) from Equation (8.191). Forbidden partitions are marked by an asterisk (from Itzykson and Nauenberg 1966).

	SU(3)	SU(4)	SU(5)	SU(6)	SU(7)	SU(8)	SU(9)	SU(10)	SU(11)	SU(12)
[1]	3	4	5	6	7	8	9	10	11	12
[2]	6	10	15	21	28	36	45	55	66	78
[1²]	3	6	10	15	21	28	36	45	55	66
[3]	10	20	35	56	84	120	165	220	286	364
[21]	8	20	40	70	112	168	240	330	440	572
[1³]	1	4	10	20	35	56	84	120	165	220
[4]	15	35	70	126	210	330	495	715	1001	1365
[31]	15	45	105	210	378	630	990	1485	2145	3003
[2²]	6	20	50	105	196	336	540	825	1210	1716
[21²]	3	15	45	105	210	378	630	990	1485	2145
[1⁴]	*	1	5	15	35	70	126	210	330	495
[5]	21	56	126	252	462	792	1287	2002	3003	4368
[41]	24	84	224	504	1008	1848	3168	5148	8008	12012
[32]	15	60	175	420	882	1680	2970	4950	7865	12012
[31²]	6	36	126	336	756	1512	2772	4752	7722	12012
[2²1]	3	20	75	210	490	1008	1890	3300	5445	8580
[21³]	*	4	24	84	224	504	1008	1848	3168	5148
[1⁵]	*	*	1	6	21	56	126	252	462	792
[6]	28	84	210	462	924	1716	3003	5005	8008	12376
[51]	35	140	420	1050	2310	4620	8580	15015	25025	40040
[42]	27	126	420	1134	2646	5544	10692	19305	33033	54054
[41²]	10	70	280	840	2100	4620	9240	17160	30030	50050
[3²]	10	50	175	490	1176	2520	4950	9075	15730	26026
[321]	8	64	280	896	2352	5376	11088	21120	37752	64064
[31³]	*	10	70	280	840	2100	4620	9240	17160	30030
[2³]	1	10	50	175	490	1176	2520	4950	9075	15730
[2²1²]	*	6	45	189	588	1512	3402	6930	13068	23166
[21⁴]	*	*	5	35	140	420	1050	2310	4620	8580
[1⁶]	*	*	*	1	7	28	84	210	462	924

fermions of spin $\frac{1}{2}$ so that SU(2) is needed in the spin space. But one can also choose as a basis the six states

$$u\uparrow, \quad u\downarrow, \quad d\uparrow, \quad d\downarrow \quad s\uparrow, \quad s\downarrow$$

which can form an invariant space for U(6). If we separate out the U(1) transformation, responsible for the conservation of particle number (Section 8.1), we remain with SU(6) transformations and use the chain

$$SU(6) \supset SU_F(3) \times SU_S(2)$$

where F stands for flavour and S stands for spin, as before. In other words, considering transformations belonging to the subgroup SU(3) × SU(2) the irreps of SU(6) are reducible to sums of products of SU(3) and SU(2) irreps.

Let us take the three-quark case—the baryons—first. The available irreps of SU(6) result from the direct product

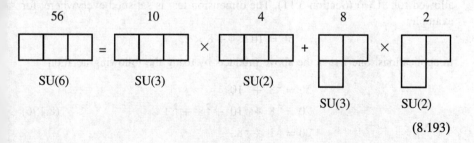

(8.192)

where the dimensions extracted from Table 8.7 are indicated above. One way to find the SU(3) × SU(2) content of these representations is by using inner products introduced in Section 4.7 because all these representations belong to S_3 inasmuch as we deal with three particles everywhere. The answer is

(8.193)

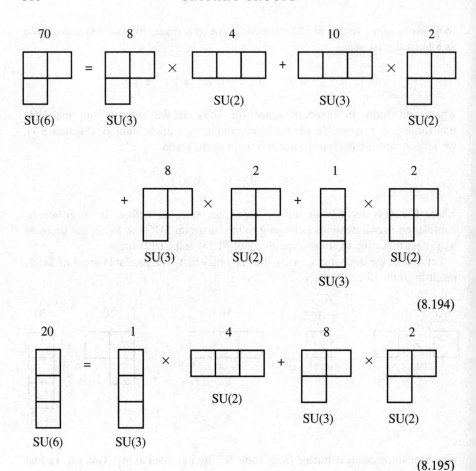

These relations are the analogues of (4.144), (4.145) and (4.148). They have extra terms allowed by SU(3). Recall that Young diagrams with at most n rows are allowed for SU(n) (Section 5.11). The dimension test is satisfied everywhere, for example:

$$56 = 10 \times 4 + 8 \times 2.$$

In applications, one writes the above products by using the following notation

$$56 = {}^2 8 + {}^4 10$$
$$70 = {}^4 8 + {}^2 10 + {}^2 8 + {}^2 1 \qquad (8.196)$$
$$20 = {}^4 1 + {}^2 8.$$

If a chosen quark Hamiltonian displays an SU(6) symmetry, then the states belonging to an SU(6) irrep would be degenerate. For example, in 56, ${}^2 8$ and ${}^4 10$ would be degenerate. Actually, the SU(6) symmetry is broken within $SU_F(3)$ by the different quark masses, as discussed above, and within SU(2) by quark–quark forces which have a spin-dependent tensor term. The spin–spin force explains the mass difference between the nucleon (essentially a $^2 8$ state) and the delta (a $^4 10$ state) as will be discussed in Chapter 10. Thus, the SU(6) breaking allows a mixture of 56, 70, and 20 multiplets (see also Chapter 10).

For mesons, described as $q\bar{q}$ systems, the corresponding SU(6) direct product is

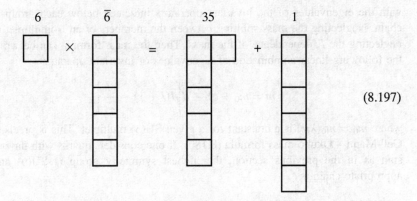

(8.197)

The decomposition of 35 or 1 into SU(3) × SU(2) representations can be obtained here most simply by dimensionality arguments. The quarks are fermions of spin $\frac{1}{2}$ so one can get $S = 0$ or $S = 1$ for a $q\bar{q}$ pair. These can be combined with the SU(3) irreps resulting from (8.160), so as to get $^1 1$, $^1 8$, $^3 1$, and $^3 8$ where the upper index stands for $2S + 1$, as for baryons. Then, the decompositions turn out to be

$$1 = {}^1 1$$
$$35 = {}^1 8 + {}^3 1 + {}^3 8.$$ (8.198)

The SU(6) multiplets like (8.196) or (8.198) are sometimes called supermultiplets and they are encountered in the study of mass spectra (Chapter 10).

Spectrum generating algebra method

An alternative way to find mass formulae, like the Gell-Mann–Okubo formula, is the *spectrum generating algebra* method. The method is currently applied both to subnuclear (Bowler et al. 1981; Iachello 1989) and nuclear physics, and it is combined with associated dynamical symmetries (Iachello and Arima 1987).

The basic idea is due to Gell-Mann and Ne'eman. One starts from the highest symmetry group G of the problem and searches for an appropriate chain

$$G \supset G' \supset G''.$$

Then, the Hamiltonian or mass operator is constructed from the invariant (Casimir) operators of the groups forming the chain. Let us illustrate the procedure for the case discussed in Section 8.6. The chain is

$$\text{SU}_F(3) \supset \text{SU}_I(2) \times \text{U}_Y(1) \supset \text{SO}_I(2) \times \text{U}_Y(1)$$
$$\downarrow \qquad \downarrow \qquad \downarrow \qquad \downarrow \qquad\qquad (8.199)$$
$$(\lambda\mu) \qquad I \qquad Y \qquad I_3$$

with the eigenvalues of the invariant operators indicated below each group of the chain. Neglecting the mass splitting between the members of an isomultiplet means neglecting the I_3 dependence of the mass. Then the mass formula can be written as the following linear combination of eigenvalues of invariant operators

$$m = m_0 + cY + d\left[I(I+1) - \frac{Y^2}{2}\right]$$

where $m_0 = m_0(\lambda\mu)$ is a constant for a given SU(3) multiplet. This is precisely the Gell-Mann – Okubo mass formula (8.180). If one considers quarks with flavour and spin as in the previous section, the highest symmetry group is SU(6) and the appropriate chain is

$$\text{SU}(6) \qquad \supset \text{SU}_F(3) \times \text{SU}_S(2) \supset \text{SU}_I(2) \times \text{U}_Y(1) \times \text{SU}_S(2)$$
$$\downarrow \qquad\qquad \downarrow \qquad \downarrow \qquad\qquad \downarrow \qquad \downarrow \qquad\qquad (8.200)$$
$$(\lambda_1\lambda_2\lambda_3\lambda_4\lambda_5) \qquad (\lambda\mu) \qquad S \qquad\qquad I \qquad Y$$

By using the eigenvalue (8.134) of the SU(3) Casimir operator, one can write the mass formula of an SU(6) multiplet as

$$m = m_0 + a(\lambda^2 + \mu^2 + \lambda\mu + 3\lambda + 3\mu) + bS(S+1) + cY + d\left[I(I+1) - \frac{Y^2}{4}\right].$$
$$(8.201)$$

This formula was proposed by Gürsey and Radicati (1964). For non-strange baryons and mesons, it can be applied by using the decompositions (8.196) and (8.198), respectively. A good fit of the $(56, 0^+)$ multiplet can be achieved by taking $m_0 = 909$ MeV, $a = 6.7$ MeV, $b = 40$ MeV, $c = -216$ MeV, and $d = -38.5$ MeV, see Fig. 8.11.

In nuclear physics, algebraic methods have been used in an attempt to describe in a unified way collective properties of nuclei. A pioneering approach is due to Elliott (1958), who used the group SU(3) to describe rotational bands of deformed nuclei. More recently, another model, called the interacting boson model, IBM, has been proposed by Arima and Iachello (1975). This is rooted in the spherical shell model developed by Mayer and Jensen and also has properties similar to the collective model of Bohr and Mottelson. The basic idea of the model is to consider dynamical symmetries and write the Hamiltonian in terms of the Casimir operators of some appropriate chains. The collective quadrupole states of nuclei have a U(6) or SU(6)

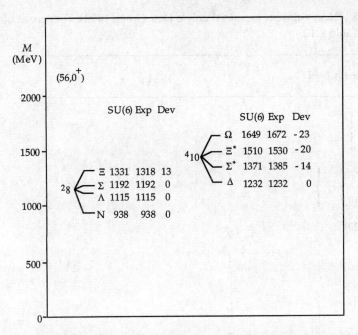

Figure 8.11 Comparison between experiment and the Gürsey–Radicati mass formula, equation (8.201), applied to the $(56, 0^+)$ multiplet. Fitted values have Dev = 0. The parameters are $m_0 = 909$ MeV, $a = 6.7$ MeV, $B = 40$ MeV, $c = -216$ MeV, and $d = -38.5$ MeV (after Iachello 1989).

structure and the chains proposed are

$$U(6) \supset U(5) \supset O(5) \supset O(3) \supset O(2)$$
$$U(6) \supset SU(3) \supset O(3) \supset O(2)$$
$$O(6) \supset O(5) \supset O(3) \supset O(2).$$

Then, instead of diagonalizing the Hamiltonian numerically, one can solve the eigenvalue problem in closed form. For further developments of the model and applications, see Iachello and Arima (1987).

8.8 HEAVY FLAVOURS

In 1974, the J/ψ particle with a mass about 3.1 GeV and a very small width of 0.063 MeV was discovered via e^+e^- collisions. To explain its properties, it was necessary to introduce a fourth quark, the charm quark c. Then, the J/ψ particle was interpreted as a $c\bar{c}$ bound state, where each quark has a mass of about 1.5 GeV. Actually, the charm quark, having a charge of $\frac{2}{3}$ units, was predicted few years before the J/ψ was

Table 8.8 Heavy-flavour mesons (Particle Data Group 1992). For masses of D and D_S, see Table 8.12.

Charmed $|C| = 1$	I	J^P
$D^\pm$ $\Big\}$ D^0	$\frac{1}{2}$	0^-
$D^*(2010)^\pm$ $\Big\}$ $D^*(2010)^0$	$\frac{1}{2}$	1^-
$D_1(2420)^0$	$\frac{1}{2}$	1^+
$D_2^*(2460)^0$	$\frac{1}{2}$	2^+
$D_S^\pm$	0	0^-
$D_S^*(2110)^\pm$	?	?
$D_{S1}(2536)^\pm$	0	1^+

Bottom $|B| = 1$	I	J^P
$B^\pm$ $\Big\}$ B^0	$\frac{1}{2}$	0^-
$B^*(5324)$	$\frac{1}{2}$	1^-

discovered (Glashow, Iliopoulos, and Maiani 1970). In 1977, a new heavy meson, the Υ particle, of mass 9.5 GeV, was discovered. The understanding of its properties required the introduction of another heavy flavour carried by the b quark with a mass of about 5 GeV, where b designates 'bottom' or 'beauty'. This allowed the interpretation of the Υ particle as a $b\bar{b}$ bound state. Based on the theory of weak interactions, it was conjectured that a sixth quark flavour should exist, the 'top' or 'true' or t quark. The t quark completes the fermion spectrum of the standard model,

Table 8.9 Heavy-flavour baryons. The number of stars represents the quality of experimental evidence: four stars—certain; three stars—very likely to certain; two stars—fair evidence; one star—poor evidence (Particle Data Group 1992). In this table, B stands for bottom quantum number.

Baryon	Mass (MeV)	J^P	I	I_3	S	C	B	Mean life τ (s)	Status
Λ_c^+	2284.9	$\frac{1}{2}^+$	0	0	0	1	0	1.9×10^{-13}	****
Σ_c^{++} Σ_c^+ Σ_c^0	2452	$\frac{1}{2}^+$	1	0	0	1	0		****
Ξ_c^+	2466.4	$\frac{1}{2}^+$	$\frac{1}{2}$	$\frac{1}{2}$	-1	1	0	3.0×10^{-13}	***
Ξ_c^0	2472.7	$\frac{1}{2}^+$	$\frac{1}{2}$	$-\frac{1}{2}$	-1	1	0	0.8×10^{-13}	***
Ω_c	2740	?	?	0	-2	1	0		*
Λ_b^0	5461	$\frac{1}{2}^+$	0	0	0	0	-1		***

so that its properties are uniquely determined once the mass is fixed. Estimates derived from precision measurements in e^+e^- annihilation and lepton–nucleon interactions indicate a top mass value between 100 and 200 GeV. In April 1994, a small number of events obtained at the Fermilab in proton-antiproton collisions gave an estimate of 174 GeV ± 17 GeV for the top quark mass. More data are required in order to confirm the existence of the t quark (CERN Courier, June 1994).

Since 1974, several baryons and mesons with charm C beauty B have been identified. Some of their properties are summarized in Tables 8.8 and 8.9. To explain the spectroscopic properties of charmed particles, one must use $SU_F(4)$, but this would require a large amount of symmetry violation. With two extra heavy flavours, c and b, one needs $SU_F(5)$, but the b quark has such a large mass that the symmetry breaking is even larger than that for the c quark, which explains the lack of attempts to use $SU_F(5)$. The $SU_F(4)$ symmetry has, however, been considered in the literature.

Table 8.10 Quark properties (Particle Data Group 1992).

Flavour	Mass (GeV)	Charge Q	I_3	S	C	B	T
u	0.002–0.008	$\frac{2}{3}$	$\frac{1}{2}$	0	0	0	0
d	0.005–0.015	$-\frac{1}{3}$	$-\frac{1}{2}$	0	0	0	0
s	0.1–0.3	$-\frac{1}{3}$	0	-1	0	0	0
c	1.3–1.7	$\frac{2}{3}$	0	0	1	0	0
b	4.7–5.3	$-\frac{1}{3}$	0	0	0	-1	0
t	> 91	$\frac{2}{3}$	0	0	0	0	1

Its use relies on the decomposition of its irreps into SU(3) irreps. From a group theory point of view, this brings us to the interesting problem of finding the SU($n-1$) content of SU(n) irreps, which will be discussed below.

Each heavy flavour is characterized by new quantum numbers: C, for charm; B, for beauty; T, for top. A list of the six quarks with the relevant quantum numbers for the present discussion can be found in Table 8.10. The C, B, and T quantum numbers are also additive like Y or I_3. Taking into account heavy flavours, the Gell-Mann–Nishijima formula (8.162) extends to

$$Q = I_3 + \frac{1}{2}(Y + C + B + T). \tag{8.202}$$

In the following, we shall present some of the group properties of SU(4) and its application to charmed particles.

The group SU(4)

The group SU(4) was first introduced in physics as a symmetry group for nuclei (Wigner and Feenberg 1941). The appropriate treatment of SU(4) is based in that case on the chain SU(4) $\supset$ SU$_I$(2) $\times$ SU$_S$(2) because for the nucleon one can define isospin I and spin S multiplets.

For elementary particles, the appropriate chain is the canonical chain

$SU(4) \supset SU(3) \supset SU(2)$

and in this case the treatment of $SU(4)$ is very similar to that of $SU(3)$.

The group $SU(4)$ is of order 15 (see Table 5.2), i.e. it has 15 generators. For the fundamental representation, they can be given as 4×4 λ_i matrices, satisfying the properties (8.11) and (8.15). They can be derived in the manner used for $SU(3)$. The result is

$$\lambda_1 = \begin{pmatrix} 0 & 1 & 0 & 0 \\ 1 & 0 & 0 & 0 \\ 0 & 0 & 0 & 0 \\ 0 & 0 & 0 & 0 \end{pmatrix}, \quad \lambda_2 = \begin{pmatrix} 0 & -i & 0 & 0 \\ i & 0 & 0 & 0 \\ 0 & 0 & 0 & 0 \\ 0 & 0 & 0 & 0 \end{pmatrix}, \quad \lambda_3 = \begin{pmatrix} 1 & 0 & 0 & 0 \\ 0 & -1 & 0 & 0 \\ 0 & 0 & 0 & 0 \\ 0 & 0 & 0 & 0 \end{pmatrix},$$

$$\lambda_4 = \begin{pmatrix} 0 & 0 & 1 & 0 \\ 0 & 0 & 0 & 0 \\ 1 & 0 & 0 & 0 \\ 0 & 0 & 0 & 0 \end{pmatrix}, \quad \lambda_5 = \begin{pmatrix} 0 & 0 & -i & 0 \\ 0 & 0 & 0 & 0 \\ i & 0 & 0 & 0 \\ 0 & 0 & 0 & 0 \end{pmatrix}, \quad \lambda_6 = \begin{pmatrix} 0 & 0 & 0 & 0 \\ 0 & 0 & 1 & 0 \\ 0 & 1 & 0 & 0 \\ 0 & 0 & 0 & 0 \end{pmatrix},$$

$$\lambda_7 = \begin{pmatrix} 0 & 0 & 0 & 0 \\ 0 & 0 & -i & 0 \\ 0 & i & 0 & 0 \\ 0 & 0 & 0 & 0 \end{pmatrix}, \quad \lambda_8 = \frac{1}{\sqrt{3}} \begin{pmatrix} 1 & 0 & 0 & 0 \\ 0 & 1 & 0 & 0 \\ 0 & 0 & -2 & 0 \\ 0 & 0 & 0 & 0 \end{pmatrix}, \quad \lambda_9 = \begin{pmatrix} 0 & 0 & 0 & 1 \\ 0 & 0 & 0 & 0 \\ 0 & 0 & 0 & 0 \\ 1 & 0 & 0 & 0 \end{pmatrix},$$

$$\lambda_{10} = \begin{pmatrix} 0 & 0 & 0 & -i \\ 0 & 0 & 0 & 0 \\ 0 & 0 & 0 & 0 \\ i & 0 & 0 & 0 \end{pmatrix}, \quad \lambda_{11} = \begin{pmatrix} 0 & 0 & 0 & 0 \\ 0 & 0 & 0 & 1 \\ 0 & 0 & 0 & 0 \\ 0 & 1 & 0 & 0 \end{pmatrix}, \quad \lambda_{12} = \begin{pmatrix} 0 & 0 & 0 & 0 \\ 0 & 0 & 0 & -i \\ 0 & 0 & 0 & 0 \\ 0 & i & 0 & 0 \end{pmatrix},$$

$$\lambda_{13} = \begin{pmatrix} 0 & 0 & 0 & 0 \\ 0 & 0 & 0 & 0 \\ 0 & 0 & 0 & 1 \\ 0 & 0 & 1 & 0 \end{pmatrix}, \quad \lambda_{14} = \begin{pmatrix} 0 & 0 & 0 & 0 \\ 0 & 0 & 0 & 0 \\ 0 & 0 & 0 & -i \\ 0 & 0 & i & 0 \end{pmatrix}, \quad \lambda_{15} = \frac{1}{\sqrt{6}} \begin{pmatrix} 1 & 0 & 0 & 0 \\ 0 & 1 & 0 & 0 \\ 0 & 0 & 1 & 0 \\ 0 & 0 & 0 & -3 \end{pmatrix}.$$

(8.203)

These matrices obey a Lie algebra of the form (8.94):

$$[\lambda_i, \lambda_j] = 2 \, i f_{ijk} \lambda_k \tag{8.204}$$

with non-vanishing, totally antisymmetric structure constants exhibited in Table 8.11.

Some of them are those of (8.95).

As for SU(3), the generators are usually denoted by

$$\widehat{F}_i = \frac{1}{2}\lambda_i. \qquad (8.205)$$

Among the 15 matrices, three are diagonal λ_3, λ_8, and λ_{15}. They correspond to generators of type H_i when the algebra is written in the standard form (Section 5.9). For SU(3), the correspondence between $\widehat{F}_i$ and the standard form was shown in Table 8.3. In the same table, the generators have been also written in terms of operators A_{ij}. These operators, defined by (8.116), result from using the harmonic oscillator in second quantization (Section 8.6). The group SU(N) is the symmetry group of an

Table 8.11 The non-vanishing structure constants f_{ijk} of SU(4).

i	j	k	f_{ijk}	i	j	k	f_{ijk}
1	2	3	1	4	10	13	$-1/2$
1	4	7	$1/2$	5	9	13	$1/2$
1	5	6	$-1/2$	5	10	14	$1/2$
1	9	12	$1/2$	6	7	8	$\sqrt{3}/2$
1	10	11	$-1/2$	6	11	14	$1/2$
2	4	6	$1/2$	6	12	13	$-1/2$
2	5	7	$1/2$	7	11	13	$1/2$
2	9	11	$1/2$	7	12	14	$1/2$
2	10	12	$1/2$	8	9	10	$1/(2\sqrt{3})$
3	4	5	$1/2$	8	11	12	$1/(2\sqrt{3})$
3	6	7	$-1/2$	8	13	14	$-\sqrt{1/3}$
3	9	10	$1/2$	9	10	15	$\sqrt{2/3}$
3	11	12	$-1/2$	11	12	15	$\sqrt{2/3}$
4	5	8	$\sqrt{3}/2$	13	14	15	$\sqrt{2/3}$
4	9	14	$1/2$				

isotropic harmonic oscillator Hamiltonian in N dimensions. So the procedure of Section 8.6 can be straightforwardly extended to any dimension N, and in particular to SU(4). Then, the group algebra takes the form (8.121) with indices running from 1 to 4. One can prove that this is consistent with (8.204). In terms of A_{ij}, one can introduce the operators

$$\widehat{I_3} = \frac{1}{2}(A_{11} - A_{22})$$
$$\widehat{Y} = \frac{1}{3}(A_{11} + A_{22} - 2A_{33}) \quad (8.206)$$
$$\widehat{Z} = \frac{1}{4}(A_{11} + A_{22} + A_{33} - 3A_{44})$$

which commute among themselves, as can be easily proved from (8.121). Therefore, these are proportional to the H_i operators. We take

$$\widehat{Y} = \frac{2}{\sqrt{3}} H_1, \quad \widehat{I_3} = H_2, \quad \widehat{Z} = H_3$$

where the first two relations are taken from Table 8.3. The eigenvalue of $\widehat{Z}$ must now be related to the charm quantum number C. The general form is (Cornwell 1984, Chapter 18, Section 4)

$$C = aZ + bB$$

where B stands for the baryon number.

Glashow, Iliopoulos and Maiani (1970) take

$$a = -1, \quad b = \frac{3}{4}, \quad (8.207)$$

which give the required properties of C, as discussed below.

Let us consider the fundamental representation $\square$ of SU(4). From (8.206), one finds that the state of maximum weight possesses the following quantum numbers

$$I_{3_{max}} = \frac{1}{2}, \quad Y_{max} = \frac{1}{3}, \quad Z_{max} = \frac{1}{4} \quad (8.208)$$

obtained by filling the box with a quantum of type 1 (see below). As for a quark, the baryonic number is $B = \frac{1}{3}$, it follows from (8.207) that the state of maximum weight has $C = 0$, and for consistency with SU(3) it must be identified with the u quark.

For SU(4), some of the ladder operators are the same as for SU(3). From Table 8.3, one can write $\widehat{I}_+ = A_{12}$, $\widehat{I}_- = A_{21}$, $\widehat{V}_+ = A_{13}$, $\widehat{V}_- = A_{31}$, $\widehat{U}_+ = A_{23}$ and $\widehat{U}_- = A_{32}$. The operators changing the eigenvalue of $\widehat{Z}$ must contain the index 4. One can easily find that

$$\widehat{Z} A_{4i} \phi = (Z - 1) A_{4i} \phi \quad (i = 1, 2, 3).$$

Thus, Z can be decreased (raised) by unity. The state with $Z = Z_{max} - 1 = -\frac{3}{4}$ has

$$C = \frac{3}{4} + \frac{3}{4}\frac{1}{3} = 1$$

and corresponds to the c quark. By using the appropriate commutation relations, one can show that the c quark is characterized by $I_3 = 0$ and $Y = \frac{1}{3}$, as expected.

The group SU(4) is of rank $\ell = 3$, therefore its weight diagrams have three dimensions. It is convenient to use C instead of Z. In Fig. 8.12, we illustrate the weight diagram of the fundamental representation. According to the general discussion of Section 5.11, any irrep of SU(4) can be described by a Young diagram of the form

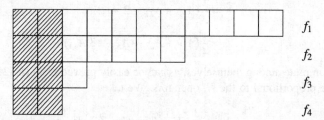

with a maximum of four rows. Row i contains f_i boxes. Columns with four boxes do not play any role (removing or adding columns of four boxes leads to equivalent representations). Then, three numbers are necessary to label an irrep of SU(4)

$$\lambda_1 = f_1 - f_2, \qquad \lambda_2 = f_2 - f_3, \qquad \lambda_3 = f_3 - f_4$$

In terms of a harmonic oscillator description like that given in section 8.6 the boxes can be filled with quanta of type 1, 2, 3, or 4. One can define the state of maximum weight by filling the ith row with f_i quanta of type i ($i = 1, 2, 3$ or 4) (see, for example, Hecht 1964).

The SU($n - 1$) content of SU(n) irreps

When the physical states correspond to basis vectors of an SU(n) irrep it is often possible to label these states by using the canonical chain

$$\text{SU}(n) \supset \text{SU}(n - 1) \supset \cdots \supset \text{SU}(2).$$

Generally, the irreps of SU(n) are reducible with respect to its subgroup SU($n - 1$). It is therefore useful to know how to find the SU($n - 1$) content of SU(n) irreps. Some applications will be considered in the next section.

There is a rule, known as Weyl's branching law, which is quite simple in practice. Let $[f_1, f_2, ..., f_{n-1}, f_n]$ be a representation of SU(n) and $[f'_1, f'_2, ..., f'_{n-1}]$ a representation of SU($n - 1$). The rule requires f'_i to satisfy the following inequalities

$$\begin{aligned} f_2 &\leq f'_1 \leq f_1 \\ f_3 &\leq f'_2 \leq f_2 \\ &\vdots \\ f_n &\leq f'_{n-1} \leq f_{n-1}. \end{aligned} \qquad (8.209)$$

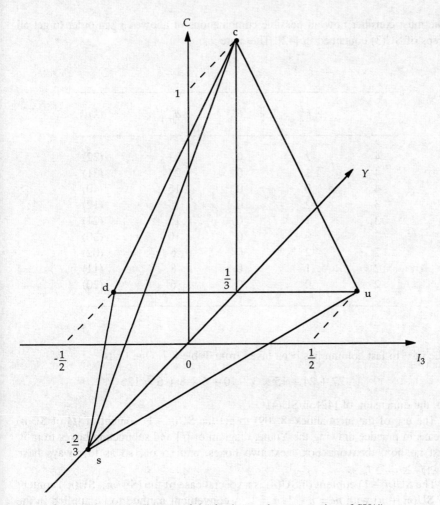

Figure 8.12 The weight diagram of the fundamental representation of SU(4).

Example 8.1 Find the SU(3) content of the irrep [42] of SU(4).
For SU(4), the partition is

$$f_1 = 4, \quad f_2 = 2, \quad f_3 = f_4 = 0.$$

Applying the inequalities (8.209), one has

$$2 \leq f'_1 \leq 4 \rightarrow f'_1 = 2, 3, 4$$
$$0 \leq f'_2 \leq 2 \rightarrow 0, 1, 2$$
$$0 \leq f'_3 \leq 0 \rightarrow f'_3 = 0.$$

One must consider now all possible combinations of allowed f'_i in order to get all irreps of SU(3) contained in [42]. These are :

f'_1	f'_2	f'_3	$d^{SU(3)}_{[f']}$	$(\lambda\mu)$
4	2	0	27	(22)
4	1	0	24	(31)
4	0	0	15	(40)
3	2	0	15	(12)
3	1	0	15	(21)
3	0	0	10	(30)
2	2	0	6	(02)
2	1	0	8	(11)
2	0	0	6	(20)

The next to last column has been taken from Table 8.7. One finds

$$27 + 24 + 15 \times 3 + 10 + 6 + 8 + 6 = 126$$

i.e. the dimension of [42] in SU(4) .

The use of the inequalities (8.209) to get the SU(n - 1) content of [f] of SU(n) means in practice drawing the Young diagram of [f] and subtracting boxes from it: first no box, then one box, next two boxes, and so on, so as to always have $f'_1 \geq f'_2 \geq ... \geq f'_{n-1}$.

The SU($n-1$) content of SU(n) is a special case of the (SU(m), SU(n)) content of SU($m+n$) with $m = n-1, n = 1$. A convenient method to be applied in the general case can be found in Itzykson and Nauenberg (1966) and the solution for Young diagrams having up to eight boxes can be found in Table C of this reference.

Charmed particles

The classification of charmed hadrons according to SU(4) is similar to the SU(3) classification. The fundamental representation of SU(4) is denoted by its dimension 4 and the complex conjugate representation by $\bar{4}$. Therefore, one must look at the decomposition of the direct product $4 \times 4 \times 4$ for baryons and of $4 \times \bar{4}$ for mesons. Applying the Littlewood rule (Section 4.6), one has

HEAVY FLAVOURS

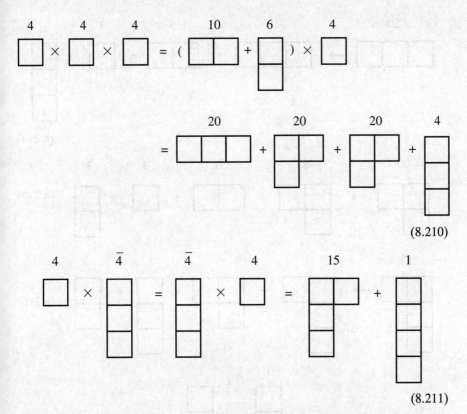

where the dimensions indicated above each diagram have been taken from Table 8.7. Applying the rule of the previous section, one can find the following decomposition into SU(3) representations

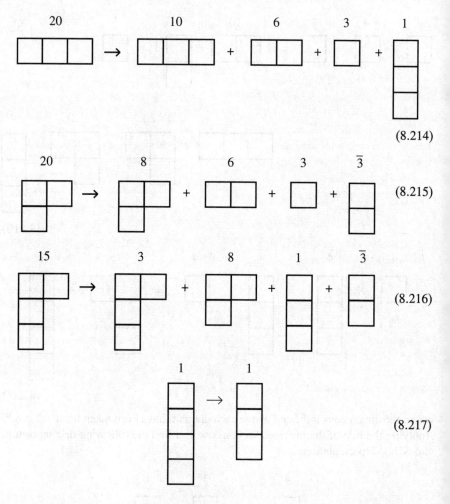

Let us first discuss the mesons for which the experimental information is more plentiful. There are mesons with *hidden* or *open* charm. Mesons with hidden charm are those made of a $c\bar{c}$ pair, i.e. with $C = 0$, while the open charm mesons are those with $|C| = 1$. To better understand the SU(4) meson classification, let us introduce an upper index, 1, 0, or –1, for the charm. Then, the decomposition of 4 and $\bar{4}$ can be written more explicitly as

$$4 = 3^0 + 1^1 \tag{8.218}$$
$$\bar{4} = \bar{3}^0 + 1^{-1}. \tag{8.129}$$

Here, 3^0 represents the u, d, and s quarks, 1^1 the c quark, $\bar{3}^0$ the $\bar{u}$, $\bar{d}$, and $\bar{s}$ antiquarks, and 1^{-1} the $\bar{c}$ antiquark. With this notation, (8.211) becomes

$$4 \times \bar{4} = (3^0 + 1^1) \times (\bar{3}^0 + 1^{-1})$$
$$= 3^0 \times \bar{3}^0 + 1^1 \times \bar{3}^0 + 3^0 \times 1^{-1} + 1^1 \times 1^{-1} \qquad (8.220)$$
$$= (8^0 + 1^0) + \bar{3}^1 + 3^{-1} + 1^0 = 15 + 1$$

We can identify, above, the meson nonets $8^0 + 1^0$ with $C = 0$ discussed in Section 8.6 as being formed of u, d, and s quarks and their antiparticles. The last SU(3) submultiplet 1^0 is a hidden charm meson formed of a $c\bar{c}$ pair and thus having $C = 0$ too. That leaves the $\bar{3}^1$ open charm mesons with $C = +1$ which contain $\bar{u}c$, $\bar{d}c$, or $\bar{s}c$ pairs and the 3^{-1} mesons also with open charm, $C = -1$, composed of pairs $u\bar{c}$, $d\bar{c}$, or $s\bar{c}$.

The charmed $|C| = 1$ mesons with $J^P = 0^-$ (pseudoscalar) and $J^P = 1^-$ (vector) are gathered together in Table 8.12. The hidden charm mesons are bound $q\bar{q}$ states:

$$\eta_c = c\bar{c} \, (J^P = 0^-), \quad J/\psi = c\bar{c} \, (J^P = 1^-).$$

By analogy with positronium, a bound e^+e^- state, the above mesons are called *charmonium*. In the ground state the J/ψ particle has a small width $\Gamma = 0.086$ MeV due to the fact that its decay into hadrons is strongly suppressed. This is a consequence of the so-called Okubo–Zweig–Iizuka (OZI) rule. Accordingly, the decay takes place by pair annihilation of the $c\bar{c}$ pair into three virtual gluons, each creating a new light $q\bar{q}$ pair. Each gluon contributes by a factor α_S, where α_S is the strong coupling constant of QCD, so that the decay amplitude is proportional to α_S^3. For $\alpha_S \sim 0.2$, required by annihilation at short separation between c and $\bar{c}$, this gives a suppression factor of about 0.01. The list of known charmonium states is given in Table 8.13. Resonances of mass larger than 3.73 GeV, the $D\bar{D}$ threshold, decay

Table 8.12 Mesons with open charm $|C| = 1$ (Particle Data Group 1992).

Name	Mass (MeV)	$^{2S+1}L_J$	Quark content
$D^0, \bar{D}^0$	1864.5	$^1S_0$	
$D^{0*}, \bar{D}^{0*}$	2007.1	$^3S_1$	$c\bar{u}$ or $\bar{c}u$
$D^\pm$	1869.3	$^1S_0$	
$D^{\pm *}$	2010.1	$^3S_1$	$c\bar{d}$ or $\bar{c}d$
$D_s^\pm$	1968.8	$^1S_0$	
$D_s^{\pm *}$	2110.3	$^3S_1$	$c\bar{s}$ or $\bar{c}s$

strongly into open charm mesons, which explains a much larger width starting with $\psi(3770)$.

Let us return to baryons and rederive (8.210) by using (8.218)

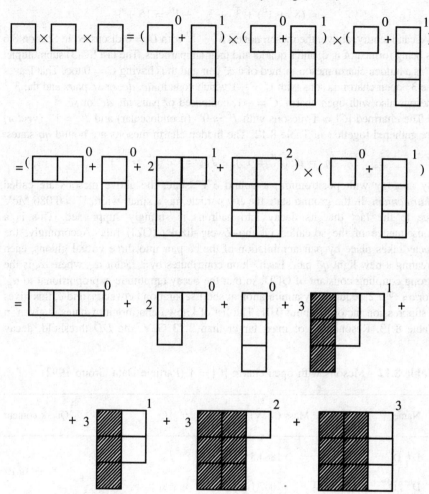

By identification, one can obtain the charm content of the SU(3) decompositions (8.213)–(8.215). The result is

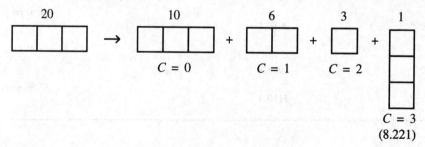

(8.221)

Table 8.13 Charmonium states (Particle Data Group 1992).

State	$I^G(J^{PC})$	Width (MeV)
$\eta_c(1S) = \eta_c(2980)$	$0^+(0^{-+})$	$10.3^{+3.8}_{-3.4}$
$J/\psi(1S) = J/\psi(3097)$	$0^-(1^{--})$	0.086 ± 0.006
$\chi_{c0}(1P) = \chi_{c0}(3415)$	$0^+(0^{++})$	14 ± 5
$\chi_{c1}(1P) = \chi_{c1}(3510)$	$0^+(1^{++})$	0.88 ± 0.14
$\chi_{c2}(1P) = \chi_{c2}(3555)$	$0^+(2^{++})$	2.00 ± 0.18
$\eta_c(2S) = \eta_c(3590)$	$?(?^{?+})$	
$\psi(2S) = \psi(3685)$	$0^-(1^{--})$	0.278 ± 0.032
$\psi(3770)$	$?^?(1^{--})$	23.6 ± 2.7
$\psi(4040)$	$?^?(1^{--})$	52 ± 10
$\psi(4160)$	$?^?(1^{--})$	78 ± 20
$\psi(4415)$	$?^?(1^{--})$	43 ± 15

$$20 \rightarrow 8 \, (C=0) + 6 \, (C=1) + 3 \, (C=2) + \bar{3} \, (C=1) \tag{8.222}$$

$$\bar{4} \rightarrow 1 \, (C=0) + 3 \, (C=1) \tag{8.223}$$

On the right-hand side, each SU(3) diagram refers to a number of uncharmed quarks. This is equal to $3 - C$. For each SU(4) representation, one can draw a three-dimensional weight diagram. For the 20-dimensional symmetric representation, the SU(3) decuplet with $C = 0$ is the basis and there are two other planes, one with $C = 1$ (six baryons) and another with $C = 3$ (three baryons) and also a point with $C = 3$ corresponding to a ccc baryon, on the top of the diagram. (See Particle Data Group (1994) p. 1321)

The 20-dimensional mixed representation has the former SU(3) baryon octet as a basis. The next plane, with $C = 1$, Fig. 8.13, contains points belonging to two different SU(3) representations, the sextet and the antitriplet. As there are only six possible combinations of three quarks, uuc, ddc, ssc, udc, usc, and dsc, containing one quark c, it means that there are only six distinct points (weights) in this plane. Therefore, the antitriplet weights are degenerate with some of the sextet representation. In the third plane, with $C = 2$, there are three points (weights) corresponding to ucc, dcc, and scc. There is no triple charmed baryon in the 20-dimensional mixed representation of SU(4). A list of the 20 possible baryon states

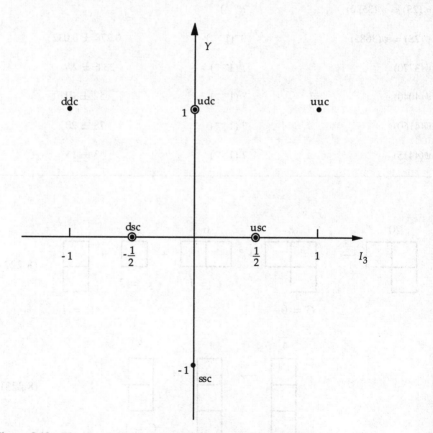

Figure 8.13 The $C = 1$ plane of the [21] representation of SU(4).

belonging to this representation is shown in Table 8.14. Among the charmed baryons, some of those with $C = 1$ have already been discovered (Table 8.9).

This classification helps to better understand the structure of the charmed baryon wave functions. In the sextet ($C = 1$), the light quarks are in a symmetric state ☐☐. In order to get a totally antisymmetric state of this subsystem, the light quarks must couple to an $S = 1$ spin state, which is also symmetric. The colour state is antisymmetric so the total state is antisymmetric. In the antitriplet ($C = 1$) the light quarks are in an antisymmetric flavour state ⊟. Then, their total spin must be zero for the same reasons as above. To distinguish between these two situations, the baryon which is mainly $S = 0$ intermediate coupling is denoted by Ξ_c and the one with $S = 1$ intermediate coupling is denoted by Ξ'_c as indicated in the table. The states with $S = 1$ and $S = 0$ intermediate coupling have been denoted by χ^λ and χ^ρ respectively, as in Chapter 4. In real charmed baryons, they should appear with different coefficients, in contrast to p or n (equal coefficients) because now the charmed quark breaks the S_3 symmetry, in as much as the three quarks are no longer identical.

8.9 THE ADJOINT REPRESENTATION

Here, we introduce the adjoint representation of SU(N). It is important in physical applications, because its basis transforms like the generators of the group.

Let us consider all traceless Hermitian $N \times N$ matrices as a linear vector space. The Hermitian traceless matrices λ_ρ of (8.14) form a basis in this ($N^2 - 1$)-dimensional space, and a general traceless Hermitian matrix h_0 is a linear combination of λ_ρ, i.e. it has the form (8.14).

The adjoint representation is defined by the transformation

$$h'_0 = u h_0 u^{-1}$$

where u is an element of SU(N). If u is an infinitesimal transformation, one has

$$u h_0 u^{-1} = (1 + i\delta\theta_i \frac{\lambda_i}{2}) h_0 (1 - i\delta\theta_i \frac{\lambda_i}{2})$$

$$= h_0 + i\delta\theta_i \left[\frac{\lambda_i}{2}, h_0\right].$$

Thus, the generators F_i of the adjoint representation are given by

$$F_i h_0 = \left[\frac{\lambda_i}{2}, h_0\right].$$

In particular, acting with F_i on $h_0 = \frac{\lambda_j}{2}$, one has

$$F_i \frac{\lambda_j}{2} = \left[\frac{\lambda_i}{2}, \frac{\lambda_j}{2}\right] = i f_{ijk} \frac{\lambda_k}{2}.$$

Table 8.14 Classification of $S = \frac{1}{2}$ baryons in SU(3) submultiplets of the 20-dimensional mixed representation of SU(4) according to (8.222).

SU(3)	Particle	Quark content	I
(C = 0)	p	u u d	$\frac{1}{2}$
	n	u d d	$\frac{1}{2}$
	Σ^+	u u s	1
	Λ^0	u d s	0
	Σ^0	u d s	1
	Σ^-	d d s	1
	Ξ^0	u s s	$\frac{1}{2}$
	Ξ^-	d s s	$\frac{1}{2}$
(C = 1)	Σ_c^{++}	u u c	1
	Σ_c^0	d d c	1
	Ω_c	s s c	0
	Σ_c^+	u d c	1
	$\Xi_c^{+'}$	u s c	$\frac{1}{2}$
	$\Xi_c^{0'}$	d s c	$\frac{1}{2}$
(C = 10)	Λ_c^+	u d c	0
	Ξ_c^+	u s c	$\frac{1}{2}$
	Ξ_c^0	d s c	$\frac{1}{2}$
(C = 2)	Ω_{cc}	u c c	$\frac{1}{2}$
		d c c	$\frac{1}{2}$
		s c c	0

THE ADJOINT REPRESENTATION

Hence, the matrix elements of these generators are precisely the structure constants

$$(F_i)_{kj} = \mathrm{i} f_{ijk} = -\mathrm{i} f_{ikj}. \tag{8.224}$$

The three generators of the adjoint representation of SU(2) are 3×3 matrices. They represent the generators of a particle of spin $S = 1$ and their form is given by (6.93) (see also Exercise 8.6).

The eight generators of the adjoint representation of SU(3) are 8×8 matrices, the elements of which are given by the structure constants (8.95).

In general, the dimension of the adjoint representation of SU(N) is equal to $N^2 - 1$, i.e. to the group order.

Another way to define the adjoint representation of SU(N) is through their Young diagrams. The representation with $f_1 = 2, f_2 = 1, ..., f_{N-1} = 1$ is called the adjoint representation. Here, we illustrate the adjoint representation of

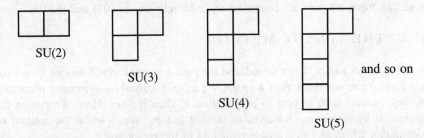

With the help of Table 8.7, one can check that the dimension is equal to the group order in each case. An adjoint representation can be defined for any Lie group (see Chen 1989, Section 5.12.2).

Exercise 8.6 Derive the matrix form of the generators of the adjoint representation of SU(2).

Solution One can choose as a basis either the Cartesian basis $i = x, y, z$ or the spherical basis $\mu = 1, 0, -1$. In the first case, one obtains the matrices (6.93) as a direct application of (8.224). In the spherical basis, one introduces the generators (Equation (6.109))

$$S_\mu = \begin{cases} \mp \frac{1}{\sqrt{2}}(S_x \pm \mathrm{i} S_y) & \mu = \pm 1 \\ S_z & \mu = 0. \end{cases} \tag{1}$$

To construct the matrix of S_y, for example, one needs the commutation relations

$$[S_y, S_+] = \frac{\mathrm{i}}{\sqrt{2}} S_0$$

$$[S_y, S_0] = -\frac{\mathrm{i}}{\sqrt{2}}(S_+ - S_-) \tag{2}$$

$$[S_y, S_-] = -\frac{\mathrm{i}}{\sqrt{2}} S_0.$$

Using Equation (8.224), one finds

$$(S_y)_{01} = \frac{i}{\sqrt{2}}, \quad (S_y)_{10} = -\frac{i}{\sqrt{2}}, \quad (S_y)_{-10} = \frac{i}{\sqrt{2}}, \quad (S_y)_{0-1} = -\frac{i}{\sqrt{2}}.$$

Thus, the matrix is

$$S_y = \frac{1}{\sqrt{2}} \begin{pmatrix} 0 & -i & 0 \\ i & 0 & -i \\ 0 & i & 0 \end{pmatrix}.$$

In an analogous way, one can find

$$S_x = \frac{1}{\sqrt{2}} \begin{pmatrix} 0 & 1 & 0 \\ 1 & 0 & 1 \\ 0 & 1 & 0 \end{pmatrix}, \quad S_z = \frac{1}{\sqrt{2}} \begin{pmatrix} 1 & 0 & 0 \\ 0 & 0 & 0 \\ 0 & 0 & -1 \end{pmatrix}.$$

Note that these matrices are consistent with the relations (6.107) and (6.108).

8.10 THE TENSOR METHOD

In Sections 4.5 and 5.11, we introduced the tensor method where tensors were used as a basis for representations of a group. We have discussed the symmetry properties of these tensors with respect to permutation of their indices. Here, we pursue the discussion by restricting ourselves to unitary groups. We describe the method of *contraction* which allows a further reduction of the tensor space.

The general principle is to first define the so-called *invariant* (or *isotropic*) tensors. Their components remain unchanged under a given transformation. For unitary transformations, we use contravariant $T^{i_1,\ldots,i_n}$ and covariant $T_{i_1,\ldots,i_n}$ tensors as in Exercise 8.4, by setting the indices as superscripts or subscripts. Let T be a tensor which transforms into T' under a unitary transformation. One has

$$T' = (u \times u \times \cdots \times u)\, T \tag{8.225}$$

where u is an $N \times N$ unitary matrix.

Its complex conjugate transforms as

$$(T^*)' = (u^* \times u^* \times \cdots \times u^*)\, T^*. \tag{8.226}$$

The relation (8.225) is used to define contravariant tensors by setting the indices as superscripts instead of subscripts as in (5.179). The relation (8.226) introduces covariant tensors with indices as subscripts as in (5.179). The relation between the two types of tensors is obtained through the Levi-Civita symbol (5.185). One can also define tensors with superscripts and subscripts. Then,

$$T^{i_1 \ldots i_n}_{a_1 \ldots a_m}$$

is a component of a tensor of rank (n,m) which transforms as

$$(T')^{i_1 \ldots i_n}_{a_1 \ldots a_m} = u^{i_1}_{j_1} \cdots u^{i_n}_{j_n}\, (u^+)^{b_1}_{a_1} \cdots (u^+)^{b_m}_{a_m}\, T^{j_1 \ldots j_n}_{b_1 \ldots b_m}. \tag{8.227}$$

For SU(N), all indices vary from 1 to N and the tensor has N^{n+m} components.

In the following, we restrict the discussion to SU(3). Its extension to SU(N) is straightforward. There are two types of invariant tensors:

1. The Kronecker symbol

$$T^i_a = \delta^i_a = \begin{cases} 1 & \text{for } i = a \\ 0 & \text{for } i \neq a. \end{cases} \tag{8.228}$$

One can immediately see that this is an invariant because

$$(\delta')^i_a = u^i_j (u^+)^b_a \delta^j_b = u^i_j (u^+)^j_a = \delta^i_a.$$

2. The Levi-Civita tensor defined as in (5.185) or with superscript indices

$$T_{ijk} = \varepsilon_{ijk}, \qquad T^{ijk} = \varepsilon^{ijk}. \tag{8.229}$$

One has, for example,

$$(\varepsilon')^{ijk} = u^i_a u^j_b u^k_c \varepsilon^{abc} = \det u \, \varepsilon^{ijk} = \varepsilon^{ijk}$$

because $\det u = 1$. The invariance of ε_{ijk} comes from $\det u^+ = 1$.

Starting from a tensor of rank (n,m) and using one of the above invariant tensors, one can construct a new tensor, of smaller rank, by the operation of contraction. This consists in multiplying an (n,m) tensor by one of the invariant tensors and summing over the repeated indices. There are three possibilities.

(a) Contraction with δ^j_i which is a trace operation:

$$T^{i_1 \cdots i_n}_{j_1 \cdots j_m} \delta^{i_1}_{j_1}.$$

Here, the first upper and lower indices have been contracted, but the operation can be applied to any pair. This leads to a tensor of rank $(n-1, m-1)$. For example,

$$T^{ij}_i \delta^i_i = T^j. \tag{8.230}$$

(b) Contraction with ε_{ijk}

$$T^{i_1 \cdots i_n}_{j_1 \cdots j_m} \varepsilon_{j_{m+1} i_\alpha i_\beta} \quad \text{with } 1 \leq \alpha \leq n, 1 \leq \beta \leq n.$$

To allow the summation, the initial tensor must have $n \geq 2$ upper indices. The outcome is a tensor of rank $(n-2, m+1)$. For example,

$$T^{ij} \varepsilon_{kij} = T_k. \tag{8.231}$$

(c) Contraction with ε^{ijk}:

$$T^{i_1 \cdots i_n}_{j_1 \cdots j_m} \varepsilon^{i_{n+1} j_\alpha j_\beta} \quad \text{with } 1 \leq \alpha \leq m, 1 \leq \beta \leq m.$$

Here, the initial tensor must have at least $m \geq 2$ lower indices to allow summation and the resulting tensor is of rank $(n + 1, m - 2)$.

The contraction operation is related to reducibility, as already mentioned. By definition, a tensor is reducible if, by one of the above contractions, one can get a new non-vanishing tensor of rank smaller than $n + m$. If that is not possible, the tensor is irreducible.

As a consequence of (a), an irreducible tensor must satisfy

$$T^{i_1 \cdots i_n}_{j_1 \cdots j_m} \delta^{i_1}_{j_1} = 0, \quad \text{etc.} \tag{8.232}$$

and from (b) and (c), it follows that the tensor must be symmetric with respect to any pair $i_\alpha\, i_\beta$ or $j_\alpha\, j_\beta$ on which the sum has been performed. These are constraints which imply that the number of independent components of an irreducible tensor of rank (n, m) may be smaller than 3^{n+m}. Examples are given below.

Tensor	Rank	Young diagram
1	0	⬚
T^i	(1,0)	▭
T_j	(0,1)	⬚
T^i_j	(1,1)	⊞
T^{ij}	(2,0)	▭▭
T_{ij}	(0,2)	⊞
T^{ijk}	(3,0)	▭▭▭
T_{ijk}	(0,3)	⊞⊞
T^{ij}_k	(2,1)	⊞▭
T^i_{jk}	(1,2)	⊞▭
$T^{ij}_{k\ell}$	(2,2)	⊞▭

The importance of irreducible tensors consists in the fact that there is a one-to-one correspondence between an irreducible tensor and an irrep. For the irreps of SU(3), the identification is

$$n = \lambda \qquad m = \mu.$$

Some examples are shown in the table opposite.

The tensor of rank (0, 0) is a scalar and it can be formed by contraction with ε^{ijk}

$$S = \varepsilon_{ijk} T^i T^{jk}. \tag{8.233}$$

T^{ij} must be antisymmetric in order to get a non-vanishing contribution. As in Exercise 8.4, one can choose

$$T^1 = u, \qquad T^2 = d, \qquad T^3 = s$$
$$T^{23} = ds - sd, \qquad T^{31} = su - us, \qquad T^{12} = ud - ds.$$

Then S is an $SU_F(3)$ scalar, a Slater determinant made up of u, d, and s quarks.

The number of independent components of each irreducible tensor gives the dimension of the corresponding irrep. For example, T^i_j ($i, j = 1, 2, 3$) has nine components, but the trace condition (8.232) introduces one constraint

$$T^1_1 + T^2_2 + T^3_3 = 0 \tag{8.234}$$

so there are only eight independent components and this corresponds to the SU(3) octet.

Due to the symmetry property

$$T^{ij} = T^{ji}$$

this tensor has only $9 - 3 = 6$ independent components and is associated to the diagram ☐☐ .

Then, the tensor $T^{ij}_{k\ell}$ must have $6 \times 6 - 9 = 27$ independent components, where the nine constraints are

$$T^{ij}_{k\ell} \delta^\ell_j = 0 \qquad (i, k = 1, 2, 3).$$

Application to mesons

The mesons are $q\bar{q}$ pairs. Let us associate the contravariant tensor T^i to quarks and the covariant tensor T_j to antiquarks:

$$T^i = q^i, \qquad T_j = \bar{q}_j.$$

Then, $q\bar{q}$ pairs are described by

$$T^i_j = q^i \bar{q}_j.$$

To belong to an irrep, T^i_j must obey (8.234). One can then define traceless tensors by

$$\widetilde{T}^i_j = q^i \bar{q}_j - \frac{1}{3} \delta^i_j \sum_k q^k \bar{q}_k \qquad (8.234a)$$

which correspond to the pseudoscalar meson octet as follows

$$\widetilde{T}^1_2 = u\bar{d} \sim \pi^+, \qquad \widetilde{T}^2_1 = d\bar{u} \sim \pi^-$$
$$\widetilde{T}^2_3 = d\bar{s} \sim K^0, \qquad \widetilde{T}^3_2 = s\bar{d} \sim \bar{K}^0$$
$$\widetilde{T}^1_3 = u\bar{s} \sim K^+, \qquad \widetilde{T}^3_1 = s\bar{u} \sim K^-,$$
$$\widetilde{T}^3_3 \sim 2s\bar{s} - u\bar{u} - d\bar{d} \sim \eta^0.$$

For π^0, the relation is not so direct:

$$\widetilde{T}^1_1 - \widetilde{T}^2_2 = T^1_1 - T^2_2 = u\bar{u} - d\bar{d} \sim \pi^0.$$

The baryons can also be regarded as forming an irreducible tensor of rank (1,1), see for example Lee (1981), Chapter 12.

8.11 CLEBSCH–GORDAN COEFFICIENTS FOR SU(3)

In Chapter 5, we discussed the Clebsch–Gordan coefficients of the permutation group S_n and in Chapter 6 those of the rotation group R_3. Here, we present the Clebsch–Gordan coefficients of SU(3). The procedure is entirely similar to the other groups. First, we need to recall that in Section 8.6 we introduced a complete labelling of basis vectors of SU(3) irreps based on a canonical chain. The principle is that the operators $\hat{I}^2, \hat{I}_3$, and $\hat{Y}$ commute among themselves and also with the invariants F^2 and G^3 of SU(3). Then, using the eigenvalues of these operators, one can label a basis vector of an irrep $(\lambda\mu)$ by $|(\lambda\mu) I I_3 Y >$. Let us consider two irreps $(\lambda'\mu')$ and $(\lambda''\mu'')$. Their Kronecker product gives rise to a Clebsch–Gordan series of type (4.121)

$$(\lambda'\mu') \times (\lambda''\mu'') = \sum_{(\lambda\mu)} m_{\lambda\mu}(\lambda\mu).$$

The direct product of two basis vectors of $(\lambda'\mu')$ and $(\lambda''\mu'')$ is a linear combination of basis vectors of $(\lambda\mu)$ with CG coefficients of SU(3):

$$|\lambda' \mu' I' I'_3 Y' > |\lambda'' \mu'' I'' I''_3 Y'' >$$
$$= \sum_{\substack{\lambda\mu \\ I I_3 Y}} < \lambda \mu I I_3 Y | \lambda' \mu' I' I'_3 Y'; \lambda'' \mu'' I'' I''_3 Y'' > |\lambda \mu I I_3 Y >. \qquad (8.235)$$

This definition of SU(3) CG coefficients is the analogue of (6.68) for the rotation group. The irrep label $\lambda\mu$ here corresponds to j there and the set $I I_3 Y$ here to m there. The sum runs over $\lambda\mu$ values which can be obtained by applying the Littlewood rule

CLEBSCH–GORDAN COEFFICIENTS FOR SU(3)   345

(Section 4.6) for the direct product $(\lambda'\mu') \times (\lambda''\mu'')$ and the dimensions can be checked with formula (4.126).

Similar to (6.68) and its inverted form (6.69), the relation (8.235) can also be inverted to give

$$|\lambda\mu I I_3 Y\rangle = \sum_{\substack{I'I'_3Y'\\I''I''_3Y''}} \langle \lambda'\mu'I'I'_3Y'; \lambda''\mu''I''I''_3Y''|\lambda\mu I I_3 Y\rangle \qquad (8.236)$$

$$\times |\lambda'\mu'I'I'_3Y'\rangle|\lambda''\mu''I''I''_3Y''\rangle$$

because both (8.235) and (8.236) represent orthogonal transformations between two orthogonal bases. The orthogonality relations have the same form as (6.71) or (6.72) of the rotation group where $j \to \lambda\mu$, $m \to II_3Y$ for SU(3).

Let us adopt the notation (de Swart 1963)

$$\langle \lambda'\mu'I'I'_3Y'; \lambda''\mu''I''I''_3Y''|\lambda\mu I I_3 Y\rangle = \begin{pmatrix} \lambda'\mu' & \lambda''\mu'' & \lambda\mu \\ I'I'_3Y' & I''I''_3Y'' & II_3Y \end{pmatrix}. \qquad (8.237)$$

Similarly to the CG coefficients of S_n (Section 4.9), the above coefficients also have a factorization property. Every coefficient can be written as a product of an SU(2) CG coefficient and an isoscalar factor

$$\begin{pmatrix} \lambda'\mu' & \lambda''\mu'' & \lambda\mu \\ I'I'_3Y' & I''I''_3Y'' & II_3Y \end{pmatrix} = C \begin{pmatrix} I' & I'' & I \\ I'_3 & I''_3 & I_3 \end{pmatrix} \begin{pmatrix} \lambda'\mu' & \lambda''\mu'' & \lambda\mu \\ I'Y' & I''Y'' & IY \end{pmatrix}. \qquad (8.238)$$

The SU(2) CG coefficients in the isospin space are the same as those in the spin space. For details, see Chapter 6. The isoscalar factors are tabulated by de Swart (1963). Note that they are independent of the projection of the isospin on the z-axis.

Corresponding to the symmetry properties (6.77)–(6.79) of SU(2) CG coefficients, here we have

$$\begin{pmatrix} \lambda'\mu' & \lambda''\mu'' & \lambda\mu \\ I'I'_3Y' & I''I''_3Y'' & II_3Y \end{pmatrix} = \zeta_1 \begin{pmatrix} \lambda''\mu'' & \lambda'\mu' & \lambda\mu \\ I''I''_3Y'' & I'I'_3Y' & II_3Y \end{pmatrix} \qquad (8.239)$$

$$\begin{pmatrix} \lambda\mu' & \lambda''\mu'' & \lambda\mu \\ I'I'_3Y' & I''I''_3Y'' & II_3Y \end{pmatrix} = \zeta_3 \begin{pmatrix} \overline{\lambda'\mu'} & \overline{\lambda''\mu''} & \overline{\lambda\mu} \\ I'I'_3Y' & I''I''_3Y'' & II_3Y \end{pmatrix} \qquad (8.240)$$

where $\overline{\lambda\mu}$ is the complex conjugate representation of $\lambda\mu$ and the label $\overline{II_3Y}$ means I, $-I_3$, $-Y$ (see Section 8.6 for complex conjugate representations), and

$$\begin{pmatrix} \lambda'\mu' & \lambda''\mu'' & \lambda\mu \\ I'I'_3Y' & I''I''_3Y'' & II_3Y \end{pmatrix}$$

$$= \zeta_2 (-)^{I'_3 + \frac{Y'}{2}} \left(\frac{d_{[\lambda\mu]}}{d_{[\lambda''\mu'']}}\right)^{\frac{1}{2}} \begin{pmatrix} \lambda'\mu' & \overline{\lambda\mu} & \lambda''\mu'' \\ I'I'_3Y' & \overline{II_3Y} & I''I''_3Y'' \end{pmatrix} \qquad (8.241)$$

where $d_{[\lambda\mu]}$ is the dimension of the $(\lambda\mu)$ irrep from Equation (8.139). The phases factors ζ_i ($i = 1, 2, 3$) depend on the overall phase convention; see de Swart (1963).

Another overall phase convention is that of Baird and Biedenharn (1963, 1964, 1965a, b). This is a generalization of the Condon and Shortley phase convention for SU(2) according to which one takes

$$C^{j_1 \; j_2 \; J}_{j_1 \; J-j_1 \; J} > 0. \tag{8.242}$$

The generalization to SU(3) is

$$\begin{pmatrix} \lambda'\mu' & \lambda''\mu'' & \lambda\mu \\ I'_H I'_{3H} Y'_H & I'' I''_3 Y'' & I_H I_{3H} Y_H \end{pmatrix} > 0 \tag{8.243}$$

where the subscript H means highest weight as defined by (8.131)–(8.133). For details, see Chen (1989), Section 7.7.3.

The CG coefficients of a group depend on the particular basis used. Another well-known application of SU(3) is the Elliott model (Elliott 1958) proposed to describe rotational bands of nuclei. In this model, SU(3) serves to describe the orbital part of the wave function. Thus, the chain used to label the irreps must be

$$\text{SU}(3) \supset R_3$$

which leads to states of given angular momentum L, required by nuclear spectroscopy. This chain introduces an ambiguity, because a representation $(\lambda\mu)$ may contain several representations with the same angular momentum. An additional label K has been introduced in the classification of states. The eigenvalues corresponding to different values of K, but the same L, are interpreted as belonging to different rotational bands. It turns out that states with different K are not orthogonal, which complicates the procedure of deriving CG coefficients. For the representations $(\lambda 0)$ and $(\lambda 1)$, only one value of K appears which allowed the derivation of analytic expressions of CG coefficients for some particular decompositions related to 2s-1d shell-model calculations (Iosifescu and Stancu 1967). Vergados (1968) gave a systematic recipe for obtaining an orthogonal basis when several values of K appear. This allowed him to derive explicit algebraic expressions for most of the SU(3) CG coefficients necessary in the 1p and 2s-1d shell-model calculations.

SUPPLEMENTARY EXERCISES

8.7 Prove that

$$v e^{ih} v^{-1} = e^{ivhv^{-1}}$$

where v and h are $n \times n$ matrices. This identity has been used in Equation (8.4).

8.8 Construct the analogue of the Table 8.3 for SU(4).

8.9 Find the SU(2) content of [3] and [21] irreps of SU(3) and the SU(3) content of [3] of SU(4).

8.10 Using Equation (8.220), construct the weight diagrams of the SU(4) irreps associated to the direct product $4 \times \bar{4}$.

9

GAUGE GROUPS

The continuous symmetries we have discussed until now are called *global* symmetries. The parameters a^ρ, Equation (5.51), of a Lie group are space-time independent. In a field theory, this implies that fields at different space-time points transform by the same amount. There are theories where the symmetry transformation contains space-time-dependent parameters $a^\rho(x)$. These transformations are called *local* or *gauge* (Weyl 1929) and form gauge groups. They generate the gauge interactions; electromagnetic, weak, and strong interactions are all gauge interactions. Therefore, the present theory of elementary particle physics is based on the principle of gauge symmetries.

In a gauge theory, one turns a global transformation acting on a Dirac field into a local transformation, by replacing a^ρ by $a^\rho(x)$ where x is a point in Minkowski space-time. The procedure is known as 'gauging' a symmetry. According to the gauge principle, the Lagrangian must remain invariant under the local gauge transformation. But the derivative acting on the $a^\rho(x)$ in the free Lagrangian spoils the invariance. To recover it, the theory is enlarged by introducing new vector fields, the *gauge fields*. Their number is equal to the number of generators of the gauge group and they generate dynamics.

The prototype gauge theory is quantum electrodynamics (QED). It is based on a local Abelian invariance, U(1) as discovered by Weyl (1929). In 1954, Yang and Mills extended the gauge principle to non-abelian symmetries. The non-abelian gauge theories introduced at that time form the basis of the modern theory of elementary particles. The first application of the Yang–Mills theory was the unification of the weak and electromagnetic interaction as a SU(2) × U(1) gauge-invariant theory. Quantum chromodynamics (QCD), created in the 1980s, is the gauge theory of the strong interactions based on the SU(3)-colour gauge group.

In this chapter, we briefly discuss the application of the gauge principle for abelian and non-abelian gauge groups.

9.1 ABELIAN GAUGE SYMMETRY

QED is an Abelian gauge theory. It can be derived by requiring the Dirac free Lagrangian to be gauge invariant under U(1) transformations. The action of a global U(1) transformation is to change a wave function by a phase factor as in (8.9),

$\psi \to e^{ia}\psi(x)$. The Lagrangian density of a free electron

$$\mathcal{L}_0 = \overline{\psi}(i\gamma^\mu \partial_\mu - m)\psi \tag{9.1}$$

is invariant under such transformation. 'Gauging' this symmetry means replacing a by $a(x)$. Then,

$$\psi(x) \to e^{ia(x)}\psi(x), \qquad \overline{\psi}(x) \to e^{-ia(x)}\overline{\psi}(x) \tag{9.2}$$

and

$$\overline{\psi}(x)\partial_\mu \psi(x) \to \overline{\psi}(x)\partial_\mu \psi(x) + i\,\overline{\psi}\Big(\partial_\mu a(x)\Big)\psi(x). \tag{9.3}$$

Therefore, the derivative introduces an extra term which must be cancelled by another one in order to preserve the gauge invariance. This can be realized by introducing a vector field $A_\mu(x)$, the gauge field, and by changing the ordinary derivative ∂_μ to the *covariant* derivative

$$D_\mu = \partial_\mu - ieA_\mu \tag{9.4}$$

provided the gauge field has the transformation property

$$A_\mu \to A_\mu + \frac{1}{e}\partial_\mu a(x). \tag{9.5}$$

To make the gauge field A_μ a dynamical field, one has to add to $\mathcal{L}_0$ the free Lagrangian for A_μ

$$\mathcal{L}_A = -\frac{1}{4}F_{\mu\nu}F^{\mu\nu} \tag{9.6}$$

where the field-strength tensor

$$F_{\mu\nu} = \partial_\mu A_\nu - \partial_\nu A_\mu \tag{9.7}$$

is obviously gauge invariant by itself. Then, the total Lagrangian

$$\mathcal{L}_{\text{QED}} = \overline{\psi}(i\gamma^\mu D_\mu - m)\psi - \frac{1}{4}F_{\mu\nu}F^{\mu\nu} \tag{9.8}$$

is gauge invariant. From this construction, one can say that in QED the photon coupling appears as a consequence of the gauge principle or, according to Weyl, the gauge invariance can be regarded as a reason for the existence of the photon.

There are a few important remarks to be made.

1. $\mathcal{L}_A$ does not contain a term of the form

    $$-\frac{1}{2}m^2 A_\mu(x)A^\mu(x)$$

    because such a term is not gauge invariant. Therefore, the photon is massless and such a statement remains valid for any gauge field.

2. $D_\mu \psi$ contains the minimal coupling to the photon, the form of which is determined by the transformation properties of the electron field under the gauge group. This feature remains valid for any gauge theory, abelian or non-abelian, and is referred to as universality.
3. The photon does not carry a charge (the U(1) quantum number) and as a consequence there are no self-coupling terms for the gauge field. The situation is different for non-abelian theories, because there the gauge fields carry charges.

9.2 NON-ABELIAN GAUGE SYMMETRY

The non-abelian symmetries are related to internal symmetries of elementary particles. The non-abelian case is illustrated here for the simplest case, the $SU_I(2)$. The global SU(2) symmetry was discussed in Section 8.2. In isospin space, the fundamental representation is spanned by an isospin doublet

$$\psi(x) = \begin{pmatrix} \psi_1(x) \\ \psi_2(x) \end{pmatrix} \tag{9.9}$$

which, under an $SU_I(2)$ transformation described by (8.59) becomes

$$\psi'(x) = \exp(-i \frac{\boldsymbol{\omega} \cdot \boldsymbol{\tau}}{2}) \psi(x) \tag{9.10}$$

where $\omega_i (i = 1, 2, 3)$ are the SU(2) parameters and $\tau_i (i = 1, 2, 3)$ the Pauli matrices (8.58). To the extent that u and d quarks have equal masses, they form a doublet like (9.9).

As in the abelian case, the free Lagrangian $\mathscr{L}_0$, where $\psi(x)$ is now given by (9.9), is invariant under global $SU_I(2)$ transformations. Let us introduce a local transformation

$$S = \exp(-i \frac{\boldsymbol{\omega}(x) \cdot \boldsymbol{\tau}}{2}). \tag{9.11}$$

The derivative term in $\mathscr{L}_0$ becomes

$$\overline{\psi}'(x) \partial_\mu \psi'(x) = \overline{\psi}(x) \partial_\mu \psi(x) + \overline{\psi}(x) S^{-1} (\partial_\mu S) \psi(x) \tag{9.12}$$

which is no longer gauge invariant. To remedy the situation, one has to introduce three vector fields $A_\mu^i (i = 1, 2, 3)$, one for each generator, and a covariant derivative D_μ containing the minimal coupling by analogy with QED

$$D_\mu = \partial_\mu - i g \frac{\boldsymbol{\tau} \cdot \boldsymbol{A}_\mu}{2} \tag{9.13}$$

where g plays the role of a 'charge'. The transformation properties of A_μ^i are established from the requirement that

$$(D_\mu \psi)' = S D_\mu \psi \tag{9.14}$$

i.e. $D_\mu \psi$ and ψ have the same transformation property. The relation (9.14) is equivalent to

$$D'_\mu = S D_\mu S^{-1} \tag{9.15}$$

or, based on (9.13), to

$$\frac{\boldsymbol{\tau} \cdot \boldsymbol{A}'_\mu}{2} = S \frac{\boldsymbol{\tau} \cdot \boldsymbol{A}_\mu}{2} S^{-1} - \frac{i}{g}(\partial_\mu S)S^{-1}. \tag{9.16}$$

where, also, use has been made of the equality $(\partial_\mu S^{-1})S = -(\partial_\mu S)S^{-1}$.

To find $A^{i'}_\mu$ in terms of A^i_μ, it is enough to consider an infinitesimal transformation

$$S \simeq 1 - i \frac{\boldsymbol{\omega}(x) \cdot \boldsymbol{\tau}}{2} \tag{9.17}$$

for which (9.16) becomes

$$\begin{aligned}\frac{\boldsymbol{\tau} \cdot \boldsymbol{A}'_\mu}{2} &= \frac{\boldsymbol{\tau} \cdot \boldsymbol{A}_\mu}{2} - i\,\omega^j A^k_\mu \left[\frac{\tau^j}{2}, \frac{\tau^k}{2}\right] - \frac{1}{g}\frac{\boldsymbol{\tau}}{2}\cdot\partial_\mu \boldsymbol{\omega} \\ &= \frac{\boldsymbol{\tau} \cdot \boldsymbol{A}_\mu}{2} + \frac{1}{2}\varepsilon^{ijk}\tau^i \omega^j A^k_\mu - \frac{1}{g}\frac{\boldsymbol{\tau}}{2}\cdot\partial_\mu \boldsymbol{\omega}\end{aligned} \tag{9.18}$$

where the su(2) algebra given by Equation (6.38) has been used to evaluate the commutator. Equation (9.18) is equivalent to

$$A^{i'}_\mu = A^i_\mu + \varepsilon^{ijk}\omega^j A^k_\mu - \frac{1}{g}\partial_\mu \omega^i \tag{9.19a}$$

or to

$$\boldsymbol{A}'_\mu = \boldsymbol{A}_\mu + \boldsymbol{\omega} \times \boldsymbol{A}_\mu - \frac{1}{g}\partial_\mu \boldsymbol{\omega}. \tag{9.19b}$$

This is a generalization of (9.5). The second term in (9.19) is not surprising because here we deal with an (iso)vector which, under a rotation, changes according to (6.31). Thus, the second term would appear even if ω were a constant.

The next stage is to construct the Lagrangian of the gauge field A^i_μ so that it is gauge invariant like (9.6). The second-rank tensor $F^i_{\mu\nu}$ is defined by

$$(D_\mu D_\nu - D_\nu D_\mu)\psi = -i g \left(\frac{\tau^i}{2}F^i_{\mu\nu}\right)\psi \tag{9.20}$$

with

$$\frac{\tau^i}{2}F^i_{\mu\nu} = \partial_\mu \frac{\boldsymbol{\tau} \cdot \boldsymbol{A}_\nu}{2} - \partial_\nu \frac{\boldsymbol{\tau} \cdot \boldsymbol{A}_\mu}{2} - i g \left[\frac{\boldsymbol{\tau} \cdot \boldsymbol{A}_\mu}{2}, \frac{\boldsymbol{\tau} \cdot \boldsymbol{A}_\nu}{2}\right] \tag{9.21}$$

or

$$F^i_{\mu\nu} = \partial_\mu A^i_\nu - \partial_\nu A^i_\mu + g\,\varepsilon^{ijk}A^j_\mu A^k_\nu. \tag{9.22}$$

The transformation property (9.14) leads to

$$[(D_\mu D_\nu - D_\nu D_\mu)\psi]' = S(D_\mu D_\nu - D_\nu D_\mu)\psi. \tag{9.23}$$

Using (9.20) on both sides, one has

$$\tau^i F^{i'}_{\mu\nu} S\psi = S\tau^i F^i_{\mu\nu} \psi$$

or

$$\boldsymbol{\tau} \cdot \boldsymbol{F}'_{\mu\nu} = S\boldsymbol{\tau} \cdot \boldsymbol{F}_{\mu\nu} S^{-1}. \tag{9.24}$$

Introducing the infinitesimal transformation (9.17), one obtains

$$F^{i'}_{\mu\nu} = F^i_{\mu\nu} + \varepsilon^{ijk}\omega^j F^k_{\mu\nu}. \tag{9.25}$$

From (9.25), one can easily see that

$$\text{tr}\left[(\boldsymbol{\tau} \cdot \boldsymbol{F}_{\mu\nu})(\boldsymbol{\tau} \cdot \boldsymbol{F}^{\mu\nu})\right] = F^i_{\mu\nu} F^{i\mu\nu}$$

is gauge invariant. Then, the total gauge invariant Lagrangian describing the interaction between the gauge fields A^i_μ and the $SU_I(2)$ doublet ψ is

$$\mathcal{L} = \overline{\psi}(i\gamma^\mu D_\mu - m)\psi - \frac{1}{4}F^i_{\mu\nu}F^{i\mu\nu} \tag{9.26}$$

with D_μ given by (9.13) and $F^i_{\mu\nu}$ by (9.22).

The Lagrangian (9.26) appears in the Weinberg–Salam theory of the electro-weak interaction which has $SU(2) \times U(1)$ as a gauge group, so that there are four gauge bosons. In a pure gauge theory, these are massless in order to ensure the gauge invariance as discussed above. If one wants to avoid this feature and obtain massive vector bosons, the gauge symmetry must be broken. A convenient possibility is spontaneous symmetry breaking. This means that the Lagrangian is still invariant under gauge symmetry but the ground state (the vacuum) is not a scalar of the symmetry group. The net result is that three of the gauge bosons acquire mass and have been identified with the W^+, W^- and Z^0 bosons. The $U(1)$ symmetry remains unbroken and is mediated by the massless photon.

One can generalize these results to any $SU(n)$. The $n^2 - 1$ generators have the commutation relations

$$[T^a, T^b] = if^{abc}T^c \tag{9.27}$$

where f^{abc} are totally antisymmetric structure constants and ψ entering the Lagrangian is supposed to belong to the fundamental representation. The covariant derivative is analogous to (9.13):

$$D_\mu = \partial_\mu - igT^a A^a_\mu \tag{9.28}$$

and the second rank tensor $F^a_{\mu\nu}$ is analogous to (9.22)

$$F^a_{\mu\nu} = \partial_\mu A^a_\nu - \partial_\nu A^a_\mu + gf^{abc}A^b_\mu A^c_\nu. \tag{9.29}$$

The Lagrangian

$$\mathscr{L} = \overline{\psi}(i\gamma^\mu D_\mu - m)\psi - \frac{1}{4}F^a_{\mu\nu}F^{a\mu\nu} \qquad (9.30)$$

is then invariant under the local gauge transformation

$$S = e^{-i\omega^a(x)T^a} \qquad (9.31)$$

and the analogue of (9.16) is

$$\boldsymbol{T} \cdot \boldsymbol{A}'_\mu = S\,\boldsymbol{T} \cdot \boldsymbol{A}_\mu\,S^{-1} - \frac{i}{g}(\partial_\mu S)S^{-1} \qquad (9.32)$$

where instead of the structure constant ε^{ijk} of SU(2) we have now the structure constant f^{abc} of SU(n) leading to

$$A^{a'}_\mu = A^a_\mu + f^{abc}\omega^b A^c_\mu - \frac{1}{g}\partial_\mu \omega^a \qquad (9.33)$$

which shows that the gauge fields A^a_μ transform among themselves like the generators, or like members of the adjoint representation of SU(n) (Section 8.9).

The term $-\frac{1}{4}F^a_{\mu\nu}F^{a\mu\nu}$ is not in fact a free Lagrangian. It contains self-interacting terms brought about by the non-linear terms in $F^a_{\mu\nu}$, which are a consequence of the non-abelian character of the gauge symmetry. This is fundamentally different from the abelian case.

For SU(3), one has

$$T^a = \frac{\lambda^a}{2} \quad (a = 1, 2, \cdots, 8) \qquad (9.34)$$

where λ^a are the Gell-Mann matrices. Quantum chromodynamics (QCD) has emerged as the theory of strong interactions in an SU(3)-colour gauge theory. The massless gauge bosons are the gluons which have self-interactions among themselves and couple to the colour quantum number.

In Yang–Mills theories, in contrast to abelian gauge theory, the coupling constant decreases at short distances. It is the case for QCD. It is called the anti-screening effect, and is quite well understood. It leads to asymptotic freedom at short distances, where perturbative QCD is applicable. The perturbative regime has been tested by several experiments. However, in the non-perturbative regime, related to long distances and quark confinement, only lattice calculations provide a solution for the ground state of hadrons. The study of the excitation spectrum of hadrons is beyond present means. In this situation, one must construct models inspired by QCD and some of them will be considered in the next chapter.

Before ending this Chapter, we add two useful tables. The first one, Table 9.1, lists the three known families of fundamental fermions forming the matter. Each fermion can be described by a wave function ψ as in (9.1). The first two rows represent leptons and the following two, quarks, as described in the previous Chapter.

Table 9.1 Families of fundamental fermions and their electrical charge.

Families			Charge
ν_e	ν_μ	ν_τ	0
e	μ	τ	-1
u	c	t	$\frac{2}{3}$
d	s	b	$-\frac{1}{3}$

In the nature, elementary particles can interact with each other through four known types of forces: a) gravitational b) weak c) electromagnetic or 4) strong forces. In fact, gravitation is believed to play a negligible role in elementary particle physics. Also, the weak interaction is so feeble that it is usually swamped by the much stronger electromagnetic and strong interactions, unless these are forbidden by selection rules. Above, it has been illustrated how the electromagnetic, weak, and strong interactions can be generated as gauge interactions via boson exchange. Table 9.2 lists these interactions together with some of their characteristics, relevant for the discussion in this Chapter.

Table 9.2 Fundamental interactions between elementary particles.

Interaction	Exchanged boson	Number of bosons	Mass	Interacting fermions
strong	gluon g	8	$m_g = 0$	quarks
electromagnetic	photon γ	1	$m_\gamma = 0$	quarks and charged leptons
weak	intermediate boson $W^\pm, Z^0$	3	$m_{W^\pm} = 80.26$ GeV $m_{Z^0} = 91.91$ GeV	quarks and leptons

*10

MULTIQUARK SYSTEMS

To date, quarks have not been observed as isolated particles. They are thought to exist only in bound states (confined). They carry a new degree of freedom, called colour, but the bound states they form are always colourless (i.e. $SU_C(3)$ singlets, cf. Exercise 10.2).

The present chapter is devoted to spectroscopic studies of multiquark systems based on their symmetries. It is far from being intended as an exhaustive review of multiquark system results. It is rather meant to present the intermediate steps of a group theoretical nature which are often omitted in research papers. In particular, we shall discuss three-quark systems, qqq, in relation to baryons, two-quark–two-antiquark systems, $2q$-$2\bar{q}$, in relation to hadronic molecules, and six-quark systems, $6q$, in relation to the short-range part of the nucleon–nucleon interaction.

The classification of hadrons, considered in the previous chapter, was related to their composite aspect only and the quarks were nearly independent. In the present chapter, we look at quark systems from a dynamical point of view, i.e. we discuss how their mass can be calculated from the interaction between quarks. In terms of quarks, hadron spectroscopy is very much like that in atomic and nuclear physics. A major difference comes from the additional degree of freedom, the colour. Quarks are fermions, so that a system of identical quarks must be described by a totally antisymmetric wave function ψ. If we denote the orbital part by R, the spin part by χ, the flavour part by ϕ and the colour part by C, then ψ must be an appropriate combination of terms $R\chi\phi C$ such as to be antisymmetric under all transpositions of the constituent particles. The extra function C makes the procedure of deriving ψ more complicated than in atomic and nuclear physics, especially if the number of quarks is greater than three, but the techniques are similar and are based on the procedures described in Chapter 4.

Therefore, for a multiquark system, the major problems to be settled are related to the dynamics and the classification of multiquark states. Without loss of generality, the approach to multiquark systems will be illustrated by using the constituent quark model. Although the foundations of the model are not well understood, it is very useful in practice. Most of the physics and group theoretical discussion holds beyond the particular model. This relies on the fact that all models have a common feature, the confinement. In the non-relativistic model, this is achieved by a spin-independent interaction potential between quarks, in the MIT bag model by appropriate boundary conditions, and in the soliton bag model by an additional scalar field or through a chromodielectric function mediated by a scalar field (Lee 1981).

In the next section we shall present a prototype of the constituent quark model. The subsequent sections will refer to the description of $3q$, $2q$-$2\bar{q}$, and $6q$ systems.

10.1 THE DYNAMICS

In a constituent quark model, a system of N confined quarks is described by a Hamiltonian H_0 of the form

$$H_0 = T + V^{\text{s.i.}}(r_1, r_2, \cdots, r_N) \tag{10.1}$$

where T is the intrinsic kinetic energy which can have a non-relativistic form for heavy quarks (c, b, t) or a relativistic or non-relativistic form for light quarks (u, d, s). The potential $V^{\text{s.i.}}$ contains a confinement interaction between quarks which stems from the chromoelectric part of the QCD Hamiltonian. Therefore, it is spin independent (s.i.) and it is also assumed to be flavour independent (Section 8.6). The potential term can contain two- or N-body terms when $N > 2$. The latter can be introduced on the basis of QCD-inspired flux tube models (Fig. 10.1). They are $SU_C(3)$ invariant. In the case of baryons, it has been shown (Carlson, Kogut, and Pandharipande, 1983a) that three-body terms contribute a few percent to the binding energy.

A good approximation to $V^{\text{s.i.}}$ is a sum of two-body terms

$$V^{\text{s.i.}} = \sum_{i<j}^{N} V^{\text{s.i.}}_{ij}. \tag{10.2}$$

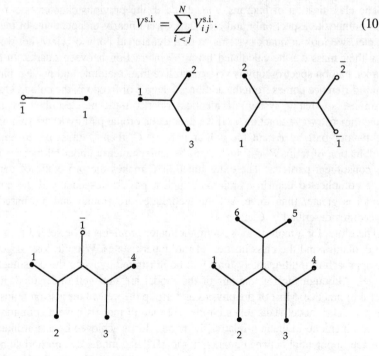

Figure 10.1 Flux-tube configurations for confined states of 2–6 quarks. For $N > 2$, they correspond to N-body forces.

THE DYNAMICS 357

Lattice gauge calculations for a heavy quark–antiquark pair (Stack 1984) give $V_{ij}^{\text{s.i.}}$ as a Coulomb type plus a linear confining potential

$$V_{ij}^{\text{s.i.}} = -\left(-\frac{\alpha_s}{r} + \frac{3}{4}c + \frac{3}{4}\sqrt{\sigma}\,r\right)\widehat{C}_i \cdot \widehat{C}_j \tag{10.3}$$

where r is the relative distance between the particles i and j, α_s the strong coupling constant, $\sqrt{\sigma}$ the string-tension constant, and c a constant to be determined from the data. The factor

$$\widehat{C}_i \cdot \widehat{C}_j = \sum_{c=1}^{8} \frac{\lambda_i^c}{2} \cdot \frac{\lambda_j^c}{2} \tag{10.4}$$

is the colour operator expressed in terms of $SU_C(3)$ Gell-Mann matrices λ_i^c ($i = 1, \ldots, 8$) acting in the colour space. Its expectation value in a two-body colour state is (Exercise 10.1)

$$<\widehat{C}_i \cdot \widehat{C}_j> = \begin{cases} -\frac{2}{3} & \text{for } [1^2] \ (qq \text{ colour antisymmetric}) \\ \frac{1}{3} & \text{for } [2] \ (qq \text{ colour symmetric}) \\ -\frac{4}{3} & \text{for } [1^3] \ (\bar{q}q \text{ colour singlet}) \\ \frac{1}{6} & \text{for } [21] \ (\bar{q}q \text{ colour octet}). \end{cases} \tag{10.5}$$

In a baryon, two quarks can be in only an antisymmetric colour state because adding a third quark must yield a colour singlet $3q$ state. According to (10.5), the interaction strength V_{qq} for a pair of quarks in a baryon is half of the strength $V_{q\bar{q}}$ for a $q\bar{q}$ pair in a meson, once the integration in the colour space has been made:

$$V_{qq} = \frac{1}{2} V_{q\bar{q}}. \tag{10.6}$$

The potential of the form (10.3) does not respect colour gauge invariance but, despite this flaw, its use is justifiable by its good results in the baryon spectra.

In (10.3), typical values of the constants are

$$\frac{4}{3}\frac{\alpha_s}{\hbar c} = 0.5, \qquad \sqrt{\sigma} = 1 \text{ GeV fm}^{-1}, \qquad c \simeq 0.8 \text{ GeV}. \tag{10.7}$$

They are consistent with parameters used to fit the spectrum of charmonium. The levels in this spectrum are given by the masses of the heavy mesons listed in Table 8.13. Although on a quite different energy scale, this spectrum is similar to that of positronium (see, for example, Nachtmann (1990), Chapter 13).

For calculation purposes, it is sometimes advantageous to fit the potential (10.3) by a functional form of the type

$$V_{ij}^{\text{s.i.}} = -\left[e_0 + \frac{1}{2}kr^2 + A\exp\left(-\frac{r^2}{\alpha^2}\right)\right]\widehat{C}_i \cdot \widehat{C}_j \tag{10.8}$$

where the parameters e_0, k, A, and α^2 can be adjusted from data on meson and/or baryon spectra. The harmonic oscillator potential ($A = 0$) has been used extensively in the calculation of baryon spectra (Isgur and Karl 1978, 1979a,b). It is a convenient first approximation to realistic potentials and its parameters have to be adjusted for each band in order to optimize the approximation. A form of the type (10.8) has been used by Harvey (1981b) in the study of the N-N problem and the same form with $A = 0$ will be considered below in the study of $2q$-$2\bar{q}$ systems. This form is convenient for use with a harmonic oscillator basis.

The total Hamiltonian of a constituent quark model is

$$H = H_0 + H^{\text{hyp}} \tag{10.9}$$

where H^{hyp} is a spin-dependent term which represents the chromomagnetic field. This term has the form

$$H^{\text{hyp}} = \sum_{i<j} V_{ij}^{\text{hyp}} \tag{10.10}$$

with

$$V_{ij}^{\text{hyp}} = \left(V_{ij}^{\text{SS}} + V_{ij}^{\text{T}}\right) \widehat{C}_i \cdot \widehat{C}_j \tag{10.11}$$

$$V_{ij}^{\text{SS}} = -\frac{8\pi\alpha_S}{3m_i m_j} \frac{\varepsilon^3}{\pi^{3/2}} e^{-\varepsilon^2 r^2} S_i \cdot S_j \tag{10.12}$$

$$V_{ij}^{\text{T}} = -\frac{3\alpha_S}{m_i m_j r^3} \left[\text{erf}(\varepsilon r) - \frac{4\varepsilon^3 r^3}{3\sqrt{\pi}} \left(1 + \frac{3}{2\varepsilon^2 r^2}\right) e^{-\varepsilon^2 r^2}\right]$$
$$\times \left(\frac{1}{r^2} S_i \cdot r \, S_j \cdot r - \frac{1}{3} S_i \cdot S_j\right). \tag{10.13}$$

Here, V_{ij}^{SS} and V_{ij}^{T} are regularized expressions of the spin–spin and tensor parts of the Fermi–Breit interaction (De Rújula, Georgi, and Glashow 1975). The quantities ε and the quark masses m_i are assumed to be free parameters. For meson spectra, the spin-orbit term of the Fermi–Breit interaction should also be retained. It can take the form

$$V_{ij}^{\text{SO}} = -\frac{6\alpha_s \varepsilon^3}{m_i m_j} \left(\frac{\text{erf}(\varepsilon r)}{4\varepsilon^3 r^3} - \frac{1}{3\pi^{1/2}} e^{-\varepsilon^2 r^2}\right) \widehat{C}_i \cdot \widehat{C}_j. \tag{10.14}$$

Its contribution cancels out in the baryon spectra and that is why it is usually neglected (Isgur and Karl 1978, 1979a,b). Note also that the major contribution to baryon spectra comes from the spin–spin term. The above regularized forms are taken from Carlson, Kogut, and Pandharipande (1983b).

Exercise 10.1 Prove the relations (10.5).

Solution For a system of two quarks i and j, the generators $\widehat{C}^c$ of SU$_C$(3) are

$$\widehat{C}^c = \widehat{C}_i^c + \widehat{C}_j^c \quad (c = 1, \cdots, 8). \tag{1}$$

Then, by analogy with angular momentum, one can write (10.4) as

$$\widehat{C}_i \cdot \widehat{C}_j = \frac{1}{2}\left(\widehat{C}^2 - \widehat{C}_i^2 - \widehat{C}_j^2\right) \quad (2)$$

where $\widehat{C}^2$, $\widehat{C}_i^2$ and $\widehat{C}_j^2$ are Casimir operators related to the two-body system, and the particles i and j, respectively. For an irrep $(\lambda\mu)$, the eigenvalue of the Casimir operator is given by (8.134):

$$\widehat{C}^2 \psi_{(\lambda\mu)} = \frac{1}{3}\left(\lambda^2 + \mu^2 + \lambda\mu + 3\lambda + 3\mu\right)\psi_{(\lambda\mu)}. \quad (3)$$

In the colour space, a quark is described by $(\lambda\mu) = (10)$ and an antiquark by $(\lambda\mu) = (01)$. Then, Equation (3) gives

$$\widehat{C}_i^2 \psi_{(10)} = \frac{4}{3}\psi_{(10)}, \qquad \widehat{C}_i^2 \psi_{(01)} = \frac{4}{3}\psi_{(01)}. \quad (4)$$

A symmetric state of two quarks ☐☐ has $(\lambda\mu) = (20)$ and its Casimir eigenvalue equation is

$$C^2 \psi_{(20)} = \frac{10}{3}\psi_{(20)}. \quad (5)$$

An antisymmetric state ☐/☐ has $(\lambda\mu) = (01)$, so that the eigenvalue associated to this representation is the same as that for an antiquark, Equation (4).

It follows immediately that

$$\langle \widehat{C}_i \cdot \widehat{C}_j \rangle_{[1^2]} = \frac{1}{2}\left(\frac{4}{3} - \frac{4}{3} - \frac{4}{3}\right) = -\frac{2}{3} \quad (6)$$

$$\langle \widehat{C}_i \cdot \widehat{C}_j \rangle_{[2]} = \frac{1}{2}\left(\frac{10}{3} - \frac{4}{3} - \frac{4}{3}\right) = \frac{1}{3}. \quad (7)$$

In a meson, a quark and an antiquark form a colour singlet state for which the Casimir eigenvalue is zero, so that we have

$$\langle \widehat{C}_i \cdot \widehat{C}_j \rangle_{[1^3]} = \frac{1}{2}\left(0 - \frac{4}{3} - \frac{4}{3}\right) = -\frac{4}{3}. \quad (8)$$

A colour octet $q\bar{q}$ pair has $(\lambda\mu) = (11)$ and a Casimir eigenvalue equation

$$C^2 \psi_{(11)} = 3\psi_{(11)}.$$

Hence, finally, one has

$$<\widehat{C}_i \cdot \widehat{C}_j>_{[21]} = \frac{1}{2}\left(3 - \frac{4}{3} - \frac{4}{3}\right) = \frac{1}{6}. \quad (9)$$

Colour octets can appear as subsystems of a larger system (see Section 10.3).

10.2 THE BARYONS

Here, we discuss non-strange baryons only and take $m_u = m_d$. For equal quark masses, the internal (Jacobi) coordinates $\boldsymbol{\rho}$ and $\boldsymbol{\lambda}$ (cf. Fig. 3.2) are defined by

$$\boldsymbol{\rho} = \frac{1}{\sqrt{2}}(\boldsymbol{r}_1 - \boldsymbol{r}_2), \qquad \boldsymbol{\lambda} = \frac{1}{\sqrt{6}}(\boldsymbol{r}_1 + \boldsymbol{r}_2 - 2\boldsymbol{r}_3). \tag{10.15}$$

The wave function ψ depends on $\boldsymbol{\rho}$ and $\boldsymbol{\lambda}$ via its orbital part ψ_{LM} which, combined with its spin χ and flavour ϕ parts, must give a totally symmetric function because the colour part C, which is a singlet, is also a totally antisymmetric function in this case (see Exercise 10.2).

According to Section 8.7, the classification of baryons can be made by using the irreps of SU(6). The decomposition of the SU(6) multiplets denoted by their dimension (56), (70), and (20), into those of $SU_F(3) \times SU_S(2)$ was given in Equations (8.196). It follows that the eigenstates of H_0, Equation (10.1), can also be classified according to SU(6). The meaning of the labels (56), (70), and (20) is that ψ_{LM} must have an S_3 symmetry indicated by one of the Young diagrams in Table 10.1, such as to give a totally symmetric function together with the associated $\chi\phi$ part. The possible combinations are shown in Table 10.1, which represents a transcription of Equations (8.196) in terms of S_3 symmetries.

But the hyperfine interaction (10.10)–(10.13) breaks the SU(2) symmetry and hence the SU(6) symmetry. Then, states belonging to distinct SU(6) irreps can be

Table 10.1 Possible S_3 symmetries for the orbital ψ_{LM}, spin χ, and flavour ϕ parts of the baryon wave functions. S = symmetric, M = mixed, A = antisymmetric.

	Orbital		Spin	Flavour
SU(6) multiplet	Young diagram	ψ_{LM}	χ	ϕ
56	▭▭▭	S	M	M
		S	S	S
70		M	S	M
		M	M	S
		M	M	M
		M	M	A
20		A	M	M
		A	S	A

THE BARYONS

mixed by diagonalizing H^{hyp} in a space spanned by SU(6) multiplets. To do the diagonalization, one needs the eigenstates of H_0. Usually, the Schrödinger equation for H_0—the unperturbed Hamiltonian—cannot be solved exactly and one looks for a variational solution. For a potential of type (10.3) plus three-body forces, a variational solution for the ground state has been proposed by Carlson, Kogut, and Pandharipande (1983a). It consists of products of two- and three-body correlation functions, which have an appropriate asymptotic behaviour for a linear potential. Let us denote by F the ground state solution. Then, it is convenient to choose variational wave functions ψ_{LM} for the excited states to be orthogonal on F. They must also satisfy the correct permutation symmetry specified in Table 10.1. They carry an amount of radial or orbital momentum excitations and are usually classified by the number N of oscillator quanta. There is a systematic procedure which can be applied to obtain these states (Moshinski 1969). It leads to functions which are polynomials in ρ and λ. The classification of orbital states for a three-quark system carrying up to $N = 2$ excitations is shown in Table 10.2, both in terms of the internal coordinates ρ and λ and in terms of harmonic oscillator states. Note that the harmonic oscillator states of symmetries $[21]_\lambda$ and $[21]_\rho$ were constructed according to Section 4.3. For example,

$$|[21]_\rho (0s)^2(0p)> = \frac{1}{\sqrt{2}}(|0s>|0p>|0s> - |0p>|0s>|0s>)$$

where the particle order 1, 2, 3 is understood in both terms on the right-hand side.

In Table 10.3, the ψ^μ_{LM} states of the $N = 3$ band are indicated. We exhibit only the $M = 0$ states because $M \neq 0$ states can be obtained from ψ^μ_{L0} through ladder operators $L_\pm$. For S_3 mixed symmetry representations, only the state with symmetry $\mu = \rho$ is given. The associated $\mu = \lambda$ state can be obtained from $\mu = \rho$ by interchanging the ρ and λ coordinates. The coefficients c, α_0, α_1 and α_2 ensure the orthogonality of the 70, 70′, and 70″ states. The SU(6) multiplets denoted by prime or double prime carry radial excitations.

Once the ψ^μ_{LM} are known, one can construct a totally antisymmetric wave function $\psi^\mu_{LM} \chi \phi C$. The colour part C does not need to be written explicitly because the integration in the colour space is given by (10.5), which says that for two quarks in a baryon, the colour operator brings a factor $-\frac{2}{3}$. It remains to construct the totally symmetric product $\psi^\mu_{LM} \chi \phi$. This is achieved with the help of the inner products explained in Section 4.7. Section 4.8. gives the required Clebsch–Gordan coefficients.

Example 10.1 Construct the three-quark totally symmetric function $|^2N(70, 2^+)\frac{3}{2}^+ \frac{3}{2}>$.

This is a positive parity state where N stands for the nucleon, in order to make a distinction with the Δ particle. The upper index 2 stands for $2S$ where S is the spin, which is coupled to $L = 2$ to give $J = m_J = \frac{3}{2}$. This function belongs to the SU(6) multiplet 70 and because it represents a nucleon, it must have the structure ψ^μ_{LM}-mixed, χ-mixed, ϕ-mixed as indicated in Table 10.1. Here, χ-mixed is required by $S = \frac{1}{2}$, the spin of the nucleon (see Equations (4.208)–(4.211)), and ϕ-mixed by its

Table 10.2 Orbital states for three-quark systems with $N = 0, 1,$ or 2 oscillator quanta in terms of intrinsic coordinates ρ and λ and the corresponding harmonic oscillator configurations. The ground state ψ_{00}^S in a harmonic oscillator basis is $\psi_{00} = \pi^{-3/2}\exp[-\frac{1}{2}(\rho^2+\lambda^2)]$, $\rho_\pm = \rho_x \pm i\rho_y$, $\lambda_\pm = \lambda_x \pm i\lambda_y$. The upper index μ specifies the S_3 symmetry (cf. Section 4.3) and P the parity.

SU(6) multiplet	N	LM	P	μ	$\psi_{L0}^\mu / \psi_{00}^S$	Harmonic oscillator configuration
56	0	00	+	S	1	$\|[3](0s)^3\rangle$
70	1	10	−	ρ	$\sqrt{2}\rho_0$	$\|[21]_\rho(0s)^2 0p\rangle$
				λ	$\sqrt{2}\lambda_0$	$\|[21]_\lambda(0s)^2 0p\rangle$
56′	2	00	+	S	$\sqrt{\frac{1}{3}}(3-\rho^2-\lambda^2)$	$\sqrt{\frac{2}{3}}\|[3](0s)^2 1s\rangle - \sqrt{\frac{1}{3}}\|[3](0s)(0p)^2\rangle$
70	2	00	+	ρ	$\sqrt{\frac{4}{3}}\,\boldsymbol{\rho}\cdot\boldsymbol{\lambda}$	$\sqrt{\frac{1}{3}}\|[21]_\rho(0s)^2 1s\rangle + \sqrt{\frac{2}{3}}\|[21]_\rho(0s)(0p)^2\rangle$
				λ	$\sqrt{\frac{1}{3}}(\rho^2-\lambda^2)$	$\sqrt{\frac{1}{3}}\|[21]_\lambda(0s)^2 1s\rangle + \sqrt{\frac{2}{3}}\|[21]_\lambda(0s)(0p)^2\rangle$
56	2	20	+	S	$\sqrt{\frac{1}{6}}[3(\rho_0^2+\lambda_0^2)-(\rho^2+\lambda^2)]$	$\sqrt{\frac{2}{3}}\|[3](0s)^2 0d\rangle - \sqrt{\frac{1}{3}}\|[3](0s)(0p)^2\rangle$
70	2	20	+	ρ	$\sqrt{\frac{2}{3}}(3\rho_0\lambda_0 - \boldsymbol{\rho}\cdot\boldsymbol{\lambda})$	$\sqrt{\frac{1}{3}}\|[21]_\rho(0s)^2 0d\rangle + \sqrt{\frac{2}{3}}\|[21]_\rho(0s)(0p)^2\rangle$
				λ	$\sqrt{\frac{1}{6}}[3(\rho_0^2-\lambda_0^2)-(\rho^2-\lambda^2)]$	$\sqrt{\frac{1}{3}}\|[21]_\lambda(0s)^2 0d\rangle + \sqrt{\frac{2}{3}}\|[21]_\lambda(0s)(0p)^2\rangle$
20	2	10	+	A	$\sqrt{\frac{1}{2}}(\rho_-\lambda_+ - \rho_+\lambda_-)$	$\|[1^3](0s)(0p)^2\rangle$

Table 10.3 The orbital states ψ_{LO}^μ of the $N=3$ negative parity band where μ denotes the S_3 symmetry ρ, S, or A. $\rho_\pm = \rho_x \pm i\rho_y$, $\lambda_\pm = \lambda_x \pm i\lambda_y$. F is a general ground state wave function. The first column indicates the corresponding SU(6) multiplet and the angular momentum. (From Stancu and Stassart 1991).

Multiplet	ψ_{LO}^μ
$(70,1^-)$	$N_{10}^\rho \rho_0 F$
$(70',1^-)$	$N_{10}^{\rho'}[1 - c(\rho^2 + \lambda^2)\rho_0]F$
$(70'',1^-)$	$N_{10}^{\rho''}[\boldsymbol{\rho}\cdot\boldsymbol{\lambda}\lambda_0 + (\alpha_0 + \alpha_1\rho^2 + \alpha_2\lambda^2)\rho_0]F$
$(56,3^-)$	$N_{30}^S\{[5(\lambda_0^2 - 3\rho_0^2) + 3(\rho^2 - \lambda^2)]\lambda_0 + 6\boldsymbol{\rho}\cdot\boldsymbol{\lambda}\rho_0\}F$
$(20,3^-)$	$N_{30}^A\{[5(\rho_0^2 - 3\lambda_0^2) - 3(\rho^2 - \lambda^2)]\rho_0 + 6\boldsymbol{\rho}\cdot\boldsymbol{\lambda}\lambda_0\}F$
$(70,3^-)$	$N_{30}^\rho\{[5(\rho_0^2 + \lambda_0^2) - 3\rho^2 - \lambda^2]\rho_0 - 2\boldsymbol{\rho}\cdot\boldsymbol{\lambda}\lambda_0\}F$
$(70,2^-)$	$N_{20}^\rho(\rho_+\lambda_- - \rho_-\lambda_+)\lambda_0 F$
$(56,1^-)$	$N_{10}^S[-(\rho^2 - \lambda^2)\lambda_0 - 2\boldsymbol{\rho}\cdot\boldsymbol{\lambda}\rho_0]F$
$(20,1^-)$	$N_{10}^A[(\rho^2 - \lambda^2)\rho_0 - 2\boldsymbol{\rho}\cdot\boldsymbol{\lambda}\lambda_0]F$

flavour structure (cf. Table 4.1). This means that spin and flavour have to be combined to give mixed S_3 symmetry states. According to Table 4.7, these are

$$(\chi\phi)_{\frac{1}{2}m_s}^\lambda = -\frac{1}{\sqrt{2}}\left(\chi_{\frac{1}{2}m_s}^\rho \phi^\rho - \chi_{\frac{1}{2}m_s}^\lambda \phi^\lambda\right)$$

and

$$(\chi\phi)_{\frac{1}{2}m_s}^\rho = -\frac{1}{\sqrt{2}}\left(\chi_{\frac{1}{2}m_s}^\lambda \phi^\rho + \chi_{\frac{1}{2}m_s}^\rho \phi^\lambda\right).$$

One can combine these functions with ψ_{LM}^λ and ψ_{LM}^ρ to give a totally symmetric state as (4.159). For $J = m_J = \frac{3}{2}$, the desired wave function is

$$|^2N(70, 2^+)\frac{3^+}{2}\frac{3}{2}\rangle$$

$$= \frac{1}{2}\sum C^2_{m_L\,m_S\,\frac{1}{2}\,\frac{3}{2}} \left[\psi_{2m_L}^\lambda \left(\chi_{\frac{1}{2}m_S}^\rho \phi^\rho - \chi_{\frac{1}{2}m_S}^\lambda \phi^\lambda\right) + \psi_{2m_L}^\rho \left(\chi_{\frac{1}{2}m_S}^\rho \phi^\lambda + \chi_{\frac{1}{2}m_S}^\lambda \phi^\rho\right)\right]$$

$$= \frac{1}{\sqrt{5}}[\psi_{22}^\lambda(\chi_-^\rho\phi^\rho - \chi_-^\lambda\phi^\lambda) + \psi_{22}^\rho(\chi_-^\rho\phi^\lambda + \chi_-^\lambda\phi^\rho)]$$

$$- \frac{1}{2\sqrt{5}}[\psi_{21}^\lambda(\chi_+^\rho\phi^\rho - \chi_+^\lambda\phi^\lambda) + \psi_{21}^\rho(\chi_+^\rho\phi^\lambda + \chi_+^\lambda\phi^\rho)].$$

The orbital states $\psi_{Lm_L}^\mu$ can be obtained from ψ_{L0}^μ of Table 10.2 by applying ladder operators $L_\pm$. By the same procedure, one can construct all SU(6) multiplet states of Table 10.1. In Table 10.4, we exhibit the detailed expression of the states of the $N = 0, 1$, and 2 bands for all allowed J values. For convenience, we chose $m_J = \frac{1}{2}$.

In a similar way, the states belonging to the $N = 3$ band can also be obtained.

Including the $N = 0, 1, 2$, and 3 bands, the number of calculated eigenvalues is larger than the number of observed resonances. Their identification with the calculated spectrum sometimes raises problems. For a proper identification of the resonances, the calculated spectrum must be complemented with a study of decay widths. An example of a spectrum for the lowest positive and negative parity states calculated within the model described in Section 10.1 and restricted to the $N = 0, 1$, and 2 bands is shown in Table 10.5.

A useful remark is that the nucleon ground state wave function is mainly a $(56, 0^+)$ multiplet, i.e. a positive parity state with total angular momentum $L = 0$ where the three quarks are located in the s-shell.

In fitting the spectrum, a key quantity is the Δ-N splitting which can be tuned through the parameters m and Λ of H^{hyp}. Most masses fall within or near the experimental interval. There are a few exceptions which are typical problems for a constituent quark model. The Roper resonance appears usually too high by about 100–150 MeV. A better description of the radial wave function brings some improvement (Stancu and Stassart 1990). The lowest masses in the $N\frac{1}{2}^-$ and $N\frac{3}{2}^-$ sectors are too low by about 50 MeV (see also Capstick and Isgur 1986). It has been shown by Høbogaasen and Richard (1983) that in a potential model based on the most general two-body interaction the first negative parity states ($N = 1$ band) must always appear below the Roper resonance ($N = 2$ band) which is at odds with experiment. Three-body interactions may change the situation. However, discrepancies of a few tens of MeV should not be regarded as a definite failure of constituent quark models because the coupling of baryons, as bound states, to baryon–meson channels could produce important shifts (Kumano 1990; Silvestre-Brac and Gignoux 1991). For a recent review of the non-relativistic quark models for baryons, the reader is referred to Richard (1992). In this article, the baryons containing quarks with unequal masses are also discussed.

Table 10.4 Three-quark wave functions including orbital $\psi^\mu_{Lm_L}$, spin χ_{Sm_S}, and flavour ϕ^μ parts for the $N = 0, 1$ and 2 bands with $m_J = \frac{1}{2}$ for all allowed J values. χ_{Sm_S} are defined in Section 4.11 and ϕ^μ are given in Tables 4.1 and 4.2. The notation of column 1 is explained in Example 10.1.

Notation	Wave function
$\|^4 N(70, 2^+) \frac{7}{2}^+ \frac{1}{2}\rangle$	$\sqrt{\frac{1}{70}}\left(\psi^\rho_{22}\phi^\rho + \psi^\lambda_{22}\phi^\lambda\right)\chi_{\frac{3}{2}-\frac{3}{2}} + \sqrt{\frac{6}{35}}\left(\psi^\rho_{21}\phi^\rho + \psi^\lambda_{21}\phi^\lambda\right)\chi_{\frac{3}{2}-\frac{1}{2}} + \sqrt{\frac{9}{35}}\left(\psi^\rho_{20}\phi^\rho + \psi^\lambda_{20}\phi^\lambda\right)\chi_{\frac{3}{2}\frac{1}{2}} + \sqrt{\frac{2}{35}}\left(\psi^\rho_{2-1}\phi^\rho + \psi^\lambda_{2-1}\phi^\lambda\right)\chi_{\frac{3}{2}\frac{3}{2}}$
$\|^4 \Delta(56, 2^+) \frac{7}{2}^+ \frac{1}{2}\rangle$	$\left(\sqrt{\frac{1}{35}}\psi^S_{22}\chi_{\frac{3}{2}-\frac{3}{2}} + \sqrt{\frac{12}{35}}\psi^S_{21}\chi_{\frac{3}{2}-\frac{1}{2}} + \sqrt{\frac{18}{35}}\psi^S_{20}\chi_{\frac{3}{2}\frac{1}{2}} + \sqrt{\frac{4}{35}}\psi^S_{2-1}\chi_{\frac{3}{2}\frac{3}{2}}\right)\phi^S$
$\|^2 N(56, 2^+) \frac{5}{2}^+ \frac{1}{2}\rangle$	$\sqrt{\frac{2}{10}}\psi^S_{21}\left(\chi^\rho_-\phi^\rho + \chi^\lambda_-\phi^\lambda\right) + \sqrt{\frac{3}{10}}\psi^S_{20}\left(\chi^\rho_+\phi^\rho + \chi^\lambda_+\phi^\lambda\right)$
$\|^2 N(70, 2^+) \frac{5}{2}^+ \frac{1}{2}\rangle$	$\sqrt{\frac{2}{20}}\psi^\lambda_{21}\left(\chi^\rho_-\phi^\lambda - \chi^\lambda_-\phi^\rho\right) + \psi^\rho_{21}\left(\chi^\rho_-\phi^\rho + \chi^\lambda_-\phi^\lambda\right)\right] + \sqrt{\frac{3}{20}}\left[\psi^\lambda_{20}\left(\chi^\rho_+\phi^\lambda - \chi^\lambda_+\phi^\rho\right) + \psi^\rho_{20}\left(\chi^\rho_+\phi^\rho + \chi^\lambda_+\phi^\lambda\right)\right]$
$\|^4 N(70, 2^+) \frac{5}{2}^+ \frac{1}{2}\rangle$	$\sqrt{\frac{12}{140}}\left(\psi^\rho_{22}\phi^\rho + \psi^\lambda_{22}\phi^\lambda\right)\chi_{\frac{3}{2}-\frac{3}{2}} + \sqrt{\frac{25}{140}}\left(\psi^\rho_{21}\phi^\rho + \psi^\lambda_{21}\phi^\lambda\right)\chi_{\frac{3}{2}-\frac{1}{2}} - \sqrt{\frac{6}{140}}\left(\psi^\rho_{20}\phi^\rho + \psi^\lambda_{20}\phi^\lambda\right)\chi_{\frac{3}{2}\frac{1}{2}} - \sqrt{\frac{27}{140}}\left(\psi^\rho_{2-1}\phi^\rho + \psi^\lambda_{2-1}\phi^\lambda\right)\chi_{\frac{3}{2}\frac{3}{2}}$
$\|^4 \Delta(56, 2^+) \frac{5}{2}^+ \frac{1}{2}\rangle$	$\left(\sqrt{\frac{12}{70}}\psi^S_{22}\chi_{\frac{3}{2}-\frac{3}{2}} + \sqrt{\frac{25}{70}}\psi^S_{21}\chi_{\frac{3}{2}-\frac{1}{2}} - \sqrt{\frac{6}{70}}\psi^S_{20}\chi_{\frac{3}{2}\frac{1}{2}} - \sqrt{\frac{27}{70}}\psi^S_{2-1}\chi_{\frac{3}{2}\frac{3}{2}}\right)\phi^S$
$\|^2 \Delta(70, 2^+) \frac{5}{2}^+ \frac{1}{2}\rangle$	$\left[\sqrt{\frac{2}{10}}\left(\psi^\rho_{21}\chi^\rho_- + \psi^\lambda_{21}\chi^\lambda_-\right) + \sqrt{\frac{3}{10}}\left(\psi^\rho_{20}\chi^\rho_+ + \psi^\lambda_{20}\chi^\lambda_+\right)\right]\phi^S$
$\|^4 N(70, 0^+) \frac{3}{2}^+ \frac{1}{2}\rangle$	$\sqrt{\frac{1}{2}}\left(\psi^\rho_{00}\phi^\rho + \psi^\lambda_{00}\phi^\lambda\right)\chi_{\frac{3}{2}\frac{1}{2}}$
$\|^2 N(56, 2^+) \frac{3}{2}^+ \frac{1}{2}\rangle$	$\sqrt{\frac{3}{10}}\psi^S_{21}\left(\chi^\rho_-\phi^\rho + \chi^\lambda_-\phi^\lambda\right) - \sqrt{\frac{2}{10}}\psi^S_{20}\left(\chi^\rho_+\phi^\rho + \chi^\lambda_+\phi^\lambda\right)$

Table 10.4 *cont.*

Notation	Wave function
$\|^2N(70, 2^+), \frac{3}{2}^+ \frac{1}{2}\rangle$	$\sqrt{\frac{3}{20}}\left[\psi_{21}^\lambda\left(\chi_-^\rho\phi^\rho - \chi_-^\lambda\phi^\lambda\right) + \psi_{21}^\rho\left(\chi_-^\rho\phi^\lambda + \chi_-^\lambda\phi^\rho\right)\right] - \sqrt{\frac{2}{20}}\left[\psi_{20}^\lambda\left(\chi_+^\rho\phi^\rho - \chi_+^\lambda\phi^\lambda\right) + \psi_{20}^\rho\left(\chi_+^\rho\phi^\lambda + \chi_+^\lambda\phi^\rho\right)\right]$
$\|^4N(70, 2^+), \frac{3}{2}^+ \frac{1}{2}\rangle$	$\sqrt{\frac{2}{10}}\left(\psi_{22}^\lambda\phi^\lambda + \psi_{22}^\rho\phi^\rho\right)\chi_{\frac{3}{2}-\frac{1}{2}} - \sqrt{\frac{1}{10}}\left(\psi_{20}^\lambda\phi^\lambda + \psi_{20}^\rho\phi^\rho\right)\chi_{\frac{3}{2}\frac{1}{2}} + \sqrt{\frac{2}{10}}\left(\psi_{2-1}^\lambda\phi^\rho + \psi_{2-1}^\lambda\phi^\lambda\right)\chi_{\frac{3}{2}\frac{3}{2}}$
$\|^2N(20, 1^+), \frac{3}{2}^+ \frac{1}{2}\rangle$	$\sqrt{\frac{1}{6}}\psi_{11}^A\left(\chi_-^\rho\phi^\lambda - \chi_-^\lambda\phi^\rho\right) + \sqrt{\frac{2}{6}}\psi_{10}^A\left(\chi_+^\rho\phi^\lambda - \chi_+^\lambda\phi^\rho\right)$
$\|^4\Delta(56, 0^+), \frac{3}{2}^+ \frac{1}{2}\rangle$	$\psi_{00}^S\chi_{\frac{3}{2}\frac{1}{2}}\phi^S$
$\|^4\Delta(56', 0^+), \frac{3}{2}^+ \frac{1}{2}\rangle$	$\psi_{00}^{S'}\chi_{\frac{3}{2}\frac{1}{2}}\phi^S$
$\|^4\Delta(56, 2^+), \frac{3}{2}^+ \frac{1}{2}\rangle$	$\left(\sqrt{\frac{2}{5}}\psi_{21}^S\chi_{\frac{3}{2}-\frac{1}{2}} - \sqrt{\frac{1}{5}}\psi_{20}^S\chi_{\frac{3}{2}\frac{1}{2}} + \sqrt{\frac{2}{5}}\psi_{2-1}^S\chi_{\frac{3}{2}\frac{3}{2}}\right)\phi^S$
$\|^2\Delta(70, 2^+), \frac{3}{2}^+ \frac{1}{2}\rangle$	$\left[\sqrt{\frac{3}{10}}\left(\psi_{21}^\rho\chi_-^\rho + \psi_{21}^\lambda\chi_-^\lambda\right) - \sqrt{\frac{2}{10}}\left(\psi_{20}^\rho\chi_+^\rho + \psi_{20}^\lambda\chi_+^\lambda\right)\right]\phi^S$
$\|^2N(56, 0^+), \frac{1}{2}^+ \frac{1}{2}\rangle$	$\sqrt{\frac{1}{2}}\psi_{00}^S\left(\chi_+^\rho\phi^\rho + \chi_+^\lambda\phi^\lambda\right)$
$\|^2N(56', 0^+), \frac{1}{2}^+ \frac{1}{2}\rangle$	$\sqrt{\frac{1}{2}}\psi_{00}^{S'}\left(\chi_+^\rho\phi^\rho + \chi_+^\lambda\phi^\lambda\right)$

Table 10.4 *cont.*

Notation	Wave function
$\|^2N(70,0^+)\tfrac{1}{2}^+\tfrac{1}{2}\rangle$	$\tfrac{1}{2}\left[\psi_{00}^\lambda\left(\chi_+^\rho\phi^\rho-\chi_+^\lambda\phi^\lambda\right)+\psi_{00}^\rho\left(\chi_+^\lambda\phi^\rho+\chi_+^\rho\phi^\lambda\right)\right]$
$\|^4N(70,2^+)\tfrac{1}{2}^+\tfrac{1}{2}\rangle$	$\sqrt{\tfrac{4}{20}}\left(\psi_{22}^\rho\phi^\rho+\psi_{22}^\lambda\phi^\lambda\right)\chi_{\tfrac{3}{2}-\tfrac{3}{2}}-\sqrt{\tfrac{3}{20}}\left(\psi_{21}^\rho\phi^\rho+\psi_{21}^\lambda\phi^\lambda\right)\chi_{\tfrac{3}{2}-\tfrac{1}{2}}+\sqrt{\tfrac{2}{20}}\left(\psi_{20}^\rho\phi^\rho+\psi_{20}^\lambda\phi^\lambda\right)\chi_{\tfrac{3}{2}\tfrac{1}{2}}-\sqrt{\tfrac{1}{20}}\left(\psi_{2-1}^\rho\phi^\rho+\psi_{2-1}^\lambda\phi^\lambda\right)\chi_{\tfrac{3}{2}\tfrac{3}{2}}$
$\|^2N(20,1^+)\tfrac{1}{2}^+\tfrac{1}{2}\rangle$	$\sqrt{\tfrac{2}{6}}\psi_{11}^A\left(\chi_-^\rho\phi^\lambda-\chi_-^\lambda\phi^\rho\right)-\sqrt{\tfrac{1}{6}}\psi_{10}^A\left(\chi_+^\rho\phi^\lambda-\chi_+^\lambda\phi^\rho\right)$
$\|^2\Delta(70,0^+)\tfrac{1}{2}^+\tfrac{1}{2}\rangle$	$\sqrt{\tfrac{1}{2}}\left(\psi_{00}^\rho\chi_+^\rho+\psi_{00}^\lambda\chi_+^\lambda\right)\phi^S$
$\|^4\Delta(56,2^+)\tfrac{1}{2}^+\tfrac{1}{2}\rangle$	$\left(\sqrt{\tfrac{4}{10}}\psi_{22}^S\chi_{\tfrac{3}{2}-\tfrac{3}{2}}-\sqrt{\tfrac{3}{10}}\psi_{21}^S\chi_{\tfrac{3}{2}-\tfrac{1}{2}}+\sqrt{\tfrac{2}{10}}\psi_{20}^S\chi_{\tfrac{3}{2}\tfrac{1}{2}}-\sqrt{\tfrac{1}{10}}\psi_{2-1}^S\chi_{\tfrac{3}{2}\tfrac{3}{2}}\right)\phi^S$
$\|^4N(70,1^-)\tfrac{5}{2}^-\tfrac{1}{2}\rangle$	$\sqrt{\tfrac{3}{20}}\left(\psi_{11}^\rho\phi^\rho+\psi_{11}^\lambda\phi^\lambda\right)\chi_{\tfrac{3}{2}-\tfrac{1}{2}}+\sqrt{\tfrac{6}{20}}\left(\psi_{10}^\rho\phi^\rho+\psi_{10}^\lambda\phi^\lambda\right)\chi_{\tfrac{3}{2}\tfrac{1}{2}}+\sqrt{\tfrac{1}{20}}\left(\psi_{1-1}^\rho\phi^\rho+\psi_{1-1}^\lambda\phi^\lambda\right)\chi_{\tfrac{3}{2}\tfrac{3}{2}}$
$\|^2N(70,1^-)\tfrac{3}{2}^-\tfrac{1}{2}\rangle$	$\sqrt{\tfrac{1}{12}}\left[\psi_{11}^\lambda\left(\chi_-^\rho\phi^\rho-\chi_-^\lambda\phi^\lambda\right)+\psi_{11}^\rho\left(\chi_-^\lambda\phi^\rho+\chi_-^\rho\phi^\lambda\right)\right]+\sqrt{\tfrac{2}{12}}\left[\psi_{10}^\lambda\left(\chi_+^\lambda\phi^\rho-\chi_+^\lambda\phi^\lambda\right)+\psi_{10}^\rho\left(\chi_+^\rho\phi^\lambda+\chi_+^\lambda\phi^\rho\right)\right]$
$\|^4N(70,1^-)\tfrac{3}{2}^-\tfrac{1}{2}\rangle$	$\sqrt{\tfrac{8}{30}}\left(\psi_{11}^\rho\phi^\rho+\psi_{11}^\lambda\phi^\lambda\right)\chi_{\tfrac{3}{2}-\tfrac{1}{2}}-\sqrt{\tfrac{1}{30}}\left(\psi_{10}^\rho\phi^\rho+\psi_{10}^\lambda\phi^\lambda\right)\chi_{\tfrac{3}{2}\tfrac{1}{2}}-\sqrt{\tfrac{6}{30}}\left(\psi_{1-1}^\rho\phi^\rho+\psi_{1-1}^\lambda\phi^\lambda\right)\chi_{\tfrac{3}{2}\tfrac{3}{2}}$

Table 10.4 *cont.*

Notation	Wave function
$\|^2\Delta(70, 1^-)\frac{3}{2}^-\frac{1}{2}\rangle$	$\left[\sqrt{\frac{1}{6}}\left(\psi_{11}^\rho \chi_-^\rho + \psi_{11}^\lambda \chi_-^\lambda\right) + \sqrt{\frac{2}{6}}\left(\psi_{10}^\rho \chi_+^\rho + \psi_{10}^\lambda \chi_+^\lambda\right)\right]\phi^s$
$\|^2N(70, 1^-)\frac{1}{2}^-\frac{1}{2}\rangle$	$\sqrt{\frac{2}{12}}\left[\psi_{11}^\lambda\left(\chi_-^\rho \phi^\rho - \chi_-^\lambda \phi^\lambda\right) + \psi_{11}^\rho\left(\chi_-^\rho \phi^\lambda + \chi_-^\lambda \phi^\rho\right)\right] - \sqrt{\frac{1}{12}}\left[\psi_{10}^\lambda\left(\chi_+^\rho \phi^\rho - \chi_+^\lambda \phi^\lambda\right) + \psi_{10}^\rho\left(\chi_+^\rho \phi^\lambda + \chi_+^\lambda \phi^\rho\right)\right]$
$\|^4N(70, 1^-)\frac{1}{2}^-\frac{1}{2}\rangle$	$\sqrt{\frac{1}{12}}\left(\psi_{11}^\rho \phi^\rho + \psi_{11}^\lambda \phi^\lambda\right)\chi_{\frac{3}{2}-\frac{1}{2}} - \sqrt{\frac{2}{12}}\left(\psi_{10}^\rho \phi^\rho + \psi_{10}^\lambda \phi^\lambda\right)\chi_{\frac{3}{2}\frac{1}{2}} + \sqrt{\frac{3}{12}}\left(\psi_{1-1}^\rho \phi^\rho + \psi_{1-1}^\lambda \phi^\lambda\right)\chi_{\frac{3}{2}\frac{3}{2}}$
$\|^2\Delta(70, 1^-)\frac{1}{2}^-\frac{1}{2}\rangle$	$\left[\sqrt{\frac{2}{6}}\left(\psi_{11}^\rho \chi_-^\rho + \psi_{11}^\lambda \chi_-^\lambda\right) - \sqrt{\frac{1}{6}}\left(\psi_{10}^\rho \chi_+^\rho + \psi_{10}^\lambda \chi_+^\lambda\right)\right]\phi^s$

THE BARYONS

Table 10.5 (a) Non-strange baryons of positive parity: results obtained in a semi-relativistic model with linear confinement and the hyperfine interaction (10.10)–(10.13) with $m = 324$ MeV and $\varepsilon = 5.5$ fm^{-1} (from Sartor and Stancu 1986).

State	Mass (MeV)	Mixing angles				
$^4N(70, 2^+)\frac{7^+}{2}$	1980	(1)				
$^4\Delta(56, 2^+)\frac{7^+}{2}$	1952	(1)				
$^2N(56, 2^+)\frac{5^+}{2}$	1754	0.833	−0.553	1.7×10^{-4}		
$^2N(70, 2^+)\frac{5^+}{2}$	1970	−0.544	−0.821	−0.174		
$^4N(70, 2^+)\frac{5^+}{2}$	2033	−0.096	−0.145	0.985		
$^4\Delta(56, 2^+)\frac{5^+}{2}$	1962	0.408	0.913			
$^2\Delta(70, 2^+)\frac{5^+}{2}$	1985	0.913	−0.408			
$^4N(70, 0^+)\frac{3^+}{2}$	1752	0.098	−0.824	0.558	−0.013	-1.8×10^{-3}
$^2N(56, 2^+)\frac{3^+}{2}$	1914	−0.760	−0.298	−0.296	0.469	0.164
$^2N(70, 2^+)\frac{3^+}{2}$	1979	0.614	−0.309	−0.556	0.332	0.329
$^4N(70, 2^+)\frac{3^+}{2}$	1985	−0.082	−0.359	−0.530	−0.491	−0.585
$^2N(20, 1^+)\frac{3^+}{2}$	2046	−0.173	−0.084	−0.108	−0.655	0.723
$^4\Delta(56, 0^+)\frac{3^+}{2}$	1285	0.977	−0.185	−0.088	0.058	
$^4\Delta(56', 0^+)\frac{3^+}{2}$	1904	0.126	0.902	−0.319	0.261	
$^4\Delta(56, 2^+)\frac{3^+}{2}$	1964	0.169	0.388	0.643	−0.638	
$^2\Delta(70, 2^+)\frac{3^+}{2}$	1979	−0.025	−0.031	−0.691	−0.722	
$^2N(56, 0^+)\frac{1^+}{2}$	941	0.969	0.174	−0.172	−0.034	1.6×10^{-3}
$^2N(56', 0^+)\frac{1^+}{2}$	1607	0.155	−0.980	−0.124	−0.014	-1.7×10^{-3}
$^2N(70, 0^+)\frac{1^+}{2}$	1795	0.172	−0.095	0.934	−0.290	0.066

Table 10.5(a) *cont.*

State	Mass (MeV)	Mixing angles				
$^4N(70,2^+)\frac{1^+}{2}$	1930	0.080	-9.7×10^{-3}	0.274	0.825	-0.488
$^2N(20,1^+)\frac{1^+}{2}$	2042	-0.030	4.7×10^{-4}	-0.083	-0.484	-0.870
$^2\Delta(70,0^+)\frac{1^+}{2}$	1910	0.908	0.419			
$^4\Delta(56,2^+)\frac{1^+}{2}$	1935	-0.419	0.908			

Table 10.5(b) Same as Table 10.5 (a), but for non-strange baryons of negative parity.

State	Mass (MeV)	Mixing angles	
$^4N(70,1^-)\frac{5^-}{2}$	1653	(1)	
$^2N(70,1^-)\frac{3^-}{2}$	1496	0.997	0.079
$^4N(70,1^-)\frac{3^-}{2}$	1714	-0.079	0.997
$^2\Delta(70,1^-)\frac{3^-}{2}$	1631	(1)	
$^2N(70,1^-)\frac{1^-}{2}$	1475	0.923	-0.384
$^4N(70,1^-)\frac{1^-}{2}$	1627	0.384	0.923
$^2\Delta(70,1^-)\frac{1^-}{2}$	1631	(1)	

THE BARYONS

An alternative approach to hadron structure is to use the algebraic methods of the kind described in Section 8.7 which lead to mass formulae. There they were related to the spin–flavour part. Recently, such methods have been extended by Bijker, Iachello, and Leviatan (1994) to include the space (orbital) part where the complete algebraic structure is

$$G = U(7) \times SU_F(3) \times SU_S(2) \times SU_C(3).$$

In this method, the three constituent quarks move in a correlated way giving rise to collective states of vibrational and rotational type. Then, the Roper resonance is described as a symmetric stretching vibration and is correctly located in the spectrum. However, in such a description there are problems with the rotational spectrum which does not reproduce a characteristic feature of hadronic spectra, namely the occurrence of linear Regge trajectories.

Exercise 10.2 Show that the colour antisymmetric three-particle state is a colour singlet for SU(3).

Solution An antisymmetric normalized function of colour quantum numbers a, b, and c is (see Chapter 4)

$$\psi_A = \frac{1}{\sqrt{6}} \sum_P \delta_P P \varphi_a(1) \varphi_b(2) \varphi_c(3). \tag{1}$$

The action of an SU(3) transformation U gives

$$U \psi_A = \psi'_A = \frac{1}{\sqrt{6}} \sum_{a'b'c'} u_{a'a} u_{b'b} u_{c'c} \sum_P \delta_P P \varphi_{a'}(1) \varphi_{b'}(2) \varphi_{c'}(3). \tag{2}$$

As the sum over P is a Slater determinant, one can return to the initial order abc by exchanging its rows (or columns). This produces a Levi–Civita factor $\varepsilon_{a'b'c'}$. Thus,

$$\begin{aligned} U \psi_A &= \frac{1}{\sqrt{6}} \sum_{a'b'c'} u_{a'a} u_{b'b} u_{c'c} \varepsilon_{a'b'c'} \sum_P \delta_P \varphi_a(1) \varphi_b(2) \varphi_c(3) \\ &= \det u \, \frac{1}{\sqrt{6}} \sum_P \delta_P \varphi_a(1) \varphi_b(2) \varphi_c(3) = \psi_A \end{aligned} \tag{3}$$

because $\det u = 1$. This shows that ψ_A remains unchanged under an SU(3) transformation, or in other words, it is a singlet.

The proof can easily be generalized to SU(N) to show that an antisymmetric N-particle wave function is an SU(N) singlet. Note that if the number of particles n is larger than the dimension N of the unitary unimodular transformation, one can still construct singlets, but they are no longer antisymmetric states. In particular for SU(3) the necessary condition is that $n/3$ must be an integer which represents the number of three-box columns in a singlet representation. Thus, SU(3) singlets are

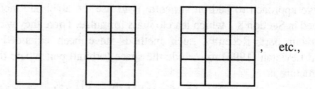, etc.,

but only the first one corresponds to a totally antisymmetric wave function.

10.3 DIQUONIA

As illustrated by Exercise 10.2, it is possible to construct colour singlet states from more than three quarks. These states would contain two or more hadrons, either free or loosely bound. Their existence is not forbidden by colour confinement. Multiquark hadron systems with more than three quarks located in a finite volume, within which colour fields are confined, form a class of so-called exotics. The simplest possible candidate for a multiquark hadron is the $2q$-$2\bar{q}$ system, also called diquonium. Its properties were first studied by Jaffe (1977a,b, 1978) with the MIT bag model including a hyperfine interaction. He found that the scalar $J^{PC} = 0^{++}$ states have lowest mass and interpreted the $I = 0, f_0(975)$ and the $I = 1, a_0(980)$ resonances as such $2q$-$2\bar{q}$ states. Jaffe's results were confirmed and improved in a non-relativistic quark model also including a hyperfine interaction by Weinstein and Isgur (1983, 1990). If $f_0(975)$ or $a_0(980)$ resonances were $J^{PC} = 0^{++}$ $q\bar{q}$ systems, they would have carried an $L = 1$ excitation as shown in Section 8.6 and this would have raised their masses to higher values than is found experimentally. They have several other properties which makes people think that they are not $q\bar{q}$ systems, but $K\bar{K}$ molecules. A list of these properties can be found, for example, in Weinstein and Isgur (1990).

The number of candidates to be interpreted as meson molecules is increasing. For example, recent proposals were made to interpret the $f_0(1720)$ resonance with $I, J^{PC} = 0, 0^{++}$ as a vector meson molecule $K^*\bar{K}^*$ or $K^*\bar{K}^* + \omega\phi$. These interpretations are somewhat controversial and more theoretical and experimental work is required (For a review, see, for example Karl, 1992(b)).

The interpretation of these resonances as hadronic molecules raises the theoretical question of whether $2q$-$2\bar{q}$ systems can form bound or quasi-bound states. The main focus of this subsection is on the symmetry properties of the $2q$-$2\bar{q}$ states in relation to their spin and colour structure. To illustrate the problem, let us consider a very simple Hamiltonian

$$H = \sum_{i=1}^{4}\left(m_i + \frac{p_i^2}{2m_i}\right) + \sum_{i<j}^{4}\left(V_{ij}^{\text{conf}} + V_{ij}^{\text{hyp}}\right) \tag{10.16}$$

where V_{ij}^{conf} has the form (10.8) with the last term dropped ($A = 0$) and V_{ij}^{hyp} is given by (10.11) with V_{ij}^{T} also dropped. We also choose a set of parameters among those proposed by Weinstein and Isgur (1983):

$$m = m_u = m_d = 330 \text{ MeV}$$
$$k = 339 \text{ MeV fm}^{-2}$$
$$e_0 = -352 \text{ MeV} \qquad (10.16a)$$
$$\alpha_S = 2.7$$
$$\varepsilon = 2 \text{ fm}^{-1}.$$

Here, the discussion is restricted to equal-mass quarks, u and d. The resulting Schrödinger equation cannot be solved exactly. One can apply the Rayleigh–Ritz variational principle to search for an approximate solution. The procedure is based on a trial wave function expanded in a chosen basis. This generates a matrix eigenvalue equation of the form (10.53). The variational parameters contained in the basis functions are determined by minimizing the ground state solution E_0. To understand the content of Equation (10.53), let us discuss the basis functions, in which one must incorporate the orbital, spin, flavour, and colour degrees of freedom. First, suppose that particles 1 and 2 are quarks and 3 and 4 are antiquarks and choose a state of total isospin $I = 0$. The appropriate flavour function is

$$\phi = \frac{1}{2}(ud - du)(\bar{u}\bar{d} - \bar{d}\bar{u}). \qquad (10.17)$$

This function is antisymmetric under the permutation of quarks or antiquarks. The total wave function must be antisymmetric under permutations (12) and (34). Then, the remaining product of orbital, spin, and colour space parts must be symmetric under permutations (12) and (34). The discussion can be easily generalized to $I \neq 0$.

Orbital wave functions

One can introduce three alternative coordinate sets depicted in Fig. 10.2:

$$\boldsymbol{\sigma} = \frac{1}{\sqrt{2}}(r_1 - r_2), \quad \boldsymbol{\sigma}' = \frac{1}{\sqrt{2}}(r_3 - r_4), \quad \boldsymbol{\lambda} = \frac{1}{2}(r_1 + r_2 - r_3 - r_4) \quad (10.18)$$

$$\boldsymbol{\rho} = \frac{1}{\sqrt{2}}(r_1 - r_3), \quad \boldsymbol{\rho}' = \frac{1}{\sqrt{2}}(r_2 - r_4), \quad \boldsymbol{x} = \frac{1}{2}(r_1 + r_3 - r_2 - r_4) \quad (10.19)$$

$$\boldsymbol{\alpha} = \frac{1}{\sqrt{2}}(r_1 - r_4), \quad \boldsymbol{\alpha}' = \frac{1}{\sqrt{2}}(r_2 - r_3), \quad \boldsymbol{y} = \frac{1}{2}(r_1 + r_4 - r_2 - r_3) \quad (10.20)$$

with the following permutation properties

$$(12)\boldsymbol{\rho} = \boldsymbol{\alpha}', \quad (12)\boldsymbol{\rho}' = \boldsymbol{\alpha}, \quad (34)\boldsymbol{\rho} = \boldsymbol{\alpha}, \quad (34)\boldsymbol{\rho}' = \boldsymbol{\alpha}'$$

$$(12)\boldsymbol{x} = -\boldsymbol{y}, \quad (34)\boldsymbol{x} = \boldsymbol{y}, \quad (23)\boldsymbol{x} = \boldsymbol{\lambda}, \quad (14)\boldsymbol{x} = -\boldsymbol{\lambda} \quad (10.21)$$

$$(12)\boldsymbol{\lambda} = \boldsymbol{\lambda}, \quad (34)\boldsymbol{\lambda} = \boldsymbol{\lambda}, \quad (23)\boldsymbol{y} = \boldsymbol{y}, \quad (14)\boldsymbol{y} = \boldsymbol{y}.$$

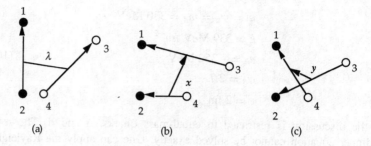

Figure 10.2 Three possible ways to define relative coordinates for a $2q$-$2\bar{q}$ system. Darkened and open circles represent quarks and antiquarks, respectively.

The orbital wave function can be expressed in terms of one of these coordinate sets. The most general orbital wave function with $L = 0$ is a function of six scalar quantities. For the asymptotic meson–meson channel with particles 1 and 3 in one meson and 2 and 4 in the other, the six variables are $\rho^2, \rho'^2, x^2, \boldsymbol{\rho} \cdot \boldsymbol{\rho}', \boldsymbol{\rho} \cdot \boldsymbol{x}$ and $\boldsymbol{\rho}' \cdot \boldsymbol{x}$ and we write

$$R(x) = R(\rho^2, \rho'^2, x^2, \boldsymbol{\rho} \cdot \boldsymbol{\rho}', \boldsymbol{\rho} \cdot \boldsymbol{x}, \boldsymbol{\rho}' \cdot \boldsymbol{x}). \tag{10.22}$$

In the exchange channel, with particles 1 and 4 in one meson and 2 and 3 in the other, one has

$$R(y) = R(\alpha^2, \alpha'^2, y^2, \boldsymbol{\alpha} \cdot \boldsymbol{\alpha}', \boldsymbol{\alpha} \cdot \boldsymbol{y}, \boldsymbol{\alpha}' \cdot \boldsymbol{y}). \tag{10.23}$$

Under permutations (12) and (34), one has

$$R(y) = (12)R(x) = (34)R(x). \tag{10.24}$$

Spin wave functions

The situation is analogous in the spin space. Let us consider $S = 0$ states. There are three orthonormal basis sets denoted by

$$|A_{12}A_{34}>, \qquad |S_{12}S_{34}> \tag{10.25}$$
$$|V_{13}V_{24}>, \qquad |P_{13}P_{24}> \tag{10.26}$$
$$|V_{14}V_{23}>, \qquad |P_{14}P_{23}> \tag{10.27}$$

which can be associated to the orbital function written in the coordinates (10.18), (10.19), and (10.20), respectively. The basis vectors (10.25) can be defined through their corresponding Young tableaux:

$$|A_{12}A_{34}> = \begin{array}{|c|c|} \hline 1 & 2 \\ \hline 3 & 4 \\ \hline \end{array} \qquad |S_{12}S_{34}> = \begin{array}{|c|c|} \hline 1 & 3 \\ \hline 2 & 4 \\ \hline \end{array} \tag{10.28}$$

As explained in Section 4.3, these vectors form the Young–Yamanouchi basis of the two-dimensional representation [22] of S_4 and have the following permutation properties

$$(12)|A_{12}A_{34}> = (34)|A_{12}A_{34}> = |A_{12}A_{34}>$$
$$(12)|S_{12}S_{34}> = (34)|S_{12}S_{34}> = -|S_{12}S_{34}>.$$

The explicit form of (10.28) in terms of quark spin states $\uparrow$ ($s=\frac{1}{2}, s_z=\frac{1}{2}$) and $\downarrow$ ($s=\frac{1}{2}, s_z=-\frac{1}{2}$) can be obtained by taking $\alpha=\uparrow$ and $\beta=\downarrow$ in the orthonormalized basis vectors (4.71) and (4.72). One gets

$$|A_{12}A_{34}> = \sqrt{\frac{1}{12}}(2\uparrow\uparrow\downarrow\downarrow + 2\downarrow\downarrow\uparrow\uparrow - \uparrow\downarrow\uparrow\downarrow - \downarrow\uparrow\uparrow\downarrow - \uparrow\downarrow\downarrow\uparrow - \downarrow\uparrow\downarrow\uparrow)$$
(10.29)

$$|S_{12}S_{34}> = \frac{1}{2}(\uparrow\downarrow\uparrow\downarrow + \downarrow\uparrow\downarrow\uparrow - \uparrow\downarrow\downarrow\uparrow - \downarrow\uparrow\uparrow\downarrow).$$
(10.30)

The basis vectors (10.26) correspond to asymptotic pseudoscalar–pseudoscalar (PP) and vector–vector (VV) direct channels, and (10.27) correspond to exchange channels. On the one hand, they can be obtained by first introducing two-body spin states from which one can build $S=0$ four-particle states. For the PP channels, the two-body states have spin 0 and for VV channels spin 1. On the other hand, they result from the action of various transpositions on the Young–Yammanouchi basis (10.28) as follows

$$|V_{13}V_{24}> = (23)|A_{12}A_{34}> = (14)|A_{12}A_{34}>$$
$$= -\frac{1}{2}|A_{12}A_{34}> + \frac{\sqrt{3}}{2}|S_{12}S_{34}>$$
(10.31)

$$|P_{13}P_{24}> = (23)|S_{12}S_{34}> = (14)|S_{12}S_{34}> = \frac{\sqrt{3}}{2}|A_{12}A_{34}> + \frac{1}{2}|S_{12}S_{34}>$$
(10.32)

$$|V_{14}V_{23}> = (13)|A_{12}A_{34}> = (24)|A_{12}A_{34}>$$
$$= -\frac{1}{2}|A_{12}A_{34}> - \frac{\sqrt{3}}{2}|S_{12}S_{34}>$$
(10.33)

$$|P_{14}P_{23}> = -(13)|S_{12}S_{34}> = -(24)|S_{12}S_{34}>$$
$$= \frac{\sqrt{3}}{2}|A_{12}A_{34}> - \frac{1}{2}|S_{12}S_{34}>.$$
(10.34)

From the above, one can see that $|V_{13}V_{24}>$ and $|V_{14}V_{23}>$ are not orthogonal to each other. The same is valid for $|P_{13}P_{24}>$ and $|P_{14}P_{23}>$.

Colour wave functions

Again, by analogy with coordinate space, one can construct a colour singlet $2q$-$2\bar{q}$ state in three ways. Using the notation of Section 8.6, these are

$$|\bar{3}_{12}3_{34}>, \quad |6_{12}\bar{6}_{34}> \tag{10.35}$$

$$|1_{13}1_{24}>, \quad |8_{13}8_{24}> \tag{10.36}$$

$$|1_{14}1_{23}>, \quad |8_{14}8_{23}>. \tag{10.37}$$

Recall that particles 1 and 2 are quarks and 3 and 4 are antiquarks. The 3 and $\bar{3}$ states are antisymmetric under the transposition (ij). In the notation of Section 8.10, they can be written as

$$|\bar{3}_{ij}^{\alpha}> = \frac{1}{\sqrt{2}}\varepsilon^{\alpha\beta\gamma}q^{\beta}(i)q^{\gamma}(j) \tag{10.38}$$

$$|3_{\alpha,ij}> = \frac{1}{\sqrt{2}}\varepsilon_{\alpha\beta\gamma}\bar{q}_{\beta}(i)\bar{q}_{\gamma}(j). \tag{10.39}$$

The states 6 and $\bar{6}$ are symmetric under the same transposition, and can be defined as

$$|6_{ij}^{\alpha}> = d^{\alpha\beta\gamma}q^{\beta}(i)q^{\gamma}(j) \tag{10.40}$$

$$|\bar{6}_{\alpha,ij}> = d_{\alpha\beta\gamma}\bar{q}_{\beta}(i)\bar{q}_{\gamma}(j) \tag{10.41}$$

where the non-vanishing $d^{\alpha\beta\gamma}$ and $d_{\alpha\beta\gamma}$ constants are

$$d^{111} = d_{111} = d^{222} = d_{222} = d^{333} = d_{333} = 1$$
$$d^{412} = d_{412} = d^{421} = d_{421} = d^{523} = d_{523} = d^{532} = d_{532} \tag{10.42}$$
$$= d^{613} = d_{613} = d^{631} = d_{631} = \frac{1}{\sqrt{2}}.$$

The definitions (10.38)–(10.41) show that $|\bar{3}_{12}3_{34}>$ and $|6_{12}\bar{6}_{34}>$ are orthogonal.

The bases (10.36) and (10.37) contain a singlet–singlet and an octet–octet colour state, which can be expressed in terms of the basis vectors (10.35). Using the tensor method (Section 8.10), one can show that

$$|1_{13}1_{24}> = \sqrt{\frac{1}{3}}|\bar{3}_{12}3_{34}> + \sqrt{\frac{2}{3}}|6_{12}\bar{6}_{34}> \tag{10.43}$$

$$|8_{13}8_{24}> = -\sqrt{\frac{2}{3}}|\bar{3}_{12}3_{34}> + \sqrt{\frac{1}{3}}|6_{12}\bar{6}_{34}> \tag{10.44}$$

for the direct channel and

$$|1_{14}1_{23}> = -\sqrt{\frac{1}{3}}|\bar{3}_{12}3_{34}> + \sqrt{\frac{2}{3}}|6_{12}\bar{6}_{34}> \tag{10.45}$$

$$|8_{14}8_{23}> = \sqrt{\frac{2}{3}}|\bar{3}_{12}3_{34}> + \sqrt{\frac{1}{3}}|6_{12}\bar{6}_{34}> \tag{10.46}$$

for the exchange channel (see Exercise 10.3).

Therefore, one has

$$<1_{13}1_{24}|1_{14}1_{23}> \neq 0, \quad <8_{13}8_{24}|8_{14}8_{23}> \neq 0. \tag{10.47}$$

Channel wave functions

Including orbital, spin, and colour degrees of freedom, one can now construct four channel wave functions with total spin $S = 0$:

$$\psi_{PP} = \frac{N_{PP}}{\sqrt{2}}[R_P(x)|1_{13}1_{24}> |P_{13}P_{24}> + R_P(y)|1_{14}1_{23}> |P_{14}P_{23}>] \quad (10.48)$$

$$\psi_{VV} = \frac{N_{VV}}{\sqrt{2}}[R_V(x)|1_{13}1_{24}> |V_{13}V_{24}> + R_V(y)|1_{14}1_{23}> |V_{14}V_{23}>] \quad (10.49)$$

$$\psi_{C_0 C_0} = \frac{N_{C_0 C_0}}{\sqrt{2}}[R_{C_0}(x)|8_{13}8_{24}> |P_{13}P_{24}> + R_{C_0}(y)|8_{14}8_{23}> |P_{14}P_{23}>] \quad (10.50)$$

$$\psi_{C_1 C_1} = \frac{N_{C_1 C_1}}{\sqrt{2}}[R_{C_1}(x)|8_{13}8_{24}> |V_{13}V_{24}> + R_{C_1}(y)|8_{14}8_{23}> |V_{14}V_{23}>]. \quad (10.51)$$

From the discussion above, one can see that they are symmetric under permutations (12) and (34) and the antisymmetry of the total wave function is established through the flavour part (10.17). Each wave function has a direct and an exchange part. The wave function ψ_{PP} corresponds to the pseudoscalar-meson–pseudoscalar-meson (PP) channel and ψ_{VV} to the vector-meson–vector-meson (VV) channel. The wave functions (10.50) and (10.51) represent closed (hidden) colour channels, because they are formed by coupling colour octet $q\bar{q}$ pairs to a singlet $2q$-$2\bar{q}$ state. Asymptotically, their energy rises to infinity. The notation is such that when the spin of each $q\bar{q}$ pair is zero the channel is denoted by $C_0 C_0$ and when it is one, the notation is $C_1 C_1$. The wave functions (10.48)–(10.51) are not orthogonal to each other as inferred from the above discussion. This raises the question as to whether or not they are linearly independent. It has been shown that (10.50) and (10.51) are linearly dependent on (10.48) and (10.49) if the orbital parts are expressed in terms of a complete set of functions (Brink and Stancu 1994). Otherwise, if the orbital wave functions are linear combinations of a finite set of functions, then (10.48)–(10.51) are linearly independent. This is usually the case in variational calculations where they all have to be taken into account.

The wave functions (10.48)–(10.51) show the general spin–colour structure of asymptotic channel wave functions of a $2q$-$2\bar{q}$ system of equal-mass quarks. They can be used as variational wave functions in the Rayleigh–Ritz method to find approximate eigenvalues and eigenstates for a given Hamiltonian. Some results with the Hamiltonian (10.16) are presented below.

To apply the variational method, the trial wave functions ψ_n must be expanded in the basis (10.48)–(10.51). The simplest choice of (10.22) is to take in all channels

$$R_i(x, b) = e^{-a^2 \rho^2} e^{-a^2 \rho'^2} e^{-b^2 x^2} \quad (i = P, V, C_0, C_1) \quad (10.52)$$

where the variational parameter a can be fixed by determining the pseudoscalar meson mass m_π from a Hamiltonian like (10.16) which describes a $q\bar{q}$ pair only. The value of m_π is obtained by minimizing the expectation value $<\phi_0|H|\phi_0>$ with respect to a where $\phi_0 \sim e^{-a^2\rho^2}$. The resulting matrix eigenvalue equation is

$$<H> C_n = E_n B C_n \qquad (10.53)$$

where $<H>$ is a 4×4 matrix obtained by integrating out the spin and colour parts of the wave functions (10.48)–(10.51), B is the overlap matrix, which includes the non-orthogonality of the basis vectors, and C_n is a column matrix containing the coefficients of the variational solution ψ_n.

To see if the $2q$-$2\bar{q}$ system binds, we need the ground state solution E_0 from which one must subtract $2m_\pi$. In Fig. 10.3, the quantity $E_0 - 2m_\pi$ is plotted as a function of the variational parameter b. From (10.52), one can see that $1/b$ plays the role of a deformation parameter. At small b, the two pairs are far apart and in the limit $b \to 0$, one gets $E_0 \to 2m_\pi$. At large b, the two $q\bar{q}$ pairs get close together and the interaction becomes repulsive. In the limit of zero separation, $r_{ij} = 0$, V_{ij}^{conf} reduces to the constant term $-e_0 \widehat{C}_i \cdot \widehat{C}_j$ which cancels out in $E_0 - 2m_\pi$. Therefore, the repulsion seen in both curves of Fig. 10.3 is due entirely to the kinetic energy. In one case, the spin–spin term has been suppressed and in that case, $E_0 - 2m_\pi$ increases steadily from zero to infinity. When the spin–spin is included, $E_0 - 2m_\pi$ gets an attractive pocket which proves that the spin–spin interaction is crucial in binding the system (for details, see Brink and Stancu 1994).

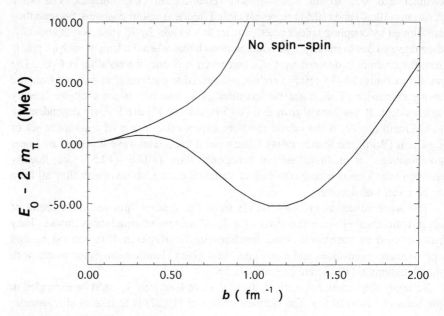

Figure 10.3 Interaction energy of a $2q$-$2\bar{q}$ system calculated with the trial wave function (10.52) in the four-channel basis (10.48)–(10.51). The parameter $a = 2fm^{-1}$ is fixed. The upper (lower) curve is calculated without (with) the hyperfine interaction represented by Equation (10.12). (From Brink and Stancu 1994).

DIQUONIA

Exercise 10.3 Prove the relations (10.43)–(10.46).

Solution Following the prescriptions of Section 8.10, one can construct from (10.38) and (10.39) a normalized singlet colour state (scalar) as follows

$$|\bar{3}_{12}3_{34}>= \frac{1}{\sqrt{12}} \varepsilon^{\alpha\beta\gamma}\varepsilon_{\alpha\lambda\sigma} q^\beta(1)q^\gamma(2)\bar{q}_\lambda(3)\bar{q}_\sigma(4). \qquad (1)$$

Its explicit form is

$$\begin{aligned}|\bar{3}_{12}3_{34}>= \frac{1}{\sqrt{12}}(&q^2q^3\bar{q}_2\bar{q}_3 - q^2q^3\bar{q}_3\bar{q}_2 - q^3q^2\bar{q}_2\bar{q}_3 + q^3q^2\bar{q}_3\bar{q}_2 \\ +& q^3q^1\bar{q}_3\bar{q}_1 - q^3q^1\bar{q}_1\bar{q}_3 - q^1q^3\bar{q}_3\bar{q}_1 + q^1q^3\bar{q}_1\bar{q}_3 \\ +& q^1q^2\bar{q}_1\bar{q}_2 - q^1q^2\bar{q}_2\bar{q}_1 - q^2q^1\bar{q}_1\bar{q}_2 + q^2q^1\bar{q}_2\bar{q}_1).\end{aligned} \qquad (2)$$

Here and below, the particle order 1, 2, 3, 4 is understood whenever it is not specified.

Next, let us write explicitly the basis vectors of the sextet representation 6_{ij} of (10.40):

$$\begin{aligned}6^1_{ij} &= q^1(i)q^1(j) \\ 6^2_{ij} &= q^2(i)q^2(j) \\ 6^3_{ij} &= q^3(i)q^3(j) \\ 6^4_{ij} &= \frac{1}{\sqrt{2}}[q^1(i)q^2(j) + q^2(i)q^1(j)] \\ 6^5_{ij} &= \frac{1}{\sqrt{2}}[q^2(i)q^3(j) + q^3(i)q^2(j)] \\ 6^6_{ij} &= \frac{1}{\sqrt{2}}[q^3(i)q^1(j) + q^1(i)q^3(j)].\end{aligned} \qquad (3)$$

Based on these components and their analogues for $\bar{6}$, one can construct a $6\bar{6}$ colour singlet as follows

$$\begin{aligned}|6_{12}\bar{6}_{34}> &= \frac{1}{\sqrt{6}}|6^\alpha_{12}>|\bar{6}^\alpha_{34}> \\ &= \frac{1}{\sqrt{6}} d^{\alpha\beta\gamma} d_{\alpha\lambda\sigma} q^\beta(1)q^\gamma(2)\bar{q}_\lambda(3)\bar{q}_\sigma(4)\end{aligned} \qquad (4)$$

Its detailed form is

$$\begin{aligned}|6_{12}\bar{6}_{34}>=\frac{1}{\sqrt{6}}[&q^1q^1\bar{q}_1\bar{q}_1 + q^2q^2\bar{q}_2\bar{q}_2 + q^3q^3\bar{q}_3\bar{q}_3 \\ &+\frac{1}{2}(q^1q^2\bar{q}_1\bar{q}_2 + q^2q^1\bar{q}_1\bar{q}_2 + q^2q^1\bar{q}_2\bar{q}_1 + q^1q^2\bar{q}_2\bar{q}_1) \\ &+\frac{1}{2}(q^2q^3\bar{q}_2\bar{q}_3 + q^3q^2\bar{q}_2\bar{q}_3 + q^3q^2\bar{q}_3\bar{q}_2 + q^2q^3\bar{q}_3\bar{q}_2) \\ &+\frac{1}{2}(q^3q^1\bar{q}_3\bar{q}_1 + q^1q^3\bar{q}_3\bar{q}_1 + q^1q^3\bar{q}_1\bar{q}_3 + q^3q^1\bar{q}_1\bar{q}_3)].\end{aligned} \qquad (5)$$

A colour singlet constructed from $q\bar{q}$ colour singlets is

$$|1_{13}1_{24}> = \frac{1}{3}q^i(1)\bar{q}_i(3)q^j(2)\bar{q}_j(4)$$
$$= \frac{1}{3}(q^1q^1\bar{q}_1\bar{q}_1 + q^2q^2\bar{q}_2\bar{q}_2 + q^3q^3\bar{q}_3\bar{q}_3 \qquad (6)$$
$$+ q^1q^2\bar{q}_1\bar{q}_2 + q^2q^1\bar{q}_2\bar{q}_1 + q^3q^1\bar{q}_3\bar{q}_1$$
$$+ q^1q^3\bar{q}_1\bar{q}_3 + q^2q^3\bar{q}_2\bar{q}_3 + q^3q^2\bar{q}_3\bar{q}_2).$$

Putting together (2), (5), and (6), one obtains the transformation (10.43), i.e.

$$|1_{13}1_{24}> = \sqrt{\frac{1}{3}}|\bar{3}_{12}3_{34}> + \sqrt{\frac{2}{3}}|6_{12}\bar{6}_{34}>. \qquad (7)$$

To construct a colour singlet from $q\bar{q}$ octets, one uses the traceless tensor (8.234a). A normalized singlet state formed by the quarks a, c, and the antiquarks b, d reads

$$|8_{ab}8_{cd}> = \frac{1}{2\sqrt{2}}\tilde{T}^i_j(a,b)\tilde{T}^j_i(c,d). \qquad (8)$$

In particular, one obtains

$$|8_{13}8_{24}> = \frac{1}{2\sqrt{2}}\left[\frac{2}{3}(q^1q^1\bar{q}_1\bar{q}_1 + q^2q^2\bar{q}_2\bar{q}_2 + q^3q^3\bar{q}_3\bar{q}_3)\right.$$
$$- \frac{1}{3}(q^1q^2\bar{q}_1\bar{q}_2 + q^2q^1\bar{q}_2\bar{q}_1 + q^2q^3\bar{q}_2\bar{q}_3 + q^3q^2\bar{q}_3\bar{q}_2 \qquad (9)$$
$$+ q^1q^3\bar{q}_1\bar{q}_3 + q^3q^1\bar{q}_3\bar{q}_1) + q^1q^2\bar{q}_2\bar{q}_1 + q^2q^1\bar{q}_1\bar{q}_2 + q^2q^3\bar{q}_3\bar{q}_2$$
$$\left. + q^3q^2\bar{q}_2\bar{q}_3 + q^1q^3\bar{q}_3\bar{q}_1 + q^3q^1\bar{q}_1\bar{q}_3 \right].$$

One can see that (2), (5), and (9) satisfy

$$|8_{13}8_{24}> = -\sqrt{\frac{2}{3}}|\bar{3}_{12}3_{34}> + \sqrt{\frac{1}{3}}|6_{12}\bar{6}_{34}>.$$

i.e. the transformation (10.44). In a similar way, one can prove the relations (10.45) and (10.46) useful for the exchange channel.

Exercise 10.4 Find the matrix elements of the colour operators $\widehat{C}_i \cdot \widehat{C}_j$ ($i < j = 1,\ldots,4$) in the basis ($|\bar{3}_{12}3_{34}>, |6_{12}\bar{6}_{34}>$).

Solution The problem is similar to that of Exercise 10.1, i.e. it is based on the eigenvalue (8.134) of the Casimir operator $\widehat{C}^2$ acting in the colour space.

For example, in the present context, the relation (6) of Exercise 10.1 leads to

$$<\bar{3}_{12}3_{34}|\widehat{C}_1 \cdot \widehat{C}_2|\bar{3}_{12}3_{34}> = <\bar{3}_{12}3_{34}|\widehat{C}_3 \cdot \widehat{C}_4|\bar{3}_{12}3_{34}> = -\frac{2}{3}. \qquad (1)$$

It is useful to introduce the two-body operator

$$\widehat{C}_{12} \cdot \widehat{C}_{34} = \frac{1}{2}\left(\widehat{C}^2 - \widehat{C}_{12}^2 - \widehat{C}_{34}^2\right) \tag{2}$$

where $\widehat{C}^2$ is associated to the whole $2q\text{-}2\bar{q}$ system and $\widehat{C}_{12}^2$ and $\widehat{C}_{34}^2$ to the quark and the antiquark pairs, respectively.

(a) The channel $|\bar{3}_{12}3_{34}>$.
The expectation value of (2) is

$$<\bar{3}_{12}3_{34}|\widehat{C}_{12} \cdot \widehat{C}_{34}|\bar{3}_{12}3_{34}> = \frac{1}{2}\left(0 - \frac{4}{3} - \frac{4}{3}\right) = -\frac{4}{3}. \tag{3}$$

The operator (2) can also be written as

$$\widehat{C}_{12} \cdot \widehat{C}_{34} = \left(\widehat{C}_1 + \widehat{C}_2\right) \cdot \left(\widehat{C}_3 + \widehat{C}_4\right) = \widehat{C}_1 \cdot \widehat{C}_3 + \widehat{C}_2 \cdot \widehat{C}_3 + \widehat{C}_1 \cdot \widehat{C}_4 + \widehat{C}_2 \cdot \widehat{C}_4. \tag{4}$$

We recall that 1 and 2 are quarks and 3 and 4 are antiquarks. For symmetry reasons, it follows that the operators on the right-hand side of (4) have equal expectation values, equal to a quarter of the value appearing on the right-hand side of (3). Therefore, one has

$$\begin{aligned}<\bar{3}_{12}3_{34}|\widehat{C}_1 \cdot \widehat{C}_3|\bar{3}_{12}3_{34}> &= <\bar{3}_{12}3_{34}|\widehat{C}_2 \cdot \widehat{C}_3|\bar{3}_{12}3_{34}> \\ &= <\bar{3}_{12}3_{34}|\widehat{C}_1 \cdot \widehat{C}_4|\bar{3}_{12}3_{34}> \\ &= <\bar{3}_{12}3_{34}|\widehat{C}_2 \cdot \widehat{C}_4|\bar{3}_{12}3_{34}> = -\frac{1}{3}.\end{aligned} \tag{5}$$

(b) The channel $|6_{12}\bar{6}_{34}>$.
In this channel, the relation (7) of Exercise 10.1 gives

$$<6_{12}\bar{6}_{34}|\widehat{C}_1 \cdot \widehat{C}_2|6_{12}\bar{6}_{34}> = <6_{12}\bar{6}_{34}|\widehat{C}_3 \cdot \widehat{C}_4|6_{12}\bar{6}_{34}> = \frac{1}{3}. \tag{6}$$

On the other hand, the expectation value of (2) is

$$<6_{12}\bar{6}_{34}|\widehat{C}_{12} \cdot \widehat{C}_{34}|6_{12}\bar{6}_{34}> = \frac{1}{2}\left(0 - \frac{10}{3} - \frac{10}{3}\right) = -\frac{10}{3}. \tag{7}$$

Using (4) and symmetry arguments as above, one obtains

$$\begin{aligned}<6_{12}\bar{6}_{34}|\widehat{C}_1 \cdot \widehat{C}_3|6_{12}\bar{6}_{34}> &= <6_{12}\bar{6}_{34}|\widehat{C}_2 \cdot \widehat{C}_3|6_{12}\bar{6}_{34}> \\ &= <6_{12}\bar{6}_{34}|\widehat{C}_1 \cdot \widehat{C}_4|6_{12}\bar{6}_{34}> \\ &= <6_{12}\bar{6}_{34}|\widehat{C}_2 \cdot \widehat{C}_4|6_{12}\bar{6}_{34}> = -\frac{5}{6}.\end{aligned} \tag{8}$$

i.e. a quarter of the value given by the Equation (7).

(c) The non-diagonal matrix elements.

To calculate the non-diagonal matrix elements, we need the inverse of the transformations (10.43), (10.44) and (10.45), (10.46). In the first case, these are

$$|\bar{3}_{12}\,3_{34}> = \sqrt{\frac{1}{3}}|1_{13}\,1_{24}> - \sqrt{\frac{2}{3}}|8_{13}\,8_{24}>$$
$$|6_{12}\,\bar{6}_{34}> = \sqrt{\frac{2}{3}}|1_{13}\,1_{24}> + \sqrt{\frac{1}{3}}|8_{13}\,8_{24}>.$$
(9)

For any colour-invariant operator O, these transformations give

$$<\bar{3}_{12}\,3_{34}|O|6_{12}\,\bar{6}_{34}> =$$
$$= \frac{\sqrt{2}}{3}[<1_{13}\,1_{24}|O|1_{13}\,1_{24}> - <8_{13}\,8_{24}|O|8_{13}8_{24}>].$$
(10)

In particular, one has

$$<1_{13}\,1_{24}|\widehat{C}_1\cdot\widehat{C}_3|1_{13}\,1_{24}> \;=\; <1_{13}|\widehat{C}_1\cdot\widehat{C}_3|1_{13}>$$
$$= \frac{1}{2}\left(0 - \frac{4}{3} - \frac{4}{3}\right) = -\frac{4}{3}$$
$$<8_{13}\,8_{24}|\widehat{C}_1\cdot\widehat{C}_3|8_{13}\,8_{24}> \;=\; <8_{13}|\widehat{C}_1\cdot\widehat{C}_3|8_{13}>$$
$$= \frac{1}{2}\left(3 - \frac{4}{3} - \frac{4}{3}\right) = \frac{1}{6}.$$

Using these values, the matrix element (10) becomes

$$<\bar{3}_{12}3_{34}|\widehat{C}_1\cdot\widehat{C}_3|6_{12}\bar{6}_{34}> = -\frac{\sqrt{2}}{2}.$$
(11)

By analogy, one has

$$<\bar{3}_{12}3_{34}|\widehat{C}_2\cdot\widehat{C}_4|6_{12}\bar{6}_{34}> = -\frac{\sqrt{2}}{2}.$$
(12)

Using the inverse transformations of (10.45) and (10.46) in a similar way, one obtains

$$<\bar{3}_{12}3_{34}|\widehat{C}_1\cdot\widehat{C}_4|6_{12}\bar{6}_{34}> = <\bar{3}_{12}3_{34}|\widehat{C}_2\cdot\widehat{C}_3|6_{12}\bar{6}_{34}> = \frac{\sqrt{2}}{2}.$$
(13)

Note, of course, that

$$<\bar{3}_{12}3_{34}|\widehat{C}_1\cdot\widehat{C}_2|6_{12}\bar{6}_{34}> = <\bar{3}_{12}3_{34}|\widehat{C}_3\cdot\widehat{C}_4|6_{12}\bar{6}_{34}> = 0$$

due to the orthogonality relations $<\bar{3}_{12}|6_{12}> = 0$, etc., and that the sum of matrix elements (11), (12), and (13) leads to

$$<\bar{3}_{12}3_{34}|\widehat{C}_{12}\cdot\widehat{C}_{34}|6_{12}\bar{6}_{34}> = 0$$

where $\widehat{C}_{12}.\widehat{C}_{34}$ is defined by (4).

10.4 SIX-QUARK SYSTEMS

The concept of quarks as fundamental particles has implications not only for the study of hadron properties, but also for their interactions. This section is devoted, in particular, to the study of the nucleon–nucleon (NN) interaction. When two nucleons overlap, one expects that their structure will influence their interaction in a range up to about 1 fm and it seems natural to try to understand the nature of the repulsive core of the NN potential from the quark–quark interaction. The problem is analogous to the derivation of molecular forces from the interaction between electrons. Two strongly overlapped nucleons can be viewed as a six-quark system confined in a region of space where the quarks interact via gluon exchange. There have been many attempts to derive the NN interaction within non-relativistic quark models and only a few attempts in which relativistic models have been used, obviously due to their practical difficulties. A description of various results can be found in review articles as, for example, by Myhrer and Wroldsen (1988) or Shimizu (1989).

The classification and construction of six-quark states is a central problem in any derivation of the NN interaction, irrespective of the underlying quark model. The purpose of this section is to mainly present some group theory aspects of the presently available classification schemes. The discussion here is relevant for any non-relativistic or relativistic quark model which provides the single-particle states necessary to build six-quark states.

As for the baryon, the symmetries of six-quark states are related to the degrees of freedom characterizing the quarks: orbital (O), colour (C), flavour (F) or isospin (I), and spin (S). Therefore, the construction of six-quark states is similar to that of the baryon states but more elaborate because, instead of the group S_3, one must use the group S_6. The use of S_6 is based on the properties of S_n described in Sections 4.7–4.10.

There are several distinct aspects related to the construction of six-quark states which will be successively discussed. These are the coupling schemes, the physical states, and the orbital part of the wave function.

The coupling schemes

From Section 10.2, one can see that the orbital symmetry of the nucleon ground state is mostly [3], or equivalently (56, 0^+) in the language of SU(6) multiplets. Therefore, the most important orbital symmetries of six-quark states are obtained from the outer product (Section 4.6)

$$[3] \times [3] = [6] + [51] + [42] + [33]. \tag{10.54}$$

On the other hand, in order to be a colour singlet, the six-quark states must have the symmetry [222] (see Exercise 10.2). The problem of finding totally antisymmetric states by incorporating the flavour and spin degrees of freedom is not so simple. Several coupling schemes can be adopted. Up to now, two coupling schemes have been used in the literature.

In the one which we call CS, one first combines the SU(3)-colour singlet [222] with the SU(2)-spin representations $[f]_S$ to a given SU(6) symmetry $[f]_{CS}$. Subsequently, one can couple this to an SU(2)-isospin representation $[f]_I$ to obtain a representation $[f]_{CSI}$ of SU(12) (see, for example, Oka and Yazaki 1984). This must be the conjugate of the partition $[f]_O$ of the orbital symmetry in order to obtain a totally antisymmetric state as given by Equation (4.184). In this scheme one can, of course, permute the orbital and the isospin symmetries, i.e. couple first $[f]_{CS}$ to $[f]_O$ and then to $[f]_I$ (see, for example, Ji and Brodsky 1986).

One can also build totally antisymmetric states based on an intermediate representation $[f]_{IS}$ of the SU(4) isospin–spin group mentioned in Section 8.8. The other intermediate coupling is between the orbital part and the colour part. Then, the scheme can be called either CO or IS. One can choose, for example, any of the orbital symmetries on the right-hand side of (10.54) and couple it to $[222]_C$. The conjugate isospin–spin symmetries that can combine with them to form totally antisymmetric states are listed in Table 10.6. This result can be obtained by using inner products of S_6 (Section 4.7) and the answer can be read from Table 4.6. For example, there one can find the Clebsch–Gordan series

$$[42] \times [2^3] = [21^4] + [31^3] + [2^3] + [321] + [42] \tag{10.55}$$

and the conjugates of the representations on the right-hand side are precisely those listed in Table 10.6 for the orbital symmetry [42].

Among the SU(4) irreps, some are denoted with an asterisk. They correspond to dibaryon states and are the most important ones. The others can safely be neglected

Table 10.6 SU(4) spin–isospin symmetry states required by the orbital symmetries of (10.54) and $[222]_C$ to form colourless totally antisymmetric six-quark states. The asterisk corresponds to dibaryon states NN, NΔ or $\Delta\Delta$.

Orbital	SU(4) spin–isospin
[6]	[33]*
[51]	[42]*
	[321]
[42]	[51]*
	[411]
	[33]*
	[321]
	[2211]
[33]	[6]*
	[42]*
	[222]
	[3111]

SIX-QUARK SYSTEMS

(Stancu and Wilets 1988). By dibaryon states, we mean NN, NΔ, or $\Delta\Delta$ systems. As explained in Section 10.2, the free nucleon N or the Δ resonance are both described by the symmetry $[3]_{IS}$. Hence, dibaryons will be characterized by the SU(4) symmetries appearing on the right-hand side of (10.54).

Each of the above schemes has its advantage. The CS scheme is convenient if one reduces the hyperfine interaction H^{hyp} (10.11) to the spin–spin term V_{ij}^{SS} which is actually the major contribution of this interaction in the baryon spectra. In this case, the expectation value of H^{hyp} can be easily calculated in the CS scheme.

In the IS scheme, one advantage is that one can interchange I and S at the group theory level. This implies that for the kinetic and confinement terms (no spin–spin) of the six-quark Hamiltonian, the matrix elements are identical for IS = (01) or (10), etc. Also the shell model literature provides fractional parentage coefficients which can be used in the quark physics too, provided the same phase convention is used everywhere.

Each of the coupling schemes provides a basis and it is useful to find the unitary transformation between bases in different schemes. For the orbital symmetries $[42]_O$ in the IS = (01) sector and $[33]_O$ in the IS = (00) sector, such transformations have been obtained (Stancu 1989). They allow us to determine the correct CS composition of the NN state introduced below.

Physical states

The states obtained by one of the coupling schemes described in the previous section form a basis called a symmetry basis. It is useful to transform them into the so-called physical basis. Here, we follow the work of Harvey (1981a, 1988). The physical basis is defined to contain all those dibaryon states which, for large separations between the baryons, can be represented in terms of NN, NΔ, or $\Delta\Delta$, i.e. in terms of pairs of colourless particles. Despite the name, this basis also contains the remaining orthogonal states which are asymptotically colour octet–octet states and are referred to as hidden-colour states (CC). The physical states are introduced by using fractional parentage coefficients. Let us illustrate the discussion by taking as an example the symmetry state $|[6]_O[222]_C[33]_{IS}>$. Its orbital part is symmetric, so that the rest must be antisymmetric. Using the fractional parentage coefficients of Harvey's (1981a) Table 5 and the formula (4.184), its antisymmetric part can be written in a compact form as

$$|[222]_C[33]_{IS}> = \sqrt{\frac{2}{5}}\,\begin{array}{c}\text{Young}\\\text{C}\end{array}\,\begin{array}{c}\text{Young}\\\text{IS}\end{array} + \sqrt{\frac{3}{5}}\,\begin{array}{c}\text{Young}\\\text{C}\end{array}\,\begin{array}{c}\text{Young}\\\text{IS}\end{array}$$

(10.56)

where only the position of the particles 5 and 6 has been marked, the other particles being irrelevant for the discussion. The argument is that in a physical state the colour

Table 10.7 Unitary transformation between the symmetry basis and the physical basis for the sector IS = (10) or (01) (from Harvey 1981a).

	$[6]_O[33]_{IS}$	$[42]_O[33]_{IS}$	$[42]_O[51]_{IS}$
NN	$\sqrt{\dfrac{1}{9}}$	$\sqrt{\dfrac{4}{9}}$	$-\sqrt{\dfrac{4}{9}}$
$\Delta\Delta$	$\sqrt{\dfrac{4}{45}}$	$\sqrt{\dfrac{16}{45}}$	$\sqrt{\dfrac{25}{45}}$
CC	$\sqrt{\dfrac{4}{5}}$	$-\sqrt{\dfrac{1}{5}}$	0

state of the pairs 5 and 6 must be antisymmetric when they are both in a symmetric orbital state as above, because at large separation they must form a baryon which is a colour singlet. Thus, the first component of (10.56) is unphysical and must be removed by cancellation in a linear combination of symmetry states.

From the fractional parentage expansion of the state $|[42]_O[33]_{IS}>$, it is found that it also contains the same undesirable colour component. This can be eliminated by taking a linear combination of $|[6]_O[33]_{IS}>$ and $|[42]_O[33]_{IS}>$ to give a physical state. The undesirable component is collected in a state orthogonal to the physical states and is called a CC state. The unitary transformation between symmetry states and physical NN, $\Delta\Delta$, and the CC states for IS = (01) or (10) is shown in Table 10.7. For other sectors, see Harvey (1981a, 1988).

Any calculation of the NN interaction requires a diagonalization of the Hamiltonian describing the six-quark system either in the physical or in the symmetry basis. In either case, the lowest state which describes the NN system contains some CC component. This is important at short distances and vanishes in the asymptotic region. The eigenvalues of H also depend on the separation between the nucleons, each nucleon being described as a cluster or bag of three quarks. At large separations, the lowest eigenvalue goes to $2m_N$, where m_N is the nucleon mass, and the CC channel eigenvalue rises to infinity because coloured objects do not exist. Note that the chosen Hamiltonian must also describe the three-quark system correctly.

As a first estimate, the NN interaction V_{NN} can be obtained in the Born–Oppenheimer approximation. The diagonalization of the Hamiltonian will provide a linear combination of channel states

$$\psi = \Sigma C_\alpha \psi_\alpha \qquad (10.57)$$

where $\alpha = $ NN, $\Delta\Delta$, etc., and in the Born–Oppenheimer approximation V_{NN} is

$$V_{NN} = <\psi|H|\psi> -2m_N \qquad (10.58)$$

where ψ corresponds to the lowest eigenvalue. This is a static treatment. In a dynamic description of V_{NN}, one must use the resonating group or the generator coordinate method (Pepin et al. 1996).

It has been shown (Harvey 1981b) that the state $|[42]_O[51]_{IS}>$ plays a very important role in the six-body problem. In the spherical limit (zero separation) its energy is comparable to that of the $|[6]_O[33]_{IS}>$ state which contains the most symmetric orbital configuration. This is the effect of the colour operator, non-existent in shell-model calculations.

The orbital part

In most applications the orbital wave functions of the six quarks are constructed from combinations of products of single-particle wave functions with the appropriate symmetry. Two approaches have been used in NN studies. The first is based on the cluster model and the second on the molecular orbitals (independent particle model). In this section, we discuss the construction of the two kinds of wave functions and the relation between them.

In the cluster model, there are two independent confining potentials located at $\pm\frac{1}{2}Z$, where Z may be regarded as a separation coordinate between the two interacting nucleons. The single-particle wave functions are eigenstates of one of these potentials and can be defined as right (R) and left (L) orbitals:

$$R_n = \psi_n\left(r - \frac{Z}{2}\right), \qquad L_n = \psi_n\left(r + \frac{Z}{2}\right). \qquad (10.59)$$

These functions do not have a definite parity with respect to the common centre and are not orthogonal:

$$<R_n|L_n> \neq 0 \qquad (10.60)$$

except for $Z \to \infty$.

In the independent-particle model, the single-particle wave functions are eigenstates of a deformed single-particle Hamiltonian which has axial and reflection symmetry. These eigenstates, called molecular orbitals, have good parity with respect to the common centre and good projection of the angular momentum. Here, we consider the lowest positive parity state which is denoted by σ and the lowest negative parity state denoted by π. When Z tends to zero, they become s ($\ell = 0$) and p ($\ell = 1$) states. At large deformation, they become degenerate.

Both the cluster model states and the molecular states form a complete basis set. The discussion here is restricted to the lowest ones, used in the calculation of the NN interaction. The lowest-order $n = 0$ cluster model states of (10.59) will be denoted simply by R and L. At $Z = 0$, one has $<R|L> = 1$ and at a finite Z, $<R|L> \neq 0$.

One can construct good parity, orthonormal states from them as

$$\sigma = \frac{R+L}{[2(1+ <R|L>)]^{1/2}}, \quad \pi = \frac{R-L}{[2(1- <R|L>)]^{1/2}}. \qquad (10.61)$$

In the following, we show that in the limit $Z \to 0$, relevant for the short-distance NN interaction, the molecular basis gives a richer composition of the six-quark states than the cluster model basis. Of course, in each case the same number of orbits are considered, σ and π in the molecular case and R and L in the cluster model. In the cluster model, we introduce normalized pseudoright and pseudoleft states r and ℓ by

$$\begin{bmatrix} r \\ \ell \end{bmatrix} = 2^{-1/2}(\sigma \pm \pi) \qquad (10.62)$$

where, by using (10.61), one gets

$$\begin{bmatrix} r \\ \ell \end{bmatrix} = \frac{1}{2}\left[\frac{R+L}{(1+ <R|L>)^{1/2}} \pm \frac{R-L}{(1- <R|L>)^{1/2}} \right]. \qquad (10.63)$$

Contrary to R and L, these functions have the property

$$<r|\ell> = 0 \quad \text{for all } Z. \qquad (10.64)$$

Let us now construct two-particle RL states. The symmetric S and antisymmetric A functions are

$$\begin{bmatrix} S \\ A \end{bmatrix} = \left[2(1\pm <R|L>^2)\right]^{-1/2}(R(1)L(2) \pm L(1)R(2)). \qquad (10.65)$$

By using the inverse of (10.61), one obtains

$$S = \left[2(1+ <R|L>^2)\right]^{-1/2}[(1+ <R|L>)\sigma(1)\sigma(2) - (1- <R|L>)\pi(1)\pi(2)] \qquad (10.66)$$

and

$$A = -\frac{1}{\sqrt{2}}[\sigma(1)\pi(2) - \pi(1)\sigma(2)]. \qquad (10.67)$$

In the limit $Z \to 0$, where $\sigma \to s$ and $\pi \to p$

$$S \to s^2 \qquad (10.68)$$
$$A \to sp - ps \qquad (10.69)$$

where the normal particle order is understood.

Let us now introduce two-body states starting from the orbitals r and ℓ instead of R and L. By analogy with (10.65), the symmetric and antisymmetric combinations, $\tilde{S}$ and $\tilde{A}$, respectively, are

$$\begin{bmatrix} \tilde{S} \\ \tilde{A} \end{bmatrix} = 2^{-1/2}[r(1)\ell(2) \pm \ell(1)r(2)]. \qquad (10.70)$$

Using (10.62), one obtains

$$\tilde{S} = 2^{-1/2}[\sigma(1)\sigma(2) - \pi(1)\pi(2)] \tag{10.71}$$

$$\tilde{A} = -2^{-1/2}[\sigma(1)\pi(2) - \pi(1)\sigma(2)]. \tag{10.72}$$

One can see that the $\tilde{A}$ and A states are identical but the $\tilde{S}$ and S are different and now in the limit $Z \to 0$ one obtains

$$\tilde{S} \to s^2 - p^2. \tag{10.73}$$

This shows that in the cluster model two-body state (10.68), the configuration p^2 is missing. This makes an essential difference between the cluster and the independent-particle models. Actually, any independent-particle model yields eigenstates σ and π and, by using (10.62), one can show that a two-body symmetric state of type (10.70) has the behaviour (10.73). The discussion can be generalized to six-quark states to find out which are the configurations missing in the cluster model basis. For this purpose, one must write six-quark states both in terms of (r, ℓ) and (σ, π) orbits. The transformation between these two representations is shown in Table 10.8. The chosen $[f]_O$ symmetries are those of (10.54). The six-quark states are divided into positive and negative parity states. Each state has a label indicated in the first column. This label described the structure of that state. For example, $51^+[6]$ indicates that the orbital symmetry is $[6]$, the parity is $+$, and the state contains five quarks in the left and one in the right bag (cluster) or vice versa.

The construction of Table 10.8 relies on the material presented in Section 4.3. One can consider the symmetry [51], for example, and use the table of that section where the explicit form of the basis vectors of [51] are given for the configuration $\alpha^3\beta^3$. One can make the replacement $\alpha \to r, \beta \to \ell$ to get $|r^3\ell^3[51]>$ or $\alpha \to \sigma, \beta \to \pi$ to obtain $|\sigma^3\pi^3[51]>$.

The states $|\sigma^5\pi[51]>$ and $|\sigma\pi^5[51]>$ are simpler and can be constructed by the technique presented in Exercise 4.2. After all that, one can use the transformation (10.62) to check the correspondence between the (r, ℓ) and (σ, π) representations indicated in Table 10.8. Using this table, the limiting structure $Z \to 0$ of each symmetry can automatically be obtained by the replacement $\sigma \to s$ and $\pi \to p$. The result can be compared to the cluster model configurations as given by Harvey (1981a).

Cluster model basis	Molecular orbital basis
$s^6[6]$	$\sqrt{5}(s^6[6]-p^6[6])-\sqrt{3}(s^4p^2[6]+s^2p^4[6])$
$s^4p^2[42]$	$s^4p^2[42] - s^2p^4[42]$
$s^3p^3[33]$	$s^3p^3[33]$

Table 10.8 Transformation from (r, ℓ) to (σ, π) configurations for six-quark states of S_6 permutation symmetries [6], [51], [42], and [33] (from Stancu and Wilets 1987).

Name	$r^n \ell^{6-n}$ configuration	$\sigma^n \pi^{6-n}$ configuration
Even parity		
$60^+[6]$	$\frac{1}{\sqrt{2}}\left[\left(r^6[6]\right) + \left(\ell^6[6]\right)\right]$	$\frac{1}{4\sqrt{2}}\left[\left(\sigma^6[6]\right) + \left(\pi^6[6]\right) + \sqrt{15}\left(\sigma^4\pi^2[6]\right) + \sqrt{15}\left(\sigma^2\pi^4[6]\right)\right]$
$51^+[6]$	$\frac{1}{\sqrt{2}}\left[\left(r^5\ell[6]\right) + \left(r\ell^5[6]\right)\right]$	$\frac{1}{4}\left[\sqrt{3}\left(\sigma^6[6]\right) - \sqrt{3}\left(\pi^6[6]\right) + \sqrt{5}\left(\sigma^4\pi^2[6]\right) - \sqrt{5}\left(\sigma^2\pi^4[6]\right)\right]$
$42^+[6]$	$\frac{1}{\sqrt{2}}\left[\left(r^4\ell^2[6]\right) + \left(r^2\ell^4[6]\right)\right]$	$\frac{1}{4\sqrt{2}}\left[\sqrt{15}\left(\sigma^6[6]\right) + \sqrt{15}\left(\pi^6[6]\right) - \left(\sigma^4\pi^2[6]\right) - \left(\sigma^2\pi^4[6]\right)\right]$
$33[6]$	$\left(r^3\ell^3[6]\right)$	$\frac{1}{4}\left[\sqrt{5}\left(\sigma^6[6]\right) - \sqrt{5}\left(\pi^6[6]\right) - \sqrt{3}\left(\sigma^4\pi^2[6]\right) + \sqrt{3}\left(\sigma^2\pi^4[6]\right)\right]$
$51^+[51]$	$\frac{1}{\sqrt{2}}\left[\left(r^5\ell[51]\right) + \left(r\ell^5[51]\right)\right]$	$-\frac{1}{\sqrt{2}}\left[\left(\sigma^4\pi^2[51]\right) - \left(\sigma^2\pi^4[51]\right)\right]$
$42^+[51]$	$\frac{1}{\sqrt{2}}\left[\left(r^4\ell^2[51]\right) + \left(r^2\ell^4[51]\right)\right]$	$-\frac{1}{\sqrt{2}}\left[\left(\sigma^4\pi^2[51]\right) + \left(\sigma^2\pi^4[51]\right)\right]$
$42^+[42]$	$\frac{1}{\sqrt{2}}\left[\left(r^4\ell^2[42]\right) + \left(r^2\ell^4[42]\right)\right]$	$\frac{1}{\sqrt{2}}\left[\left(\sigma^4\pi^2[42]\right) + \left(\sigma^2\pi^4[42]\right)\right]$
$33[42]$	$\left(r^3\ell^3[42]\right)$	$\frac{1}{\sqrt{2}}\left[\left(\sigma^4\pi^2[42]\right) - \left(\sigma^2\pi^4[42]\right)\right]$

Table 10.8 *cont.*

Odd parity		
51⁻[51]	$\frac{1}{\sqrt{2}}\left[\left(r^5\ell[51]\right)-\left(r\ell^5[51]\right)\right]$	$-\frac{1}{4}\left[\sqrt{2}\left(\sigma^5\pi[51]\right)-\sqrt{2}\left(\sigma\pi^5[51]\right)+2\sqrt{3}\left(\sigma^3\pi^3[51]\right)\right]$
42⁻[51]	$\frac{1}{\sqrt{2}}\left[\left(r^4\ell^2[51]\right)-\left(r^2\ell^4[51]\right)\right]$	$-\frac{1}{\sqrt{2}}\left[\left(\sigma^5\pi[51]\right)+\left(\sigma\pi^5[51]\right)\right]$
33[51]	$\left(r^3\ell^3[51]\right)$	$-\frac{1}{4}\left[\sqrt{6}\left(\sigma^5\pi[51]\right)-\sqrt{6}\left(\sigma\pi^5[51]\right)-2\left(\sigma^3\pi^3[51]\right)\right]$
42⁻[42]	$\frac{1}{\sqrt{2}}\left[\left(r^4\ell^2[42]\right)-\left(r^2\ell^4[42]\right)\right]$	$\left(\sigma^3\pi^3[42]\right)$
33[33]	$\left(r^3\ell^3[33]\right)$	$-\left(\sigma^3\pi^3[33]\right)$

Here, we can see that the replacement of R, L states of the cluster model by r, ℓ states related to a molecular basis brings new configurations in the spherical limit. In the molecular basis, the configuration $s^n p^{6-n}$ also appears with s and p interchanged and for higher symmetries such as [6] extra configurations are present too. Moreover, Table 10.8 shows that in the molecular basis one can also construct states of a given orbital symmetry $[f]$ not only through occupying the r and ℓ states with three quarks each, but also from clustering the quarks, two in one bag, four in the other, or five in one bag and one in the other. These are configurations natural in the molecular basis and are neglected in cluster model calculations. It has been shown that they give a substantial lowering of the upper bound (Stancu and Wilets 1988, 1989, 1994) and represent an important part of the total wave function of the NN system, as described by (10.57).

Matrix elements

Let us consider a Hamiltonian consisting of one-body and two-body terms as, for example, that described in Section 10.1. One has to evaluate matrix elements of this Hamiltonian between antisymmetric six-quark states. Remember that in the NN problem all six quarks are identical. The matrix elements of a one-body operator between two antisymmetric n-particle states ψ_a and ψ_b can be reduced to the calculation of a single matrix element

$$<\psi_a| \sum_{i=1}^{n} O(i)|\psi_b> = n <\psi_a|O(n)|\psi_b> \qquad (10.74)$$

(see Exercises 4.8 and 4.9). The matrix elements of a two-body operator, like (10.2), may be written as

$$<\psi_a| \sum_{i<j}^{n} V_{ij}|\psi_b> = \frac{n(n-1)}{2} <\psi_a|V_{n-1,n}|\psi_b> . \qquad (10.75)$$

To be more specific, let us consider the SU(4)-isospin-spin classification scheme, the IS scheme. As mentioned above, one first couples the colour singlet $[222]_C$ to a specific orbital symmetry $[f']_O$ to give a state of symmetry $[f]$ which has to be combined with a conjugate symmetry state $[\tilde{f}]_{IS}$ in order to give an antisymmetric $[1^6]$ state as defined by the general form (4.184):

$$\psi_6 \bigg(([f']_O[222]_C)_{[f]}[\tilde{f}]_{IS}\bigg)_{[1^6]}$$

$$= \sqrt{\frac{1}{d_{[f]}}} \sum_Y (-)^{n_Y^f} |[f']_O[222]_C ; [f]Y > |[\tilde{f}]_{IS}\tilde{Y}> \qquad (10.76)$$

$$= \sqrt{\frac{1}{d_{[f]}}} \sum_{YY'Y''} (-)^{n_Y^f} S([f']_O Y' [222]_C Y'' \| [f]_{IS} Y)$$

$$\times |[f']_O Y' > |[222]_C Y'' > |[\tilde{f}]_{IS} \tilde{Y} >$$

SIX-QUARK SYSTEMS

where, in the last equality, (4.135) has been used, with $S([f']Y'[f'']Y''\|[f]Y)$ an S_6 Clebsch–Gordan coefficient. In calculating the matrix elements (10.74), it is necessary to expand the states (10.76) in terms of sums of products of antisymmetric states of the first $n - 1$ particles and the last particle. To calculate (10.75), a similar expansion is required in terms of products of antisymmetric states of the first $n - 2$ particles and the last pair of particles. To carry out this procedure, it is necessary to write the orbital, colour, and spin–isospin functions as linear combinations of products $\psi_5\phi_1$, with ψ_5 a state of five quarks, and ϕ_1 of one quark. The required linear combinations are obtained with the help of fractional parentage coefficients (see Harvey 1981a). In order to maintain the antisymmetry of the five-particle wave functions, one can use the factorizability property (4.186) of the CG coefficient of S_6 into the isoscalar factor K and a CG coefficient for S_5. With the help of the S_5 CG coefficients, one can recombine the orbital, colour, and the spin–isospin functions into a totally antisymmetric five-particle state. In the one-body matrix elements, this part of the six-particle wave function is unaffected and gives an overlap. Its evaluation is simplified by the orthogonality property (4.167) of the CG coefficients of S_n. For example, in a simplified notation, one has

$$< \psi_5([f']_O\alpha'[f'']_C\alpha''[f]_{IS}\alpha)|\psi_5([\bar{f}']_O\bar{\alpha}'[\bar{f}'']_C\bar{\alpha}''[\bar{f}]_{\overline{IS}}\bar{\alpha}) >$$
$$= \delta_{f'\bar{f}'}\delta_{f''\bar{f}''}\delta_{f\bar{f}}\delta_{\alpha''\bar{\alpha}''}\delta_{\alpha\bar{\alpha}}\delta_{I\bar{I}}\delta_{S\bar{S}} < \psi_5([f']_O\alpha')|\psi_5([f']_O\bar{\alpha}') > \qquad (10.77)$$

where α', α'' and α represent possible structures of $[f']$, $[f'']$ and $[f]$ respectively. In using molecular orbitals, one can have, for example, $\alpha' = r^3\ell^2$, $\bar{\alpha}' = r^4\ell$ etc., when α' and $\bar{\alpha}'$ are associated to distinct deformations of the six-quark system. Such situations appear in the study of the NN interaction based on the generator coordinate method. Otherwise, for identical deformations, the property (10.64) yields $\delta_{\alpha'\bar{\alpha}'}$.

For two-body matrix elements, the discussion is similar. In this case, one has to evaluate matrix elements of an operator acting on the $(n - 1)$th and the nth particles. Then, (10.76) has to be expanded in terms of sums of products $\psi_4\phi_2$. For the CG coefficients appearing in (10.76), one must use the factorization property (4.186) twice. If one wishes to have the last two particles with a definite permutation symmetry, one has to use the diagonalized Young–Yamanouchi–Rutherford representation (4.109). This leads to the so-called $\overline{K}$-matrix developed in Section 4.10. Then, ψ_6 can be written as

$$\psi_6 = \sum_{[f]\alpha,\beta} P^f_{\alpha\beta}\psi_4([f]\alpha)\phi_2([11]\beta) \qquad (10.78)$$

where α and β represent different possible structures. Note that ϕ_2 must always be antisymmetric. Then, the two-body matrix element on the right-hand side of (10.75) becomes

$$< \psi_a|V_{56}|\psi_b > = \sum_{\substack{[f] \\ \alpha\beta, \bar{\alpha}\bar{\beta}}} P^f_{\alpha\beta}P^f_{\bar{\alpha}\bar{\beta}} < \psi_4[f]\alpha|\psi_4[f]\bar{\alpha} >< \phi_2[11]\beta|V_{56}|\phi_2[11]\bar{\beta} >$$

$$(10.79)$$

where the coefficients P are products of $\overline{K}$ matrix elements and two-body fractional parentage coefficients associated to the orbital, colour, and isospin–spin space and can be found in tables. In practice, in the colour space, it is necessary to know only whether the last pair is in a symmetric or antisymmetric state in order to use (10.5) and the fractional parentage coefficients do not need to be known explicitly. They add up together into a normalization constant.

The procedure discussed above can and has been applied to the derivation of the NN interaction at short distances, both for non-relativistic models (Harvey 1981b; Harvey, Le Tourneux, and Lorazo 1984) and relativistic models (Koepf, Wilets, Pepin, and Stancu 1994). In all cases a hard core is found, but the interpretation of its nature is not always the same.

In relativistic models, a part of the quark–quark interaction is associated with one-gluon exchange. Exercise 10.5 describes a detailed but important aspect of the calculations based on relativistic models.

Exercise 10.5 Prove that the total colour chromoelectrostatic energy of a configuration a^6 of six quarks in a colour singlet state is zero. By a we understand σ or π states which, in the spherical limit, tend to s and p states, respectively.

Solution Note that this is a generalization of the well-known result that for a colour singlet state of three quarks, each in an s state, the mutual and the self-interaction chromoelectrostatic energies cancel out. The proof for a^6 is entirely analogous to that for s^3.

The chromoelectrostatic energy of six quarks, each in an a state, can be written as

$$H_{CE}(a^6) = 2\pi\alpha_S \left(\sum_{i=1}^{6} \boldsymbol{E}_a(i)\right)^2 \qquad (1)$$

where α_S is the strong coupling constant and $\boldsymbol{E}_a(i)$ is the colour electric field of the quark i. In quantum chromodynamics, $\boldsymbol{E}_a(i)$ has eight colour components:

$$E_a^c(i) \sim \widehat{C}^c(i) \quad (c = 1, 2, \cdots, 8) \qquad (2)$$

where $\widehat{C}^c(i)$ are the SU(3)-colour generators (9.34).

Using (10.74) and (10.75), one can write the expectation value of H_{CE} as

$$<H_{CE}(a^6)> \,=\, 2\pi\alpha_S[6 < \boldsymbol{E}_a(6) \cdot \boldsymbol{E}_a(6) > \,+\, 30 < \boldsymbol{E}_a(5) \cdot \boldsymbol{E}_a(6) >] \qquad (3)$$

where the first term on the right-hand side represents the self-interaction and the second term the mutual interaction. The above matrix elements can be written in factorized form as

$$< \boldsymbol{E}_a(6) \cdot \boldsymbol{E}_a(6) > \,=\, \mathscr{S}_a < \widehat{C}(6) \cdot \widehat{C}(6) > \qquad (4)$$
$$< \boldsymbol{E}_a(5) \cdot \boldsymbol{E}_a(6) > \,=\, \mathscr{M}_a < \widehat{C}(5) \cdot \widehat{C}(6) > \qquad (5)$$

where $\mathscr{S}_a$ and $\mathscr{M}_a$ are numerical coefficients resulting from integration in the coordinate space. If one calculates the self-energy by using the diagram depicted in

SIX-QUARK SYSTEMS

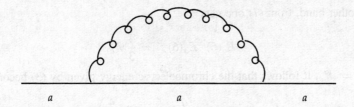

Figure 10.4 Self-energy diagram with identical incoming, intermediate, and outgoing quark states.

Fig. 10.4, as required by the minimal energy prescription of the MIT bag model, then the numerical coefficients $\mathscr{S}_a$ and $\mathscr{M}_a$ are identical and represent an electric monopole when $a = s$. For the MIT bag model, one has (see, for example, Wroldsen and Myhrer 1984)

$$\mathscr{S}_s = \mathscr{M}_s = \frac{\alpha_S}{8}\frac{0.278}{R} \tag{6}$$

where R is the bag radius

The colour matrix element of (4) is just the eigenvalue of the SU(3) Casimir operator as given by (8.134), by setting $\lambda = 1, \mu = 0$

$$<\widehat{C}(6)\cdot\widehat{C}(6)> = \frac{4}{3}. \tag{7}$$

The colour matrix element of (5) is obtained from the first and second rows of Equations (10.5):

$$<\widehat{C}(5)\cdot\widehat{C}(6)> = \begin{cases} -\frac{2}{3} & \text{for } [1^2] \\ \frac{1}{3} & \text{for } [2]. \end{cases} \tag{8}$$

For a colour singlet state of six quarks (see Exercise 10.2), there are five independent colour wave functions corresponding to the following Young tableaux (see Section 4.3)

1	2
3	4
5	6

1	3
2	4
5	6

1	2
3	5
4	6

1	3
2	5
4	6

1	4
2	5
3	6

One can see that the symmetric colour state $\boxed{5\,6}$ appears with the weight $\frac{2}{5}$ and the antisymmetric colour state $\begin{smallmatrix}\boxed{5}\\\boxed{6}\end{smallmatrix}$ with the weight $\frac{3}{5}$. These considerations, together with Equation (8), give

$$<\boldsymbol{E}_a(5)\cdot\boldsymbol{E}_a(6)> = \mathscr{M}_a\left(\frac{2}{5}\cdot\frac{1}{3} - \frac{3}{5}\cdot\frac{2}{3}\right) = -\frac{4}{15}\mathscr{M}_a.$$

On the other hand, from (7) one gets

$$< E_a(6) \cdot E_a(6) > = \frac{4}{3}\mathscr{S}_a.$$

For $\mathscr{S}_a = \mathscr{M}_a$, it follows that the chromoelectric energy given by (3) becomes

$$< H_{\text{CE}}(a^6) > = 2\pi\alpha_S\left(6 \cdot \frac{4}{3} - 30 \cdot \frac{4}{15}\right)\mathscr{S}_a = 0$$

SUPPLEMENTARY EXERCISES

10.6 Derive ψ_{L0}^μ of Table 10.2 by writing explicitly the basis vectors of column 7 in terms of harmonic oscillator states 0s, 0p, 1s, and 0d.

10.7 Using the results of Exercise 10.4 and the transformations (10.43) and (10.44), show that the confining potential

$$V^c = -\sum_{i<j}^{4}\left(e_0 + \frac{1}{2}k\, r_{ij}^2\right)\widehat{C}_i \cdot \widehat{C}_j$$

takes the following matrix form in the basis $(|1_{13}1_{24}>, |8_{13}8_{24}>)$

$$< V^c > = \frac{8}{3}e_0 \mathbf{1} + \frac{k}{6}\begin{bmatrix} 8(\rho^2 + \rho'^2) & 4\sqrt{2}\boldsymbol{\rho}\cdot\boldsymbol{\rho}' \\ 4\sqrt{2}\boldsymbol{\rho}\cdot\boldsymbol{\rho}' & 9x^2 + \frac{7}{2}(\rho^2 + \rho'^2) + 5\boldsymbol{\rho}\cdot\boldsymbol{\rho}' \end{bmatrix}$$

where $\boldsymbol{\rho}, \boldsymbol{\rho}'$, and $\mathbf{x}$ are defined by (10.19).

10.8 Prove row 1 of Table 10.8.

Appendix A

CONSERVATION LAWS

In this appendix we illustrate the conservation laws of physics for two cases. The first is the conservation of energy in classical mechanics. The second is Noether's theorem in the case of internal symmetries. Both are relevant to the discussion in Chapter 1, but the second can be understood only after reading Chapter 6.

A.1 ENERGY CONSERVATION IN CLASSICAL MECHANICS

In many cases in classical mechanics one can obtain a number of *first integrals* which are relations of the type

$$f(q_1, q_2, \ldots, q_n, \dot{q}_1, \dot{q}_2, \ldots, \dot{q}_n, t) = \text{const.}$$

where q_i are the generalized coordinates of a classical system and $\dot{q}_i$ their time derivatives. These relations are first-order differential equations and they in fact include conservation laws.

Here we consider the conservation of *total energy* for conservative systems. A system is conservative if $F = -\nabla V$, where V, the potential energy, is independent of velocities and the constraints are time independent. This implies that the Lagrangian $L = T - V$ does not depend explicitly on time. Then its total derivative is

$$\frac{dL}{dt} = \sum_i \frac{\partial L}{\partial q_i} \frac{dq_i}{dt} + \sum_i \frac{\partial L}{\partial \dot{q}_i} \frac{d\dot{q}_i}{dt}. \tag{A.1}$$

Using the Lagrange equations

$$\frac{\partial L}{\partial q_i} = \frac{d}{dt} \frac{\partial L}{\partial \dot{q}_i} \tag{A.2}$$

one can write

$$\frac{d}{dt}\left(L - \sum_i \dot{q}_i \frac{\partial L}{\partial \dot{q}_i}\right) = 0 \tag{A.3}$$

or that

$$\sum_i \dot{q}_i \frac{\partial L}{\partial \dot{q}_i} - L = H = \text{const.} \tag{A.4}$$

is a constant of motion. In terms of the generalized momenta

$$p_i = \frac{\partial L}{\partial \dot{q}_i} \tag{A.5}$$

one has

$$H = \sum_i \dot{q}_i p_i - L \tag{A.6}$$

and in fact H is precisely the total energy of the system. To see this, one can write

$$p_i = \frac{\partial L}{\partial \dot{q}_i} = \frac{\partial T}{\partial \dot{q}_i} \tag{A.7}$$

because V is independent of $\dot{q}_i$ for a conservative system. On the other hand, when the constraints are time independent, the kinetic energy T is a homogeneous quadratic function of the generalized velocities. From Euler's theorem it follows that

$$\sum_i \dot{q}_i \frac{\partial T}{\partial \dot{q}_i} = 2T. \tag{A.8}$$

Hence

$$H = T + V \tag{A.9}$$

i.e. the constant of motion H of Equation (A.4) is the total energy of the system.

A.2 NOETHER'S THEOREM

The formal connection between invariance and conservation laws is contained in Noether's theorem. This theorem can be formulated in classical mechanics starting from a Lagrangian $L(q_i, \dot{q}_i, t)$ where q_i are generalized coordinates and $\dot{q}_i$ their time derivatives, or equivalently in the four-dimensional space x^μ ($\mu = 0, 1, 2, 3$) of classical field theory. There, the starting point is a Lagrangian density $\mathscr{L}(\phi_\alpha, \partial_\mu \phi_\alpha)$ which is a functional of the field amplitudes ϕ_α and their first derivatives $\partial_\mu \phi_\alpha$. This functional must be invariant under some symmetry group. For most field theories, the Lagrangian approach and Noether's theorem can be carried over directly to the quantum domain without difficulty (Bjorken and Drell 1965, Chapter 11).

Noether's theorem states that for a system described by the Lagrangian

$$L = \int d^3x \, \mathscr{L}\left(\phi_\alpha(x), \, \partial_\mu \phi_\alpha(x)\right) \tag{A.10}$$

and satisfying the Euler–Lagrange equations

$$\partial_\mu \frac{\partial \mathscr{L}}{\partial\left(\partial_\mu \phi_\alpha\right)} - \frac{\partial \mathscr{L}}{\partial \phi_\alpha} = 0 \tag{A.11}$$

any *continuous* symmetry transformation which leaves the action $S = \int L\,dt$ invariant implies the existence of a conserved current J_μ

$$\partial^\mu J_\mu = 0 \tag{A.12}$$

where J_μ is given by equation (A.22) below. The charge defined by

$$Q(t) = \int d^3x\, J_0(x) \tag{A.13}$$

is a constant of motion

$$\frac{dQ}{dt} = 0 \tag{A.14}$$

because the surface integral at infinity is zero.

Here, the discussion is restricted to classical field theory and Noether's theorem is illustrated in the case of internal symmetries. If, for example, the internal degree of freedom is the spin, the group of transformations under which $\mathscr{L}$ is invariant is the rotation group which is a subgroup of the Lorentz group. In the rotation case, the infinitesimal transformation is

$$x' = x - \boldsymbol{\omega} \times x \tag{A.15}$$

where $\boldsymbol{\omega} = (\omega^1, \omega^2, \omega^3)$ with ω^a the infinitesimal parameters. The effect of this transformation on the field component is

$$\phi'_\alpha = U(\omega)\phi_\alpha(x) = \left(\delta_{\alpha\beta} - i\omega^a S^a_{\alpha\beta}\right)\phi_\beta(x)$$

or

$$\delta\phi_\alpha = -i\omega^a S^a_{\alpha\beta}\,\phi_\beta(x) \tag{A.16}$$

where $S^a_{\alpha\beta}$ are the matrix elements of the generators S^a in the space given by ϕ_α (see Section 6.1). Note that the form (A.16) is valid in general, i.e. for any group of transformations under which the Lagrangian density is invariant and the generators of that group satisfy the corresponding Lie algebra

$$\left[S^a, S^b\right] = f^{abc} S^c \tag{A.17}$$

where f^{abc} are the structure constants.

The corresponding change in the Lagrangian density is

$$\delta\mathscr{L} = \frac{\partial\mathscr{L}}{\partial\phi_\alpha}\delta\phi_\alpha + \frac{\partial\mathscr{L}}{\partial(\partial_\mu \phi_\alpha)}\delta(\partial_\mu \phi_\alpha). \tag{A.18}$$

Using the Euler–Lagrange equations and the relation

$$\delta(\partial_\mu \phi_\alpha) = \partial_\mu(\delta\phi_\alpha) \tag{A.19}$$

one can write

$$\delta \mathcal{L} = \partial_\mu \frac{\partial \mathcal{L}}{\partial\left(\partial_\mu \phi_\alpha\right)} \delta\phi_\alpha + \frac{\partial \mathcal{L}}{\partial\left(\partial_\mu \phi_\alpha\right)} \partial_\mu\left(\delta\phi_\alpha\right)$$

$$= \partial_\mu \left[\frac{\partial \mathcal{L}}{\partial\left(\partial_\mu \phi_\alpha\right)} \delta\phi_\alpha \right] \qquad (A.20)$$

$$= \omega^a \, \partial_\mu \left[\frac{\partial \mathcal{L}}{\partial\left(\partial_\mu \phi_\alpha\right)} \left(-\mathrm{i}\, S^a_{\alpha\beta}\, \phi_\beta\right) \right]$$

where (A.16) has been used in the last row.

For an invariant Lagrangian one has $\delta \mathcal{L} = 0$ which implies that for any ω^a one has

$$\partial^\mu J^a_\mu = 0 \qquad (A.21)$$

with the conserved current J^a_μ defined by

$$J^a_\mu = -\mathrm{i}\, \frac{\partial \mathcal{L}}{\partial\left(\partial^\mu \phi_\alpha\right)}\, S^a_{\alpha\beta}\, \phi_\beta. \qquad (A.22)$$

The *charge* is defined by

$$Q(t) = \int \mathrm{d}^3 x\, J^a_0 \qquad (A.23)$$

and is conserved

$$\frac{\mathrm{d}Q}{\mathrm{d}t} = 0 \qquad (A.24)$$

because integrating (A.21) on a sphere with radius $R \to \infty$ the surface term goes to zero.

If the internal degrees of freedom are not affected by the group of transformations which leave $\mathcal{L}$ invariant, the current (A.22) takes a simpler form because in that case the matrix S becomes

$$S = 1 \qquad (A.25)$$

as, for example, in the case of a particle with spin zero.

Example. The Lagrangian density of a free Dirac particle is

$$\mathcal{L} = \overline{\psi}\left(\mathrm{i}\, \partial^\mu \gamma_\mu - m\right)\psi \qquad (A.26)$$

with ψ a spinor of components ψ_α ($\alpha = 1, 2, 3, 4$) so that the identification is

$$\phi_\alpha = \psi_\alpha.$$

Then one has

$$\frac{\partial \mathscr{L}}{\partial\left(\partial^\mu \psi_\alpha\right)} = i\overline{\psi}_\alpha \gamma_\mu$$

and the current takes the form

$$J_\mu^a = \overline{\psi}_\alpha \gamma_\mu S^a_{\alpha\beta} \psi_\beta. \tag{A.27}$$

For a rotation, the matrix S^a associated with the Lorentz group is identical to the component Σ^a of the vector operator defined by (7.86)

$$S^a = \Sigma^a = \frac{i}{4}\left[\gamma^b, \gamma^c\right] \tag{A.26}$$

or else

$$\mathbf{S} = \frac{1}{2}\begin{pmatrix}\boldsymbol{\sigma} & 0 \\ 0 & \boldsymbol{\sigma}\end{pmatrix}. \tag{A.29}$$

Then the conserved charge is

$$J_0 = \int d^3x\, \psi^+ S\, \psi. \tag{A.30}$$

Remark. Noether's theorem proves that the invariance under a continuous symmetry group of a Lagrangian leads to a conservation law. The converse is not always true. Examples are given by Goldstein (1980, Section 12.7).

Appendix B

THE REARRANGEMENT THEOREM, SCHUR'S LEMMAS, AND THE ORTHOGONALITY THEOREM

In the following, we present the rearrangement theorem and prove both Schur's lemmas and the orthogonality theorem.

The rearrangement theorem If all elements $g \in G$ are arranged in a given order and are multiplied on the left (or on the right) by an arbitrarily fixed element g_i, the sequence of products $g_i g$ (or $g g_i$) form the same group but with its elements arranged in a different order.

Proof The proof is trivial. One has to show that the products $g_i g$ reproduce each group element only once. Suppose that two elements g_1 and g_2 reproduce the same element of G by multiplication with g_i, that is

$$g_i g_1 = g_i g_2.$$

it then follows that $g_1 = g_2$, i.e. each group element is reproduced only once.

Schur's lemma 1 Let S and S' be two irreducible representations of a group G defined in the linear vector spaces L and L' of dimension n and n' respectively. If a linear operator A transforming vectors from L to L' satisfies

$$S'(g) A = A S(g) \tag{B.1}$$

for any g then either A gives a one-to-one correspondence between L and L' or $A = 0$.

Proof Let us take a vector $x \in L$. The operator A gives its corresponding vector in L'

$$x' = A x. \tag{B.2}$$

1. *Case* $n < n'$. Let M' be the image of L in L' obtained through A as illustrated in Fig. B.1. Then $\{0\} \leq M' \leq L'$, i.e. M' can have any dimension between 0 and n' by construction. On the other hand, using (B.1) one gets

$$S'(g) A x = A S(g) x = A y \tag{B.3}$$

where $y = S(g) x$ is another vector of L. But both Ax and Ay belong to L' which shows that M' is an invariant subspace of L'. In that case there are only two possibilities: $M' = \{0\}$ or L'. The second is excluded because $n < n'$ and A mapping vectors of L onto L' implies that M' can have at most dimension n. It follows that $M' = \{0\}$ or alternatively $A = 0$.

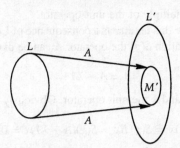

Figure B.1 The image of the mapping A.

2. *Case* $n = n'$. There are two possibilities: (a) A does or (b) does not establish a one-to-one correspondence. In case (a) the operator A is non-singular and the two representations are equivalent. In case (b) one has $\dim M' < \dim L'$. But as before $M' = \{0\}$ or L'. The only possibility is $M' = \{0\}$ i.e. $A = 0$.

3. *Case* $n > n'$. Let us define in L the subspace N of vectors which map the null vector in L' i.e. $x_0 \in N$ if $Ax_0 = 0$, as shown in Fig. B.2. The subspace N is called the kernel of the mapping, defined in Section 2.3. As $n > n'$ there are indeed distinct vectors in L which map into the same vector in L'. Take, for example $x_1 = x$ and $x_2 = x + x_0$. To have $Ax_1 = Ax_2 = Ax$ requires

$$Ax_0 = 0 \tag{B.4}$$

which shows that $N \neq \{0\}$. But according to (B.1) and (B.4) one gets

$$A S(g) x_0 = S'(g) A x_0 = 0 \tag{B.5}$$

i.e. $S(g) x_0$ also belongs to N, i.e. N is an invariant subspace of L. Then $N = L$ which implies $A = 0$.

Schur's Lemma 2 If $S(g)$ is a finite dimensional irreducible representation of a group G and A is an arbitrary linear operator such that

$$A S(g) = S(g) A \tag{B.6}$$

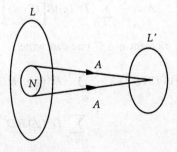

Figure B.2 The kernel of the mapping A.

for all $g \in G$ then A is a multiple of the unit operator.

Proof We choose to prove this lemma as a consequence of Lemma 2. In the space L of the irreducible representation $S(g)$ the operator A can be diagonalized as $Ax = \lambda x$. Let us build the operator

$$B = A - \lambda \mathbf{1} \tag{B.7}$$

with λ an eigenvalue of A and $\mathbf{1}$ the unit operator. Obviously B also commutes with $S(g)$ which gives

$$B\,S(g)x = S(g)\,Bx = S(g)(A - \lambda\mathbf{1})x = 0, \tag{B.8}$$

i.e. any eigenvector $S(g)x$ belonging to λ maps to the null vector. This means that B does not establish a one-to-one correspondence and according to case 2 of the previous lemma it follows that $B = 0$ or

$$A = \lambda \mathbf{1}. \tag{B.9}$$

Orthogonality theorem for finite groups Let D^μ and D^ν be two unitary irreducible representations of a finite group G of order N. If D^μ and D^ν are not equivalent one has

$$\frac{1}{N}\sum_{g \in G} D^\mu_{ij}(g)\left[D^\nu_{k\ell}(g)\right]^* = 0 \tag{B.10}$$

and if they are equivalent they satisfy the relation

$$\frac{1}{N}\sum_{g \in G} D^\mu_{ij}(g)\left[D^\nu_{k\ell}(g)\right]^* = \frac{\delta_{\mu\nu}}{n_\mu} M_{ik}\left(M^{-1}\right)_{\ell j} \tag{B.11}$$

where n_μ is the dimension of D^μ and M is a similarity transformation (Section 3.4) which transforms the matrices of the representation μ into those of ν and vice versa.

Proof One can introduce a matrix B of dimension $n_\mu \times n_\nu$ and construct out of it the matrix A as

$$A = \frac{1}{N}\sum_{g_i \in G} D^\mu(g_i) B\, D^\nu(g_i^{-1}). \tag{B.12a}$$

identical to (see Equation (3.21))

$$A = \frac{1}{N}\sum_{g_i \in G} D^\mu(g_i) B\left[D^\nu(g_i)\right]^{-1}. \tag{B.12b}$$

Choosing an arbitrary element $g \in G$ one can write

$$D^\mu(g) A\, D^\nu(g^{-1}) = \frac{1}{N}\sum_{g_i \in G} D^\mu(g\,g_i)\, B\, D^\nu\left((g g_i)^{-1}\right)$$

$$= \frac{1}{N}\sum_{g'_i \in G} D^\mu(g'_i)\, B\, D^\nu\left(g'^{-1}_i\right) \tag{B.13}$$

$$= A$$

where the second equality holds due to the rearrangement theorem. The sum now runs over the elements $g'_i = g\,g_i$ of the same group but arranged in a different order. Using Equation (3.21) in matrix form

$$D^\mu(g^{-1}) = \left[D^\mu(g)\right]^{-1} \tag{B.14}$$

one can write (B.13) as

$$D^\mu(g)A = A D^\nu(g). \tag{B.15}$$

This is the matrix form of the relation (B.1) and we can now apply Schur's lemmas. According to Lemma 1 if D^μ and D^ν are not equivalent this implies $A = 0$ which, for a unitary representation reads

$$\sum_{g \in G} D^\mu(g) B \left[D^\nu(g)\right]^{+} = 0. \tag{B.16}$$

One has the freedom to choose the matrix B with all but one of its elements equal to zero. Let us take

$$B_{j\ell} = 1. \tag{B.17}$$

Then the matrix element (i,k) of (B.16) leads to the relation (B.10) which proves the first part of the orthogonality theorem.

Now suppose D^μ and D^ν are equivalent

$$D^\mu = M D^\nu M^{-1} \tag{B.18}$$

where M is a non-singular matrix of dimension $n_\mu \times n_\mu$. The relation (B.15) becomes

$$D^\nu(g) M^{-1} A = M^{-1} A D^\nu(g). \tag{B.18}$$

Then according to Lemma 2 $M^{-1}A$ is a multiple λ of the unit matrix:

$$M^{-1}A = \lambda \mathbf{1} \tag{B.19}$$

or $A = \lambda M$. Replacing (B.18) in (B.12b) and multiplying it on the left by M^{-1} one has

$$M^{-1}A = \frac{1}{N}\sum_{g \in G} D^\nu(g) M^{-1} B \left[D^\nu(g)\right]^{-1}. \tag{B.20}$$

Due to the trace property $\text{tr}(ABCD) = \text{tr}(DABC)$, etc., and definition (B.17) it follows that

$$\text{tr}(M^{-1}A) = \text{tr}(M^{-1}B) = (M^{-1})_{\ell j}. \tag{B.21}$$

Using (B.19) in (B.21) one obtains the value of λ in terms of n_μ and $(M^{-1})_{\ell j}$, from which one can write

$$A = \lambda M = \frac{(M^{-1})_{\ell j}}{n_\mu} M. \tag{B.22}$$

The matrix element (i, k) of (B.12b) with A defined by (B.22) and B by (B.17) becomes, for unitary matrices D,

$$\frac{(M^{-1})_{\ell j}}{n_\mu} M_{ik} = \frac{1}{N} \sum_{g \in G} D^\mu_{ij}(g) \left[D^\mu_{k\ell}(g) \right]^*$$

which is just (B.11) with $\mu = \nu$ and this proves the second part of the theorem.

Appendix C

INVARIANT INTEGRATION

In Section 3.10, we introduced invariant integration for compact Lie groups in terms of the invariant measure dg.

Invariant integration is an extension of the invariant sum for finite groups. There, one has

$$\sum_{g \in G} f(g'g) = \sum_{g \in G} f(g) \tag{C.1}$$

and

$$\sum_{g \in G} f(gg') = \sum_{g \in G} f(g) \tag{C.2}$$

i.e. the sum is left- and right-invariant because $g'g$ or gg' also cover the whole group with its elements arranged in a different order (the rearrangement theorem, Appendix B). For continuous groups, the sum is replaced by an integral over the group parameters $\varepsilon_1, \varepsilon_2, \ldots, \varepsilon_r$. One should be able to define a left- and a right-invariant integral. For linear Lie groups, one can show that there always exists a left- and a right-invariant integral. Their definitions are

$$\int_G f(g'g) \, d_\ell g = \int_G f(g) \, d_\ell g \tag{C.3}$$

$$\int_G f(gg') \, d_r g = \int_G f(g) \, d_r g \tag{C.4}$$

where $d_\ell g$ and $d_r g$ are left- and right-invariant measures.

If G is a compact Lie group, one can take $d_\ell g = d_r g = dg$ and one has

$$\int_G f(g'g) \, dg = \int_G f(g) \, dg \tag{C.5}$$

$$\int_G f(gg') \, dg = \int_G f(g) \, dg. \tag{C.6}$$

These integrals are finite for every continuous function $f(g)$ if the group is compact. Then, one can write the invariant measure dg in terms of a weight function $\rho(\varepsilon_1, \varepsilon_2, \ldots, \varepsilon_r)$ where ε_i are the group parameters defined in closed, finite intervals $a_i \leq \varepsilon_i \leq b_i$ $(i = 1, 2, \ldots, r)$. One can choose ρ such that

$$\int_G dg = \int_{a_1}^{b_1} d\varepsilon_1 \int_{a_2}^{b_2} d\varepsilon_2 \cdots \int_{a_r}^{b_r} d\varepsilon_r \rho(\varepsilon_1, \varepsilon_2, \cdots, \varepsilon_r) = 1. \tag{C.7}$$

In terms of the weight ρ, any of the integrals (C.5) or (C.6) becomes

$$\int_{\varepsilon \in G} f(\varepsilon')\rho(\varepsilon)\, d\varepsilon = \int_{\varepsilon \in G} f(\varepsilon)\rho(\varepsilon)\, d\varepsilon \tag{C.8}$$

where $\varepsilon = (\varepsilon_1, \varepsilon_2, \ldots, \varepsilon_r)$ and $\varepsilon' = (\varepsilon'_1, \varepsilon'_2, \ldots, \varepsilon'_r)$ denote the parameters of g and $g'g$ or gg', respectively. One can also write

$$\int_{\varepsilon \in G} f(\varepsilon)\rho(\varepsilon)\, d\varepsilon = \int_{\varepsilon' \in G} f(\varepsilon')\rho(\varepsilon')\, d\varepsilon' = \int_{\varepsilon \in G} f(\varepsilon')\rho(\varepsilon')\left|\frac{\partial \varepsilon'}{\partial \varepsilon}\right|\, d\varepsilon \tag{C.9}$$

where $\left|\frac{\partial \varepsilon'}{\partial \varepsilon}\right|$ is the Jacobian. Comparison of (C.8) and (C.9) leads to

$$\rho(\varepsilon')\left|\frac{\partial \varepsilon'}{\partial \varepsilon}\right| = \rho(\varepsilon) \tag{C.10}$$

for all ε. Here, one can choose $\varepsilon = 0$. Then,

$$\rho(\varepsilon') = \rho(0)\left|\frac{\partial \varepsilon'}{\partial \varepsilon}\right|^{-1}_{\varepsilon=0} = \left|\frac{\partial \varepsilon'}{\partial \varepsilon}\right|^{-1}_{\varepsilon=0} \tag{C.11}$$

where the second equality holds if $\rho(0) = 1$. Then (C.11) defines ρ up to a norm which is fixed by (C.7).

APPLICATION TO UNITARY GROUPS

The elements of a unitary group are unitary matrices. Each unitary matrix can be brought to a diagonal form (see Section 8.1):

$$u = \begin{vmatrix} \varepsilon_1 & & & \\ & \varepsilon_2 & & \\ & & \ddots & \\ & & & \varepsilon_n \end{vmatrix} \tag{C.12}$$

where

$$\varepsilon_i = e^{ih'_{ii}} = e^{-i\theta_i}. \tag{C.13}$$

The parameters θ_i are real and $0 \le \theta_i \le 2\pi$. They are the only ones needed to define the characters, although $U(n)$ depends on n^2 parameters. For $SU(n)$, there is an additional constraint

$$\sum_{i=1}^{n} \theta_i = 0 \tag{C.14}$$

in agreement with Equation (8.7). The expression of the character of an irrep of $SU(n)$ described by the partition $[f_1, f_2, \ldots, f_n = 0]$ is given in Appendix D.

One can show that the weight $\rho(\varepsilon)$ of a unitary group can be expressed in terms of ε_i by the following formula (for a proof, see for example Elliott and Dawber 1979, Section 18.12)

$$\rho(\varepsilon) = \prod_{i<j} |\varepsilon_i - \varepsilon_j|^2 = |\Delta|^2 \tag{C.15}$$

where Δ is defined by (D.7).

Example U(2) and SU(2)

$$\rho = |\varepsilon_1 - \varepsilon_2|^2 = |e^{-i\theta_1} - e^{-i\theta_2}|^2 = 4\sin^2\frac{\theta_1 - \theta_2}{2}.$$

According to (C.14), for SU(2) one has $\theta_1 + \theta_2 = 0$. Thus, there is only one free parameter, which can be $\phi = \theta_1 - \theta_2$. Then, the weight for SU(2) is

$$\rho = 4\sin^2\frac{\phi}{2}. \tag{C.16}$$

The integral on the left hand side of (C.7) becomes

$$\int_0^{4\pi} d\phi \, \rho(\phi) = 8\pi.$$

Thus, in order to satisfy (C.7), one must divide (C.16) by 8π.

For the rotation group parametrized in terms of the Euler angles (Section 6.1), the expression of the weight function is

$$\rho(\alpha, \beta, \gamma) = \frac{1}{8\pi^2} \sin\beta.$$

A proof can be found in Gel'fand and Shapiro (1956).

Appendix D

DIMENSION OF AN SU(N) IRREP

In this appendix we derive the formula (8.191) giving the dimension of an SU(n) irrep.

Usually the dimension of an irrep is determined from the character of the identity element (Section 4.3). The character of an irrep described by a Young tableau $f_1, f_2, \ldots, f_{n-1}, f_n = 0$ is given by the following formula (Weyl 1946, p. 201)

$$\chi^{[f]} = \frac{\begin{vmatrix} \varepsilon_1^{f_1+n-1} & \varepsilon_1^{f_2+n-2} & \cdots & \varepsilon_1^{0} \\ \varepsilon_2^{f_1+n-1} & \varepsilon_2^{f_2+n-2} & \cdots & \varepsilon_2^{0} \\ \vdots & \vdots & & \vdots \\ \varepsilon_n^{f_1+n-1} & \varepsilon_n^{f_2+n-2} & \cdots & \varepsilon_n^{0} \end{vmatrix}}{\begin{vmatrix} \varepsilon_1^{n-1} & \varepsilon_1^{n-2} & \cdots & \varepsilon_1^{0} \\ \varepsilon_2^{n-1} & \varepsilon_2^{n-2} & \cdots & \varepsilon_2^{0} \\ \vdots & \vdots & & \vdots \\ \varepsilon_n^{n-1} & \varepsilon_n^{n-2} & \cdots & \varepsilon_n^{0} \end{vmatrix}}. \tag{D.1}$$

This is a symmetric function of the quantities

$$\varepsilon_i = e^{-i\theta_i} \quad (i = 1, 2, \cdots, n) \tag{D.2}$$

called *eigenvalues* of an element u of U(n), see Appendix C.

For SU(n), we recall that columns filled with n boxes in a Young diagram can be neglected because they give equivalent representations (Section 5.11). Removing these columns, an irrep can be relabelled by

$$f_1 - f_n, \; f_2 - f_n, \; \cdots, \; f_{n-1} - f_n, \; 0$$

or one can take from the outset

$$f_1, \; f_2, \; \cdots, \; f_{n-1}, 0$$

as was done in Equation (D.1).

First, it is convenient to introduce the notation

$$\begin{aligned} \ell_1 &= f_1 + n - 1 \\ \ell_2 &= f_2 + n - 2 \\ &\vdots \\ \ell_n &= 0. \end{aligned} \tag{D.3}$$

Next, one must be careful about taking the limit $\varepsilon_i \to 1$ because both the numerator and denominator vanish in this limit. One can choose

$$\varepsilon_1 = \varepsilon^{n-1}, \ \varepsilon_2 = \varepsilon^{n-2}, \ \cdots, \ \varepsilon_n = \varepsilon^0 \tag{D.4}$$

with

$$\varepsilon = e^{-i\theta} \quad \theta \to 0 \tag{D.5}$$

depending on a single parameter θ. Then, the dimension $d^{[f]}_{SU(n)}$ of an irrep represented by the partition $[f]$ is

$$d^{[f]}_{SU(n)} = \frac{\begin{vmatrix} (\varepsilon^{\ell_1})^{n-1} & (\varepsilon^{\ell_2})^{n-1} & \cdots & (\varepsilon^{\ell_n})^{n-1} \\ (\varepsilon^{\ell_1})^{n-2} & (\varepsilon^{\ell_2})^{n-2} & \cdots & (\varepsilon^{\ell_n})^{n-2} \\ \vdots & \vdots & & \vdots \\ (\varepsilon^{\ell_1})^{0} & (\varepsilon^{\ell_2})^{0} & \cdots & (\varepsilon^{\ell_n})^{0} \end{vmatrix}}{\begin{vmatrix} (\varepsilon^{n-1})^{n-1} & (\varepsilon^{n-2})^{n-1} & \cdots & (\varepsilon^{0})^{n-1} \\ (\varepsilon^{n-1})^{n-2} & (\varepsilon^{n-2})^{n-2} & \cdots & (\varepsilon^{0})^{n-2} \\ \vdots & \vdots & & \vdots \\ (\varepsilon^{n-1})^{0} & (\varepsilon^{n-2})^{0} & \cdots & (\varepsilon^{0})^{0} \end{vmatrix}}. \tag{D.6}$$

Both determinants have the same general form (Hamermesh 1962, Equation (7.21)):

$$\Delta(x_1, \cdots, x_n) = \begin{vmatrix} x_1^{n-1} & x_2^{n-1} & \cdots & x_n^{n-1} \\ x_1^{n-2} & x_2^{n-2} & \cdots & x_n^{n-2} \\ \vdots & \vdots & & \vdots \\ x_1^0 & x_2^0 & \cdots & x_n^0 \end{vmatrix} = \prod_{i<j}(x_i - x_j) \tag{D.7}$$

and at this level one can take the limit

$$\varepsilon^{\ell_i} - \varepsilon^{\ell_j} \underset{\theta \to 0}{\sim} -i\theta(\ell_i - \ell_j)$$

which leads to

$$d^{[f]}_{SU(n)} = \frac{\Delta(\ell_1, \ell_2, \cdots, \ell_n = 0)}{\Delta(n-1, n-2, \cdots, 0)}. \tag{D.8}$$

Using (D.7) and the notation (D.3), one obtains the desired formula

$$d^{[f]}_{SU(n)} = \prod_{i<j} \frac{f_i - f_j + j - i}{j - i}. \tag{D.9}$$

One can use an alternative notation. Let λ_i be the number of columns of length i. Then, one can write

$$\ell_1 = \lambda_1 + \lambda_2 + \cdots + \lambda_{n-1} + n - 1$$
$$\ell_2 = \lambda_2 + \cdots + \lambda_{n-1} + n - 2$$
$$\vdots$$
$$\ell_{n-1} = \lambda_{n-1} + 1$$
$$\ell_n = 0$$

from which

$$\ell_1 - \ell_2 = \lambda_1 + 1$$
$$\ell_1 - \ell_3 = \lambda_1 + \lambda_2 + 2$$
$$\vdots$$
$$\ell_1 - \ell_n = \lambda_1 + \lambda_2 + \cdots + \lambda_{n-1} + n - 1$$
$$\ell_2 - \ell_3 = \lambda_2 + 1$$
$$\vdots$$

Noting that

$$\Delta(n-1, n-2, \cdots, 0) = (n-1)!(n-2)! \cdots 1!$$

one can write

$$d_{SU(n)}^{[f]} = \frac{(\lambda_1+1)(\lambda_2+1)..(\lambda_{n-1}+1)(\lambda_1+\lambda_2+2)..(\lambda_{n-2}+\lambda_{n-1}+2)..(\lambda_1+\lambda_2+...+\lambda_{n-1}+n-1)}{1!2!...(n-1)!}.$$

(D.10)

For example, for SU(3) we used the notation $\lambda_1 = \lambda$, $\lambda_2 = \mu$ which gave us the formula (8.139) as a particular case of (D.10).

References

Alder, S.L. (1969). *Phys. Rev.* **177**, 2426.
Alvarez-Gaumé, L., Gomez, C., and Sierra, G. (1990). *Nucl. Phys.* **B330**, 347.
Arima, A. and Iachello, F. (1975). *Phys. Rev. Lett.* **35**, 1069.
Bailin, D. and Love, A. (1994) *Supersymmetric gauge field theory and string theory.* Institute of Physics Publishing, Bristol.
Baird, G.E. and Biedenharn, L.C. (1963). *J. Math. Phys.* **4**, 1449.
Baird, G.E. and Biedenharn, L.C. (1964). *J. Math. Phys.* **5**, 1723.
Baird, G.E. and Biedenharn, L.C. (1965a). *J. Math. Phys.* **5**, 1730.
Baird, G.E. and Biedenharn, L.C. (1965b). *J. Math. Phys.* **6**, 1847.
Bargmann, V. (1936). *Z. Phys.* **99**, 576.
Baxter, R.J. (1982). *Exactly solved models in statistical mechanics.* Academic Press, New York.
Belavin, A.A., Polyakov, A.M., and Zamolodchikov, A.M. (1984). *Nucl. Phys.* **B241**, 333.
Biedenharn, L.C. (1989). *J. Math.* **A22**, L873.
Biedenharn, L.C. (1990). *Quantum groups* (ed. H.-D. Doebner and J.D. Hennig). Springer-Verlag, Berlin, p. 67.
Bijker, R., Iachello, F., and Leviatan, A. (1994). *Ann. Phys. (N.Y.)*, **236**, 69.
Birman, J. (1974). *Braids, links and mapping class groups.* Princeton University Press, Princeton, N.J.
Bjorken, J.D. and Drell, S.D. (1964). *Relativistic quantum mechanics.* McGraw-Hill, New York.
Bjorken, J.D. and Drell, S.D. (1965). *Relativistic quantum fields.* McGraw-Hill, New York.
Boerner, H. (1963). *Representations of groups.* North-Holland, Amsterdam.
Bohr, A. and Mottelson, B.R. (1969). *Nuclear structure*, vol. 1. Benjamin, New York.
Bonatsos, D. (1994). *Proceedings of the international conference on many-body physics, Coïmbra, Portugal* (ed. C. Fiolhais, M. Fiolhais, C. Sousa, and J.N. Urbano). World Scientific, Singapore.
Bonatsos, D. and Daskaloyannis, C. (1993). *Phys. Rev.* **A48**, 3611.
Bonatsos, D., Brito, L., Menezes, D., Providência, C., and da Providência, J. (1993a). *J. Phys.* **A26**, 895.
Bonatsos, D., Daskaloyannis, C. and Kokkotas, K. (1993b). *Phys. Rev.* **A48**, R3407.
Bowler, K.L., Corvi, P.J., Hey, A.J.G., Jarvis, P.D., and King, R.C. (1981). *Phys. Rev.* **D24**, 197.
Brink, D.M. and Satchler, R. (1968). *Angular momentum*, 2nd edn. Oxford University Press, Oxford.
Brink, D.M. and Stancu, Fl (1994). *Phys. Rev.* **D49**, 4665.
Capstick, S. and Isgur, N. (1986). *Phys. Rev.* **D34**, 2809.
Carlson, J. and Pandharipande, V.R. (1991). *Phys. Rev.* **D43**, 1652.
Carlson, J., Kogut, J., and Pandharipande. V.R. (1983a). *Phys. Rev.* **D27**, 233.
Carlson, J., Kogut, J., and Pandharipande, V.R. (1983b). *Phys, Rev.* **D28**, 2807.
Chen, Jin-Quan (1989). *Group representation theory for physicists.* World Scientific, Singapore.
Chen, Jin-Quan, and Gao, Mei-Juan (1981). *Reduction coefficients of the permutation group and their applications.* Beijing Science Press, Beijing.

Chew, G.F. and Frautschi, S.C. (1961). *Phys. Rev. Lett.* **7**, 394.
Close, F.E. (1979). *An introduction to quarks and partons.* Academic Press, London.
Constantinescu, F. and Lüdde, M. (1992). *J. Math. Phys. Gen.* **25**, L1273– L1280.
Cornwell, J.F. (1984). *Group theory in physics,* Vols. 1 and 2, Third edition. Academic Press, London.
Cornwell, J.F. (1989). *Group theory in physics,* Vol. 3. Academic Press, London.
Courant, R. and Hilbert, D. (1953). *Methods of mathematical physics.* Interscience, New York.
Courrier CERN (1994). **34**(5), 1.
De Rújula, A., Georgi, H., and Glashow, S.L. (1975). *Phys. Rev.* **D12**, 147.
de Swart, J.J. (1963). *Rev. Mod. Phys.* **35**, 916.
de Swart, J.J. (1966). *Proceedings of the 1966 CERN School of Physics, Noordwijk ann Zee, June 6–18, 1966,* CERN Preprint 66–29.
Drinfeld, V.G. (1986). *Quantum groups* (Proceedings of the International Congress of Mathematics). MSRI, Berkeley, Ca.
Dynkin, E.B. (1947). *Uspeki Mat. Nauk. (N.S.)* **2**, no. 4 (20), 59–127 (Translation 1950: *Amer. Math. Soc. Trans. Series*) **1**, 9, 328–469.
Eckart, C. (1930). *Rev. Mod. Phys.* **2**, 305.
Edmonds, A.R. and Flowers B.H. (1952). *Proc. Roy. Soc.* **214**, 515.
Eisenhart, L.P. (1933). *Continuous groups of transformations.* Princeton University Press, Princeton, N.J.
Elliott, J.P. (1958). *Proc. Roy. Soc. (London)* **A245**, 128. *Ibid.* **A245**, 562.
Elliott, J.P. and Dawber, P.G. (1979). *Symmetry in physics.* MacMillan, London.
Elliott, J.P. and Harvey, M. (1963). *Proc. Roy. Soc.* **A272**, 557.
Elliott, J.P., Hope, J., and Jahn, H.A. (1953). *Phil. Trans. Roy. Soc.* **246**, 241.
Fadeev, L.D. (1984). *Integrable models in $(l + 1)$-dimensional quantum field theory* (Les Houches XXXIX), Course 8 (ed. J.-B. Zuber and R. Stora). Elsevier, Amsterdam.
Fadeev, L.D., Reshetikhin, N. Yu., and Takhtajan, L.A. (1989). *Algebra i Analiz* **1**, 193 (in Russian).
Feng Pan (1991). *J. Phys.* **A24**, L803.
Flowers, B.H. (1952). *Proc. Roy. Soc.* **212**, 248.
Fock, V. (1935). *Z. Phys.* **98**, 145.
Frame, J.S., Robinson, G. de B., and Thrall, R.M. (1954). *Can. J. Math.* **6**, 316.
Frobenius, G. (1900). *Über die Charaktere der Symmetrische Gruppe.* Sitz. Ber. Preuss. Akad., Berlin, p. 516.
Gel'fand, I.M. and Shapiro, A. Ya. (1956). *Amer. Math. Soc. Trans. Series 2,* **2**, 207, translated from *Uspeki Mat. Nauk.* (N.S.) (1952) 7, no. 1 (47) 3.
Gel'fand, I.M., Minlos, R.A., and Shapiro, Z. Ya. (1963). *Representations of the rotation and Lorentz groups and their applications.* Pergamon Press, London.
Gell-Mann, M. (1961) *Caltech Report* CTSL-20. Caltech, Ca.
Gell-Mann, M. (1962). *Phys. Rev.* **125**, 1067.
Gell-Mann, M. and Ne'eman, Y. (1964). *The eightfold way.* Benjamin, New York.
Glashow, S.L., Iliopoulos, J., and Maiani, L. (1970). *Phys. Rev.* **D2**, 1285.
Goldstein, H. (1980). *Classical mechanics.* Addision Wesley.
Greiner, W. and Müller, B. (1989). *Quantum mechanics, symmetries.* Springer-Verlag, Berlin.
Gürsey, F. and Radicati, L.A. (1964). *Phys. Rev. Lett.* **13**, 173.
Hamermesh, M. (1962). *Group theory.* Addison-Wesley, Reading, UK.
Harvey, M. (1981*a*). *Nucl. Phys.* **A352**, 301.

Harvey, M. (1981b). *Nucl. Phys.* **A352**, 326.
Harvey, M. (1988). *Nucl. Phys.* **A481**, 834.
Harvey, M., LeTourneux, J., and Lorazo, B. (1984). *Nucl. Phys.* **A424**, 428.
Hecht, K.T. (1964). 'Collective models', in *Selected topics in nuclear spectroscopy* (ed. B.J. Verhaar). North-Holland, Amsterdam.
Higgs, P.W. (1979). *J. Phys.* **A12**, 309.
Høgaasen, H. and Richard, J.-M. (1983). *Phys. Lett.* **B124**, 520.
Iachello, F. (1989). *Nucl. Phys.* **A497**, 23c.
Iachello, F. and Arima, A. (1987). *The interacting boson model.* Cambridge University Press.
Iosifescu, M. and Stancu, Fl. (1967). *Nucl. Phys.* **B1**, 471.
Isgur, N. and Karl, G. (1978). *Phys. Rev.* **D18**, 4187.
Isgur, N. and Karl, G. (1979a). *Phys. Rev.* **D19**, 2653.
Isgur, N. and Karl, G. (1979b). *Phys. Rev.* **D20**, 1191.
Itzykson, C. and Nauenberg, M. (1966). *Rev. Mod. Phys.* **38**, 95.
Itzykson, C. and Zuber, J.-B. (1980). *Quantum Field Theory.* McGraw-Hill, New York.
Jackson, J.D. (1975). *Classical electrodynamics*, Second edition. Wiley, New York.
Jaffe, R.L. (1977a). *Phys. Rev.* **D15**, 267.
Jaffe, R.L. (1977b). *Phys. Rev.* **D15**, 281.
Jaffe, R.L. (1978). *Phys. Rev.* **D17**, 1444.
Jahn, H.A. (1950). *Proc. Roy. Soc.* **A201**, 516.
Jahn, H.A. (1951). *Proc. Roy. Soc.* **A205**, 192.
Ji, Cheung-Ryong, and Brodsky, S.J. (1986). *Phys. Rev.* **D34**, 1460.
Jimbo, M. (1985). *Lett. Math. Phys.* **10**, 63. *Ibid.*, **11**, 247.
Jimbo, M. (1989). *Intern. J. Modern Phys.* **A4**, 3759.
Jones, V.F.R. (1987). *Ann. Math.* **126**, 335.
Kac, V.G. (1990). *Infinite dimensional Lie algebras.* Cambridge University Press.
Karl, G. (1992a). *Phys. Rev.* **D45**, 247.
Karl, G. (1992b). *Intern. J. of Mod. Phys.* **E1**, 491.
Kibler, M. and Winternitz, P. (1987). *J. Phys.* **A20**, 4097.
Koepf, W., Wilets, L., Pepin, S., and Stancu, Fl. (1994). *Phys. Rev.* **C50**, 614.
Kulish, P.P. and Reshetikin, Yu. N. (1981). *Zap. Nauchn. Sem. LOMI* **101**, 101 [(1983) *J. Sov. Math.* **23**, 2435].
Kumano, S. (1990). *Phys. Rev.* **D41**, 195.
Landau, L.D. and Lifshitz, E.M. (1965). *Quantum mechanics, nonrelativistic theory*, second edition. Pergamon Press, Oxford.
Lawrence, R.J. (1990). *Comm. Math. Phys.* **135**, 141.
Lee, T.D. (1981). *Particle physics and introduction to field theory.* Harwood Academic Publishers, Chur, Switzerland
Lee, T.D. and Yang, C.N. (1956). *Nuovo Cimento* **3**, 749.
Leemon, H.I. (1979). *J. Phys.* **A12**, 489.
Lichtenberg, D.B. (1970). *Unitary symmetry and elementary particles.* Academic Press, New York.
Littlewood, D.E. (1958). *The theory of group characters*, second edition. Oxford University Press.
Macfarlane, A.J. (1989). *J. Math.* **A22**, 4581.
Messiah, A. (1964). *Quantum mechanics.* North-Holland, Amsterdam.
Moshinski, M. (1969). *The harmonic oscillator in modern physics: from atoms to quarks.* Gordon and Breach, London.

Moshinski, M., Quesne, C., and Loyola, G. (1990). *Annals of Physics* **198**, 103.
Myhrer, F. and Wroldsen, J. (1988). *Rev. Mod. Phys.* **60**, 629.
Nachtmann, O. (1990). *Elementary particle physics, concepts and phenomena.* Springer-Verlag, Berlin.
Ne'eman, Y. (1961). *Nucl. Phys.* **26**, 222.
Ohnuki, Y. (1960). *Proceedings of the International High-Energy Conference*, CERN, Geneva, p. 843.
Oka, M. and Yazaki, K. (1984). *Quarks and nuclei*, (ed. W. Weise). World Scientific, Singapore, p. 489.
Particle Data Group (1992). *Phys. Rev.* **D45**, (11–II). Review of particle properties.
Particle Data Group (1994). *Phys. Rev.* **D50** (3–I). Review of particle properties.
Pauli, W. (1926). *Z. Phys.* **36**, 336.
Pennington, M.E. (1991). *In search for hadrons beyond the quark model. Proceedings of the Dalitz conference* (ed. I.J.R. Aitchison, C.H. Llewellyn Smith, and J.E. Paton). World Scientific, Singapore.
Pepin, S., Stancu, Fl. (1996). Liège University, preprint.
Pepin, S., Stancu, Fl., Koepf, W., and Wilets L. (1996). *Phys. Rev.* **C53**, 1368.
Peter, F. and Weyl, H. (1927). *Math. Ann.* **97**, 137.
Racah, G. (1942). *Phys. Rev.* **62**, 438.
Racah, G. (1943). *Phys. Rev.* **63**, 367.
Racah, G. (1949). *Phys. Rev.* **76**, 1352.
Racah, G. (1950). *Lincei Rend. Sci. Fis. Mat. Nat.* **8**, 108.
Racah, G. (1965). 'Group theory and spectroscopy', in *Ergebnisse der Exakten Naturwissenschaften*, vol. 37. Springer-Verlag, Berlin.
Richard, J.-M. (1992). *Physics Reports* **212**, 1.
Richtmyer, R.D. (1981). *Principles of advanced mathematical physics.* Springer-Verlag, Berlin.
Robinson, G. de B. (1961). *Representation theory of the symmetric group.* Edinburgh University Press.
Rose, M.E. (1955). *Multipole fields.* John Wiley, New York.
Rose, M.E. (1957). *Elementary theory of angular momentum.* John Wiley, New York.
Rotenberg, M., Bivius, R., Metropolis, N., and Wooten, J.K., Jr. (1959). *The 3-j and 6-j symbols.* Technology Press, Cambridge, Mass.
Rutherford, D.E. (1948). *Substitutional analysis.* Edinburgh University Press.
Sakurai, J.J. (1967). *Advanced quantum mechanics.* Addison-Wesley, London.
Sartor, R. and Stancu, Fl. (1986). *Phys. Rev.* **D34**, 3405.
Schiff, L. (1968). *Quantum mechanics*, third edition. McGraw-Hill, New York.
Schindler, S. and Mirman, R. (1978). *Computer Phys. Comm.* **15**, 131.
Schweber, S.S. (1961). *An introduction to relativistic quantum field theory.* Harper and Row, New York.
Schwinger, J. (1952). *On angular momentum.* US Atomic Energy Commission, NYO-3071.
Shimizu, K. (1989). *Rep. Prog. Phys.* **52**, 1.
Silvestre-Brac, B. and Gignoux, C. (1991). *Phys. Rev.* **D43**, 3699.
Sklyanin, E.K. (1980). *Zap. Nauchn. Sem. LOMI* **95**, 55 [(1982) *J. Sov. Math.* **19**, 1532].
Smirnov, V.I. (1964). *A course of higher mathematics*, vol. 4, § 121. Pergamon Press, Oxford.
Smirnov, Yu., Tolstoy, V.N., and Kharitonov, Yu.I. (1993). *Yad. Phys.* **56**, 223 ; *Phys. At. Nucl.* **56**, 690.
Stack, J. (1984). *Phys. Rev.* **D29**, 1213.

Stancu, Fl. (1989). *Phys. Rev.* **C39**, 2030.
Stancu, Fl. and Stassart, P. (1990). *Phys. Rev.* **D41**, 916.
Stancu, Fl. and Stassart, P. (1991). *Phys. Lett.* **B269**, 243.
Stancu, Fl. and Wilets, L. (1987). *Phys. Rev.* **C36**, 726.
Stancu, Fl. and Wilets, L. (1988). *Phys. Rev.* **C38**, 1145.
Stancu, Fl. and Wilets, L. (1989). *Phys. Rev.* **C40**, 1901.
Stancu, Fl. and Wilets, L. (1994). *Proceedings of the international conference on many-body physics, Coïmbra, Portugal* (ed. C. Fiolhais, M. Fiolhais, C. Sousa, and J.N. Urbano). World Scientific, Singapore.
Sun, Chang-Pu and Fu, Hong-Chen (1989). *J. Phys.* **A22**, L983.
Talmi, I. (1993). *Simple models of complex nuclei.* Harwood Academic Publishers, Chur, Switzerland.
Tinkham, M. (1964). *Group theory and quantum mechanics.* McGraw-Hill, New York.
Tung, Wu-Ki. (1985). *Group theory in physics.* World Scientific, Singapore.
Vergados, J.D. (1968). *Nucl. Phys.* **A111**, 681.
Weinstein, J. and Isgur, N. (1983). *Phys. Rev.* **D27**, 588.
Weinstein, J. and Isgur, N. (1990). *Phys. Rev.* **D41**, 2236.
Weyl, H. (1929). *Z. Phys.* **56**, 330.
Weyl, H. (1946). *The classical groups.* Princeton University Press.
Weyl, H. (1950). *The theory of groups and quantum mechanics.* Dover, New York.
Wigner, E.P. (1927). *Z. Phys.* **43**, 624.
Wigner, E.P. (1932). *Göttinger Nachr.* **31**, 546.
Wigner, E.P. (1959). *Group theory and its application to quantum mechanics of atomic spectra.* Academic Press, London.
Wigner, E.P. and Feenberg, E. (1941). *Rep. Progr. Phys.* **8**, 274.
Wroldsen, J. and Myhrer, F. (1984). *Z. Phys.* **C25**, 59.
Wybourne, B.G. (1970). *Symmetry principles and atomic spectroscopy.* Wiley-Interscience, New York.
Wybourne, B.G. (1974). *Classical groups for physicists.* Wiley Interscience, New York.
Yamanouchi, T. (1937). *Proc. Phys. Math. Soc. Japan,* 3rd series, **19**, 436.
Yang, C.N. and Mills, R. (1954). *Phys. Rev.* **96**, 191.
Young, A. (1928). *Proc. London Math. Soc.* **28**, 255.

INDEX

abelian group 9
abstract group 10
accidental degeneracy 233
anomalies 8 n.

braid groups 144–7
branching law of S_n 90

Cartan 154, 171, 181, 183
  criterion for semi-simple groups 161, 174
  integers 179
  matrix 189
  subalgebra 172
Cartan–Weyl basis 174
Cayley's theorem 12, 20
character, *see* representation
charmonium 333, 357
Chevalley basis 174
class (conjugacy) 22
Clebsch–Gordan
  coefficients for R_3 220
  coefficients for S_n 114
  coefficients for SU(3) 344
  series 57, 112, 219
compact group 169
continuous groups 15, 149
  order 16
  see also Lie groups
coset 17
cyclic group 10

diquonia 372
direct product group 27, 182
direct sum of Lie algebras 170, 182
dynamical symmetry 209; *see also* accidental degeneracy
Dynkin diagrams 186–9

Euclidean groups 161, 169, 237–40, 259
  unitary representations 239

finite group 9
  C_n, D_n 11
  order 9
  Vierergruppe 11

fractional parentage coefficients 139, 385, 394
Frobenius formula 82

gauge symmetries 347–52
Gell-Mann–Nishijima relation 304, 324
Gell-Mann–Okubo mass formula 313
global symmetries 347
G-parity 271–4
Gürsey–Radicati mass formula 320

hadron classification 300–11
heavy flavours 321
Heine–Borel theorem 169
helicity 240, 259
homomorphism 19, 246, 266
  kernel of 19, 247
hyperchange 301

infinite groups 148
inner product 108, 112–22; *see also* permutation group
invariant operators 193, 195–7
  Casimir 195
invariant subspace 32, 68, 114, 123, 128, 211
irrep, *see* representation
isomorphism 18
  isomorphic algebras 170, 185, 230
  isomorphic groups 165, 209, 230
isoscalar factor 128

J/ψ meson 321; *see also* charmonium

Kac–Moody algebra 163
Killing form 154; *see also* metric tensor
K-matrix, *see* isoscalar factor

420 INDEX

Lagrange theorem 17
Lie algebra 149, 154
  classical 183
  exceptional 186
  ideal or invariant subalgebra 162
  semi-simple 163
  simple 163
  standard form 171–5
  subalgebra 162
Lie groups
  general linear 164–5, 195, 197–8, 202
  generators 154
  orthogonal 165, 210
  orthogonal $O(n,m)$ 165, 167
  special linear 165–6, 202
  structure constants 151
  symplectic 165, 167, 185
  unitary 165–6, 202
linear operator 31
  vector space 31
little group 260
Littlewood's rule 109
local (or gauge) symmetries 347
Lorentz group 230, 241–56
  boost 248
  Dirac particle 250–3
  generators 249
  homogeneous 246
  non-compact 244
  non-unitary representations 255
  see also Lie groups

magnetic moments of baryons 141
metric tensor 153, 174–6

Noether's theorem 7, 398–401

orthogonality theorem 45, 404
outer product 108, 383; see also permutation group

parity 6
partition 23
  conjugate 61, 119
Pauli–Lubanski vector 258
Pauli's fundamental theorem 251
permutation group S_n 59–143; see also symmetric group
Poincaré group 169, 237, 256–61
point group, see finite group

quantum groups 146, 204–8, 234

Racah 128, 177, 194, 196
rank of a semi-simple group 172
rearrangement theorem 20, 47, 48, 64, 401
representation (linear)
  adjoint 50, 337–9
  character of 34
  complex conjugate 203, 371, 295, 345
  contragradient 203
  definition 31
  direct product of 42, 103
  equivalent 32, 203
  faithful 32
  fully irreducible 40
  fundamental 193, 216
  induced 171, 239, 260
  irreducible or irrep 40
  matrix representation 32
  pseudoreal 204
  real 204
  regular 45, 48–50, 89
  unitary 33
Robinson formula 87
root 172
  diagram 181
  positive 186
  properties 176
  simple 186
rotation group 210–30
  Euler angles 213, 216
  improper 212
  proper 212
  representations 216–25
  see also Lie groups
Runge–Lenz vector 234

Schur
  functions (or S-functions) 89
  lemmas 43, 196, 286, 305, 402–4
semi-direct product group 27, 238, 241, 257
semi-direct sum of Lie algebras 170
semi-simple group 26, 158
similarity transformation 33; see also equivalent representations
simple group 26, 158
spectrum generating algebra 319, 371
spin of a Dirac particle, see Lorentz group
spin of a vector particle 225
structure constants
  of a finite group 51
  of a Lie group 153
subgroup 16
  index of 17
  invariant or normal 26, 159
  regular subgroups 20
symmetric group 2, 12
  adjoint transposition 14

INDEX

alternating group 14
  cycle 13
  degree 12
  order 12
  parity of a permutation 14

tensor method 103–8, 198–202
  contraction 199, 340
  invariant tensors 201
  irreducible tensor operator 227
  see also Wigner–Eckart theorem
translation group
  T_3 237
  T_4 257

unitary groups 262–346
  SU(2) 265–70
  SU(3) 274–315
  SU(4) 324–6
  see also Lie groups
universal covering group 170
Υ meson 322

Van der Waerden 181
Virasoro algebra 163

weight 193
  dominant 194
  equivalent 194
  highest 194
  positive 194
  simple 193, 194
Weinberg–Salam theory 351
Weyl 1, 2, 5, 59, 219
  branching law 328
  group 179
  reflection 179, 194
  tableau 65, 71, 199, 201
  theorems 193
Wigner 1, 4, 222
  rotation 260
Wigner–Eckart theorem 226

Yang–Baxter equation 146, 204
Yang–Mills theory 5, 345
Yamanouchi
  phase convention 70, 74, 94
  symbol 63, 91
Young diagram 60, 199
Young tableau 62, 199
  axial distance 91
  conjugate 125
  normal 63
  standard 62
Young–Yamanouchi
  basis 89, 119
  representation 92
Young–Yamanouchi–Rutherford
    representation 101, 393